Selected Titles in This Series

257 **Peter A. Cholak, Steffen Lempp, Manuel Lerman, a Editors,** Computability theory and its applications: Currei 2000

256 **Irwin Kra and Bernard Maskit, Editors,** In the tradition of Ahlfors and Bers: Proceedings of the first Ahlfors-Bers colloquium, 2000

255 **Jerry Bona, Katarzyna Saxton, and Ralph Saxton, Editors,** Nonlinear PDE's, dynamics and continuum physics, 2000

254 **Mourad E. H. Ismail and Dennis W. Stanton, Editors,** q-series from a contemporary perspective, 2000

253 **Charles N. Delzell and James J. Madden, Editors,** Real algebraic geometry and ordered structures, 2000

252 **Nathaniel Dean, Cassandra M. McZeal, and Pamela J. Williams, Editors,** African Americans in Mathematics II, 1999

251 **Eric L. Grinberg, Shiferaw Berhanu, Marvin I. Knopp, Gerardo A. Mendoza, and Eric Todd Quinto, Editors,** Analysis, geometry, number theory: The Mathematics of Leon Ehrenpreis, 2000

250 **Robert H. Gilman, Editor,** Groups, languages and geometry, 1999

249 **Myung-Hwan Kim, John S. Hsia, Yoshiyuki Kitaoka, and Rainer Schulze-Pillot, Editors,** Integral quadratic forms and lattices, 1999

248 **Naihuan Jing and Kailash C. Misra, Editors,** Recent developments in quantum affine algebras and related topics, 1999

247 **Lawrence Wasson Baggett and David Royal Larson, Editors,** The functional and harmonic analysis of wavelets and frames, 1999

246 **Marcy Barge and Krystyna Kuperberg, Editors,** Geometry and topology in dynamics, 1999

245 **Michael D. Fried, Editor,** Applications of curves over finite fields, 1999

244 **Leovigildo Alonso Tarrío, Ana Jeremías López, and Joseph Lipman,** Studies in duality on noetherian formal schemes and non-noetherian ordinary schemes, 1999

243 **Tsit Yuan Lam and Andy R. Magid, Editors,** Algebra, K-theory, groups, and education, 1999

242 **Bernhelm Booss-Bavnbek and Krzysztof Wojciechowski, Editors,** Geometric aspects of partial differential equations, 1999

241 **Piotr Pragacz, Michał Szurek, and Jarosław Wiśniewski, Editors,** Algebraic geometry: Hirzebruch 70, 1999

240 **Angel Carocca, Víctor González-Aguilera, and Rubí E. Rodríguez, Editors,** Complex geometry of groups, 1999

239 **Jean-Pierre Meyer, Jack Morava, and W. Stephen Wilson, Editors,** Homotopy invariant algebraic structures, 1999

238 **Gui-Qiang Chen and Emmanuele DiBenedetto, Editors,** Nonlinear partial differential equations, 1999

237 **Thomas Branson, Editor,** Spectral problems in geometry and arithmetic, 1999

236 **Bruce C. Berndt and Fritz Gesztesy, Editors,** Continued fractions: From analytic number theory to constructive approximation, 1999

235 **Walter A. Carnielli and Itala M. L. D'Ottaviano, Editors,** Advances in contemporary logic and computer science, 1999

For a complete list of titles in this series, visit the AMS Bookstore at **www.ams.org/bookstore/**.

Computability Theory and Its Applications

Current Trends and Open Problems

CONTEMPORARY MATHEMATICS

257

Computability Theory and Its Applications

Current Trends and Open Problems

Proceedings of a 1999 AMS-IMS-SIAM
Joint Summer Research Conference
Computability Theory and Applications
June 13–17, 1999
University of Colorado, Boulder

Peter A. Cholak
Steffen Lempp
Manuel Lerman
Richard A. Shore
Editors

American Mathematical Society
Providence, Rhode Island

The AMS-IMS-SIAM Joint Summer Research Conference in the Mathematical Sciences on Computability Theory and Applications was held at the University of Colorado, Boulder, CO, June 13–17, 1999, with support from the National Science Foundation, Grant DMS-9618514.

2000 *Mathematics Subject Classification.* Primary 03C57, 03D25, 03D28, 03D30, 03D45, 03D80, 03E15, 03E35, 03F35, 03H15.

Any opinions, findings, and conclusions or recommendations expressed in this material are those of the authors and do not necessarily reflect the views of the National Science Foundation.

Library of Congress Cataloging-in-Publication Data

Computability theory and its applications : current trends and open problems : proceedings of a 1999 AMS-IMS-SIAM, joint summer research conference, computability theory and applications, June 13–17, 1999, University of Colorado, Boulder / Peter A. Cholak... [et al.], editors.

p. cm. — (Contemporary mathematics, ISSN 0271-4132 ; 257)

Includes bibliographical references.

ISBN 0-8218-1922-4 (alk. paper)

1. Computable functions—Congresses. I. Cholak, Peter, 1962– II. Contemporary mathematics (American Mathematical Society) ; v. 257.

QA9.59. C66 2000

511.3—dc21 00-036278

∞ The paper used in this book is acid-free and falls within the guidelines
established to ensure permanence and durability.
Visit the AMS home page at URL: `http://www.ams.org/`

10 9 8 7 6 5 4 3 2 1 05 04 03 02 01 00

Contents

A sunny summer day in July, 1997.
A walk through the woods outside Kazan,
Along the Volga river.
Ideas for a conference emerge.
Focus only on open problems.
Provide sufficient background for
Fruitful discussions about the problems.
Allow enough free time to
Immerse ourselves in the problems.
Assemble a mix of people with diverse, but connected interests.
Provide new ideas and insights.
A consensus emerges.
A list of focus areas is proposed, discussed, modified.
Potential speakers are matched with topics.
A successful proposal to the AMS.
A conference of a different nature sees the light of day.

Preface

This volume is a faithful and expanded reflection of most of the talks which were presented at the Conference on Computability Theory and Applications, as part of the Joint Summer Research Conferences in the Mathematical Sciences, Boulder, Colorado, June 13–17, 1999. The meeting focused on open problems in Computability Theory and some related areas in which the ideas, methods and/or results of Computability Theory play a role. Some talks delved in depth into a narrowly focused group of problems, providing a description of what had been done and delineating the obstacles to solution. Others covered a wider area, providing the rationale for interest in the area and the directions pursued, and a broad cross-section of central open problems. Discussions ensued. Some problems were solved quickly at the meeting, and others since that time. The result is, we hope, a snapshot of the status of Computability Theory at the end of the millennium, and a list of fruitful directions for research early in the next millennium.

All papers in this volume reflect invited talks, are written by the invited speakers, sometimes with a co-author, and are refereed. Alekos Kechris and George Odifreddi were unable to speak, but submitted papers. Anil Nerode spoke on Computable Analysis and Topology and Gerald Sacks on Higher Recursion Theory, but there are no follow-up papers for these talks. Yiannis Moschovakis was, unfortunately, not able to attend and talk about Recursion in Computer Science. In addition, two three-hour blocks of time were allocated for Barry Cooper to describe, in detail, his work on constructing automorphisms of the degrees.

Of particular note was the intersection of this meeting with one on Topology which was being held simultaneously. In particular, schedules were arranged so that both groups could meet to hear the talk by Shmuel Weinberger on joint work[1] with Alex Nabutovsky which uses results about computably enumerable sets and degrees to obtain results on manifolds with Riemannian metrics in differential topology. This application of Computability Theory was brought to our attention by Robert Soare who had a part in developing some of the computability-theoretic results used in this work. It added yet another direction of application beyond those which were represented by our original list of topics.

We thank the AMS and NSF for funding the conference. Special thanks go to Donna Salter who handled the organization and logistics superbly, and to Christine Thivierge who assisted with the preparation of this volume. Of course, the success of the Conference can be mainly attributed to the speakers and participants, whose active involvement brought many of the initial goals to fruition.

Peter A. Cholak
Steffen Lempp
Manuel Lerman
Richard A. Shore

Editors and Conference Co-organizers
December 15, 1999

[1]These results are presented in A. Nabutovsky and S. Weinberger: *Variational problems for Riemannian functionals and arithmetic groups*, and *The fractal nature of Riem/Diff*, both to appear. For background on this material, see also A. Nabutovsky: *Disconnectedness of sublevel sets of some Riemannian functionals*, Geometric and Functional Analysis, **6** (1996), 703-725, and *Geometry of the space of triangulations of a compact manifold*, Communications in Mathematical Physics, **181** (1996), 303-330.

Program of Invited Talks, AMS Summer Research Conference on Computability Theory and Applications

Sunday, June 13, 1999

- Sergey Goncharov, Novosibirsk, Computable model theory
- Julia Knight, Notre Dame, Models of arithmetic
- Serikzhan Badaev, Almaty, Numeration theory
- Jeffrey Remmel, San Diego, Computable algebra
- Bakhadyr Khoussainov, Auckland, Issues in computable presentations of models
- Mikhail Peretyat'kin, Almaty, Finitely axiomatizable theories and Lindenbaum algebras

Monday, June 14, 1999

- Peter Cholak, Notre Dame, The lattice of computably enumerable sets
- Robert Soare, Chicago, The lattice of computably enumerable sets
- Klaus Ambos-Spies, Heidelberg, Genericity and randomness
- Barry Cooper, Leeds, Proof of the automorphism theorem, parts 1,2

Tuesday, June 15, 1999

- André Nies, Chicago, Definability and coding
- Richard Shore, Cornell, Natural definability in degree structures
- Gerald Sacks, Harvard & MIT, Higher recursion theory

Wednesday, June 16, 1999

- Stephen Simpson, Penn State, Reverse mathematics
- Carl Jockusch, Urbana, Π^0_1 classes and computable combinatorics
- C. T. Chong, Singapore, Reverse computability theory
- Harvey Friedman, Ohio State, Reverse mathematics
- Alexandra Shlapentokh, East Carolina, Issues related to Hilbert's Tenth Problem
- Anil Nerode, Cornell, Computable analysis and topology

Thursday, June 17, 1999

- Theodore Slaman, Berkeley, Applications of recursion theoretic methods in set theory
- Marat Arslanov, Kazan, D.c.e. and n-c.e. degrees
- Marcia Groszek, Dartmouth, Independence results (from ZFC) in recursion theory
- Manuel Lerman, Connecticut, Lattice embeddings into the computably enumerable degrees
- Andrea Sorbi, Siena, Enumeration degrees

Conference Participants

Ambos-Spies, Klaus
Universität Heidelberg
Mathematisches Institut
Im Neuenheimer Feld 294
D-69120 Heidelberg, Germany
ambos@math.uni-heidelberg.de

Arana, Andrew
University of Notre Dame
Department of Mathematics
Mail Distribution Center
Notre Dame, IN 46556-5683, USA
E-mail: andrew.arana.1@nd.edu

Arslanov, Marat
Kazan State University
Department of Mechanics and Math.
ul. Kremlevskaya 18
420008 Kazan, Russia
marat.arslanov@ksu.ru

Badaev, Serikzhan
Kazakh Academy of Sciences
Institute of Mathematics
125 Pushkin Street
Almaty 480100, Kazakhstan
badaev@math.kz

Calhoun, William C.
Bloomsburg University
Department of Math., CS & Stat.
Bloomsburg, PA 17815, USA
wcalhoun@bloomu.edu

Cenzer, Douglas
University of Florida
Department of Mathematics
310 Little Hall
Gainesville, FL 32611-8105, USA
cenzer@math.ufl.edu

Cholak, Peter
University of Notre Dame
Department of Mathematics
Mail Distribution Center
Notre Dame, IN 46556-5683, USA
peter.cholak.1@nd.edu

Chong, Chi Tat
National University of Singapore
Department of Mathematics
Lower Kent Ridge Road
Singapore 119260, Singapore
chongct@math.nus.edu.sg

Coles, Richard
30 Slade Rd Four Oaks
Sutton Coldfield
West Midlands B75 5PG, England
coles@cs.auckland.ac.nz

Cooper, Barry
University of Leeds
School of Mathematics
Leeds LS2 9JT, England
pmt6sbc@leeds.ac.uk

Davis, Martin
3360 Dwight Way
Berkeley, CA 94704-2523, USA
martin@eipye.com

Ealy, Jr. Clifton E.
Western Michigan University
Department of Mathematics & Statistics
Kalamazoo, MI 49008-5152, USA
Clifton.E.Ealy@wmich.edu

Englert, Burkhard
University of Connecticut
Department of Mathematics
U-9 Rm 111, 196 Auditorium Rd.
Storrs, CT 06269-3009, USA
englert@math.uconn.edu

Ershov, Yuri
Academy of Sciences
Siberian Branch
Mathematical Institute
630090 Novosibirsk, Russia
root@ershov.nsu.ru

Fejer, Peter
University of Massachusetts
Department of Mathematics & CS
Boston, MA 02125-3393, USA
fejer@cs.umb.edu

Friedman, Harvey
The Ohio State University
Department of Mathematics
231 West 18th Avenue
Columbus, OH 43210-1101, USA
friedman@math.ohio-state.edu

Galminas, Lisa
Northwestern State Univ of LA
Department of Mathematics
Natchitoches, LA 71497, USA
galminas@alpha.nsula.edu

Giorgi, Matthew B.
School of Mathematics
University of Leeds
Leeds LS2 9JT, United Kingdom
matt@amsta.leeds.ac.uk

Goncharov, Sergey
Academy of Sciences
Siberian Branch
Mathematical Institute
630090 Novosibirsk, Russia
gonchar@imi.nsu.ru

Griffor, Edward R.
ComAdvisors, Inc.
19111 West Ten Mile Road
Suite 167
Southfield, MI 48075, USA
egriffor@comadvisors.com

Groszek, Marcia
Dartmouth College
Department of Mathematics
6188 Bradley Hall
Hanover, NH 03755-3551, USA
marcia.groszek@dartmouth.edu

Harizanov, Valentina
George Washington University
Department of Mathematics
Funger Hall
2201 G Street NW
Washington, DC 22052
harizanv@gwis2.circ.gwu.edu

Harrington, Leo
University of California
Department of Mathematics
Berkeley, CA 94720-0001, USA
leo@math.berkeley.edu

Herrmann, Eberhard
Institut für Mathematik
Humboldt-Universität zu Berlin
Math.-Naturwiss. Fakultät II
Unter den Linden 6
D-10099 Berlin, Germany
herrmann@mathematik.hu-berlin.de

Hirschfeldt, Denis
Cornell University
Department of Mathematics
Malott Hall
Ithaca, NY 14853-7901, USA
drh@math.cornell.edu

Hirst, Jeff
Appalachian State University
Department of Mathematical Sciences
Boone, NC 28608, USA
jlh@cs.appstate.edu

Ho, Kejia (Joyce)
University of Illinois
Department of Mathematics
1409 West Green Street
Urbana, IL 61801-2975, USA
k-ho1@math.uiuc.edu

Hummel, Tamara
Allegheny College
Department of Mathematics
Meadville, PA 16335, USA
thummel@alleg.edu

Jockusch, Carl
University of Illinois
Department of Mathematics
1409 W. Green St.
Urbana, IL 61801-2975, USA
jockusch@math.uiuc.edu

Khoussainov, Bakh
University of Auckland
Department of Computer Science
Private Bag 92019
Auckland, New Zealand
bmk@cs.auckland.ac.nz

Knight, Julia
University of Notre Dame
Department of Mathematics
Mail Distribution Center
Notre Dame, IN 46556-5683, USA
Julia.F.Knight.1@nd.edu

Kučera, Antonín
Charles University
Department of Theoretical CS
Faculty of Mathematics & Physics
Malostranské nám. 25
CZ-110 00 Prague 1, Czech Republic
kucera@ktisun2.ms.mff.cuni.cz

Kudinov, Oleg
Academy of Sciences
Siberian Branch
Mathematical Institute
630090 Novosibirsk, Russia
kud@math.nsc.ru

LaForte, Geoff
University of West Florida
Institute for Human & Machine Cog.
11000 University Parkway
Pensacola, FL 32514, USA
glaforte@coginst.uwf.edu

Lawton, Linda
University of Illinois
Department of Mathematics
1409 W. Green Street
Urbana, IL 61801-2975, USA
lawton@math.uiuc.edu

Lempp, Steffen
University of Wisconsin
Department of Mathematics
480 Lincoln Drive
Madison, WI 53706-1388, USA
lempp@math.wisc.edu

Leonhardi, Steven
Winona State University
Department of Mathematics & Statistics
Winona, MN 55987, USA
leonhardi@vax2.winona.msus.edu

Lerman, Manuel
University of Connecticut
Department of Mathematics
U-9 Rm 111, 196 Auditorium Rd
Storrs, CT 06269-3009, USA
mlerman@math.uconn.edu

Li, Angsheng
Academica Sinica
Institute of Software
Post Office Box 8718
Beijing 100080, PR of China
liang@ox.ios.ac.cn

Marcone, Alberto
Universita di Udine
Dip. di Matematica e Inform.
Via delle Scienze 206
33100 Udine, Italy
marcone@dimi.uniud.it

McAllister, Alex
Centre College
Department of Mathematics
600 W. Walnut Street
Danville, KY 40422-1394, USA
alexmcal@centre.edu

McNicholl, Timothy H.
University of Dallas
Department of Mathematics
Irving, TX 75062, USA
tmcnicho@acad.udallas.edu

Miller, Russell
University of Chicago
Department of Mathematics
5734 University Avenue
Chicago, IL 60637-1514, USA
russell@math.uchicago.edu

Morozov, Andrei
Academy of Sciences
Siberian Branch
Mathematical Institute
630090 Novosibirsk, Russia
morozov@math.nsc.ru

Nerode, Anil
Cornell University
Department of Mathematics
Malott Hall
Ithaca, NY 14853-4201, USA
anil@math.cornell.edu

Nies, André
University of Chicago
Department of Mathematics
5734 University Avenue
Chicago IL 60637-1514, USA
nies@math.uchicago.edu

Peretyat'kin, Mikhail
Kazakh Academy of Sciences
Institute of Mathematics
125 Pushkin Street
Almaty 480021, Kazakhstan
peretya@math.kz

Remmel, Jeffrey
University of California
Department of Mathematics
La Jolla, CA 92093-0112, USA
remmel@kleene.ucsd.edu

Sacks, Gerald
Massachusetts Institute of Technology
Department of Mathematics
Cambridge, MA 02139-4358, USA
sacks@math.harvard.edu

Shlapentokh, Alexandra
East Carolina University
Deptartment of Mathematics
Greenville, NC 27858-4353, USA
shlapentokh@math.ecu.edu

Shore, Richard A.
Cornell University
Department of Mathematics
Malott Hall
Ithaca, NY 14853-4201, USA
shore@math.cornell.edu

Simpson, Stephen
Pennsylvania State University
Department of Mathematics
McAllister Building
University Park, PA 16802-6401, USA
simpson@math.psu.edu

Slaman, Theodore
University of California
Department of Mathematics
Berkeley, CA 94720-0001, USA
slaman@math.berkeley.edu

Smuga-Otto, Maciej
University of Wisconsin
Department of Mathematics
480 Lincoln Drive
Madison, WI 53706-1388, USA
smug-ot@math.wisc.edu

Soare, Robert
University of Chicago
Department of Mathematics
5734 University Avenue
Chicago, IL 60637-1514, USA
soare@cs.uchicago.edu

Solomon, Reed
University of Wisconsin
Department of Mathematics
480 Lincoln Drive
Madison. WI 53706-1388, USA
rsolomon@math.wisc.edu

Sorbi, Andrea
Dip. di Matematica
Via del Capitano 15
I-53100 Siena, Italy
sorbi@unisi.it

Stephan, Frank
Universität Heidelberg
Mathematisches Institut
Im Neuenheimer Feld 294
D-69120 Heidelberg, Germany
fstephan@math.uni-heidelberg.de

Thurber, John
Eastern Oregon University
Department of Mathematics
La Grande, OR 97850, USA
jthurber@eou.edu

Wald, Kevin
University of Chicago
Department of Mathematics
5734 University Ave
Chicago, IL 60637-1514, USA
wald@math.uchicago.edu

Walk, Steve
University of Notre Dame
Department of Mathematics
Mail Distribution Center
Notre Dame, IN 46556-5683, USA
Stephen.M.Walk.1@nd.edu

Wang, Dejia
University of Wisconsin
Department of Mathematics
480 Lincoln Drive
Madison, WI 53706-1388, USA
dwang@math.wisc.edu

White, Walker
Cornell University
Department of Mathematics
Malott Hall
Ithaca, NY 14853-4201, USA
wmwhite@math.cornell.edu

Contemporary Mathematics
Volume **257**, 2000

Randomness in Computability Theory

KLAUS AMBOS-SPIES AND ANTONÍN KUČERA

ABSTRACT. We discuss some aspects of algorithmic randomness and state some open problems in this area. The first part is devoted to the question "What is a computably random sequence?" Here we survey some of the approaches to algorithmic randomness and address some questions on these concepts. In the second part we look at the Turing degrees of Martin-Löf random sets. Finally, in the third part we deal with relativized randomness. Here we look at oracles which do not change randomness.

1. Introduction

Formalizations of the intuitive notions of computability and randomness are among the major achievements in the foundations of mathematics in the 20th century.

It is commonly accepted that various equivalent formal computability notions – like Turing computability or μ-recursiveness – which were introduced in the 1930s and 1940s adequately capture computability in the intuitive sense. This belief is expressed in the well known Church-Turing thesis (see Soare [**30**] for a recent account of the history of formal computability).

The process of formalizing randomness took a long time. In part this was due to the fact that randomness in an absolute sense does not exist. So one faced the problem to decide which form of restricted randomness adequately captures the intuitive notion. In 1940 Church suggested that intuitive randomness should be viewed as algorithmic randomness and he proposed a formal computable randomness notion. His thesis became widely accepted but his formal concept had some flaws from the statistical point of view. In 1966, however, Martin-Löf [**22**] introduced a new formal concept of algorithmic randomness which eliminated these insufficiencies and which is widely considered to be adequate.

In this note we discuss some aspects of algorithmic randomness focusing on some open problems. In the first part (Section 2) we review some of the important formal randomness notions introduced in the literature and pose some still open problems arising from these notions. In particular we discuss the shift from computable randomness (Church) to computably enumerable randomness (Martin-Löf) which still leaves open the question whether there is an adequate formalization of computable randomness. The second part (Section 3) is devoted to (Martin-Löf)

1980 *Mathematics Subject Classification.* Primary 03D80; Secondary 03D28.

randomness in computability theory. Here we focus on some questions on the (Turing) degrees of random sets which are related to the diagonalization strength of random sets. In the final part (Section 4) we look at relativizations (in the computational sense) of randomness. Here we discuss some recent work and problems on oracles which do not help to strengthen randomness.

We should emphasize that we do not intend to give a complete treatment of algorithmic randomness here but that we only look at some selected topics which are of current interest in computability theory. In particular we do not address the important areas of Kolmogorov complexity and of resource-bounded randomness. For a more complete treatment of algorithmic randomness in general we refer the reader to the monographs of Li and Vitanyi [**20**] and van Lambalgen [**17**].

We assume the reader to be familiar with the basic concepts and results of computability theory. Our notation is standard. For unexplained concepts and notation, see [**24**] or [**29**]. Following Soare [**30**] we use the terms *computable* and *computably enumerable (c.e.)* also in a formal sense, i.e. in place of the previously used terms *recursive* and *recursively enumerable*. We identify an infinite binary sequence $A(0)A(1)A(2)\ldots \in \{0,1\}^\omega$ with a set $A = \{n : A(n) = 1\} \subseteq \omega$ and with the characteristic function $A : \omega \to \{0,1\}$ of A. At some places we also identify finite binary strings with numbers. The initial segment $A(0)A(1)A(2)\ldots A(n)$ of a sequence A is denoted by $A[n]$. A *set* will be a set of natural numbers, a *class* will be a set of sets of natural numbers.

In order to minimize the prerequisites from measure theory, we will only look at the uniform distribution over $\{0,1\}$ under which the events 0 and 1 are equiprobable. The corresponding product measure on $\{0,1\}^\omega$ is denoted by μ and called the *Lebesgue measure.*

2. What is a computably random sequence?

Intuitively we can view a random sequence $A = A(0)A(1)A(2)\ldots$ as an infinite binary sequence where the individual members $A(n)$ of the sequence are determined by independent chance experiments like the tossing of a fair coin. We will expect that such a sequence is *chaotic*, i.e. does not show any regularities, and that it is *typical* in the sense that it does not have any particular rare properties. Each of these observations, which will be made more precise below, can be used to give a definition of the most popular algorithmic randomness concept due to Martin-Löf.

First attempts to define randomness, however, were based on a much simpler and more specific observation: Since we assume that the events 0 and 1 are equiprobable, the frequency of these events in a random sequence A will agree in the limit, i.e. for $i = 0, 1$,

$$\limsup_{n\to\infty} \frac{||\{m < n : A(m) = i\}||}{n} = \frac{1}{2} \tag{1}$$

will hold. This property of *frequency stability* (one also says that A satisfies the *law of large numbers*) alone, however, obviously does not suffice to guarantee randomness as the example of the sequence $01010101\ldots$ demonstrates. In order to avoid such trivial counter examples, in 1919 von Mises [**21**] proposed to call a sequence A random if frequency stability is preserved by the operation of taking infinite subsequences using "admissible place selection rules". Here, a place selection rule in the sense of von Mises can be formalized by a total function $f : \{0,1\}^{<\omega} \to \{0,1\}$, called a *selection function.* Then the subsequence $A_f = A(n_0)A(n_1)A(n_2)\ldots$ of

A selected by f is determined by the numbers $n_0 < n_1 < n_2 < \dots$ for which $f(A[n-1]) = 1$. In other words, the question whether a place n is selected for the subsequence is decided on the base of the values $A(0), \dots, A(n-1)$ of the sequencce A at the previous places.

Now it is easy to see that there is no sequence A such that all infinite subsequences of A selected by some selection function satisfy the law of large numbers. This reflects the more general observation that randomness in an absolute sense does not exist. Von Mises, however, left open the question of what selection functions should be admitted. In 1939 Wald [**34**] proved that random sequences exist relative to *any countable* class of selection functions. Based on Wald's work, in 1940 Church [**5**] suggested that randomness in the intuitive sense should be viewed as algorithmic randomness and he proposed to call a sequence random if it is random relative to the class of computable selection functions. This gives the following first nontrivial and sound definition of randomness, where, following a suggestion by Kolmogorov (see [**32**]) we call randomness in the sense of von Mises *stochasticity.*

Definition 2.1 (von Mises, Wald, Church). *Let F be a countable class of selection functions. A sequence A is F-*stochastic *if, for any selection function f in F which selects an infinite subsequence A_f of A, the subsequence A_f satisfies the law of large numbers. The sequence A is* computably stochastic *or* Church random *if A is F-stochastic for the class F of computable selection functions.*

The concept of stochasticity can be illustrated by the following alternative game theoretic characterization in terms of prediction functions (see e.g. [**32**]). A *prediction function* p is a partial function from $\{0,1\}^{<\omega}$ to $\{0,1\}$. This function p should be viewed as the strategy of a player P in the following game, which proceeds in infinitely many rounds and in which P attempts to predict the bits of an initially completely hidden infinite binary sequence A. This sequence is revealed to the player bit by bit, the $(n+1)$th bit $A(n)$ being revealed at the end of round n. In this round, based on the previously seen bits $A(0), \dots, A(n-1)$, P may make a guess $i \in \{0,1\}$ at the value of $A(n)$ ($p(A[n-1]) = i$) or he may decide not to dare a guess ($p(A[n-1]) \uparrow$). The player P will *win* this game (formally, p will *succeed on* A) if he makes infinitely many guesses and the frequency of his correct guesses surpasses that of the wrong guesses, i.e.

$$\limsup_{n\to\infty} \frac{||\{m < n : p(A[m]) = A(m+1)\}||}{||\{m < n : p(A[m]) \downarrow\}||} > \frac{1}{2}.$$

Then a sequence A is Church random if and only if no player using an effective strategy will win the above game for A, i.e. if no partially computable prediction function with computable domain succeeds on A. (If we admit arbitrary partially computable prediction functions, we obtain a strictly stronger stochasticity concept, which we may call *partially computable stochasticity.*)

From the statistical point of view, however, stochasticity has some shortcomings whence it is not considered an adequate formalization of randomness. This was observed first in 1939 by Ville [**33**]. Ville pointed out that a typical sequence does not only have the property (1) of frequency stability but, in addition, the rate of convergence of the limit in (1) has to obey a certain law, known as the *law of the iterated logarithm.* In particular, in a typical sequence A it cannot happen that in all initial segments $A[n]$ the number of the ones is higher than that of the zeroes.

For any countable F, however, Ville constructed an F-stochastic sequence with the latter property.

Since by this time the measure theoretical foundation of probability theory by Kolmogorov [**9**] was widely accepted, the interest in random sequences in probability theory decreased. This might explain why a more satisfying randomness concept was obtained only in the 60s by Martin-Löf.

Martin-Löf explained the shortcomings of stochasticity as arising from its being based on a single statistical law, namely the law of large numbers. Consequently, he proposed that an algorithmic random sequence should obey all possible effective statistical laws and by identifying (the sequences obeying some) laws with classes of sequences of measure 1, he formalized the notion of a statistical law.

For this purpose Martin-Löf used the fact that a *null class* $\mathbf{C}$, i.e. a class $\mathbf{C}$ of sequences of measure 0, can be described by a sequence of open covers $\{\mathbf{C}_n : n \geq 0\}$ where the measure $\mu(\mathbf{C}_n)$ of the nth cover $\mathbf{C}_n$ is bounded by 2^{-n}. In order to make this more precise, for a given binary string σ let $\mathbf{B}_\sigma$ denote the basic open class consisting of the infinite binary strings A which extend σ. Then a class $\mathbf{C}$ has measure 0 iff $\mathbf{C}$ is contained in the intersection of classes $\mathbf{C}_n$, $n \geq 0$, where each $\mathbf{C}_n$ is the union of countably many basic open classes $\mathbf{B}_{\sigma_{n,m}}$ $(m \geq 0)$ such that the infinite sum of the measures of the individual classes $\mathbf{B}_{\sigma_{n,m}}$ $(m \geq 0)$ is bounded by 2^{-n}. Martin-Löf calls such a cover of a null class a *sequential test*. Then, by defining effective sequential tests, one is led to the notion of an effective null class and thereby to a new notion of effective randomness.

Definition 2.2 (Martin-Löf). *An* effective sequential test *or a* Martin-Löf test *is a computably enumerable set* $T \subseteq \omega \times \{0,1\}^{<\omega}$ *such that, for*

$$T_n = \{\sigma : (n, \sigma) \in T\}$$

and

$$\mathbf{C}(T, n) = \bigcup\{\mathbf{B}_\sigma : \sigma \in T_n\},$$

$\mu(\mathbf{C}(T,n)) \leq 2^{-n}$ *holds for all* n. *A sequence* A fails *the test* T *if* $A \in \mathbf{C}(T,n)$ *for all* $n \geq 0$. *Otherwise* A passes *the test* T. *A class* $\mathbf{C}$ *is a* Martin-Löf null class, *if there is a Martin-Löf test* T *such that all members of* $\mathbf{C}$ *fail* T. *A sequence* A *is* Martin-Löf random *if* A *passes all effective sequential tests.*

No statistical deficiencies are known for Martin-Löf randomness whence this concept is widely considered to be an adequate formalization of algorithmic randomness. Solovay has given an alternative measure theoretic definition of randomness and has shown it to be equivalent to Martin-Löf randomness (see Li and Vitanyi [**20**], page 152): A sequence A is *Solovay random* if, for any effective sequence $\{S_n\}_{n\in\omega}$ of Σ^0_1 subsets of $\{0,1\}^\omega$ with $\sum_{n\in\omega} \mu(S_n)$ finite, there are only finitely many numbers n with $A \in S_n$. Moreover, as Schnorr [**26**] has shown, Martin-Löf's notion, which is based on an effectivization of the notion of typicalness, can also be characterized in terms of effectively chaotic sequences. Here, roughly speaking, a sequence A is considered to be chaotic if its initial segments $A[n]$ are *incompressible*, i.e. the initial segments do not allow any descriptions which are shorter than the initial segments themselves. This idea is effectivized (and thereby made sound) by considering algorithmic incompressibility in the sense of (prefix-free) Kolmogorov complexity. Historically, Kolmogorov's attempt to define randomness in terms of Kolmogorov complexity preceded and motivated Martin-Löf's work. (See the monograph of Li and Vitanyi [**20**] for more details.)

Despite these facts supporting Martin-Löf randomness one should note, however, that Martin-Löf randomness strengthens Church randomness not only from the statistical point of view but also from a computational point of view by passing from computability to (the more general concept of) computable enumerability. So – in contrast to Church randomness which is a (from the probabilistical point of view insufficient) formalization of *computable randomness* – Martin-Löf randomness formalizes *computably enumerable randomness.* Consequently, Martin-Löf random sequences are commonly called Σ^0_1-*random* or 1-*random* sequences.

This – or actually some related observations – led Schnorr [**26**] to criticize Martin-Löf's concept as being too restrictive, and he proposed an alternative concept which overcame the deficiencies of stochasticity but which is tied to computability instead of computable enumerability. For this purpose he refined the notion of a Martin-Löf test T by requiring, in addition, that the measure of the classes $\mathbf{C}(T,n)$ be computable. (This is more than just requiring that the test be computable. It is easy to see that any effective sequential test T can be converted into an equivalent computable test T.)

Definition 2.3 (Schnorr). *A* totally effective sequential test *or* Schnorr test *is a Martin-Löf test T for which the function $m(n) = \mu(\mathbf{C}(T,n))$ is computable. A sequence A is* Schnorr random *if A passes all totally effective sequential tests.*

As Schnorr pointed out, this concept corresponds to the constructive measure introduced by Brouwer [**3**], and he has shown that – in contrast to Church randomness – Schnorr random sequences satisfy the law of the iterated logarithm, whence there are Church random sequences which are not Schnorr random. As recently shown by Wang [**35**], however, Schnorr randomness is *not* a strengthening of Church randomness, namely there are Schnorr random sequences which are not Church random. So, if we agree that computable randomness should entail Church randomness, Schnorr randomness does not formalize computable randomness.

We can clarify the situation somewhat by giving game theoretical characterizations of Martin-Löf randomness and Schnorr randomness which are due to Schnorr [**26**]. These characterizations use effectivizations of martingales, certain betting games which Ville [**33**] had already shown can be used for defining null classes (and the classical measure μ in general). The betting game considered here refines the above prediction game by allowing the player to express different degrees of confidence in his guesses.

The game, in which the player P makes bets on the hidden bits of a sequence A and which again proceeds in infinitely many rounds, is as follows: Initially the player P has a certain capital $d(\lambda)$ and the sequence A he bets against is completely hidden. In round n, after having seen the initial segment $A[n-1]$ of A, the player P will bet a certain part s of his current capital $d(A[n-1])$ on a possible value i of $A(n)$. If the bet is correct the stake is doubled, otherwise the stake is lost. (I.e. $d(A[n]) = d(A[n-1]) + s$ if $A(n) = i$ and $d(A[n]) = d(A[n-1]) - s$ if $A(n) = 1-i$.) The player *wins* if his capital is unbounded in the course of the game.

Formally this game is captured by the following definition.

Definition 2.4 (Ville). *A* martingale *is a function* $d : \{0,1\}^{<\omega} \to \mathbf{R}_+$ *which maps strings to nonnegative real numbers and which satisfies the so-called* martingale property

$$d(\sigma 0) + d(\sigma 1) = 2d(\sigma) \tag{2}$$

for all strings σ. *The martingale* succeeds *on a sequence* A *if*

$$\limsup_{n\to\infty} d(A[n]) = \infty$$

and d *succeeds on a class* $\mathbf{C}$ *of sequences if* d *succeeds on every sequence* A *in* $\mathbf{C}$.

Ville [**33**] has shown that a class $\mathbf{C}$ of sequences has measure 0 if and only if there is a martingale succeeding on $\mathbf{C}$. So typicalness can be defined in terms of martingales and – using this approach – it seems to be natural to define computable randomness as follows.

Definition 2.5. *A sequence* A *is* computably random *if there is no computable martingale which succeeds on* A.

This notion was introduced by Schnorr [**26**] (but he did not use the term computably random). The importance of this notion was stressed by Lutz [**19**] who developed a resource-bounded measure theory based on resource-bounded variants of this randomness concept. Schnorr has also shown that every computably random sequence is computably stochastic, i.e. Church random. Namely every computable prediction function which succeeds on a sequence A can be transformed to a computable martingale which succeeds on A too. In fact, Ambos-Spies et. al. [**1**] have given a characterization of stochasticity in terms of martingales: A sequence A is computably stochastic iff no *simple* computable martingale succeeds on A where a martingale d is simple if the set $\{d(\sigma 0)/d(\sigma) : \sigma \in \{0,1\}^{<\omega}\}$ is finite (i.e. intuitively, if the player has only a finite number of possible choices for the fraction of his current capital which he is going to bet in any round of the game).

The martingale characterizations of Martin-Löf randomness and Schnorr randomness are as follows. A sequence is Martin-Löf random iff no computably enumerable (i.e. computably approximable from below) martingale succeeds on A. This implies that every Martin-Löf random sequence is computably random. The converse, however, fails as shown by Schnorr too.

For a characterization of Schnorr randomness in terms of martingales one has to effectivize the notion of success: A martingale d *effectively succeeds* on a sequence A if there is a computable function $f : \omega \to \omega$ which is nondecreasing and unbounded such that

$$\limsup_{n\to\infty} (d(A[n]) - f(n)) > 0.$$

Then a sequence is Schnorr random iff no computable martingale effectively succeeds on A.

The uniform characterization of the algorithmic randomness concepts above in terms of martingales immediately implies the following implications among these concepts:

$$\begin{array}{ccc}
\Sigma^0_1\text{-randomness (Martin-Löf)} & & \\
\Downarrow & & \\
\text{computable randomness} & \Rightarrow & \text{computable stochasticity (Church)} \\
\Downarrow & & \\
\text{Schnorr randomness} & &
\end{array}$$

(By the negative results mentioned above, no other implications hold.) It is an interesting open question, whether the measure theoretic approach underlying the definitions of Martin-Löf randomness and Schnorr randomness also yields a characterization of computable randomness.

Open Problem 2.6. *Is there a characterization of computable randomness in terms of effective statistical tests?*

An answer to this question might help to solve the problem of giving an adequate formalization of the intuitive notion of computable randomness.

Open Problem 2.7. *Give a formal concept capturing the intuitive notion of computable randomness (if there is any). Is computable randomness in the formal sense of Definition 2.5 adequate?*

Answers to this question may prove controversial but a discussion of it should be of interest. There seems to be a broad consensus that Martin-Löf randomness is an adequate formalization of computably enumerable randomness, not of computable randomness (unless one will argue that, due to missing closure properties of the class of computable functions, both concepts will coincide). If we compare Schnorr randomness with computable randomness then the former captures randomness in the sense of constructive mathematics while the latter is related to randomness in the sense of computability theory. This point of view is supported by the above observation that computable randomness but not Schnorr randomness implies computable stochasticity. Since in addition – in contrast to stochasticity – the martingale approach is satisfying from the statistical point of view in the classical (noneffective) case, we can consider computable randomness as a reasonable candidate for formalizing the underlying intuitive notion.

Yet it is conceivable that classically equivalent concepts might have different strengths when being effectivized whence computable randomness in the sense of Definition 2.5 might be too weak a concept.

In fact – in an unsuccessful attempt to give a characterization of Martin-Löf randomness in the style of stochasticity – more flexible selection rules and betting games have been introduced. The concepts considered so far are monotone, i.e. the selection of a place or the bet on the value of the sequence at this place only depends on the values of the sequence at the previous places. In the 1960s Kolmogorov [**10**] and, independently, Loveland [**18**] gave a more general interpretation of admissible selection rules than that by computable selection functions. A *Kolmogorov-Loveland (KL) selection rule* consists of a pair of partial computable functions that, given a sequence A, depending on the information on A already obtained, first will choose the place n of the next bit $A(n)$ of A to be revealed and second, before examining $A(n)$ will determine whether this bit will be added to the selected subsequence or not. The sequence A is *Kolmogorov-Loveland stochastic* (*KL-stochastic* for short) if all subsequences of it obtained by a Kolmogorov-Loveland selection rule satisfy the law of large numbers. It has been shown by Muchnik (see [**32**]) and Shen' [**27**] that the following implications are strict.

$$\Sigma^0_1\text{-randomness} \Rightarrow \text{KL-stochasticity} \Rightarrow \text{computable stochasticity.}$$

KL-stochasticity extends computable stochasticity in two ways: First it allows partial selection functions (which makes it rather a concept of partially computable randomness than of computable randomness) and second the places of the selected bits are not necessarily chosen in increasing order.

It should be noted, however, that it is questionable whether nonmonotone selection is adequate for characterizing computable randomness. It is easy to see that for any given computably enumerable set B we can give a KL-selection rule which selects the places in B from any given sequence A. So one might argue that – like Martin-Löf randomness – KL-stochasticity is a notion of computably enumerable randomness. An interesting open question about robustness of KL-stochasticity is the following. While, for computably stochastic (random) sequences, every infinite subsequence selected by a computable selection function is computably stochastic (random) again, the corresponding fact is not known for KL-stochasticity (see [**32**]).

Open Problem 2.8. *Is every infinite subsequence of a KL-stochastic sequence which is selected by a KL-selection rule KL-stochastic again?*

Recently, corresponding nonmonotone extensions of the martingale concept have been given by Muchnik, Semenov and Uspensky [**23**] and, independently (in the resource-bounded setting), by Buhrmann et. al. [**2**]. We call a sequence *nonmonotone computably random* if no nonmonotone partially computable martingale succeeds on it (see [**23**] for a formal definition). Muchnik et. al. have shown that nonmonotone computable randomness is a proper strengthening of computable randomness (in fact of partially computable randomness defined in the straightforward way). They left open, however, the question of whether this new concept coincides with Martin-Löf randomness.

Open Problem 2.9. *Is every nonmonotone computably random sequence Σ^0_1-random?*

We conclude our discussion of algorithmic randomness concepts with this interesting question. In the next section we look at Martin-Löf randomness in the context of computability theory.

3. Randomness and degrees of unsolvability

In this section we will look at the distribution of the Martin-Löf random sets from the point of view of computability theory, i.e. we will look at the (Turing) degrees of Martin-Löf random sets. Our presentation here will not be complete. Summaries of further results in this area can be found in [**8**], [**13**], and [**16**]. Here we will focus on the question of which coding techniques are compatible with Martin-Löf randomness.

In this section, as is common in computability theory, we use the term 1-randomness for Martin-Löf randomness and we call a degree 1-random if it contains a 1-random set. The classes of the 1-random sets and 1-random degrees are denoted by RAND and **RAND** respectively.

Since the class RAND of 1-random sets has measure 1, Sacks' theorem that any upper cone $\mathbf{D}(\geq \mathbf{a})$ of Turing degrees with a nonzero base $\mathbf{a}$ has measure 0 (see [**25**]) implies that, for any nonzero degree $\mathbf{a}$ there is a 1-random degree incomparable with $\mathbf{a}$. Moreover, it is immediate by definition that a 1-random set cannot be c.e. hence not computable, whence $\mathbf{0} \notin \mathbf{RAND}$. On the other hand, Schnorr [**26**] has shown that there are 1-random sets in Δ^0_2.

The above and some other elementary results on the distribution of the 1-random sets follow from the fact that the 1-random sets form a Σ^0_2 class, i.e. RAND is the effective union of Π^0_1 classes. This is an immediate consequence of the existence of *universal* Martin-Löf tests ([**22**]): There is a Martin-Löf test U such that a set A is 1-random if and only if A passes the test U, i.e. if and only if $A \notin \mathbf{C}(U,n)$ for some n. Since, by the computable enumerability of U, the classes $\mathbf{C}(U,n)$ are uniformly Σ^0_1, RAND is the effective union of the Π^0_1 classes $\overline{\mathbf{C}(U,n)}$.

This allows us to apply general results on Π^0_1 classes. (For recent surveys on Π^0_1 classes see Cenzer [**4**] and the article by Cenzer and Jockusch in this volume.) For instance, by the Low-Basis Theorem of Jockusch and Soare for Π^0_1 classes [**7**], there is a low 1-random degree. Similarly, by another general observation of Jockusch and Soare on Π^0_1 classes, there is a hyperimmune-free 1-random degree (see [**29**], Exercise 5.15).

The observation that RAND is a Σ^0_2 class can be improved if we are interested only in the 1-random degrees. Namely, the class **RAND** is the class of degrees of the members of some Π^0_1 class (in fact this is true even if we consider m-degrees in place of Turing degrees). This is an immediate consequence of the following *normal form* for 1-random sets.

Theorem 3.1 (Kučera [**11**]; see also [**15**]). *Let U be a universal Martin-Löf test. Then every 1-random set A is the finite shift $\sigma * B$, $\sigma \in \{0,1\}^{<\omega}$, of some member B of the Π^0_1 class $\overline{\mathbf{C}(U,n)}$.*

In order to obtain more specific results on the distribution of the 1-random degrees we have to analyze the relations between randomness and diagonalization. A 1-random set can be viewed as the product of a diagonalization over certain classes which are c.e. approximable in measure. So, as one often intuitively says, a 1-random set has certain diagonalizations built-in. For instance, it is easy to show that, for any 1-random set A, the even part $A_0 = \{n : 2n \in A\}$ and the odd part $A_1 = \{n : 2n+1 \in A\}$ of A are T-incomparable and 1-random again (whence a 1-random degree cannot be minimal among all degrees or among the 1-random degrees) and A itself is effectively bi-immune (see e.g. Theorems 5 and 6 in Kučera [**11**]).

In order to capture the diagonalization strength of 1-random sets, it is instructive to compare 1-randomness with the concept of diagonal nonrecursiveness, which is defined in terms of more explicit general diagonalizations over Σ^0_1 objects (see [**6**]).

Definition 3.2. *A total function $f : \omega \to \omega$ is* diagonally nonrecursive (DNR) *if $f(x) \neq \varphi_x(x)$ for all numbers x. If, in addition, f is 0-1-valued then f is a 0-1-DNR function. A degree* **a** *is (0-1-)DNR if it contains a (0-1-)DNR function.*

Note that the class of 0-1-DNR functions (interpreted as sets) is a Π^0_1 class. Moreover, one can easily show that $\mathbf{0}'$ is a 0-1-DNR degree and the classes of the DNR degrees and 0-1-DNR degrees are closed upwards. The 0-1-DNR degrees coincide with the degrees of complete extensions of Peano arithmetic and also with the degrees of sets which separate an effectively inseparable pair of c.e. sets (see Simpson [**28**] for details). Moreover, the DNR degrees played a central role in Kučera's priority-free solution of Post's problem (see [**12**]).

Kučera [**11**] has shown that every 1-random set computes a DNR function, whence, by upward closure of the DNR degrees, every 1-random degree is a DNR degree.

Theorem 3.3 (Kučera [**11**]). *Every 1-random degree is a DNR degree.*

Here we cannot replace DNR degrees by 0-1-DNR degrees, however, since Jockusch and Soare **[7]** have shown that the class of 0-1-DNR degrees has measure 0.

There is a general technique for effectively coding information into a 0-1-DNR function. This is based on the observation that for any nonempty Π^0_1 class of 0-1-DNR functions $\mathbf{C}$ one can effectively compute (from an index of $\mathbf{C}$) a number x_0 such that $f_0(x_0) = 0$ and $f_1(x_0) = 1$ for some functions $f_0, f_1 \in \mathbf{C}$ (this fact is an analogue of Gödel's incompleteness phenomenon). So the place x_0 can be used to code some bit of information. By iteration, this observation yields a complete computable coding tree for the class $\mathbf{C}$ which allows direct coding of any set into some member of $\mathbf{C}$. (See Kučera **[13]** for a detailed presentation and refinements of this technique.)

In **[11]** and **[13]**, based on similar ideas, Kučera gives a coding technique for 1-random sets. This technique uses the observation that any nonempty Π^0_1 subclass $\mathbf{C}$ of RAND has positive measure $\mu(\mathbf{C}) > 0$ and that a lower bound $\varepsilon > 0$ on $\mu(\mathbf{C})$ can be effectively found. To be more precise, Kučera has shown that (for a given universal Martin-Löf test U and a given number n) for any Π^0_1 class $\mathbf{C}$ which intersects $\overline{\mathbf{C}(U,n)}$ (and therefore contains 1-random sets) one can effectively compute (from an index of $\mathbf{C}$) a real $\varepsilon > 0$ such that $\mu(\mathbf{C}) > \varepsilon$. By viewing the Π^0_1 class $\mathbf{C}$ as the set of the infinite paths through some computable binary tree, this allows the computation of a bound k on the least place x at which the left-most member and the right-most member of $\mathbf{C}$ differ. From this we obtain, not a unique coding location for a bit of information, but some coding interval. Again, by iteration, this yields a coding tree but here this tree is not computable but only computable from the halting problem.

This coding technique has been used by Kučera to show that all levels of the high-low hierarchy of Δ^0_2 degrees contain 1-random sets (**[13]**, Theorem 12 and Remark 8) and that all complete degrees are 1-random:

Theorem 3.4 (Kučera **[11]**). *Every degree* $\mathbf{a} \geq \mathbf{0}'$ *is 1-random.*

This shows that above $\mathbf{0}'$ the 1-random degrees and the 0-1-DNR degrees coincide. One might interprete this as saying that – in the presence of $\emptyset'$ – the above coding techniques for 0-1-DNR functions and 1-random sets coincide. In order to find out whether coding into 1-random sets has to depend on $\emptyset'$, it will be interesting to see whether – and if so, to what extent – a coincidence of 0-1-DNR degrees and 1-random degreees can be found outside the upper cone of the complete degrees.

Open Problem 3.5. *Is there a degree* $\mathbf{a} \not\geq \mathbf{0}'$ *which is both 0-1-DNR and 1-random?*

If so, the following questions are of interest.

Open Problem 3.6. *Is there a degree* $\mathbf{a} \not\geq \mathbf{0}'$ *such that the upper cone* $\mathbf{D}(\geq \mathbf{a})$ *consists entirely of degrees which are both 0-1-DNR and 1-random? Does the upper cone* $\mathbf{D}(\geq \mathbf{a})$ *have this property if* $\mathbf{a}$ *is 0-1-DNR and 1-random?*

Some limitation on the distribution of the simultaneously 1-random and 0-1-DNR degrees has been shown by Kučera.

Theorem 3.7 (Kučera **[11]** and **[13]**). *There is a 0-1-DNR degree* $\mathbf{a}$ *below* $\mathbf{0}'$ *such that no degree* $\mathbf{b} \leq \mathbf{a}$ *is both 1-random and 0-1-DNR.*

This also shows that the class of 1-random degrees is not closed upwards since every 0-1-DNR degree bounds a 1-random degree.

4. Relativized randomness: lowness

In computability theory, one studies not only 1-randomness but also randomness relative to larger classes than the class Σ_1^0 of the c.e. sets. A straightforward extension here is n-randomness ($n \geq 1$) which is obtained by replacing Σ_1^0 by the nth level Σ_n^0 of the arithmetical hierarchy. An alternative way for strengthening randomness is relativization. A set A is *1-random relative to* a set B, *B-1-random* for short, if A passes all sequential tests which are c.e. in B. (In the following we let RAND^B denote the class of B-1-random sets.) We can use relativized randomness to describe the different levels of arithmetical randomness defined above, namely $(n+1)$-randomness coincides with 1-randomness relative to the nth jump $\emptyset^{(n)}$ of the empty set $\emptyset$. In addition, as van Lambalgen has shown, relative randomness can be used to describe independence (see [**16**] for more details):

Theorem 4.1 (van Lambalgen). *Let A be 1-random and let B be 1-random relative to A. Then A is 1-random relative to B and the join $A \oplus B$ of A and B is 1-random.*

Van Lambalgen and Zambella raised the question whether it can happen that, for a noncomputable set A, 1-randomness and 1-randomness relative to A coincide. Such sets are called low for the class RAND, or shortly ML-*low*, according to the following general definition.

Definition 4.2. *A set A is low for a (relativizable) class* $\mathbf{C}$ *if the relativization* $\mathbf{C}^A$ *of the class* $\mathbf{C}$ *relative to A coincides with* $\mathbf{C}$.

The low sets from computability theory are just the sets which are low for the class of the T-complete sets according to this definition. Note that, for an ML-low set A, $\mathrm{RAND} = \mathrm{RAND}^A$. Since RAND is defined in terms of typicalness, intuitively this means that an ML-low set A cannot detect any regularities in any ML-random sequence.

Obviously any computable set is ML-low and the class of ML-low sets is closed downwards under Turing reducibility. Moreover, since obviously no set A is A-1-random and since 1-randomness and A-1-randomness coincide for an ML-low set A, no such A can be 1-random, whence the class of ML-low sets has measure 0.

Recently Kučera and Terwijn [**15**] gave an affirmative answer to the question of van Lambalgen and Zambella by showing that there is a noncomputable ML-low set. In fact they proved that such a set can be computably enumerable:

Theorem 4.3 (Kučera and Terwijn [**15**]). *There exists a simple (hence c.e. but not computable) set A which is ML-low.*

As one can easily show, the proof of this theorem can be modified to construct a promptly simple ML-low set. Since the degrees of the promptly simple sets are just the noncappable c.e. degrees (see Soare [**29**], Chapter XIII), the downward closure of ML-lowness implies that every noncomputable c.e. set bounds a noncomputable c.e. ML-low set. By a result of Kučera ([**14**], Theorem 6), however, not every low c.e. set A is ML-low. Another limiting result on the distribution of the ML-low sets, due to Kučera and Terwijn [**15**], shows that lowness for 1-randomness implies computational lowness: Every ML-low set A is generalized low, i.e. satisfies

$A' \equiv_T A \oplus \emptyset'$. In fact, there is a function $f \leq_T \emptyset'$ which dominates all partial functions which are computable relative to some ML-low oracle.

Kučera and Terwijn's technique for constructing nontrivial ML-low sets only works below the halting problem $\emptyset'$. This leads to the question whether there are ML-low sets which cannot be reduced to $\emptyset'$ and – if so – if there are uncountably many ML-low sets.

Open Problem 4.4. *Is there an ML-low set which is not in Δ_2^0? Do uncountably many ML-low sets exist? Is there a computability theoretic characterization of the degrees of ML-low sets?*

Terwijn and Zambella [**31**] investigated lowness for the class of Schnorr random sets and got a computability theoretical characterization of these sets which surprisingly is quite different from the above results for ML-lowness. Before we describe these results we have to point out, however, that Terwijn and Zambella work with lowness in a somewhat different sense. Instead of looking at lowness for the class of random sets they look at lowness for the corresponding class of null classes. Due to the existence of a Martin-Löf universal test, this does not make a difference in the case of 1-randomness. In the case of Schnorr randomness, however, where universal tests do not exist, the situation might be different. If we call S-*lowness* and S_0-*lowness* lowness for the classes of Schnorr random sets and Schnorr null classes, respectively, then trivially S_0-lowness implies S-lowness but the converse is not known.

Open Problem 4.5. *Is every S-low set also S_0-low?*

Now Terwijn and Zambella's characterization of S_0-lowness is as follows:

Theorem 4.6 (Terwijn and Zambella [**31**]). *A set A is S_0-low if and only if A is computably traceable.*

Here a *computable trace T for a function $f : \omega \to \omega$* is a computable set $T \subseteq \omega \times \omega$ such that for the sections $T^{[k]} = \{m : (k,m) \in T\}$ of T the following holds: $f(k) \in T^{[k]}$, $T^{[k]}$ is finite, and the function which maps k to the canonical index of $T^{[k]}$ is computable. Then a set A is *computably traceable* if there is a computable bound $h : \omega \to \omega$ such that every total function $g \leq_T A$ possesses a computable trace T satisfying $||T^{[k]}|| \leq h(k)$ for all $k \geq 0$.

Terwijn and Zambella use this characterization of Schnorr lowness to show that there are uncountably many S_0-low sets and that the degrees of S_0-low sets are hyperimmune-free. The latter implies that noncomputable S_0-low sets are T-incomparable with the halting problem $\emptyset'$, hence not in Δ_2^0. So, in particular, none of the known noncomputable ML-low sets is S_0-low.

Open Problem 4.7. *Is there a noncomputable set which is both ML-low and S-low (S_0-low)?*

Since, by the above results, lowness for Martin-Löf randomness and Schnorr randomness are very different, it will be interesting to analyze lowness also for other randomness concepts, in particular for the intermediate concept of computable randomness (in the sense of Definition 2.5). Since the proofs for the existence of nontrivial ML-low sets and S-low sets use the charcterization of the corresponding randomness concepts in terms of sequential tests and since we do not know such a characterization for computable randomness (see Problem 2.6 above), the question of the existence of nontrivial low sets for the class of the computably random sets might also be of technical interest.

Open Problem 4.8. *Are there noncomputable sets which are low for the class of computably random sets? If so, what is the relation between these sets and the ML-low (S-low) sets?*

Acknowledgement

We thank an anonymous referee for helpful comments and corrections.

References

[1] K. Ambos-Spies, E. Mayordomo, Y. Wang, X. Zheng. Resource-bounded balanced genericity, stochasticity and weak randomness. In: Proceedings STACS 96, Lect. Notes Comput. Sci. 1046 (1996) 63-74.

[2] H. Buhrmann, D. v. Melkebeek, K. W. Regan, D. Sivakumar, M. Strauss. A generalization of resource-bounded measure with an application. In: Proceedings STACS 98, Lect. Notes Comput. Sci. 1373 (1998) 161-171.

[3] L. E. J. Brouwer. Begründung der Mengenlehre unabhängig vom logischen Satz vom ausgeschlossenen Dritten. Zweiter Teil. Verhdl. Nederl. Akad. Wet. afd Natuurk. Sect. 12, 7. 1919.

[4] D. Cenzer. Π^0_1 classes in recursion theory. In: Handbook of Computability Theory (E. R. Griffor, Ed.), 37-85, North Holland, 1999.

[5] A. Church. On the concept of a random sequence. Bull. Amer. Math. Soc. 46 (1940) 130-135.

[6] C. G. Jockusch, M. Lerman, R. I. Soare, R. M. Solovay. Recursively enumerable sets modulo iterated jumps and extensions of Arslanov's completeness criterion. Journal of Symbolic Logic 54 (1989) 1288-1323.

[7] C. G. Jockusch, Jr. and R. I. Soare. Π^0_1 classes and degrees of theories. Trans. Amer. Math. Soc. 173 (1972) 33-56.

[8] S. M. Kautz. Degrees of random sets. PhD thesis, Cornell University, 1991.

[9] A. N. Kolmogorov. Grundbegriffe der Wahrscheinlichkeitsrechnung. Springer, 1933.

[10] A. N. Kolmogorov. On tables of random numbers. Shankhyā. Ser. A 25 (1963) 369-376.

[11] A. Kučera. Measure, Π^0_1 classes and complete extensions of PA. In: Proceedings "Recursion Theory Week, Oberwolfach", Lect. Notes Math. 1141 (1985) 245-259.

[12] A. Kučera. An alternative, priority-free, solution to Post's problem. In: Proceedings 12th MFCS, Lect. Notes Comput. Sci. 223 (1986) 493-500.

[13] A. Kučera. On the use of diagonally nonrecursive functions, In: Proceedings Logic Colloquium '87, Stud. Logic Found. Math. 129 (1989) 219-239, North-Holland.

[14] A. Kučera. On relative randomess. Annals of Pure and Applied Logic 63 (1993) 61-67.

[15] A. Kučera and S. A. Terwijn. Lowness for the class of random sets. Journal of Symbolic Logic (to appear).

[16] M. van Lambalgen. The axiomatization of randomness. Journal of Symbolic Logic 55 (1990) 1143-1167.

[17] M. van Lambalgen. Random Sequences. PhD thesis, Univ. Amsterdam, 1987.

[18] D. W. Loveland. A new interpretation of von Mises' concept of a random sequence. Z. Math. Logik Grundlagen Math. 12 (1966) 279-294.

[19] J. H. Lutz. Almost everywhere high nonuniform complexity. Journal of Computer and System Sciences 44 (1992) 220-258.

[20] M. Li and P. Vitanyi. An Introduction to Kolmogorov Complexity and Its Applications. Second edition. Springer, 1997.

[21] R. von Mises. Grundlagen der Wahrscheinlichkeitstheorie. Math. Z. 5 (1919) 52-99.

[22] P. Martin-Löf. The definition of random sequences. Inform. Control 6 (1966) 602-619.

[23] A. A. Muchnik, A. L. Semenov, V. A. Uspensky. Mathematical metaphysics of randomness. Theor. Comput. Sci. 207 (1998) 263-317.

[24] H. Rogers, Jr. Theory of Recursive Functions and Effective Computability. McGraw-Hill, 1967.

[25] G. E. Sacks. Degrees of Unsolvability (Rev. Ed.), Ann. of Math. Studies 55, Princeton Univ. Press, 1966.

[26] C. P. Schnorr. Zufälligkeit und Wahrscheinlichkeit. Lect. Notes Math. 218, Springer, 1971.

[27] A. Kh. Shen'. The frequency approach to the definition of a random sequence. Semiotika Inform. 18 (1982) 14-42. (In Russian).

[28] S. G. Simpson. Degrees of unsolvability: a survey of results. In: Handbook of Mathematical Logic (J. Barwise, Ed.), 631-652, North-Holland, 1977.

[29] R. I. Soare. Recursively Enumerable Sets and Degrees. Springer, 1987.

[30] R. I. Soare. The history and concept of computability. In: Handbook of Computability Theory (E. R. Griffor, Ed.), 3-36, North Holland, 1999.

[31] S. A. Terwijn and D. Zambella. Algorithmic randomness and lowness. ILLC Technical Report ML–1997–07, University of Amsterdam, 1997.

[32] V. A. Uspenskii, A. L. Semenov, A. Kh. Shen'. Can an individual sequence of zeros and ones be random? Russian Math. Surveys 45 (1990) 121-189.

[33] J. Ville. Etude Critique de la Notion de Collectif. Gauthier-Villars, 1939.

[34] A. Wald. Die Widerspruchsfreiheit des Kollektivbegriffs der Wahrscheinlichkeitsrechnung. Ergebnisse eines math. Koll. 8 (1936) 38-72.

[35] Y. Wang. Randomness and Complexity. PhD thesis, Univ. Heidelberg, 1996.

MATHEMATISCHES INSTITUT, UNIVERSITÄT HEIDELBERG, IM NEUENHEIMER FELD 294, D-69120 HEIDELBERG, GERMANY

DEPARTMENT OF THEORETICAL COMPUTER SCIENCE, CHARLES UNIVERSITY PRAGUE, MALOSTRANSKÉ NÁM. 25, CZ-118 00 PRAGUE 1, CZECH REPUBLIC

Contemporary Mathematics
Volume **257**, 2000

Open questions about the n-c.e. degrees

Marat Arslanov

1. Introduction

In this paper, we discuss the problem of the similarity of the first-order theories of the partial orderings of the n-c. e. degrees for different $n > 1$. The question of whether the first-order theories of the n-c. e. degrees are elementarily equivalent, i. e., whether they satisfy the same first-order sentences, is still open. The only progress is the proof of the non-elementary equivalence of the partial orderings of the c. e. degrees ($\mathcal{R}$) on the one hand and the n-c. e. degrees for $n > 1$ ($\mathcal{D}_n$) on the other hand. This was first shown by Arslanov [1985,1988] who proved that $(\mathcal{R}, <)$ and $(\mathcal{D}_n, <)$ are different at the Σ_3^0-level. Differences between $(\mathcal{R}, <)$ and $(\mathcal{D}_n, <)$ at the Σ_2^0-level later were demonstrated by Downey [1989] and by Cooper, Harrington, Lachlan, Lempp and Soare [1991]. (Since both $\mathcal{R}$ and $\mathcal{D}_n, n > 1$, allow the embedding of any finite partial order, they are elementarily equivalent at the Σ_1^0-level.)

Downey [1989] conjectured that the n-c. e. degrees (for all $n > 1$) are elementarily equivalent.

In this paper, we describe some recent results related to this conjecture. They consist of two parts. In the first part, we consider sentences in an expanded language which distinguish $\mathcal{D}_n$ for different $n > 1$ and study the complexity of these sentences. In the second part, we describe Kalimullin's recent negative solution to Downey's Conjecture for an extended structure, namely, the enumeration degrees (or e-degrees for short) $\mathbf{D}_E$: The mapping $\iota : \mathbf{D}_T \longrightarrow \mathbf{D}_E$ defined by $\iota(\deg_T(A)) = \deg_e(A \oplus \bar{A})$ defines an order-theoretic embedding of the set of all Turing degrees $\mathbf{D}_T$ into $\mathbf{D}_E$ preserving joins and least element. (For a detailed survey of the structure of the e-degrees see Cooper [1990] and Sorbi [ta].)

Our notation is generally standard and follows Soare [1987]; thus sets will be identified with their characteristic functions. Throughout the paper we will refer to some fixed acceptable numbering $\{W_e : e \in \omega\}$ of the computably enumerable (c. e.) sets. Recall that if a set V is c. e. in another set A and $A \leq_T V$, then V is said to be A-CEA.

1991 *Mathematics Subject Classification.* Primary: 03D25.
Key words and phrases. d. c. e. degree, n-c. e. degree.
Partially supported by RFBR Grant 99-01-00174.

For $n \geq 1$, we say that a degree $\mathbf{a}$ is n-CEA if either $n = 1$ and $\mathbf{a}$ is c. e., or $n > 1$ and $\mathbf{a}$ is c. e. in some $(n-1)$-CEA degree $\mathbf{b} \leq \mathbf{a}$. We say $\mathbf{b}$ is $CEA(\mathbf{a})$, or $\mathbf{b}$ is $\mathbf{a}$-CEA, if $\mathbf{b}$ is c. e. in $\mathbf{a}$ and $\mathbf{a} \leq \mathbf{b}$. As usual, $\bar{X}$ denotes the complement $\omega - X$ of X.

A set A is n-c. e. if there is a computable function $f(s, x)$ such that for every x,

f(0,x) = 0,

$\lim_s f(s, x) = A(x)$, and

$| \{s : f(s, x) \neq f(s+1, x)\} | \leq n$.

The 2-c. e. sets are also known as the d-c. e. sets as they are the differences of c. e. sets, i. e. the ones of the form $B - C$ with both B and C c. e. Similarly the n-c. e. for $n > 1$ are the sets of the form $B - C$ with B c. e. and D is $(n-1)$-c. e. A degree $\mathbf{a}$ is called an n-c. e. degree if it contains an n-c. e. set, and it is called a *properly* n*-c. e. degree* if it contains an n-c. e. set but no $(n-1)$-c. e. set.

It follows from Proposition 1.1 that any n-c. e. degree is n-CEA. The converse fails inside $\mathcal{D}(\leq \mathbf{0}')$ even for $n = 2$ (Jockusch and Shore [1984]). Theorem 1.2 states that, inside the 2-CEA degrees, the n-c. e. degrees for $n \geq 3$ collapse to the d-c. e. degrees.

PROPOSITION 1.1. (Lachlan, unpublished) *If* $\mathbf{d} > \mathbf{0}$ *is* n*-c. e. for* $n > 1$ *then there is an* $(n-1)$*-c. e. degree* $\mathbf{a} \leq \mathbf{d}$ *such that* $\mathbf{d}$ *is c. e. in* $\mathbf{a}$.

THEOREM 1.2. (Arslanov, LaForte and Slaman [1998]) *For any 2-CEA degree* $\mathbf{a}$ *and* ω*-c. e. degree* $\mathbf{b} \leq \mathbf{a}$ *there is a d-c. e. degree* $\mathbf{d}$ *such that* $\mathbf{b} \leq \mathbf{d} \leq \mathbf{a}$. *In particular, for any 2-CEA degree* $\mathbf{a}$ *if* $\mathbf{a}$ *is* ω*-c. e. then* $\mathbf{a}$ *is d-c. e.*

2. Turing degrees

A difference between the elementary theories of the structures $\mathcal{D}_n$ and $\mathcal{D}_m$ for $n, m \geq 2, n \neq m$, in an expanded language can be obtained from the following result of Jockusch and Shore [1984].

THEOREM 2.1. (Jockusch and Shore [1984]) *For any* $n \geq 2$, *there is a* $(n+1)$*-c. e. degree which is not* n*-CEA.*

Let CE($\mathbf{x}$) denote the relation "$\mathbf{x}$ is c. e.", and CEIN($\mathbf{x}, \mathbf{y}$) the relation "$\mathbf{x}$ is c. e. in $\mathbf{y}$". Then CE($\mathbf{x}$) = CEIN($\mathbf{x}, \mathbf{0}$). It follows from Proposition 1.1 and Theorem 2.1 that for all $n, m \geq 2, m < n$, $\mathcal{D}_m \models \Psi$ and $\mathcal{D}_n \models \neg\Psi$, where Ψ denotes the following sentence:

$$\forall \mathbf{x}_1 > \mathbf{0} \exists \mathbf{x}_2 \ldots \exists \mathbf{x}_m (\mathbf{x}_1 > \ldots \mathbf{x}_m \ \& \ CE(\mathbf{x}_m) \ \&$$

$$\& \ CEIN(\mathbf{x}_1, \mathbf{x}_2) \ \& \ \ldots \ \& \ CEIN(\mathbf{x}_{m-1}, \mathbf{x}_m)).$$

Therefore, the structures $\mathcal{D}_m$ and $\mathcal{D}_n$, $n, m \geq 2, m \neq n$, are not elementarily equivalent if the relation "*c. e. in*" is uniformly definable (i. e., by the same formula) in both the m-c. e. and the n-c. e. degrees.

For another difference (but which is not definable in this expanded language) between m-c. e. and n-c. e. degrees for $n, m \geq 2, n \neq m$, consider the following well-known result (see, for instance, Jockusch and Shore [1985]): n-CEA degrees are dense for all $n \geq 1$. Since by Proposition 1.1 each n-c. e. degree is n-CEA, it follows that for all n-c. e. degrees $\mathbf{a} < \mathbf{b}$ there is an n-CEA degree $\mathbf{c}$ such that $\mathbf{a} < \mathbf{c} < \mathbf{b}$.

But there are 3-c. e. degrees $\mathbf{a} < \mathbf{b}$ such that there is no 2-CEA degree between them. To prove this, consider properly 3-c. e. degrees $\mathbf{u} < \mathbf{v}$ such that there is no *d*-c. e. degree between $\mathbf{u}$ and $\mathbf{v}$ (Hay and Lerman, see Arslanov [1997, Theorem 3.7]). Suppose that there is some 2-CEA degree $\mathbf{c}$ such that $\mathbf{u} \leq \mathbf{c} \leq \mathbf{v}$. By Theorem 1.2, we have $\mathbf{u} < \mathbf{c} < \mathbf{v}$. Now it follows from Theorem 1.2 that there is a d-c. e. degree $\mathbf{b}$ such that $\mathbf{u} < \mathbf{b} \leq \mathbf{c}$, which is a contradiction.

Similar arguments work for higher levels of the hierarchy. Namely, the following theorem is a generalized version of Theorem 2.1 and can be proved in the same way. (Note that the same result with almost the same proof was obtained also in Arslanov [1982].)

THEOREM 2.2. *The n-CEA degrees are not dense in the $(n+1)$-c. e. degrees for all $n > 1$, i. e., there exist $(n+1)$-c. e. degrees $\mathbf{u} < \mathbf{v}$ such that there is no n-CEA degree between $\mathbf{u}$ and $\mathbf{v}$.*

Theorems 1.2 and 2.1 imply the following

THEOREM 2.3. i) *The structures $(\mathcal{D}_n, CEIN, <)$ and $(\mathcal{D}_m, CEIN, <)$ are not elementarily equivalent for $n \neq m, n, m \geq 1$.*

ii) *The relations "c. e." and "d-c. e." are definable in $(\mathcal{D}_n, CEIN, <)$ for any $n \geq 3$.*

Theorem 2.3 immediately suggests the following

QUESTION 2.4. *i) Is the relation "$\mathbf{x}$ is c. e. in $\mathbf{y}$" definable in $\mathcal{D}_n$ for each $n \geq 2$?*

ii) Is the relation "$\mathbf{x}$ is c. e." definable in $\mathcal{D}_n$ for each $n \geq 2$? Is the relation "$\mathbf{x}$ is m-c. e." definable in $\mathcal{D}_n$ for each $n > m, m \geq 2$?

iii) (Lempp) *Are $(\mathcal{D}_2, CE, <)$ and $(\mathcal{D}_3, CE, <)$ elementarily equivalent?*

Concerning item ii) of this question, note the following

THEOREM 2.5. (Cooper and Yi [1996] for $n = 2$, Arslanov, LaForte and Slaman [1998] for $n > 2$) *Let $n \geq 1$. For any n-c. e. degree $\mathbf{b}$ and c. e. degree $\mathbf{a} < \mathbf{b}$, there exists some d-c. e. $\mathbf{c}$ with $\mathbf{a} < \mathbf{c} < \mathbf{b}$.*

Now it is natural to ask the following

QUESTION 2.6. *Is the converse of this statement true, i. e., for any non-c. e. n-c. e. degree $\mathbf{a}$, is there an n-c. e. degree $\mathbf{b} > \mathbf{a}$ such that there is no n-c. e. degree $\mathbf{c}$ strictly between them?*

An affirmative answer to this question would provide the following interesting definition of the c. e. degrees in $\mathcal{D}_n, n \geq 2$:

$$\mathbf{a} \text{ is c. e. iff } \forall \mathbf{b} > \mathbf{a} \exists \mathbf{c} (\mathbf{b} > \mathbf{c} > \mathbf{a}).$$

We conjecture, however, that the answer to this question is negative.

The above considerations illustrate the fundamental role of relative enumerability in the study of the n-c. e. degrees, in particular in the study of the above-mentioned open questions.

Relative enumerability is also closely related to problems in effective model theory. E. g., by an observation of Downey and Jockusch (see Coles, Downey and Slaman [ta]), there is a presentation of least Turing degree for any torsion-free abelian group of finite type and rank 1 iff for any set $X \subseteq \omega$, the set

$$\{\deg(A') \mid X \text{ is computably enumerable in } A\}$$

has a least element.

This latter condition is an instance of the problem of studying which degrees **a** have the property that a fixed degree **x** is computably enumerable in **a**. This question is trivial if **x** is computably enumerable, or if **x** is 1-generic (the answer being "all **a**" and "all $\mathbf{a} \geq \mathbf{x}$", respectively); but very little is known about the general case. We summarize the results known to us in Theorem 2.7. It states that non-c. e. Δ_2^0-sets cannot be c. e. in a minimal pair of c. e. degrees, and non-c. e. d-c. e. sets cannot be c. e. in a minimal pair of (not necessarily c. e.) degrees. On the other hand, there is a 3-c. e. set which is c. e. in a minimal pair of degrees. Finally, item (iv) of the Theorem states that (i) and (ii) do not hold for degrees instead of sets.

THEOREM 2.7. (i) *There is no non-c. e. d-c. e. set D which is c. e. in some noncomputable sets A and B such that* $\deg(A) \cap \deg(B) = \mathbf{0}$.

(ii) *There is no non-c. e. Δ_2^0-set C which is c. e. in some noncomputable c. e. sets A and B such that* $\deg(A) \cap \deg(B) = \mathbf{0}$.

(iii) (R. Coles, R. Downey, and T. Slaman [ta]) *There are a non-c. e. 3-c. e. set C and noncomputable sets A and B such that C is c. e. in A and B, and* $\deg(A) \cap \deg(B) = \mathbf{0}$.

(iv) *There are sets A, B and D such that D is of properly d-c. e. degree, A and B are c. e. ,*$\deg(A) \cap \deg(B) = \mathbf{0}$, *and D c. e. in A, and $\bar{D}$ c. e. in B.*

Proof. (i) Let $D = D_1 - D_2$ be a d-c. e. set, $X \leq_T D$ be a set such that D is c. e. in X. Then $f^{-1}(D) \leq_T X$ for any $1-1$ computable function f whose range is D_1. Indeed, let $D = \mathrm{dom}\Phi^X$ for a partial computable functional Φ. Given x, wait for a stage s such that either $f(x) \in D_{2,s} - D_{2,s-1}$, or $\Phi^X(f(x))$ is defined at stage s. If the former case holds then $x \notin f^{-1}(D)$, if the latter case holds then $x \in f^{-1}(D)$. Also $f^{-1}(D)$ is obviously co-c. e., and if D is not c. e. then $f^{-1}(D)$ is noncomputable.

(ii) Let $C = \lim_s C_s$, and $C = \mathrm{dom}\Phi^A = \mathrm{dom}\Psi^B$, where A and B are given c. e. sets. We define a c. e. set D such that $D \leq_T A, B$, and C is c. e. in D. In particular, if C is not c. e. then D is not computable.

Let

$$D_v = \{\langle s, x\rangle : x \leq s \leq v \;\&\; x \notin C_v \;\&$$

$$\forall t \leq v \forall \sigma \forall \tau (x \in C_t \;\&\; \Phi_t^\sigma(x)\downarrow \to \sigma \not\subseteq A_v) \;\&\; (x \in C_t \;\&\; \Psi_t^\tau(x)\downarrow \to \tau \not\subset eqB_v)\}.$$

We then have

a) $x \in C \leftrightarrow \exists s(\langle s, x\rangle \notin D)$, and

b) $\langle s, x\rangle \notin D \leftrightarrow \exists v \exists \sigma \subset A(x \in C_v \;\&\; \Phi_v^\sigma(x)\downarrow \;\&\; x \notin \cup_{v'<v} D_{v'})$.

It follows from a) that C is c. e. in D. It follows from b) that $\omega - D$ is c. e. in A. Since D is c. e., $D \leq_T A$. Similarly for $D \leq_T B$.

Now we briefly sketch the proof of (iv). We construct c. e. sets A, B, and a d-c. e. set D, meeting the following requirements:

$$\mathcal{R}_e : D \neq \Phi_e^{W_e} \vee W_e \neq \Psi_e^D,$$

$$\mathcal{N}_e : \Phi_e^A = \Phi_e^B = \text{total function } f \to f \text{computable},$$

where $\{(W_e, \Phi_e, \Psi_e)\}_{e\in\omega}$ is some enumeration of all possible triples consisting of a c. e. set W and partial computable functionals Φ and Ψ. In addition, we will ensure that D is c. e. in A and $\bar{D}$ c. e. in B using a common method which works as follows. When a witness x is enumerated into D (or $\bar{D}$, respectively) at stage s

then we appoint a certain marker $\alpha(x)$ ($\beta(x)$, respectively). Then we allow x to be removed from D (or $\bar{D}$, respectively) at a later stage t only if $A \upharpoonright \alpha(x) \neq A_t \upharpoonright \alpha(x)$ (or $B \upharpoonright \beta(x) \neq B_t \upharpoonright \beta(x)$, respectively).

Obviously, this ensures that D is c. e. in A and $\bar{D}$ is c. e. in B, and all requirements together prove (iv). In particular, $A \not\leq_T B$ and $B \not\leq_T A$. (If, for instance, $B \leq_T A$, then by the $\mathcal{N}$-requirements, $B \equiv_T \emptyset$ and $\bar{D}$ c. e. in B. It follows that $\deg(D)$ is c. e., which contradicts the $\mathcal{R}$-requirements.)

All the strategies are the obvious ones. To meet the $\mathcal{N}_e$- requirements, we use the usual minimal pair strategy. To meet an $\mathcal{R}_e$-requirement we choose an unused candidate x and a marker $\beta(x)$ (since at the beginning $x \in \bar{D}$) greater than any number mentioned thus far in the construction, and wait for a stage s at which

$$D(x) = \Phi_e^{W_e}(x) \;\&\; W_e \upharpoonright \varphi_e(x) = \Psi_e^{D \upharpoonright \psi_e \varphi_e(x)} \upharpoonright \varphi_e(x),$$

and $\beta(x)$ is greater than the $\mathcal{N}$-restraints of higher priority. (If the former never happens then x is a witness to the success of R, and if $\beta(x)$ is not greater than the $\mathcal{N}$-restraints of higher priority then we choose a new witness). Then preserve $D \upharpoonright \psi_{e,s}\varphi_{e,s}(x)$, enumerate x into D, $\beta(x)$ into B, define $\alpha(x)$ greater than any number mentioned thus far in the construction, and wait for a stage t at which

$$D(x) = \Phi_e^{W_e}(x) \;\&\; W_e \upharpoonright \varphi_e(x) = \Psi_e^{D \upharpoonright \psi_e \varphi_e(x)} \upharpoonright \varphi_e(x).$$

Then remove x from D, enumerate $\alpha(x)$ into A, and preserve $D \upharpoonright \psi_{e,t}\varphi_{e,t}(x)$.

It should now be clear how to carry out a construction to meet all the requirements and hence to prove the theorem. □

Finally, there are questions of decidability. We already mentioned that the existential theory of the n-c. e. degrees decidable for any $n \geq 1$. By a result of Slaman (1990, unpublished), their $\forall\exists\forall\exists$-theories are undecidable.

QUESTION 2.8. *Is the $\forall\exists$-theory of $\mathcal{D}_n$ is decidable? Is the $\forall\exists\forall$-theory of $\mathcal{D}_n$ undecidable for each $n > 1$? Is the first-order theory of $\mathcal{D}_n$ the same at the $\forall\exists$-level for each $n \geq 2$?*

3. Enumeration degrees

In this section we consider the n-c. e. enumeration degrees. We denote by $\mathcal{D}_n^E$ the set of all n-c. e. e-degrees for $n \geq 2$.

DEFINITION 3.1. *An e-degree $\mathbf{a}$ is n-splittable avoiding an e-degree $\mathbf{c}$ ($n > 1$) if there are e-degrees $\mathbf{b}_1, \mathbf{b}_2, \ldots, \mathbf{b}_n$ such that $\mathbf{a} = \mathbf{b}_1 \cup \mathbf{b}_2 \cup \cdots \cup \mathbf{b}_n$ and $\mathbf{c} \not\leq \mathbf{b}_i$ for all i, $1 \leq i \leq n$.*

THEOREM 3.2. (Kalimullin [ta]) *If $1 < m < 2n$ then every m-c. e. e-degree $\mathbf{a}$ is n-splittable avoiding any nonzero Δ_2^0 e-degree $\mathbf{c}$ by some m-c. e. e-degrees $\mathbf{b}_1, \mathbf{b}_2, \ldots, \mathbf{b}_n$.*

Proof. Choose an m-c. e. set $A \in \mathbf{a}$ and a Δ_2^0-set $C \in \mathbf{c}$. Let $A = \lim_s A_s$ and $C = \lim_s C_s$ be the appropriate computable approximations. We may assume that $A_0(x) = 0$ and $|\{s : A_s(x) \neq A_{s+1}(x)\}| \leq m$ for all x and, hence, $|\{s : x \in A_s - A_{s+1}\}| < n$. Assume also, that $|\{x : A_s(x) \neq A_{s+1}(x)\}| \leq 1$ for all s.

We construct sets B_i as $A \cup V_i$, where $1 \leq i \leq n$ and V_i are c. e. sets (therefore $B_i \leq_e A$) such that the degrees $\mathbf{b}_i = \deg(B_i)$ are as desired.

We construct V_i as $\{V_{i,s}\}_{s\in\omega}$ so that $V_i = \bigcup_s V_{i,s}$. To arrange that B_i is m-c. e. it is sufficient to achieve

$$x \in V_{i,s+1} - V_{i,s} \Longrightarrow x \in A_s - A_{s+1}$$

for all x, s and i. Furthermore, in our construction, we will have at every stage s at most one i such that $x \in V_{i,s+1} - V_{i,s}$. Since $|\{s : x \in A_s - A_{s+1}\}| < n$, this implies that $V_1 \cap V_2 \cap \cdots \cap V_n = \emptyset$.

Thus $A = B_1 \cap B_2 \cap \cdots \cap B_n$, and $A \equiv_e B_1 \oplus B_2 \oplus \cdots \oplus B_n$. To meet $\mathbf{c} \not\leq \mathbf{b}_i$, it is sufficient to satisfy for all $e \in \omega$ and $i = 1, 2, \ldots, n$ the following requirements:

$$\mathcal{N}_{e,i} : C \neq \Phi_e^{A\cup V_i}.$$

For any finite set G, define the use-function $u(G; e, x, s)$ as the finite set F with the least canonical number such that $F \subseteq G$ and $\langle x, F\rangle \in \Phi_{e,s}$. If there is no such F, then $u(G; e, x, s) = \emptyset$.

The construction.

Stage $s = 0$. Set $V_{i,0} = \emptyset$ for all i, $1 \leq i \leq n$.

Stage $s + 1$. Given $V_{i,s}$, define the following:

(length function) $l(e, i, s) = \max\{x \leq s : C_s \restriction x \subseteq \Phi_{e,s}^{A_s\cup V_{i,s}} \restriction x\}$,

(restraint function) $R(e, i, s) = \bigcup\{F : (\exists t \leq s)(\exists x < l(e, i, t))[F = u(A_t \cup V_{i,t}, e, x, t)]\}$,

where $1 \leq i \leq n$.

If $A_s \subseteq A_{s+1}$ then do nothing (i. e., $V_{i,s+1} = V_{i,s}$ for all i). If $x \in A_s - A_{s+1}$ (such x must be unique) then choose the least pair $\langle e_0, i_0\rangle \leq s$, $1 \leq i_0 \leq n$, such that $x \notin V_{i_0,s}$ and $x \in R(e_0, i_0, s)$ (if there is no such $\langle e_0, i_0\rangle$ then do nothing). Let $V_{i,s+1} = V_{i,s}$ for all $i \neq i_0$, $1 \leq i \leq n$, and $V_{i_0,s+1} = V_{i_0,s} \cup \{x\}$ (preserving the restraints of N_{e_0,i_0}).

Let $V_i = \bigcup_s V_{i,s}$ and $R(e, i) = \bigcup_s R(e, i, s)$. Assume for the sake of a contradiction that there is $\langle e, i\rangle$, $1 \leq i \leq n$, which is the least pair such that $|R(e, i)| = \infty$ or $N_{e,i}$ is not satisfied. Then $\limsup_s l(e, i, s) = \infty$. Choose a stage s' such that $A_s(x) = A_{s'}(x)$ for all $s \geq s'$, $x \in \bigcup\{R(e', i') : \langle e', i'\rangle < \langle e, i\rangle, 1 \leq i' \leq n\}$. Then

$$x \in C \Longleftrightarrow (\exists s > s')[x \in C_s \;\&\; x < l(e, i, s)]$$

for every $x \in \omega$.

Thus, $R(e, i)$ is finite and $N_{e,i}$ is satisfied for every e, i, $1 \leq i \leq n$, since C is not c. e. □

THEOREM 3.3. (Kalimullin [ta]) *For all $n > 1$, there is a $2n$-c. e. e-degree* $\mathbf{a}$ *which is not n-splittable avoiding some nonzero 3-c. e. e-degree* $\mathbf{c}$.

Theorems 3.2 and 3.3 immediately imply the following

COROLLARY 3.4. (Kalimullin [ta]) *If $1 < m < 2p \leq n$ for some integer p then the structures $(\mathcal{D}_m^E, <)$ and $(\mathcal{D}_n^E, <)$ are not elementarily equivalent.*

Proof. Let $\mathcal{F}_p$ be the formula

$$(\forall \mathbf{a})(\forall \mathbf{c} > \mathbf{0})(\exists \mathbf{b}_1)\ldots(\exists \mathbf{b}_p)[\mathbf{a} = \mathbf{b}_1 \cup \cdots \cup \mathbf{b}_p \;\&\; \mathbf{c} \not\leq \mathbf{b}_1 \;\&\; \ldots \;\&\; \mathbf{c} \not\leq \mathbf{b}_p].$$

Then by Theorem 3.2 and Theorem 3.3 we have $\mathcal{D}_m^E \models \mathcal{F}_p$ iff $m < 2p$, for all $m > 1$. □

For $n > 1$, the question of the elementary equivalence of the partial orderings $\{D_{2n}^E, \leq\}$ and $\{D_{2n+1}^E, \leq\}$ remains open. But note the following

THEOREM 3.5. (Kalimullin [ta]) *The structures* $(\mathcal{D}^E_{2n}, P, <)$ *and* $(\mathcal{D}^E_{2n+1}, P, <)$ *are not elementarily equivalent for any* $n > 1$. *(Here* $P(\mathbf{x})$ *denotes that* $\mathbf{x}$ *is a* Π^0_1*-e-degree.)*

The properties of the n-c. e. e-degrees have not been carefully studied. Below, we give a brief overview of known results and current open questions about the n-c. e. e-degrees for $n > 1$.

It follows from Proposition 1.1 that each $(n+1)$-c. e. e-degree contains a co-n-c. e. set; in particular, each d-c. e. e-degree is Π^0_1 and hence total. Below any nonzero n-c. e. e-degree $(n \geq 2)$ there is a properly 3-c. e. e-degree, therefore $(\mathcal{D}^E_n, <)$ has no minimal elements for any $n \geq 2$ (Arslanov, Kalimullin, Sorbi [ta]). On the other hand, it is easy to show that there is a nonzero ω-c. e. e-degree which bounds no nonzero n-c. e. e-degree for any $n > 2$.

The diamond lattice is embeddable into the 3-c. e. e-degrees preserving $\mathbf{0}_e$ and $\mathbf{0}'_e$ (Kalimullin [ta]). (Therefore, the structures $(\mathcal{D}^E_2, <)$ and $(\mathcal{D}^E_n, <)$, $n > 2$, are also not elementarily equivalent.)

In conclusion, we consider the following important open question:

QUESTION 3.6. *Are the* n*-c. e. e-degrees dense for each* $n \geq 3$*? Are the* ω*-c. e. e-degrees dense?*

An affirmative answer to the following partial question could help obtain a negative answer to Question 3.6:

QUESTION 3.7. *Let* $\mathbf{a} < \mathbf{0}'_e$ *be an* ω*-c. e. e-degree. Is there a total* ω*-c. e. e-degree* $\mathbf{b}$ *such that* $\mathbf{a} \leq \mathbf{b} < \mathbf{0}'_e$*?*

Indeed, let $D <_T \emptyset'$ be a d-c. e. set with no ω-c. e. set C such that $D <_T C <_T \emptyset'$ (as constructed in Cooper, Harrington, Lachlan, Lempp and Soare [1991]). Then the 3-c. e. set $D \oplus \bar{D}$ has maximal e-degree below $\mathbf{0}'_e$. Indeed, we have $D \oplus \bar{D} <_e \bar{K}$. Suppose that there is an ω-c. e. set C such that $D \oplus \bar{D} <_e C <_e \bar{K}$. Then, if the answer to Question 3.7 is "yes", there is an ω-c. e. set X such that $C <_e X \oplus \bar{X} <_e \bar{K}$. Therefore, $D \oplus \bar{D} <_e X \oplus \bar{X} <_e \bar{K}$ and $D <_T X <_T K$, which contradicts the choice of D.

Below, we briefly sketch the following observation which shows that the answer to Question 3.7 is "yes" at least for Δ^0_2-enumeration degrees. The full proof can be found in Arslanov, Cooper, Kalimullin, Li [ta].

THEOREM 3.8. *Let* $\mathbf{a} < \mathbf{0}'_e$ *be a* Δ^0_2*-e-degree. Then there is a total* Δ^0_2*-e-degree* $\mathbf{b}$ *such that* $\mathbf{a} \leq \mathbf{b} < \mathbf{0}'_e$.

Proof. Let $A <_e \bar{K}$ be a Δ^0_2-set. To ensure that $A \leq_e X \oplus \bar{X} <_e \bar{K}$ for a set X, it is enough to construct X such that $A \leq_e X$ and $X <_T K$. We need to satisfy the requirements:

$$P : x \in A \leftrightarrow \exists y(\langle x, y\rangle \in X)$$

(which obviously gives $A \leq_e X$), and

$$S_e : K = \Phi^X_e \to \bar{K} = \Psi^A_e$$

(i. e., given a Turing partial-computable functional Φ_e, we build an enumeration operator Ψ_e).

Strategy for P: Given $x \in A$ at a stage s, enumerate $\langle x, y\rangle$ into X for a big y. If later x leaves A then remove $\langle x, y\rangle$ from X. If later again x enters A then enumerate $\langle x, y'\rangle$ into X with a new big y', etc.

Strategy for S_e: For any given x, while $x \in \bar{K}$:

1) Wait for $\Phi^{X_s}_{e,s}(x) \downarrow= 0$ at a stage s with a *use* $\varphi_{e,s}(x)$. Let $F_x = \{y \leq \varphi_{e,s}(x) \mid y \in X_s\}$. By the P-strategy, for any $y \in F_x$, we have $y = \langle a, b\rangle$ for some $a \in A_s$. Define $F_{A,x} = \{a \mid \exists b \exists y(y \in F_x \;\&\; y = \langle a, b\rangle\}$.

2) Enumerate $\langle x, F_{A,x}\rangle$ into Ψ^A_e, and restrain $X \restriction \varphi_{e,s}(x)$ from other strategies from now on.

In case of any A-change at $F_{A,x}$ go back to 1).

If later $x \in K$, and again $K(x) = \Phi^{X_t}_{e,t}(x)$ at a stage $t > s$, then this means that we had an $X \restriction \varphi_{e,s}(x)$-change between stages s and t, which means that some $y \leq \varphi_{e,s}(x)$ is removed from F_x (since nothing from this interval can be enumerated into F_x after stage s). This means, by the P-strategy, that an element $a \in F_{A,x}$ is removed from A between stages s and t. Therefore, either $x \notin \Psi^A_e$, or later x enters A again. In the latter case, we may diagonalize $K(x)$ against $\Phi^X_e(x)$, enumerating back into X all elements F_x removed from X between stages s and t.

The explicit construction can now be carried out using a standard priority argument, so we will not give it here. □

References

[1] Arslanov, M. M. [1982], On a hierarchy of degrees of unsolvability, *Ver. Metodi i Kibernetika* **18**, 10-18 (In Russian).

[2] Arslanov, M. M. [1985], Structural properties of the degrees below $\mathbf{0}'$, *Sov. Math. Dokl. N.S.* **283** no. 2, 270-273.

[3] Arslanov, M. M. [1988], On the upper semilattice of Turing degrees below $\mathbf{0}'$, *Sov. Math.* **7**, 27-33.

[4] Arslanov, M. M., Cooper, S.B., Kalimullin, I. Sh., Li, A. [ta], Splitting properties of Σ^0_2-enumeration degrees, to appear.

[5] Arslanov, M. M., LaForte, G. L. and Slaman, T. A. [1998], Relative enumerability in the difference hierarchy, *J. Symb. Logic* **63**, 411-420.

[6] Arslanov, M. M., Kalimullin, I. Sh., Sorbi, A. [ta], Density results in the Δ^0_2 e-degrees, to appear.

[7] Coles, R. J., Downey, R. G., and Slaman, T. A. [ta], Every set has a least jump enumeration, to appear.

[8] Cooper, S. B. [1990] Enumeration reducibility, nondeterministic computations and relative computability of partial functions, in: K. Ambos-Spies, G. Müller, and G. E. Sacks, editors, *Recursion Theory Week, Oberwolfach 1989*, volume 1432 of *Lecture Notes in Mathematics*, pages 57-110, Heidelberg, Springer-Verlag.

[9] Cooper, S. B., Harrington, L., Lachlan, A. H., Lempp, S. and Soare, R. I. [1991], The *d*-c. e. degrees are not dense, *Ann. Pure and Applied Logic* **55**, 125-151.

[10] Cooper, S. B., Yi, X. [1996], Isolated d-r. e. degrees, Preprint.

[11] Downey, R. G. [1989], D. r. e. degrees and the nondiamond thèorem, *Bull. London Math. Soc.***21**, 43-50.

[12] Jockusch, C. G., Jr. and Shore, R. A., [1984], Pseudo-jump operators II: Transfinite iterations, hierarchies and minimal covers, *J. Symb. Logic* **49**, 1205-1236.

[13] Jockusch, C. G., Jr. and Shore, R. A. [1985], REA operators, r. e. degrees and minimal covers, in: Recursion Theory (A. Nerode and R.A. Shore, eds.), Proc. of Symposia in Pure Mathematics, **42**, Amer. Math. Soc., Providence, Rhode Island, 3-11.

[14] Kalimullin, I. Sh. [ta], Splitting properties of *n*-c. e. enumeration degrees, to appear.

[15] Soare, R. I., [1987], *Recursively Enumerable Sets and Degrees*, Springer-Verlag, Berlin.

[16] Sorbi, A. [ta] Open problems in the enumeration degrees, this volume.

Department of Mathematics, Kazan State University, Kazan 420008, RUSSIA
E-mail address: `Marat.Arslanovksu.ru`

Contemporary Mathematics
Volume **257**, 2000

The Theory of Numberings: Open Problems

Serikzhan Badaev and Sergey Goncharov

ABSTRACT. We suggest a general approach to the notion of computability for the wide variety of classes of constructive objects which admit a constructive description in formal languages with a Gödel numbering for formulas. We consider mainly problems related to the algebraic and elementary properties of Rogers semilattices of computable numberings of families of sets in Ershov's hierarchy as well as the arithmetical and the analytical hierarchy.

1. Background and preliminaries

The study of the phenomenon of computability leads to a number of very interesting directions in mathematics and applications (cf. Handbook on Recursive Mathematics [**41**], [**42**] and the textbook [**25**]). One of these very important directions is the theory of computable numberings.

In recursive mathematics and algorithm theory, we encounter various situations which naturally lead one to the study of classes of constructive objects. An examination of the algorithmic properties of classes of constructive objects fares best with the techniques and notions of the theory of computable numberings. The idea of using such numberings goes back to Gödel, who applied a computable numbering of formulas for embedding the metatheory of number theory into the theory of numbers.

The next step toward applying computable numberings, in treating the class of partial computable functions, was taken by Kleene [**43**], who constructed a universal partial computable function. The numeration theorem of Kleene has great significance in the development of the theory of algorithms and the study of properties of computable objects. It allows one to derive the basic recursion theorems and may be thought of as the starting point of the theory of computable numberings.

In the early 1950's (cf. [**44**], [**81**]), Kolmogorov provided the methodological foundation for its investigation. Later, Uspensky, his former student, carried out a systematic study of computable numberings of partial computable functions and applied them to the theory of algorithms [**78**]–[**80**]. Independently, a study of computable numberings of the family $\mathcal{F}$ of all unary partial computable functions

---

1991 *Mathematics Subject Classification.* 03D45.

S.A. Badaev was supported in part by an NSF grant for AMS Summer Research Conferences.

S.S. Goncharov was supported in part by an NSF grant for AMS Summer Research Conferences and by Grant RFBR 99-01-00485.

was initiated by Rogers [**72**]. He brought into consideration the class $\mathrm{Com}\,(\mathcal{F})$ of all possible computable numberings of $\mathcal{F}$ and a reducibility relation $\leqslant$ amongst them, which determined a preordering on that class. Factorization with respect to an equivalence relation $\equiv$, induced by the preordering, made it possible to construct a partially ordered set $\mathcal{R}(\mathcal{F}) = \langle \mathrm{Com}\,(\mathcal{F})/\!\equiv, \leqslant \rangle$, which forms an upper semilattice. Rogers concerned himself with computable numberings in the largest element of that semilattice. Its minimal elements and some other properties were later introduced by Friedberg [**27**], Mal'tsev [**59**]–[**62**], and Pour-El [**67**]–[**69**]. The papers of Rice [**70**], [**71**] and of Dekker and Myhill [**12**] also have to be added to the category of early investigations into computable numberings.

The further development of the theory of numberings was undertaken by Ershov (cf. the monograph [**21**]), Ershov and Lavrov [**20**], Lachlan [**52**]–[**56**], Khutoretsky [**45**]–[**47**], V'jugin [**85**]–[**87**], Selivanov [**74**], [**75**], Denisov [**13**], [**14**], Goncharov [**30**], [**32**]–[**36**], Badaev [**1**]–[**10**], Kummer [**48**], [**49**], [**51**] and others.

The results of the theory of computable numberings are mostly applied in recursive mathematics [**41**], [**42**], [**25**]. One of the powerful methods for constructing families of c. e. sets with a finite number of computable Friedberg numberings suggested by Goncharov [**30**] is the starting-point of the research on the algorithmic dimension of recursive models (see [**11**], [**28**], [**31**], [**39**], [**82**]–[**84**] and others).

Applications of the theory of numberings to classical recursion theory are also significant. For instance, using a theorem of Goncharov [**30**] on the number of7 computable Friedberg numberings of families of c. e. sets, Kummer [**50**] solved a famous problem on the recursive isomorphism types of partial computable functions (cf. the textbook of Rogers [**73**, Chapter 4]). Namely, he proved that for each $k \in \omega$ there exists a computable function having exactly k recursive isomorphism types. Besides, we might mention applications of the results on positive equivalences [**5**], [**57**] in proof theory [**76**], and applications of one-element Rogers semilattices of families of computable functions [**2**], [**74**] in learning theory [**51**].

While studying an algorithmic object, some derived constructive objects and structures arise, and they can usually be described in a formal language. For example, we encounter classes of $0^{(n)}$-computable sets, families of positive equivalences, and classes of Σ^0_3-distributive lattices when dealing with the problem of describing the structure of the $\mathbf{m}$-degrees. And very often, we need to study some classes of uniformly definable derived objects and structures of a given type to obtain knowledge on a separate object or a structure of that type.

The notion of a computable numbering of a family of recursive models naturally arose and became fruitful in recursive model theory [**25**]. Questions on the existence of computable and universal computable numberings are among the problems that occupy the center-stage in the area. And conversely, in order to shed light on the structure and properties of computable numberings of families of recursive models, it is important to study the computability of classes of computable numberings of families of c. e. sets [**28**], [**31**].

Computable numberings of many other classes of constructive objects are also useful in resolving various problems of algorithm theory. We might mention objects such as positive equivalences [**21**], morphisms $\mathrm{Mor}\,((S,\nu),(S_0,\nu_0))$ in the category of numbered sets [**21**], Σ^0_3-sets and Σ^0_3-structures [**25**], and others.

Goncharov and Sorbi offered in [**38**] a general approach for studying classes of objects which admit a constructive description in a formal language. Going through all the above-mentioned approaches to defining computable numberings of families of mathematical objects, they reached the general definition of a computable numbering as defined below. But first we recall some basic notions of the theory of numberings. We refer to the textbooks [**62**], [**73**], [**77**] for the basic notions of recursion theory, and to [**21**]–[**23**] for the notions of the theory of numberings.

DEFINITION 1.1. Any surjective mapping α of the set ω of natural numbers onto a nonempty set A is called a *numbering* of A. Let α and β be numberings of A. We say that a numbering α is *reducible* to a numbering β (in symbols, $\alpha \leqslant \beta$) if there exists a computable function f such that $\alpha(n) = \beta f(n)$ for any $n \in \omega$. We say that the numberings α and β are *equivalent* (in symbols, $\alpha \equiv \beta$) if $\alpha \leqslant \beta$ and $\beta \leqslant \alpha$.

DEFINITION 1.2. Let $\theta_\alpha \rightleftharpoons \{< x, y > \mid \alpha x = \alpha y\}$. A numbering α is called *decidable (or positive)* if θ_α is a decidable (or, respectively, a c. e.) set.

It is obvious that if α and β are equivalent numberings then α is decidable (or positive) if and only if β is decidable (or positive, respectively). Every decidable numbering of an infinite family is equivalent to a one-to-one or single-valued numbering [**21**], [**22**].

Let A be some set of objects. We are interested only in those objects that admit a certain constructive description. Define some language $\mathcal{L}$ and the interpretation of that language determined as a partial surjective mapping $i\colon \mathcal{L} \to A$. For any object $a \in A$, each formula in $i^{-1}(a)$ is interpreted as a description of a. For $\mathcal{L}$, we consider a Gödel numbering $G\colon \omega \to \mathcal{L}$.

DEFINITION 1.3. A numbering $\alpha\colon \omega \to A$ is called a *computable numbering of A in the language $\mathcal{L}$ with respect to the interpretation i* if there exists a computable function f for which the formula $G(f(n))$ distinguishes an element $\alpha(n)$ in $\mathcal{L}$ relative to i.

This approach allows one to unify in a very natural way various notions of computability and relative computability for different classes of constructive objects. For example, let us apply Definition 1.4 to the case of families $\mathcal{A}$ of partial computable functions. For a language $\mathcal{L}$, we take the language of Turing machines. We consider the function computed by a Turing machine M as an interpretation $i(M)$ of M. In this case, we obtain the standard notion of a computable numbering $\alpha\colon \omega \to \mathcal{A}$ of the family $\mathcal{A}$ of partial computable functions: α is computable (relative to i) if and only if there exists a partial computable function $g(n, x)$ such that the functions $\alpha(n)$ and $\lambda x g(n, x)$ coincide for any $n \in \omega$ [**21**], [**72**].

For a given language $\mathcal{L}$ and an interpretation i, we consider the class $\mathrm{Com}(A)$ of all computable numberings of a set A. The relation $\equiv$ is an equivalence relation, and the reducibility $\leqslant$ induces a partial ordering on equivalence classes of that relation. We denote the partially ordered set $\langle \mathrm{Com}(A)/{\equiv}, \leqslant \rangle$ by $\mathcal{R}(A)$. If α and β are in $\mathrm{Com}(A)$, then a numbering $\alpha \oplus \beta$ of A is defined as follows: $\alpha \oplus \beta(2n) = \alpha(n)$ and $\alpha \oplus \beta(2n+1) = \beta(n)$. The numbering $\alpha \oplus \beta$ determines the least upper bound of the pair $\alpha/{\equiv}$, $\beta/{\equiv}$ in $\mathcal{R}(A)$. Thus, $\mathcal{R}(A)$ can be regarded as an upper semilattice.

DEFINITION 1.4. The upper semilattice $\mathcal{R}(A)$ is called the *Rogers semilattice* (of the class of computable numberings of A in the language $\mathcal{L}$ with respect to the interpretation i) of A.

The Rogers semilattice $\mathcal{R}(A)$ represents the algorithmic complexity of computations of the set A as a whole, and problems in the theory of computable numberings concern mainly the algebraic and elementary properties of Rogers semilattices.

2. Algebraic properties of Rogers semilattices

We restrict ourselves to problems on computable numberings in well-known hierarchies of sets, namely the hierarchy of Ershov [**16**], [**17**], [**19**], and the arithmetical and analytical hierarchy [**73**]. First of all, let us examine in detail the approach of Goncharov and Sorbi to the notion of computability for families of sets in any class Σ^0_n or Π^0_n of the arithmetical hierarchy.

We consider formulas of arithmetic in the signature $\langle +, \cdot, 0, s, \leqslant \rangle$ as a language $\mathcal{L}$. Let $\mathbb{N}$ stand for the standard model of arithmetic. For $x \in \omega$, denote by $\mathbf{x}$ an arithmetic term defining x, i. e., the term $s(s(\dots s(0)\dots))$, in which the symbol s occurs x times. If $\overline{x} = (x_1, x_2, \dots, x_k)$ is a tuple of natural numbers, then $\overline{\mathbf{x}} = (\mathbf{x}_1, \mathbf{x}_2, \dots, \mathbf{x}_k)$ is the tuple of corresponding terms.

It is known [**73**] that any set A is in Σ^0_n if and only if there exists a formula of arithmetic $\varphi(\overline{x})$ such that $\overline{x} \in A \Leftrightarrow \mathbb{N} \models \varphi(\overline{\mathbf{x}})$ and the prenex normal form of φ begins with an existential quantifier and has at most n alternating groups of like quantifiers.

Denote the formula of arithmetic with Gödel number e by $\Psi_e(x_1, \dots, x_k)$, where $x_1, \dots, x_k$ are all the free variables of Ψ_e. One of the possible Gödel numberings of the formulas of arithmetic is suggested in [**24**]. A formula is called a $\Sigma_n(\Pi_n)$-*formula* if its prenex normal form begins with an existential quantifier (or a universal quantifier, respectively) and has at most n alternating groups of like quantifiers. We consider $\Sigma_n(\Pi_n)$-formulas of arithmetic as a language $L^{\Sigma_n}(L^{\Pi_n})$ for describing $\Sigma^0_n(\Pi^0_n)$-subsets of ω^k. Let $\mathcal{A}$ be a family of $\Sigma^0_n(\Pi^0_n)$-subsets of ω^k and $\alpha\colon \omega \to \mathcal{A}$ be a numbering of $\mathcal{A}$.

Definition 2.1. A numbering α is called Σ^0_n (or Π^0_n)-*computable* if there exists a computable function f such that, for any $m \in \omega$, $\overline{x} \in \alpha(m) \iff \mathbb{N} \models \Psi_{f(m)}(\overline{\mathbf{x}})$, where $\Psi_{f(m)}$ is a Σ_n(or Π_n)-formula with Gödel number $f(m)$.

It is worth noting that Definition 2.1 follows from Definition 1.4 if an interpretation i of the language $L^{\Sigma_n}(L^{\Pi_n})$ is chosen such that

$$i^{-1}(A) = \{\varphi \in L^{\Sigma_n}(L^{\Pi_n}) \mid \overline{x} \in A \Leftrightarrow \mathbb{N} \models \varphi(\overline{\mathbf{x}})\}.$$

Remark 2.2. It is sufficient to deal with the notion of Σ^0_n-computability only. For any numbering $\alpha\colon \omega \to \mathcal{A}$ of a class of Σ^0_n-sets, we can define a numbering β of the complements of the sets of $\mathcal{A}$ as $\beta(m) = \omega \setminus \alpha(m), m \in \omega$. Then the numbering β is Π^0_n-computable if and only if α is Σ^0_n-computable numbering [**38**].

Theorem 2.3 (S. S. Goncharov, A. Sorbi [**38**]). *Let $\mathcal{A}$ be a family of Σ^0_n-subsets of ω^k and $n \geq 1$. A numbering $\alpha\colon \omega \to \mathcal{A}$ is Σ^0_n-computable if and only if $\{(m, \overline{x}) \mid \overline{x} \in \alpha(m)\} \in \Sigma^0_n$.*

Remark 2.4. It should be mentioned that analogues of Theorem 2.3 are also valid for families of sets of both Ershov's hierarchy and the analytical hierarchy. In the case of the hierarchy of Ershov, we could consider Boolean combinations of Σ-formulas of arithmetic. And in the case of the analytical hierarchy, we need to use Σ-formulas of second-order arithmetic.

Theorem 2.3 and Remark 2.4 imply that the computability of a numbering α of a family from a given class $\mathcal{K}$ is equivalent to the definability of its *universal set* $\{< x, n >| \, x \in \alpha n\}$ in terms of the class $\mathcal{K}$. Thus, we could define this notion directly as follows.

DEFINITION 2.5. Let $\mathcal{K}$ be any class $\Sigma^i_n, n \geq 1, i \in \{-1, 0, 1\}$ of the hierarchy of Ershov ($i = -1$), the arithmetical ($i = 0$), or the analytical ($i = 1$) hierarchy, and let $\mathcal{A} \subseteq \mathcal{K}$ be a family of subsets of ω. A numbering $\alpha : \omega \to \mathcal{A}$ is called a *$\mathcal{K}$-computable* numbering of $\mathcal{A}$ if the universal set $\{< x, n >: x \in \alpha n\}$ of α is in $\mathcal{K}$.

In the case of families of c. e. sets ($\mathcal{K} = \Sigma^0_1$), Definition 2.5 coincides with the classical notion of a computable numbering (see the textbook of Ershov [**21**] or [**22**], [**23**], [**72**], [**79**]), and the universal set $\{< x, n >: x \in \alpha n\}$ is thought of as a uniformly effective procedure for computing numbers x into the sets αn.

In the case of $\mathcal{K} = \Sigma^{-1}_{n+1}$, the computability of α means that the universal set of α is $0^{(n)}$-enumerable by the strong hierarchy theorem of Kleene and Post [**73**, Chapter 14]. Therefore, many constructions of the classical theory of numberings can be relativized to computations in the arithmetical hierarchy by means of oracles.

In the case of $\mathcal{K} = \Sigma^{-1}_{n+1}$, any number x may be put into αn at some stage of a uniformly effective procedure, but, in contrast with the classical case, x may later be removed from αn. And in contrast with the case of the arithmetical hierarchy, for any x, the number of enumerations of x into, and extractions of x from, αn, is bounded by $n + 1$.

In what follows, the Rogers semilattice of $\mathcal{K}$-computable numberings of a family $\mathcal{A}$ will be denoted by $\mathcal{R}^i_n(\mathcal{A})$ if $\mathcal{K} = \Sigma^i_n$. We will omit indices i, n in the notation $\mathcal{R}^i_n(\mathcal{A})$ in the classical case $i = 0, n = 1$ (as opposed to the generalized case).In the theory of algorithms, all objects are treated modulo some computable equivalence, and the notion of equivalent numberings is just the suitable notion here. Any index n of the set $\alpha(n)$ with respect to numbering α may be thought of as a description of that set in some formal language. If a computable function f reduces a numbering α to a numbering β then f is a uniform translation of the descriptions of the objects by α to the descriptions of the same objects by β. Therefore, equivalent numberings are not distinguishable from an algorithmic point of view. This approach allows one to formulate most of the problems on numberings in terms of Rogers semilattices. And in the general setting, these problems can be formulated as follows.

- Find global algebraic properties of Rogers semilattices (such as cardinality, type of the algebraic structure, ideals, segments, covers, etc.)
- Describe invariants and among them a number of extremal elements to distinguish different Rogers semilattices.
- Classify numberings which generate special elements in Rogers semilattices (extremal elements, limit points, split elements, etc.).

REMARK 2.6. In what follows, we will deal only with computability in the arithmetical hierarchy and in the hierarchy of Ershov. We note that almost nothing is known about the Rogers semilattices of families of sets in the analytical hierarchy. So all the questions to be discussed below for the cases of Ershov's hierarchy and the arithmetical hierarchy are interesting also for the case of the analytical hierarchy.

Historically, the first two problems on Rogers semilattices of families of c. e. sets were raised by Ershov in [**15**]:

- What is the cardinality of a Rogers semilattice?

- Can a Rogers semilattice be a lattice?

Investigations of Mal'tsev [**59**], Ershov [**15**], Denisov [**13**], Khutoretsky [**47**], Marchenkov [**64**], [**65**], Badaev [**1**], [**2**], Selivanov [**74**], [**75**] and others towards the complete resolution of these problems are summarized in the following two theorems.

THEOREM 2.7 (A.B. Khutoretsky [**47**]). *Let $\mathcal{A}$ be a family of c. e. sets. If the Rogers semilattice $\mathcal{R}(\mathcal{A})$ contains two elements then it is infinite.*

Indeed, Khutoretsky embedded a linear ordering of order type ω into $\mathcal{R}(\mathcal{A})$ above each (non-greatest) element of the Rogers semilattice. The existence of infinite chains in every non-trivial Rogers semilattice was proved by Badaev [**1**], who also showed that chains can be chosen to include any given (non-greatest and non-minimal) element of $\mathcal{R}(\mathcal{A})$.

THEOREM 2.8 (V.L. Selivanov [**75**]). *Let $\mathcal{A}$ be a family of c. e. sets. If the Rogers semilattice $\mathcal{R}(\mathcal{A})$ contains two elements then it is not a lattice.*

In the case of the arithmetical hierarchy, both of Ershov's problems were completely solved by Goncharov and Sorbi [**38**]. Their solution is based on the connection of Σ^0_{n+1}-computability with computations relative to the $0^{(n)}$-oracle.

THEOREM 2.9 (S. S.Goncharov, A. Sorbi [**38**]). *If a family $\mathcal{A} \subseteq \Sigma^0_{n+1}$ contains at least two elements then the Rogers semilattice $\mathcal{R}^0_{n+1}(\mathcal{A})$ is infinite and is not a lattice.*

In the more important case of an infinite family $\mathcal{A}$, the authors construct an infinite sequence of elements of $\mathcal{R}^0_n(\mathcal{A})$, any pair of distinct elements of which has no lower bound in $\mathcal{R}^0_n(\mathcal{A})$. In the case of a finite family $\mathcal{A}$, the following simple but very useful observation was applied. We can consider any Σ^i_n-computable family $\mathcal{A}$ as a Σ^i_m-computable family for $m > n$. Then there is a natural embedding of the Rogers semilattice $\mathcal{R}^i_n(\mathcal{A})$ onto an ideal of the Rogers semilattice $\mathcal{R}^i_m(\mathcal{A})$.

At first glance, classical computability seems very close to Σ^{-1}_n-computability since the former is definable by means of Σ-formulas of arithmetic and the latter is definable by Boolean combinations of such formulas. Actually, this is not the case. In our opinion, the solution to Questions 1-4 below could lead to an explanation of the nature of Σ^{-1}_n-computability.

Question 1. *Let $\mathcal{A} \subseteq \Sigma^{-1}_{n+1}$ and let the Rogers semilattice $\mathcal{R}^{-1}_{n+1}(\mathcal{A})$ contain at least two elements. Is $\mathcal{R}^{-1}_{n+1}(\mathcal{A})$ infinite?*

Question 2. *Let $\mathcal{A} \subseteq \Sigma^{-1}_{n+1}$ and let the Rogers semilattice $\mathcal{R}^{-1}_{n+1}(\mathcal{A})$ contain at least two elements. Is $\mathcal{R}^{-1}_{n+1}(\mathcal{A})$ a lattice?*

Mal'tsev defined an important class of families of c. e. sets which have one-element Rogers semilattice, namely, the class of effectively discrete families, in [**59**], [**61**]. Later, analogous classes for families of computable functions were introduced by Ershov [**15**]. V'jugin suggested a notion of a weakly effectively discrete family in [**85**]. We recall the definitions of these classes.

Let $\mathcal{A}$ be a family c. e. sets. We introduce a topology $\tau_{\mathcal{A}}$ on $\mathcal{A}$ as follows. For any finite set G, denote by $V(G)$ the family $\{F \mid F \in \mathcal{A}, G \subseteq F\}$. We take the family $\{V(G) \mid G \textit{ is a finite subset of set in} \mathcal{A}\}$ as a base for the open sets in $\tau_{\mathcal{A}}$.

DEFINITION 2.10. A family $\mathcal{A}$ is said to be *discrete* if the topology $\tau_{\mathcal{A}}$ is discrete, i. e., for any $F \in \mathcal{A}$ there exists a finite subset G of F such that $V(G) = \{F\}$. A family $\mathcal{A}$ is said to be *weakly effectively discrete* if there exists a computable sequence $G_0, G_1, \dots, G_n, \dots$ of finite sets such that every $V(G_n), n \in \omega$, contains just one element of $\mathcal{A}$, and every element $F \in \mathcal{A}$ belongs to some $V(G_n)$. A family $\mathcal{A}$ is said to be *effectively discrete* if the above sequence $G_0, G_1, \dots, G_n, \dots$ of finite sets can be chosen strongly computable.

If we identify functions with their graphs, we obtain the corresponding notions for families of computable functions.

THEOREM 2.11 (A.I. Mal'tsev [**59**], [**61**]). *If $\mathcal{A}$ is an effectively discrete family of c. e. sets then all its computable numberings are positive. In particular, the Rogers semilattice $\mathcal{R}(\mathcal{A})$ consists of a single element.*

The same result holds for effectively discrete families of computable functions [**15**]. Note that the discreteness of a family of computable functions is necessary for its Rogers semilattice to be a one-element set.

Question 3. *Is the effective discreteness of $\mathcal{A} \subseteq \Sigma^{-1}_{n+1}$ sufficient for $\mathcal{R}^{-1}_{n+1}(\mathcal{A})$ to consist of a single element?*

A lot of examples of discrete families of computable functions and families of c. e. sets are known whose Rogers semilattices consist of one element (see [**2**], [**6**], [**8**], [**74**], [**85**]). In all these examples, every computable numbering of the family is positive.

Selivanov found an example of a non-discrete family of c. e. sets with a one-element Rogers semilattice. The family constructed by him in [**75**] is very close to the discrete one, it does not contain two sets one of which is a subset of the other, and all its computable numberings are positive also. In addition, he raised a very natural question, namely, whether the Rogers semilattice of a family of c. e. sets is always infinite if it contains two sets one of which is a subset of the other. Up to the late 1970's, there was a conjecture that the semilattice of c. e. $\mathbf{m}$-degrees can be embedded into the Rogers semilattice of any family of c. e. sets containing two sets one of which is a subset of the other. However, the following theorem holds.

THEOREM 2.12 (S. S. Goncharov, S. A. Badaev [**40**]). *There exists an infinite set of c. e. sets $\mathcal{A}$ such that $\mathcal{A}$ contains a least set under inclusion and $|\mathcal{R}(\mathcal{A})| = 1$.*

Question 4. *Find a characterization of the Σ^i_n-computable families with one-element Rogers semilattices.*

It seems that a solution to Question 4 will strongly depend on the case of families which contain two sets one of which is a subset of the other. Thus, the following question may be considered a step towards the solution of Question 4.

Question 5. *Find structural conditions for a family $\mathcal{A} \subseteq \Sigma^i_n$ containing two sets one of which is a subset of the other to have an infinite Rogers semilattice.*

We recall some natural classes of families of c. e. sets relevant to Question 5.

- finite non-discrete families [**21**];
- the family of all c. e. subsets of ω [**21**];

- families not closed under unions of computable increasing chains of its elements [**52**];
- families with a greatest element under inclusion [**3**].

It seems that the following theorem explains why often a family containing two sets one of which is a subset of the other has an infinite Rogers semilattice. The reason is a separability property, separating the small set from the large set.

DEFINITION 2.13. Let $A \subseteq \omega, B \subseteq \omega$. We say that A is *effectively separable from* B if there exists a c. e. set S such that $A \subseteq S$ and $B \cap S = \emptyset$.

THEOREM 2.14 (S. A. Badaev [**7**]). *Let a family $\mathcal{A}$ of c. e. sets contain sets $A \subset B$. Then $\mathcal{R}(\mathcal{A})$ is infinite if at least one of the following conditions holds:*
(i) the index set $\alpha^{-1}(A)$ of A with respect to α is effectively separable from the index set $\alpha^{-1}(B)$ of B for some computable numbering α of $\mathcal{A}$; or
(ii) the complement of the set B is effectively separable from the set A.

As a corollary, we obtain a very important fact: If a family $\mathcal{A}$ of c. e. sets contains two sets one of which is a subset of the other and $|\mathcal{R}(\mathcal{A})| = 1$ then $\mathcal{A}$ has no computable positive numbering.

Question 4 was raised by Ershov in [**15**] for the classical case and has not been resolved in general except for the case of $i = 0, n > 1$ (see Theorem 2.9). Two different complete solutions to Question 4 for families of computable functions are known (Goncharov [**37**], [**40**] and Kummer [**51**]). Indeed, they found different algorithmic conditions for every computable numbering of a family of computable functions to be positive. Note also that the solution of Goncharov was obtained in his early paper [**29**] in terms of constructive models.

The papers of Denisov [**14**], Ershov and Lavrov [**20**] (see also [**21**, Appendix 1]), Marchenkov [**64**], and V'jugin [**87**]–[**89**] are devoted to questions on the algebraic description of segments of Rogers semilattices of families of c. e. sets.

The first non-trivial results were obtained by Marchenkov, who showed in [**64**] that any initial segment of the semilattice of c. e. $\mathbf{m}$-degrees can be embedded as a partially ordered set above any non-greatest element of the Rogers semilattice $\mathcal{R}(\mathcal{A})$ if the family $\mathcal{A}$ of c. e. sets satisfies some restrictions.

Question 6. *Describe partially ordered sets which can be embedded above non-greatest element of Rogers semilattices $\mathcal{R}^i_n(\mathcal{A})$. In particular, do there exist minimal covers above any non-greatest element of $\mathcal{R}^i_n(\mathcal{A})$?*

The characterization of Lachlan [**56**] of the initial segments of the c. e. $\mathbf{m}$-degrees played a crucial role in the investigation of the algebraic structure of Rogers semilattices. His description was in terms of direct limits of chains of increasing finite distributive lattices which satisfy some natural condition on the uniform effectiveness of the operations and relations on these lattices. (We refer for details to [**56**] or to [**21**, Appendix 1]). Such direct limits form upper semilattices which, for brevity, we will call *Lachlan semilattices.*)

In [**88**], V'jugin proved that arbitrary segments of the c. e. $\mathbf{m}$-degrees have the same description as the initial ones, and in [**89**] he showed that the initial segments of the Rogers semilattice of the family $\mathcal{C}$ of all c. e. sets are exactly the co-ideals of Lachlan semilattices satisfying some additional requirements of an algorithmic nature.

The constructions of Lachlan from [**55**] and [**56**] were applied by Ershov and Lavrov [**20**] to the study of Rogers semilattices of finite families of c. e. sets. It was proved that all non-trivial Rogers semilattices of finite families have the same local structure since their initial segments are exactly isomorphic to Lachlan semilattices. Besides, for each of them there is an isomorphic semilattice embedding to any other non-trivial Rogers semilattice. More importantly, in [**20**], a system of invariants for Rogers semilattices was discovered. These invariants allow one to distinguish differences between some Rogers semilattices of finite families. For example, if two finite families of c. e. sets are chains of different length under inclusion then their Rogers semilattices are not isomorphic. In contrast to this, Rogers semilattices of finite families containing a least element under inclusion are isomorphic to the semilattice of c. e. $\mathbf{m}$-degrees if these families have at least one non-least element and all such elements are incomparable under inclusion. This fact was established by Denisov [**14**], and his proof is one of the most complicated proofs in the theory of numberings.

Question 7. *Is the system of invariants of Ershov and Lavrov complete?*

There is currently very little known about invariants of Rogers semilattices of infinite families of c. e. sets. We note only that V'jugin constructed in [**87**] an infinite sequence of infinite families of c. e. sets whose Rogers semilattices have different initial segments.

Question 8. *Find a characterization of the segments of the Rogers semilattice $\mathcal{R}^i_n(\mathcal{A})$ if $\mathcal{A} \subseteq \Sigma^i_n$ is infinite for $i = 0, n = 1$.*

Questions on the existence of ideals with minimal elements as well as on the existence of ideals without minimal elements are the most popular problems in the study of the ideals of Rogers semilattices. Every ideal of a Rogers semilattice $\mathcal{R}(\mathcal{A})$ contains a minimal element if $\mathcal{A}$ is a family of computable functions or if $\mathcal{A}$ is a finite family [**21**]. If $\mathcal{C}$ is the family of all c. e. sets then $\mathcal{R}(\mathcal{C})$ contains both ideals with minimal elements and ideals without them [**18**], [**46**]. And, finally, there exist families $\mathcal{A}$ such that the Rogers semilattice $\mathcal{R}(\mathcal{A})$ has no minimal element. The problem of the existence of Rogers semilattices without minimal elements was suggested by Ershov at the end of the 1960's in close connection with the search for analogues of the famous speed-up theorem of Blum. The first example was obtained by V'jugin [**87**]. He constructed a very complicated family of c. e. sets every computable numbering of which is the join of two incomparable numberings of that family. The second, very simple, example is due to Badaev [**4**], [**8**]; it was established by applying a minimality criterion of numberings (see Theorem 3.7 below). In contrast to the former example, the family constructed in the latter is discrete.

Question 9. *Find a characterization of the ideals of the Rogers semilattice $\mathcal{R}^i_n(\mathcal{A})$ if $\mathcal{A} \subseteq \Sigma^i_n$ is infinite for $i = 0, n = 1$.*

Question 10. *Do there exist Rogers semilattices $\mathcal{R}^0_n(\mathcal{A}), n > 1$, without minimal elements?*

Question 11. *Do there exist Rogers semilattices $\mathcal{R}^{-1}_{n+1}(\mathcal{A})$ without minimal elements?*

3. Problems on minimal numberings

Let us now consider problems concerning minimal elements in Rogers semilattices. We refer to the survey of Badaev and Goncharov [**10**] for a detailed list of open problems on minimal numberings in the classical case.

The starting point of the study of minimal numberings is the famous theorem of Friedberg [**27**] on the existence of a one-to-one computable numbering of the family $\mathcal{C}$ of all c. e. sets. Pour-El and Putnam gave, in [**69**], the following nice example of a family of c. e. sets without computable one-to-one numberings:

$$\{\{2x, 2x+1\} \mid x \in K\} \cup \{\{2x\}, \{2x+1\} \mid x \in \omega \setminus K\},$$

where K is the creative set.

In a result comparing Gödel numberings of $\mathcal{C}$ studied by Rogers [**72**] with the one-to-one numberings of Friedberg, Pour-El concluded in [**67**] that

- one-to-one numberings are not equivalent to Gödel numberings;
- one-to-one numberings generate minimal elements of the semilattice $\mathcal{R}(\mathcal{C})$;
- there exist at least two minimal elements of $\mathcal{R}(\mathcal{C})$ generated by one-to-one numberings,

and she suggested calling computable one-to-one numberings *Friedberg numberings.*

DEFINITION 3.1. A numbering α of a set A is called a *minimal* numbering if $\beta \leqslant \alpha$ implies $\alpha \leqslant \beta$ for any numbering β of A.

Obviously, any Σ^i_n-computable numbering of a family $\mathcal{A} \subseteq \Sigma^i_n$ generates a minimal element of the Rogers semilattice $\mathcal{R}^i_n(\mathcal{A})$.

Friedberg numberings and positive numberings are special but very important cases of minimal numberings. The main problems on minimal numberings are the following two questions raised by Yu. L. Ershov in the late 1960's for the classical case.

Question 12. *Find structural conditions under which a family $\mathcal{A} \subseteq \Sigma^i_n$ has minimal Σ^i_n-computable numberings.*

Question 13. *What numbers are realized as the number of elements in a family of minimal numberings?*

REMARK 3.2. Here and in what follows, "number of numberings" means, of course, the number of numberings modulo equivalence of numberings.

Every finite family has a unique minimal numbering which is in fact a decidable numbering [**21**], [**22**]. So problems on minimal numberings are interesting only for infinite families. We could also exclude from our considerations families of computable functions since each member of such a family has a Friedberg numbering and any minimal numbering of it is one-to-one [**61**], and the number of Friedberg numberings is either 1 or ω[**65**]. We are now going to discuss some approaches to answering Question 12; then we will consider Question 13.

Question 12 for the case of Friedberg numberings of families of c. e. sets was raised by Mal'tsev [**62**]. Some necessary structural conditions for a family of c. e. sets to have Friedberg numberings were found by Lachlan [**53**], [**54**]. Many sufficient structural conditions were suggested in the papers of Ershov [**18**], Florence [**26**], Kummer [**48**], [**49**], Mal'tsev [**62**], Pour-El and Howard [**68**] and others.

In the case of generalized computability, investigations of both Friedberg numberings and positive numberings are at a very early stage.

Conjecture 14. *For all n, the family of all* Σ^{-1}_{n+1}*-sets has* Σ^{-1}_{n+1}*-computable Friedberg numberings.*

Conjecture 15. *For all n, the family of all* Σ^{-1}_{n+1}*-sets has* Σ^{-1}_{n+1}*-computable positive non-decidable numberings.*

Question 16. *Is effective discreteness sufficient for a family* $\mathcal{A} \subseteq \Sigma^i_{n+1}$ *to have* Σ^i_{n+1}*-computable positive numberings?*

For families of Σ^0_n-computable sets, Question 12 on Friedberg numberings may be reduced to the corresponding question on positive numberings. This follows from Theorem 3.3.

THEOREM 3.3 (S. S. Goncharov and A. Sorbi [**38**]). *If an infinite family* $\mathcal{A} \subseteq \Sigma^0_{n+2}$ *has a positive* Σ^0_{n+2}*-computable numbering, then* $\mathcal{A}$ *has a Friedberg* Σ^0_{n+2}*-computable numbering.*

Question 17. *Suppose an infinite family* $\mathcal{A} \subseteq \Sigma^{-1}_{n+2}$ *has a* Σ^{-1}_{n+2}*-computable positive numbering. Does* $\mathcal{A}$ *have a* Σ^{-1}_{n+2}*-computable Friedberg numbering?*

There are families of c. e. sets which have computable positive but not Friedberg numberings [**18**], [**63**] as well as families which have minimal numberings but no positive numberings [**6**], [**66**]. There are also many structural conditions for a family of c. e. sets to have computable positive numbering [**3**], [**21**], [**61**], [**63**], [**85**], [**86**]. And there are a few structural conditions on the existence of minimal non-positive numberings [**63**], [**46**].

It does not seem realistic to search for natural structural conditions, which are both necessary and sufficient for a family of c. e. sets to have computable Friedberg (positive) numberings. Nevertheless, some algorithmic criteria were found.

THEOREM 3.4. *A family* $\mathcal{A}$ *of* Σ^0_{n+1}*-sets has a* Σ^0_{n+1}*-computable positive numbering if and only if* $\mathcal{A}$ *has a* Σ^0_{n+1}*-computable numbering* α *such that* $\theta_\alpha \rightleftharpoons \{< x, y > \mid \alpha x = \alpha y\} \in \Delta^0_2$.

In the classical case of families of c. e. sets, Theorem 3.4 was proved in [**3**]. The following criterion for the minimality of numberings does not depend on their computability.

THEOREM 3.5 (S. A. Badaev [**4**], [**6**]). *A numbering* α *is a minimal numbering of a set* A *if and only if for every c. e. set* R*, if* $\alpha(R) = A$ *then there exists a c. e. equivalence* $\eta \subseteq \theta_\alpha$ *such that the closure of* R *with respect to* η *coincides with* ω.

Problem 13 restricted to Friedberg numberings and positive numberings was completely solved by Goncharov for families of c. e. sets.

THEOREM 3.6 (S. S. Goncharov [**30**], [**36**]). *For every* $n \in \omega$ *there exists a family of c. e. sets which has exactly* n *computable Friedberg (positive) numberings.*

Applications of this theorem are well-known and were discussed in Section 1.

It is convenient to split Question 12 into some independent sub-questions to make it more detailed since, in the classical case, it was shown that the number of minimal numberings of various types is not arbitrary.

THEOREM 3.7 (S. S. Goncharov [**33**]). *If a family $\mathcal{A}$ of c. e. sets has a computable Friedberg numbering α then either α is the least numbering or $\mathcal{A}$ has an infinite number of computable positive numberings.*

Question 18. *Suppose a family $\mathcal{A} \subseteq \Sigma^i_n$ has a unique computable positive numbering which is not the least numbering. Does the Rogers semilattice $\mathcal{R}^i_n(\mathcal{A})$ contain an infinite number of minimal elements?*

In the classical case, Question 18 was asked by S. S. Goncharov [**35**]. We can answer Question 18 in particular cases. We need a notion introduced by Goncharov in [**32**].

DEFINITION 3.8. If we replace the computable function f by a $0'$-computable function f in Definition 1.1 on the reducibility of numberings then we obtain the notion of $0'$-reducibility of numberings. Two numberings are called *$0'$-equivalent* if they are $0'$-reducible each to other.

THEOREM 3.9 (S. A. Badaev [**9**]). *If all computable numberings of a family $\mathcal{A}$ of c. e. sets are $0'$-equivalent and $\mathcal{A}$ has a computable positive numbering α then either α is the least numbering or $\mathcal{A}$ has an infinite number of computable minimal numberings.*

Theorem 3.9 gives the answer to Question 13 for the following special important classes: families of finite sets; families of computable functions; and weakly effectively discrete families.

Goncharov constructed in [**34**] an example of a family $\mathcal{A}$ of finite sets which has a unique positive but non-least numbering. Theorem 3.9 implies that $\mathcal{A}$ has an infinite number of minimal numberings. Therefore, the following question is of interest.

Question 19 *Does there exist a family of Σ^i_n-sets possessing a unique Σ^i_n-computable minimal numbering which is neither a least nor a positive numbering?*

And, finally, we note that a solution to Question 13 might strongly depend on an answer to the following question.

Question 20. *Suppose a family $\mathcal{A} \subseteq \Sigma^i_n$ has two $0'$-equivalent but not equivalent Σ^i_n-computable minimal numberings. Does $\mathcal{A}$ have an infinite number of Σ^i_n-computable minimal numberings?*

References

[1] S. A. Badaev, *On incomparable numberings.* Siberian Math. J., 1974, v. 15, no. 4, p. 730–738.

[2] S. A. Badaev, *On computable enumerations of the families of total recursive functions.* Algebra and Logic, 1977, v. 16, no. 2, pp. 83-98.

[3] S. A. Badaev, *On positive enumerations.* Siberian Math. J., 1977, v. 18, no. 3, pp. 343–352.

[4] S. A. Badaev, *About one of Goncharov's problems.* Siberian Math. J., 1991, v. 32, no. 3, pp. 532–534.

[5] S. A. Badaev, *On weakly pre-complete positive equivalences.* Siberian Math. J., 1991, v. 32, no. 2, pp. 321–323.

[6] S. A. Badaev, *On minimal enumerations*, Siberian Adv. Math., 1992, v. 2, no. 1, pp. 1–30.

[7] S. A. Badaev, *On cardinality of semilattices of numberings of non-discrete families*, Sib. Math. J., 1993, v. 34, no. 5, pp. 795–800.

[8] S. A. Badaev, *Minimal enumerations.* Trudy Inst. Matem. SO RAN, 1993, v. 25, pp. 3-34. — Inst. of Mathematics, Novosibirsk, 1993 (Russian).

[9] S. A. Badaev, *Minimal numberings of positively computable families*, Algebra and Logic, 1994, v. 33, no. 2, pp. 131–141.

[10] S. A. Badaev, S. S. Goncharov, *On computable minimal enumerations*, Algebra. Proceedings of the Third International Conference on Algebra Dedicated to the Memory of M. I. Kargopolov. Krasnoyarsk, August 23–28, 1993, Walter de Gruyter, Berlin–New York, 1995, pp. 21–32.

[11] P. Cholak, S. S. Goncharov, B. Khoussainov, R. A. Shore, *Computably categorical structures and extensions by constants.* Journal of Symbolic Logic, 1999, v. 64, no. 1, pp. 13–37.

[12] J. C. E. Dekker, J. Myhill, *Some theorems on classes of recursively enumerable sets.* Trans. Amer. Math. Soc., 1958, v. 89, no. 1, p. 25–59.

[13] S. D. Denisov, *On* $\mathbf{m}$*-degrees of recursively enumerable sets.* Algebra i Logika, 1970, v. 9, no. 4, pp. 422–427 (Russian).

[14] S. D. Denisov, *Structure of the upper semilattice of recursively enumerable* $\mathbf{m}$*-degrees and allied questions.* Algebra i Logika, 1978, v. 17, no. 6, pp. 643–683 (Russian).

[15] Yu. L. Ershov, *Numberings of the families of total recursive functions*, Sib. Math. J., 1967, v. 8, no. 5, pp. 1015-1025 (Russian).

[16] Yu. L. Ershov, *On a hierarchy of sets, I.* Algebra i Logika, 1968, v. 7, no. 1, pp. 47–74 (Russian).

[17] Yu. L. Ershov, *On a hierarchy of sets, II.* Algebra i Logika, 1968, v. 7, no. 4, pp. 15–47 (Russian).

[18] Yu. L. Ershov *Computable enumerations.* Algebra and Logic, 1968, v. 7, no. 5, pp. 330-346.

[19] Yu. L. Ershov, *On a hierarchy of sets, III.* Algebra i Logika, 1970, v. 9, no. 1, pp. 20–31 (Russian).

[20] Yu. L. Ershov, I. A. Lavrov, *Upper semilattices* $L(S)$. Algebra i Logika, 1973, v. 12, no. 2, pp. 167–189 (Russian).

[21] Yu. L. Ershov, *Theory of numberings.*—Nauka, Moscow, 1977 (Russian).

[22] Yu. L. Ershov, *Theorie der Numerierungen.* Z. Math. Logik Grundlagen Math., 1977, v. 23, pp. 289–371.

[23] Yu. L. Ershov, *Theory of numberings.* — Preprint NSU, no. 18, Novosibirsk, 1996.

[24] Yu. L. Ershov, *Definability and Computability.* — Plenum Press Corp., New York, 1996.

[25] Yu. L. Ershov, S. S. Goncharov, *Constructive models.* — Plenum Press Corp., New-York, 1999.

[26] J. B. Florence, *Strong enumeration properties of recursively enumerable classes.* Z. Math. Logik Grundl. Math., 1969, Bd. 15, H. 2, S. 181–192.

[27] R. M. Friedberg, *Three theorems on recursive enumeration.* J. Symbolic Logic, 1958, v. 23, no. 3, pp. 309–316.

[28] S. S. Goncharov, *The problem of the number of non-self-equivalent constructivizations.* Sov. Math. Dokl., 1980, v. 21, pp. 411–414.

[29] S. S. Goncharov, *Autostability of models and Abelian groups.* Algebra and Logic, 1980, v. 19, no. 1, pp. 13–27.

[30] S. S. Goncharov, *Computable single-valued numerations.* Algebra and Logic, 1980, v. 19, no. 5, pp. 325–356.

[31] S. S. Goncharov, *On the problem of number of non-self-equivalent constructivizations.* Algebra and Logic, 1980, v. 19, no. 6, pp. 401–414.

[32] S. S. Goncharov, *Limit equivalent constructivizations.* Trudy Inst. Matem. SO AN SSSR, 1982, v. 2, pp. 4–12. — Nauka, Novosibirsk, 1982 (Russian).

[33] S. S. Goncharov, *Positive numerations of families with one-valued numerations.* Algebra and Logic, 1983, v. 22, no. 5, pp. 345–350.

[34] S. S. Goncharov, *The family with unique univalent but not the smallest enumeration.* Trudy Inst. Matem. SO AN SSSR, 1988, v. 8, pp. 42–48. — Nauka, Novosibirsk, 1988 (Russian).

[35] S. S. Goncharov, *A unique positive enumeration.* Siberian Adv. Math, 1994, v. 4, no. 1, pp. 52–64.

[36] S. S. Goncharov, *Positive computable enumerations.* Russian Acad. Sci. Dokl. Math., 1994, v. 48, no. 2, pp. 268–270.

[37] S. S. Goncharov, S. A. Badaev, *Classes with pairwise equivalent enumerations.* Lecture Notes in Computer Science, Springer–Verlag, 1994, v. 813, pp. 140–141.

[38] S. S. Goncharov, A. Sorbi, *Generalized computable numerations and non-trivial Rogers semilattices.* Algebra and Logic, 1997, v. 36, no. 4, pp. 359–369.

[39] S. S. Goncharov, B. Khoussainov, *On spectrum of degrees of decidable relations.* Russian Acad. Sci. Dokl. Math., 1997, v. 352, no. 3, pp. 301–303 (Russian).

[40] S. S. Goncharov, S. A. Badaev, *Families with one-element Rogers semilattice.* Algebra and Logic, 1998, v. 37, no. 1, pp. 21–34.

[41] *Handbook of Recursive Mathematics. Volume 1, Recursive model theory.* — Elsevier, Amsterdam, 1998.

[42] *Handbook of Recursive Mathematics. Volume 2, Recursive Algebra, Analysis and Combinatorics.* — Elsevier, Amsterdam, 1998.

[43] S. C. Kleene, *Introduction to Metamathematics.* — Van Nostrand, Princeton, 1952, and Elsevier, Amsterdam, 1964, 1971.

[44] A. N. Kolmogorov, V. A. Uspensky, *Defining an algorithm.* Usp. Mat. Nauk, 1958, v. 13, no. 4 (82), pp. 3–28.

[45] A. B. Khutoretsky, *The reducibility of computable enumerations.* Algebra and Logic, 1969, v. 8, no. 2, pp. 145–151.

[46] A. B. Khutoretsky, *Two existence theorems for computable numerations.* Algebra i Logika, 1969, v. 8, no. 4, pp. 484–492 (Russian).

[47] A. B. Khutoretsky, *On the cardinality of the upper semilattice of computable numberings,* Algebra and Logic, 1971, v. 10, no. 5, pp. 348–352.

[48] M. Kummer, *Recursive enumeration without repetition revisited.* Lecture Notes in Mathematics, 1990, v. 1432, pp. 255–275.

[49] M. Kummer, *An easy priority-free proof of a theorem of Friedberg.* Theoretical Computer Science, 1990, v. 74, no. 2, pp. 249–251.

[50] M. Kummer, *Some applications of computable one-one numberings.* Arch. Math. Logic, 1990, v. 30, no. 4, p. 219–230.

[51] M. Kummer, *A learning-theoretic characterization of discrete families of recursive functions,* Information Processing Letters, 1995, v. 54, pp. 205–211.

[52] A. H. Lachlan, *Standard classes of recursively enumerable sets.* Z. Math. Logik Grundl. Math., 1964, Bd. 10, H. 1, S. 23–42.

[53] A. H. Lachlan, *On recursive enumeration without repetition.* Z. Math. Logik Grundl. Math., 1965, Bd. 11, H. 3, S. 209–220.

[54] A. H. Lachlan, *On recursive enumeration without repetition: a correction.* Z. Math. Logik Grundl. Math., 1967, Bd. 13, H. 2, S. 99–100.

[55] A. H. Lachlan, *Two theorems on many-one degrees of recursively enumerable sets.* Algebra and Logic, 1972, v. 11, no. 2, pp. 216–229.

[56] A. H. Lachlan, *Recursively enumerable many-one degrees.* Algebra and Logic, 1972, v. 11, no. 3, pp. 326–358.

[57] A. H. Lachlan, *A note on positive equivalence relations.* Z. Math. Logik Grundl. Math., 1987, Bd. 33, H. 1, S. 43–46.

[58] I. A. Lavrov, *Computable numberings.* Logic, Foundations of Mathematics and Computability Theory. - Dordrecht, Holland: D. Reidel Publishing Company, 1977, pp. 195–206.

[59] A. I. Mal'tsev, *Constructible algebras. I,* Uspekhi Mat. Nauk, 1961, v. 16, no. 3, pp. 3–60.

[60] A. I. Mal'tsev, *Toward a theory of computable families of objects.* Algebra i Logika, 1964, v. 3, no. 4, pp. 5-31 (Russian).

[61] A. I. Mal'tsev, *Positive and negative enumerations.* Soviet Math. Dokl., 1965, v. 160, pp. 75-77.

[62] A. I. Mal'tsev, *Algorithms and recursive functions,*— Wolters-Noordoff Publishing, Groningen, 1970.

[63] S. S. Marchenkov, *The minimal numerations of systems of recursively enumerable sets.* Sov. Math. Dokl., 1971, v. 12, pp. 843–845.

[64] S. S. Marchenkov, *The semilattices of computable numerations.* Soviet Math. Dokl., 1971, v. 12, pp. 886–888.

[65] S. S. Marchenkov, *On computable enumerations of families of general recursive functions.* Algebra and Logic, v. 11, 1972, no. 5, pp. 326–336.

[66] S. S. Marchenkov, *The existence of families without positive numerations.* Math. Notes, 1973, v. 13, no. 4, pp. 360–363.

[67] M. B. Pour-El, *Gödel numberings versus Friedberg numberings.* Proc. Amer. Math. Soc., 1964, v. 15, no. 2, pp. 252–256.

[68] M. B. Pour-El, W. A. Howard, *A structural criterion for recursive enumeration without repetition.* Z. Math. Logik Grundl. Math., 1964, Bd. 10, H. 2, S. 105–114.

[69] M. B. Pour-El, H. Putnam, *Recursively enumerable classes and their applications to sequences of formal theories.* Arch. Math. Logik und Grundlagenforschung, 1965, v. 8, pp. 104–121.

[70] H. G. Rice, *Classes of recursively enumerable sets and their decision problems.* Trans. Amer. Math. Soc., 1953, v. 74, no. 2, pp. 358–366.

[71] H. G. Rice, *On completely recursively enumerable classes and their key arrays.* J. Symbolic Logic, 1956, v. 21, no. 3, pp. 304–308.

[72] H. Rogers, *Gödel numberings of partial computable functions.* J. Symbolic Logic, 1958, v. 23, no. 3, pp. 49–57.

[73] H. Rogers, *Theory of recursive functions and effective computability.* — McGraw-Hill, New York, 1967.

[74] V. L. Selivanov, *Enumerations of families of general recursive functions.* Algebra and Logic, 1976, v. 15, no. 2, pp. 128–141.

[75] V. L. Selivanov, *Two theorems on computable enumerations*, Algebra and Logic, 1976, v. 15, no. 4, pp. 297–306.

[76] V. Yu. Shavrukov, *Remarks on Uniformly Finitely Pre-complete Positive Equivalences.* Math. Log. Quart., 1996, v. 42, pp. 67–82.

[77] R. I. Soare, *Recursively enumerable sets and degrees. A study of computable functions and computably generated sets.* — Springer-Verlag, Berlin-New York, 1987.

[78] V. A. Uspensky, *On computable operations.* Dokl. Akad. Nauk SSSR, 1955, v. 103, no. 5, pp. 773–776 (Russian).

[79] V. A. Uspensky, *Systems of denumerable sets and their numberings.* Dokl. Akad. Nauk SSSR, 1955, v. 105, no. 6, pp. 1155–1158 (Russian).

[80] V. A. Uspensky, *Some remarks on enumerable sets.* Z. Math. Logik Grundl. Math., 1957, Bd. 3, S. 157–170.

[81] V. A. Uspensky, *Algorithms: main ideas and applications.* — Kluwer Academic Publishers Group, Dordrecht, 1993.

[82] Yu. G. Ventsov, *A family of recursively enumerable sets with finite classes of nonequivalent univalent computable numerations.* Vychisl. Sistemy, 1987, v. 120, pp. 105–142 (Russian).

[83] Yu. G. Ventsov, *On algorithmic dimension of models.* Soviet Math. Dokl., 1989, v. 39, pp. 237–239.

[84] Yu. G. Ventsov, *Computable classes of constructivizations of models of infinite algorithmic dimension.* Algebra and Logic, 1994, v. 33, no. 1, pp. 79–84.

[85] V. V. V'jugin, *Discrete classes of recursively enumerable sets.* Algebra i Logika, 1972, v. 11, no. 3, pp. 243–256 (Russian).

[86] V. V. V'jugin, *Minimal numerations of computable classes of recursively enumerable sets.* Soviet Math. Dokl., 1973, v. 14, pp. 1338–1340.

[87] V. V. V'jugin, *On some examples of upper semilattice of computable numerations.* Algebra and Logic, 1973, v. 12, no. 5, pp. 277–286.

[88] V. V. V'jugin, *Discrete classes of recursively enumerable sets.* Algebra i Logika, 1973, v. 12, no. 6, pp. 635–654 (Russian).

[89] V. V. V'jugin, *On upper semilattices of numberings.* Soviet Math. Dokl., 1974, v. 217, no. 4, pp. 749–751 (Russian).

Kazakh State University, Department of Mechanics and Mathematics, 480071 Almaty, Kazakhstan
E-mail address: badaev@math.kz

Academy of Sciences, Siberian Branch, Mathematical Institute, 630090 Novosibirsk, Russia
E-mail address: gonchar@math.nsc.ru

Contemporary Mathematics
Volume **257**, 2000

Π_1^0 Classes – Structure and Applications

Douglas Cenzer and Carl G. Jockusch, Jr.

ABSTRACT. We present results and open problems on Π_1^0 classes, which are effectively closed sets of reals. These include applications of Π_1^0 classes to various mathematical problems in analysis, combinatorics, and the theory of orderings in which the complexity of solutions to the problems is considered. This complexity can be measured in terms of Turing degrees, and in the sense of index sets for certain properties of the solution set. Such questions are closely related to the strength of the statement that solutions exist, in the sense of "reverse mathematics." Effective Ramsey theory is given special attention. The lattice structure of the Π_1^0 classes is compared with the structure of the Π_1^0 subsets of ω and several open problems are given on the lattice of Π_1^0 classes of sets.

Introduction

Effectively closed sets of functions are called Π_1^0 classes in computability theory. A Π_1^0 class may be expressed as the set of infinite paths through a computable tree. Π_1^0 classes arise naturally in the study of computability theory and its applications. The solution sets to many problems in logic, combinatorics, algebra, and other areas are Π_1^0 classes and, in some cases, these solution sets can represent arbitrary Π_1^0 classes of certain types. This leads to conclusions about the complexity of solutions as well as the structure of the solutions. Resource-bounded problems in some cases correspond to resource-bounded Π_1^0 classes with similar applications. The study of Π_1^0 classes is closely connected to the system WKL_0 of second-order number theory (based on weak König's Lemma). The inclusion lattice of Π_1^0 classes has an interesting algebraic structure, in some ways analogous to the dual of the lattice of computably enumerable subsets of ω. There are many connections between the study of the computably enumerable (c.e.) sets and degrees, the Turing degrees, and the Π_1^0 classes.

Research in computability theory has focused on the family of subsets of the set $\mathbb{N}$ of all natural numbers, and functions from $\mathbb{N}$ to $\mathbb{N}$, along with the structure of the Turing degrees of these sets and functions, and the inclusion lattice of the

---

1991 *Mathematics Subject Classification.* Primary 03D30, 03D80; Secondary 03C62, 03D25, 03E15, 03F35, 03F60.

Jockusch's research was partially supported by NSF Grant DMS-98-03073. He thanks Linda Lawton for helpful comments and corrections.

computably enumerable (c.e.) sets. These sets (not necessarily c.e.) and functions are the potential members of a Π_1^0 class. The fundamental results about Π_1^0 classes concern the complexity of the members of a class. For example, the Kreisel-Shoenfield Basis Theorem [**K53, S58**] states that every nonempty Π_1^0 class of sets contains a member computable from $\mathbf{0}'$. On the other hand, Kleene [**K55**] showed that there is a nonempty Π_1^0 class of *functions* with no hyperarithmetic member. In this paper we will focus mainly on Π_1^0 classes of sets, since these occur so frequently in the analysis of the complexity of the solutions to effectively given problems.

Section one of this paper contains some definitions and results and an open question on Π_1^0 classes of sets and functions and their members.

A key motivation for the study of Π_1^0 classes is the role they play in the study of computable aspects of numerous areas of mathematics. In many problems, the family of solutions to a given problem (such as the family of 4-colorings of a given infinite planar graph) can be viewed as a closed set under some natural topology. Since a Π_1^0 class is simply an effectively closed set, the family of solutions to a computably presented problem will often be a Π_1^0 class. Furthermore, a *compact* family of solutions will often be a *bounded* or *computably bounded* (c.b.) Π_1^0 class. (A class S of functions is *bounded* if there is a function g with $f(n) \leq g(n)$ for all $f \in S$ and $n \in \mathbb{N}$ and S is *computably bounded* (c.b.) if there is such a g which is computable.) One may also consider classes which are $\Pi_1^{0,x}$ for some oracle x, that is, are the class of infinite paths through a tree T which is computable from x. We will see that c.b. Π_1^0 classes behave like Π_1^0 classes of sets and bounded Π_1^0 classes behave like $\Pi_1^{0,0'}$ classes of sets.

For example, Bean [**B76**] observed that the set of 4-colorings of a computable graph is a computably bounded Π_1^0 class. It follows from the Kreisel-Shoenfield Basis Theorem that any computable graph which is 4-colorable must have a coloring c computable from $\mathbf{0}'$.

The more difficult *representation problem* is whether every (c.b.) Π_1^0 class represents the solution set to some computable instance of a given problem. In general, if P and Q are subsets of ω^ω we say that Q *represents* P if there is a degree-preserving bijection $\mathcal{F}$ of P onto Q, i.e. for all $f \in P$, $\mathcal{F}(f) \equiv_T f$.

For example, Remmel [**R86**] showed that every computably bounded Π_1^0 class can be represented as the family of k-colorings of some highly computable graph, provided that k-colorings which differ only by a permutation of the colors are identified. Jockusch and Soare [**JS72**] constructed a nonempty Π_1^0 class of sets such that any two of its members have incomparable Turing degree. It follows from these two results that there exists a highly computable, 4-colorable graph G such that any two 4-colorings of G which do not differ by a permutation of the colors are Turing incomparable. Furthermore, Cenzer and Remmel [**CR98b**] showed that for any computable graph G, there is a polynomial time graph G^* and a degree-preserving correspondence between the colorings of G and the colorings of G^*.

Section two of this paper contains some representation results and open questions.

The axiom **Weak König's Lemma** is the assertion in the language of second-order arithmetic that any infinite tree $T \subseteq 2^{<\omega}$ has an infinite path. Of course, this implies in the weak base theory RCA_0 that for any $k \in \omega$, any infinite tree $T \subseteq k^{<\omega}$ has an infinite path. More generally, König's Lemma states that any infinite, finite-branching tree has an infinite path. Now suppose we express the solution set of some infinite mathematical problem as the set of paths through a binary tree. Then finite restrictions of the problem correspond to finite levels of the tree. Furthermore, we place some condition on the problem such that each finite restriction will have a solution. For example, let the paths through a tree $T \subseteq 4^{<\omega}$ be the legal 4-colorings of a countably infinite planar graph. Then the (finite) 4-color theorem implies that each finite subgraph has a 4-coloring, so that T is an infinite tree. Weak König's Lemma then implies that there is a 4-coloring of the entire graph.

The *reverse mathematics* program of H. Friedman and S. Simpson is a program to analyze the logical strength of mathematical results, typically by showing that they are equivalent to one of a limited number of standard set existence principles over a certain weak subsystem RCA_0 of second order arithmetic. In connection with the above example, Hirst [**H89**] showed that the following statement is equivalent to Weak König's Lemma over RCA_0: "If every finite subset of a graph $\mathbf{G}$ has a 4-coloring, then $\mathbf{G}$ has a 4-coloring."

Section three contains some background, results and open questions on subsystems of second order arithmetic.

Ramsey's theorem has been a useful tool in many areas of mathematics. The version we will consider states that for all k and n, every k-coloring C of $[\mathbb{N}]^n$ has an infinite homogeneous set H, that is, an infinite set H such that all n-element subsets of H have the same color. For a computable coloring C of $[\mathbb{N}]^n$, the set of all homogeneous sets is a Π_1^0 class. However, this fact is of no direct use since this class includes the finite homogeneous sets. Nonetheless, Π_1^0 classes play a crucial but indirect role in the analysis of the effective content of Ramsey's theorem. Specker [**S71**] constructed a computable 2-coloring of $[\mathbb{N}]^2$ with no infinite computable homogeneous set, which implies that Ramsey's Theorem for 2-colorings of $[\mathbb{N}]^2$ is not provable in RCA_0.

Jockusch [**J72**] showed that, for any $n \geq 2$ and any k, any computable k-coloring of $[\mathbb{N}]^n$ has an infinite Π_n^0 homogeneous set, but not necessarily an infinite Σ_n^0 homogeneous set. Simpson [**S98**] showed that, for $n \geq 3$ and $k \geq 2$, Ramsey's Theorem for k-colorings of $[\mathbb{N}]^n$ is equivalent to ACA_0 (Arithmetic Comprehension) over RCA_0, but the full Ramsey's Theorem is not provable in ACA_0.

Section four contains some results and open questions on computable aspects of Ramsey Theory.

The complexity of a problem may be assessed by studying the complexity of various *index sets* associated with the problem. Here one enumerates all effectively given instances of a problem, for example, all c.e. graphs, as $G_0, G_1, \ldots$, and defines the index set associated with a certain problem and a property $\mathcal{R}$ as $\{e : G_e \text{ has property } \mathcal{R}\}$. Cenzer and Remmel [**CR98a, CR98b, CR99a**] and Gasarch [**G98**] have analyzed the complexity of many problems in this way. In particular, it is shown in [**CR98b**] that the index set for graphs which are k-colorable but have no computable k-coloring is Π_3^0 complete.

Section five contains some known results and open questions on index sets for Π_1^0 classes and associated problems.

The Π_1^0 classes form a lattice, $\mathcal{E}_\Pi$, under inclusion which has many aspects for investigation. Important issues include the definability and complexity of various properties, automorphisms of the lattice and orbits under automorphisms, and the analysis of certain substructures of the lattice. This lattice may be compared and contrasted with the lattice of Π_1^0 subsets of ω, which is of course the dual under complementation of the well-known lattice $\mathcal{E}$ of computably enumerable sets. For example, Cholak, Coles, Downey and Herrmann [**CCDH**] recently showed that the family of all perfect thin classes is definable in $\mathcal{E}_\Pi$, that any two perfect thin classes are automorphic in this lattice, and that the degrees of the perfect thin classes are precisely the c.e. *array non-computable* degrees. Here, the *degree* of a Π^0_1 class P is the degree of the set of strings extendible to elements of P, and the array non-computable degrees were defined and studied by Downey, Jockusch, and Stob in [**DJS90**] and [**DJS96**]. The study of the lattice of Π^0_1 classes can give more information about the problems represented by Π^0_1 classes which were discussed above.

Cenzer and Nies [**CN**] studied intervals of the lattice of Π^0_1 classes, showing in particular that there exist intervals which are not Boolean algebras but which have a decidable theory, in contrast with the results of Nies [**N97, Nta**] that the theory of any interval of the lattice $\mathcal{E}$ of c.e. sets which is not a Boolean algebra codes true arithmetic and is therefore undecidable.

Section six contains some known results and open questions on the lattice of Π^0_1 classes.

1. Π^0_1 classes and their members

In this section, we provide some background on Π^0_1 classes and discuss *basis* and *antibasis* results which will be needed for applications.

Some definitions are needed. Let $\omega^{<\omega}$ be the set of *strings*, that is, functions from a finite initial segment of ω into ω. Similarly, $2^{<\omega}$ is the set of strings with values in $\{0,1\}$. We write $\sigma \preceq \tau$ for $\sigma \subseteq \tau$. For $x \in \omega^\omega$ and $n \in \omega$, let $x\lceil n = (x(0), x(1), \ldots, x(n-1))$. Let $\sigma \prec x$ if $\sigma = x\lceil n$ for some n. We write $|\sigma|$ for the cardinality of the domain of the string σ and often identify σ with the sequence $(\sigma(0), \sigma(1), \ldots, \sigma(|\sigma|-1))$.

A subset T of $\omega^{<\omega}$ is a *tree* if whenever $\tau \in T$ and $\sigma \preceq \tau$, it is the case that $\sigma \in T$. For any tree T, $[T]$ denotes the set of infinite paths through T, that is,

$$[T] = \{x \in \omega^\omega : (\forall n)[x\lceil n \in T]\}.$$

The set of *extendible* nodes of T is defined by

$$Ext(T) = \{\sigma : (\exists x \in [T])[\sigma \prec x]\}.$$

The usual product topology on the space ω^ω has a sub-basis of *intervals*

$$I(\sigma) = \{x : \sigma \prec x\}.$$

With this topology, the closed subsets of ω^ω are exactly those of the form $[T]$ for some tree T. For the subspace 2^ω, the clopen sets are just finite unions of intervals.

A Π^0_1 *class* in ω^ω is an effectively closed set, i.e. one of the form $[T]$ for some *computable* tree T. It is easily seen that an equivalent definition is obtained by requiring T to be primitive recursive, or only co-c.e., instead of computable. A Π^0_1 class is called *decidable* if it has the form $[T]$ for some computable tree T such that $Ext(T)$ is computable. A Π^0_1 class $P \subseteq 2^\omega$ is called a Π^0_1 *class of sets* and clearly has the form $[T]$ for some computable tree $T \subseteq 2^{<\omega}$.

Although to show formally that a given class S is Π^0_1 one should produce a computable tree T with $S = [T]$, the intuition is even simpler. It suffices to have an effective procedure which, given an oracle for a function $f \notin S$, discovers that

$f \notin S$ within finitely many steps, and runs forever on any oracle $f \in S$. Of course, this is in complete analogy with the situation for Π_1^0 subsets of ω.

The complexity of a computable tree and its infinite paths is related to various notions of *boundedness*. For a given function $g : \omega^{<\omega} \to \omega$, a tree $T \subseteq \omega^{<\omega}$ (as well as the class $[T]$) is said to be *g-bounded* if for every $\sigma \in \omega^{<\omega}$ and every $i \in \omega$, if $\sigma^\frown i \in T$, then $i < g(\sigma)$. Thus, for example, if $g(\sigma) = 2$ for all σ, then trees in $2^{<\omega}$ are g-bounded. T is said to be *finite branching* and $[T]$ to be *bounded* if T is g-bounded for some g, that is, if each node of T has finitely many immediate successors. A computable tree T is said to be *highly computable* and the class $[T]$ is said to be *computably bounded (c.b.)* if T is g-bounded for some computable function g. Cenzer and Remmel introduced in [**CR98a**] intermediate notions of *almost bounded* and *almost computably bounded* trees and classes. Resource-bounded trees and Π_1^0 classes have been studied recently by Cenzer and Remmel in [**CR95, CR98b, CRta**] In particular, a tree T is said to be *highly p-time* if there exist p-time functions g and h such that, for all $\sigma \in T$ of length n, $|\sigma(n-1)| \leq g(1^n)$ and $h(\sigma)$ is a list of all i such that $\sigma^\frown i \in T$.

For any disjoint c.e. sets A and B, S is said to be a *separating set* for A and B if $A \subseteq S$ and $S \cap B = \emptyset$. The family of separating sets of A and B is a classic example of a Π_1^0 class.

An element x of a Π_1^0 class P is said to be *isolated* if there is some σ such that $P \cap I(\sigma) = \{x\}$. The Cantor-Bendixson derivative $D(P)$ is the set of nonisolated points of P; the αth iteration of D is D^α. The *rank* of x in P is the least α such that $x \in D^\alpha(P) \setminus D^{\alpha+1}(P)$. If P is countable, the *rank* of P is the least ordinal α such that $D^\alpha(P) = \emptyset$. (For P uncountable, the *rank* of P is the least α such that $D^\alpha(P) = D^{\alpha+1}(P)$, but we will not consider the rank of uncountable classes.)

An infinite Π_1^0 class $P \subseteq 2^\omega$ is said to be *thin* if every Π_1^0 subclass Q of P is equal to $U \cap P$ for some clopen set U. P is said to be *minimal* if every Π_1^0 subclass of P is either finite or cofinite in P.

The first example of a thin Π_1^0 class is due implicitly to D. Martin and M. Pour-El in [**MP70**]. They constructed an axiomatizable, essentially undecidable theory T such that every axiomatizable extension of T is finitely axiomatizable over T. It is easy to see that the class of complete extensions of such a theory T is a thin Π_1^0 class, and it is perfect because it contains no computable element. Information on the degrees of such theories may be found in [**D87**] and [**DJS90**]. The first explicit construction of a thin Π_1^0 class is due to S. Simpson (unpublished). Countable thin Π_1^0 classes of arbitrary computable rank, including minimal classes, were constructed in [**CDJS**].

Basis theorems say roughly that "Every nonempty simply definable class has a simply definable member." Kleene observed that, given the set of extendible nodes $Ext(T)$ as an oracle, one can easily compute an infinite path through T (if one exists) by letting $x(n)$ be the least k such that $(x(0), \ldots, x(n-1), k) \in Ext(T)$. For an arbitrary computable tree, $Ext(T)$ is Σ_1^1 and for a decidable tree, $Ext(T)$ is of course computable. The following basis results are fundamental.

THEOREM 1.1. *For any nonempty Π_1^0 class $P \subseteq \omega^\omega$,*

(a) *P has a member computable from some Σ_1^1 set;*
(b) *if P is bounded, then P has a member computable from $\mathbf{0}''$;*
(c) *if P is computably bounded, then P has a member computable from $\mathbf{0}'$;*
(d) *if P is decidable, then P has a computable member.*

(e) *if P is highly polynomial-time decidable, then P has a p-time member.*

PROOF. We will just show how to get the complexity of $Ext(T)$, where T is a computable tree with $P = [T]$. In general, $Ext(T)$ is Σ^1_1 by the form of its definition.

If T is finitely branching, then König's Lemma implies that

$$\sigma \in Ext(T) \iff (\forall n > |\sigma|)(\exists \tau)[|\tau| = n \ \& \ \sigma \preceq \tau \ \& \ \tau \in T].$$

If T is highly computable, let f be a computable function such that $\sigma(i) < f(i)$ for all $\sigma \in T$ and all $i < |\sigma|$. Then the quantifier "$\exists \tau$" above is bounded, since $\tau(i) < f(i)$ for all $i < n$. Hence, in this case, $Ext(T)$ is a Π^0_1 set. □

Part (a) of this theorem is the Kleene basis theorem [**K43**], part (c) is the Kreisel-Shoenfield basis theorem [**K53**] and part (e) is from [**CR92**].

Jockusch and Soare [**JS72**] obtained a number of further basis results for the computably bounded case.

THEOREM 1.2. *Every nonempty c.b. Π^0_1 class P contains:*

1. *an element of low degree.*
2. *an element of c.e. degree.*
3. *two elements whose degrees have infimum zero.*
4. *an element of hyperimmune-free degree.*

Part (1) is known as the *Low Basis Theorem* and is proved by forcing with nonempty Π^0_1 subclasses of P. Part (2) is obtained by considering the lexicographically least element of P, and the final two parts are proved by forcing with nonempty Π^0_1 subclasses of P.

For countable classes, we have stronger basis results, due to Kreisel [**K59**], stemming from the fact that every nonempty countable closed set must have an isolated point.

THEOREM 1.3. *Let P be a nonempty countable Π^0_1 class of functions. Then*

1. *P has a hyperarithmetic member.*
2. *If P is bounded, then P has a member computable from $\mathbf{0}'$.*
3. *If P is computably bounded, then P has a computable member.*

The following result of Jockusch and Soare [**JS72**] is relevant to the later material on second order arithmetic.

THEOREM 1.4. *For any sequence $C_0, C_1, \ldots$ of noncomputable sets and any nonempty c.b. Π^0_1 class P, P has a member x such that no C_i is Turing reducible to x.*

In general, the degree of an element x of a Π^0_1 class P is bounded in terms of the rank of x in P. We will give just the result for rank one.

THEOREM 1.5. [**CDJS**] *Let P be a Π^0_1 class of sets and suppose $x \in P$.*

1. *If x has rank 1 in P, then $x \leq_T 0''$. If, in addition, P is decidable, then $x \leq_T 0'$.*
2. *If P is thin, then $x' \leq_T x \oplus 0''$, and if P is thin and decidable, then $x' \leq_T x \oplus 0'$.*

Downey has posed the following general problem.

PROBLEM 1.6. Characterize the degrees of members of thin Π_1^0 classes.

An *antibasis* theorem states that a given family of reals is *not* a basis for a certain kind of class. Such theorems help to distinguish the various notions of boundedness and may be applied to show that all solutions to certain problems are complicated in some sense.

THEOREM 1.7. 1. *(Kleene, [**K43**]) There is a nonempty Π_1^0 class of functions with no hyperarithmetic member.*
2. *(Jockusch-McLaughlin, see also [**JLR**]) There is a bounded, nonempty Π_1^0 class of functions with no member computable from $\mathbf{0}'$.*
3. *(Jockusch, [**J74**]) There is a nonempty Π_1^0 class of sets with no member which is a finite Boolean combination of c.e. sets.*
4. *There is a disjoint pair of c.e. sets with no computable separating set.*

There is a natural, *universal* Π_1^0 class $U \subset 2^\omega$ defined by

$$U = \{x \in 2^\omega : (\forall e)[x(e) \neq \varphi_e(e)]\}.$$

The universal class U is a class of separating sets, i.e. the class of sets which separate $\{e : \varphi_e(e) = 0\}$ and $\{e : \varphi_e(e) = 1\}$. It is clear that U is nonempty and has no computable element.

Simpson [**S77**] defined a partial ordering $<<$ of degrees which is stronger than the usual ordering $<$. Let $\mathbf{b} << \mathbf{a}$ mean that every infinite tree $T \subseteq 2^{<\omega}$ of degree $\leq \mathbf{b}$ has an infinite path of degree $\leq \mathbf{a}$. Simpson showed that the degrees are densely ordered by $<<$, and much further interesting work on this ordering has been done by Kučera ([**K85**], [**K88**]). Note that $\mathbf{0} << \mathbf{a}$ means that every nonempty Π_1^0 class of sets has an element of degree $\leq \mathbf{a}$.

PROBLEM 1.8. Is $<<$ (naturally) first-order definable in $\mathcal{D}$, the usual partial ordering of the degrees?

The *universal* Π_1^0 class U defined above has the property that the degrees of the elements of U coincide with the degrees of complete extensions of Peano arithmetic. Now D. Scott showed that for any degree $\mathbf{a}$, $\mathbf{a} >> \mathbf{0}$ if and only if there exists $f \in U$ of degree $\leq \mathbf{a}$. Thus the degrees of elements of U are exactly the degrees $\mathbf{a} >> \mathbf{0}$, which means that each element of U can compute an element of any nonempty Π_1^0 class. This is the sense in which U is universal.

Here are some further results which can be viewed as antibasis theorems.

THEOREM 1.9. 1. [**JS72**] *If* $\mathbf{a} >> \mathbf{0}$, *then every countable partially ordered set can be embededded in the degrees* $\leq \mathbf{a}$.
2. *(Arslanov, see* [**S87**], *page 88) If* $\mathbf{0} << \mathbf{a}$ *and* $\mathbf{a}$ *is a c.e. degree, then* $\mathbf{a} = \mathbf{0}'$.
3. *(Kučera, see* [**S87**], *pages 116-118) If* $\mathbf{0} << \mathbf{a} \leq \mathbf{0}'$, *then there is a promptly simple degree* $\mathbf{b} \leq \mathbf{a}$. *(This result was proved without the priority method and, in combination with the low basis theorem, yields a priority-free solution to Post's problem.)*

Parts 2 and 3 of the above result were actually proved with the hypothesis that $\mathbf{0} << \mathbf{a}$ replaced by the weaker hypothesis that there is a function f of degree at most $\mathbf{a}$ which is *fixed-point free* in the sense that $(\forall e)[W_e \neq W_{f(e)}]$. In part 3 the prompt simplicity of $\mathbf{b}$ is a direct consequence of the construction. The weaker hypothesis mentioned above does not suffice for part 1, though, since M. Kumabe [**Kta**] has announced that there is a fixed-point free function of minimal degree.

COROLLARY 1.10. 1. There is a nonempty Π_1^0 class of sets with no element whose degree is both low and c.e.

2. There is a nonempty Π_1^0 class of sets having no pair of elements $\leq_T K$ whose degrees have infimum $\mathbf{0}$.

THEOREM 1.11. [**JS72**] *There is a nonempty* Π_1^0 *class* P *of sets such that any two distinct elements of* P *are Turing incomparable.*

However no nontrivial class of separating sets has the property that any two distinct elements are Turing incomparable.

2. Applications of Π_1^0 classes

In this section, we consider a few examples of Π_1^0 classes as solution sets of particular computable mathematical problems and give some open problems on the representation of Π_1^0 classes. For further details on these examples and for additional examples, the reader is referred to [**CR98b**] and [**G98**] in the *Handbook of Recursive Mathematics.*

Logical Theories. The set $P(\Gamma)$ of complete extensions of a given axiomatizable theory Γ in first-order logic is always a Π_1^0 class. (We use "complete" to mean "maximal consistent.") For the theory $\Gamma = PA$ (Peano Arithmetic), this provides the historically first example of a nonempty Π_1^0 class with no computable member. It follows from results cited in the previous section that PA does have complete extensions of low degree, c.e. degree, hyperimmune-free degree, etc.

Suppose that a thin class P is the set of complete extensions of some axiomatizable theory Γ. Then for any axiomatizable extension Δ of Γ, the set of complete consistent extensions of Δ is a Π_1^0 subclass Q of P. The fact that $Q = P \cap U$ for some clopen U means that Δ is a finitely generated (or *principal*) extension. Thus we say that Γ is a *maximal* theory. For a minimal class P, any axiomatizable extension Δ of Γ has the stronger property that either Δ has only finitely many complete extensions or all but finitely many of the complete extensions of Γ are extensions of Δ. The perfect thin set constructed by Martin and Pour-El [**MP70**] is the set of complete extensions of an axiomatizable, essentially undecidable theory T such that any axiomatizable extension of T is principal.

Ehrenfeucht [**E61**] showed that any Π_1^0 class $P \subseteq 2^\omega$ can be represented in the form $P(\Gamma)$ for some axiomatizable propositional theory Γ. It follows that there is a consistent axiomatizable theory such that any two distinct complete extensions of Γ have incomparable Turing degree.

For a *decidable* theory Γ, $P(\Gamma)$ is a decidable Π_1^0 class and every decidable Π_1^0 class of sets can be represented by the set of complete extensions of a decidable propositional theory. It follows that every decidable consistent theory has a decidable complete extension. Cenzer and Remmel [**CRta**] showed that the problem of whether every polynomial-time decidable tree represents the complete consistent extensions of a p-time decidable theory is equivalent to $P = NP$.

A minimal decidable class corresponds to a decidable theory T with exactly one undecidable complete extension and such that any axiomatizable extension of T is a principal extension of T.

Marek, Remmel and Nerode have studied natural versions of *nonmonotonic* logic corresponding to arbitrary, bounded and computably bounded Π_1^0 classes. (See Chapter 5 of [**CR98b**] for details.)

Now the complete extensions of an axiomatizable theory can be viewed as the maximal ideals of the c.e. Lindenbaum Boolean algebra. Thus every (decidable) Π_1^0 class of sets may be represented as the family of maximal ideals of a c.e. (computable) Boolean algebra.

Commutative Rings. It is easy to see that the prime ideals of a c.e. commutative ring with unity compose a c.b. Π_1^0 class. The example of Boolean rings shows that any c.b. Π_1^0 class of sets may be represented as the set of prime ideals of some c.e. commutative ring with unity and any decidable Π_1^0 class of sets may be represented as the set of prime ideals of some computable ring.

It was shown by Friedman, Simpson and Smith [**FSS**] that any Π_1^0 class of separating sets may be represented as the set of prime ideals of a computable commutative ring with unity.

PROBLEM 2.1. Does *every* c.b. Π_1^0 class represent the set of prime ideals of some computable commutative ring with unity?

Graph Theory. There are many interesting problems from computable graph theory. The *Handbook of Recursive Mathematics* ([**CR98b, G98**] is a good source. The Ramsey problem will be discussed in section 4 below.

The graph coloring problem was discussed in the introduction and demonstrates the interesting properties of thin and minimal classes. If a thin class P is the set of k-colorings of some computably enumerable graph G, then for any computable coloring f of a computable subgraph H, the set of extensions of f to G is a Π_1^0 subclass of P and thus there is a *finite* subgraph G_1 and a coloring g_1 of G_1 such that the extensions of f are exactly the extensions of g_1. For a minimal class P and a computable coloring f of a computable subgraph H, we see that either there are only finitely many k-colorings which extend f or all but finitely many k-colorings of G extend f.

Matching problems provide an example where all classes of separating sets, but not all c.b. Π_1^0 classes can be represented by a particular problem. A *computable society* $S = (B, G, K)$ consists of disjoint computable sets B (boys) and G (girls) and a computable binary relation $K \subseteq B \times G$. A solution here is a *marriage assignment*, that is, a 1-1 mapping $f : B \to G$ such that $(b, f(b)) \in K$ for all $b \in B$. S is *bounded* if the set $K(b) = \{g : K(b,g)\}$ is finite for each $b \in B$ and is *highly computable* if $|K(b)|$ can be computed effectively from b. Cenzer and Remmel [**CR98b**] showed that every class of separating sets can be represented as the set of marriage assignments of a highly computable society. On the other hand, it was shown by Misercque [**M78**] that not every c.b. Π_1^0 class can represent the set of marriage assignments of a highly computable society.

The *Stable Marriage* problem [**GI**] of Knuth refines the relation K by a *preference order*, a strict linear ordering R_b of G for each $b \in B$ and R_g of B for each $g \in G$. An instance of this problem is computable if these preference orders are uniformly computable. A solution here is a *stable marriage assignment*, that is a marriage assignment f such that for each $b \in B$ and $g \in G$, if $f(b) \neq g$, then it never happens that b prefers g to $f(b)$ and g prefers b to $f^{-1}(g)$. It is clear that the set of stable marriage assignments is a Π_1^0 class and Cenzer and Remmel showed at the meeting that any (unbounded) Π_1^0 class may be so represented.

PROBLEM 2.2. Find versions of the stable marriage problem corresponding to bounded and computably bounded Π_1^0 classes.

Partially Ordered Sets. In this section, we discuss two open problems from [**CR98a**].

Dilworth's Theorem states that every finite partial ordering (poset) of width k can be decomposed into the union of k chains, where the *width* of a finite partial ordering is the cardinality of the largest antichain. There is a natural dual to this theorem which says that every poset of height k can be covered by k antichains. The family of all decompositions of a given computable partial ordering as the union of k chains (or antichains) (k fixed) can be presented as a c.b. Π_1^0 class.

Cenzer and Remmel [**CR98a**] and, independently, J. Hirst obtained a positive answer for the special case where P is the class of separating sets of a disjoint pair of c.e. sets. The situation is analogous for antichains in place of chains.

PROBLEM 2.3. Does there exist for every c.b. Π_1^0 class P a computable partial ordering (A, R) and a natural number k such that P represents the class of decompositions of (A, R) as a union of k chains (antichains)?

The *dimension* of a poset (A, R) is defined to be the least n such that there exist n linear orderings $(A, L_1), \ldots, (A, L_n)$ such that $R = L_1 \cap \cdots \cap L_n$. Cenzer and Remmel showed that the set of solutions to the n-dimensionality problem can be expressed as a c.b. Π_1^0 class and that any class of separating sets can be represented by an instance of the dimensionality problem.

PROBLEM 2.4. Does there exist for every c.b. Π_1^0 class P a computable partial ordering (A, R) and a natural number n such that P represents the class of solutions of the n-dimensionality problem for Q?

Analysis. There are several problems related to computably continuous functions on the reals $\mathbb{R}$, the interval $[0, 1]$, on 2^ω and on ω^ω, for which the solution set is a c.b. Π_1^0 class. The key here is that the graph of a computably continuous function is always a Π_1^0 class and that, for the latter three spaces, any computably continuous function has a decidable Π_1^0 graph and any function with a Π_1^0 graph is computably continuous. See [**C93, CR98b, CR99a**] for details. For any computably continuous function, the following sets are all Π_1^0 classes: the set of zeroes, the set of points where f attains a maximum (minimum), the set of fixed points of f and the Julia set of f–that is, the complement of the basin of attraction of a computable periodic point. The first two problems can represent any Π_1^0 class in the given space, the fixed point problem can represent any Π_1^0 class except for the space [0,1] where the class must have a computable member, and the last problem can represent any class which is bounded and has computable maximum and minimum elements. Ko [**Ko98**] showed that if the Π_1^0 class P has either a polynomial time maximum element or a p-time minimum element, then there is a p-time computable function f with Julia set P.

It is also easy to see that the family of fixed points of a computably continuous mapping on the unit square, $[0, 1] \times [0, 1]$ is a Π_1^0 class. The Brouwer Fixed Point Theorem says that any such mapping must have a fixed point, but Orevkov [**O63**] showed that there need not be a computable fixed point.

PROBLEM 2.5. Does every nonempty Π_1^0 class $P \subseteq [0, 1] \times [0, 1]$ represent the set of fixed points of some computably continuous mapping?

Note that any bounded Π_1^0 class is bounded by a $\mathbf{0}'$-computable function. In the other direction, Jockusch, Lewis, and Remmel [**JLR**] showed that any $\Pi_1^{0,\mathbf{0}'}$

class bounded by a $\mathbf{0}'$-computable function is represented by some bounded Π_1^0 class. Thus, the study of bounded Π_1^0 classes is essentially reduced to the study of c.b. Π_1^0 classes via relativization to $\mathbf{0}'$. An application of bounded Π_1^0 classes to the problem of Rado selectors was given in [**JLR**].

3. Reverse Mathematics

Subsystems of second order arithmetic are used to measure the proof-theoretic strength of various theorems about the existence of solutions to mathematical problems, often by appealing to the minimum complexity of a solution to a computable problem. Second order arithmetic is the theory of natural numbers together with sets of natural numbers. An ω-model of second order arithmetic is one in which the natural numbers are standard. Let PA be the usual axioms of Peano Arithmetic and let P^- be the standard algebraic axioms for number theory (without the induction scheme).

A *Scott set* is a nonempty set $\mathcal{F} \subseteq 2^\omega$ such that whenever $T \subseteq 2^{<\omega}$ is an infinite tree computable from a finite join of elements of $\mathcal{F}$, then $[T] \cap \mathcal{F} \neq \emptyset$. The arithmetical sets form a Scott set. More generally, the sets representable in a complete extension of Peano arithmetic form a Scott set, as Scott proved in [**S62**].

The base theory RCA_0 is P^- with Δ_1^0 comprehension and Σ_1 induction. The ω-models of RCA_0 are those nonempty subsets of $\mathcal{P}(\mathbb{N})$ which are closed under join and closed downwards under Turing reducibility.

The system WKL_0 is RCA_0 together with Weak König's Lemma. The ω-models of WKL_0 are precisely the Scott sets. In WKL_0 one may prove that there is a noncomputable set by formalizing the proof that U has no computable element. However, the existence of a noncomputable c.e. set cannot be proved in WKL_0 because there is a Scott set containing no noncomputable c.e. set.

It is shown in Simpson [**S98**] that the existence of separating sets is equivalent, in a precise sense, to WKL_0 over RCA_0. Together with the representation result cited in the previous section, this can be applied to show that Dilworth's Theorem is equivalent to WKL_0 over RCA_0.

The system ACA_0 is RCA_0 together with arithmetic comprehension. The ω-models of ACA_0 are the ω-models of RCA_0 which are closed under the jump operation. Simpson [**S98**] showed that WKL_0 is a proper extension of RCA_0 and that ACA_0 is a proper extension of WKL_0.

The fact that ACA_0 does not follow from WKL_0 follows from Theorem 1.4 above.

M. Groszek and T. Slaman showed that the existence of a set of minimal degree is provable in $WKL_0 + \Sigma_3$ induction. The proof showed that there is a nonempty Π_1^0 class P such that every element of P is either of minimal degree or bounds some nonzero c.e. degree. Hence every element of P computes some set of minimal degree, so every Scott set contains a real of minimal degree.

PROBLEM 3.1. Can the existence of a minimal degree be proved in WKL_0?

H. Friedman and A. McAllister have posed the following related problems.

PROBLEM 3.2.
1. Is it provable in WKL_0 that for every noncomputable set X there is a set Y which is Turing incomparable with X?
2. Is it true that for every Scott set $\mathcal{F}$ and every noncomputable set $X \in \mathcal{F}$ there exists $Y \in \mathcal{F}$ such that X and Y are Turing incomparable?

The following related problem, which would seem to be a natural approach to part 2 of Problem 3.2, is also open.

PROBLEM 3.3. Does there exist for every nonzero degree $\mathbf{b}$ a nonempty $\Pi_1^{0,\mathbf{b}}$ class $P \subseteq 2^\omega$ such that every element of P has degree incomparable with $\mathbf{b}$?

4. Ramsey's Theorem

Let $[\mathbb{N}]^n$ denote the family of all n-element sets of natural numbers. A k-coloring C of $[\mathbb{N}]^n$ is a map from $[\mathbb{N}]^n$ into $\{1, 2, \ldots, k\}$. A set $H \subseteq \mathbb{N}$ is said to be *homogeneous* for C if all n-element subsets of H have the same color. Ramsey's Theorem states that for all k and n, every k-coloring C of $[\mathbb{N}]^n$ has an infinite homogeneous set H. For a fixed n and k, denote this theorem by RT^n_k.

Specker [**S71**] constructed a computable 2-coloring with no infinite computable homogeneous set. This result implies that the computable sets form an ω-model of RCA_0 in which RT^2_2 is false, so RT^2_2 is not provable in RCA_0.

Jockusch [**J72**] found bounds on the complexity of infinite homogeneous sets.

THEOREM 4.1. 1. *Any computable k-coloring of $[\mathbb{N}]^n$ has an infinite Π^0_n homogeneous set.*
2. *For any $n \geq 2$ there is a computable 2-coloring of $[\mathbb{N}]^n$ with no infinite Σ^0_n homogeneous set.*
3. *For any $n \geq 2$, there is a computable 2-coloring of $[\mathbb{N}]^n$ such that $\mathbf{0^{(n-2)}} \leq_T H$ for every infinite homogeneous set H.*

The proof of (i) is an induction on n using a priority construction in the base step $n = 2$ and the low basis theorem in the induction step to economize on the quantifier complexity of the infinite homogeneous set.

Simpson [**S98**] studied the proof-theoretic strength of Ramsey's Theorem and obtained the following related results.

THEOREM 4.2. 1. *For each $n \geq 3$ and $k \geq 2$, RT^n_k is equivalent to ACA_0 over RCA_0.*
2. *Ramsey's Theorem is not provable in ACA_0.*

It remained a mystery for over twenty years whether RT^2_2 is also equivalent to ACA_0 over RCA_0. This problem was resolved by D. Seetapun [**SS95**] who obtained the following result, which implies that RT^2_k does not imply ACA_0 in RCA_0.

THEOREM 4.3. *For any computable k-coloring of $[\mathbb{N}]^2$ and any noncomputable sets $C_0, C_1, \ldots$ there is an infinite homogeneous set A such that $(\forall i)[C_i \not\leq_T A]$.*

Slaman [**SS95**] then raised the question of whether RT^2_2 implies PA over RCA_0 and conjectured that a negative answer could be obtained by showing that, for some n, every computable 2-coloring of $[\mathbb{N}]^2$ has an infinite homogeneous set X which is low$_n$ (i.e. $X^{(n)} \leq_T 0^{(n)}$). This was recently done by Cholak, Jockusch, and Slaman for $n = 2$ in [**CJS**], with the following two main results.

THEOREM 4.4. *For any computable k-coloring of $[\mathbb{N}]^2$ there is an infinite homogeneous set X which is low$_2$.*

The idea of the proof is to first use the existence of a low$_2$ r-cohesive set [**JS93**] to make it possible to assume that the given coloring is *stable*, i.e. for each a the color of the pair $\{a, b\}$ is independent of b for b sufficiently large. This reduces the result to showing that, for each Δ_2^0 set A, there is an infinite low$_2$ set X such that X is contained in A or $\overline{A}$, which is the main lemma. The proof of the existence of a low$_2$ r-cohesive set and the proof of the main lemma both depend heavily on the low basis theorem. This proof is then adapted to models of second-order arithmetic to obtain the following result.

THEOREM 4.5. *Any Π_1^1 statement provable from $RCA_0 + I\Sigma_2 + RT_2^2$ is already provable from $RCA_0 + I\Sigma_2$. In particular, $B\Sigma_3$ (Σ_3 bounding) is not provable from RCA_0+ RT_2^2.*

It is also shown by J. Hirst in [**H87**] (see also [**CJS**], Corollary 11.5) that $(\forall k) RT_k^2$ *does* imply $B\Sigma_3$. This and Theorem 4.5 imply that $(\forall k) RT_k^2$ is strictly stronger than RT_2^2 over RCA_0. Here are some open questions raised in [**CJS**].

PROBLEM 4.6. Is Weak König's Lemma provable in $RCA_0 + RT_2^2$?

A corresponding problem in computable Ramsey theory is the following:

PROBLEM 4.7. Is there a computable 2-coloring of pairs having every infinite homogeneous set of degree $\mathbf{a} >> \mathbf{0}$?

It seems difficult to obtain a negative answer to the above problem by known methods because the techniques in [**SS95**] and [**CJS**] for showing that every computable 2-coloring of pairs has an infinite homogeneous set $A \not\geq_T \mathbf{0}'$ involve choosing elements of various Π_1^0 classes, and it seems easily possible that one of these classes could be a class such as U which contains only sets of degree $\mathbf{a} >> \mathbf{0}$.

PROBLEM 4.8. Is $I\Sigma_2$ provable in $RCA_0 + RT_2^2$?

Note: J. Hirst [**H87**] showed that $B\Sigma_2$ is provable in this theory.

PROBLEM 4.9. Characterize the degrees $\mathbf{d}$ such that every computable 2-coloring of pairs has an infinite homogeneous set of degree at most $\leq \mathbf{d}$.

Note: It was proved in [**CJS**] that $\mathbf{d} >> \mathbf{0}'$ iff every computable 2-coloring of pairs has an infinite homogeneous set of degree with *jump* of degree $\leq \mathbf{d}$.

PROBLEM 4.10. Is RT_2^2 provable from $RCA_0 + SRT_2^2$, where SRT_2^2 is Ramsey's theorem for stable 2-colorings of pairs?

The following problem is related to this.

PROBLEM 4.11. Does every computable stable 2-coloring of pairs have an infinite *low* homogeneous set? Equivalently, is it true that for every Δ_2^0 set A there is an infinite low set L which is contained in A or $\overline{A}$?

Downey, Lempp, Hirschfeldt, and Solomon, have recently obtained a negative answer to Problem 4.11 [private communication]. However, their method does not appear to answer Problem 4.10.

A c.e. partition of exponent n is a c.e. subset of $[\mathbb{N}]^n$. Hummel and Jockusch showed in [**HJ99**] that every c.e. partition of exponent n has an infinite Π_n^0 homogeneous set. Key ingredients of the proofs were the concept of *retraceability* and the existence of a low$_2$ cohesive set, due to Jockusch and Stephan [**JS93**].

An infinite set A is *n-cohesive* if A is almost homogeneous for each c.e. partition of exponent n. Thus, the 1-cohesive sets are exactly the cohesive sets.

The following results are from [**HJ99**].

THEOREM 4.12. 1. *There is a* Π^0_2 *2-cohesive set.*
2. *For all* n *there is a an* n*-cohesive set* $A \leq_T 0^{(n)}$, *so* $A \in \Delta^0_{n+1}$.

Here are two open problems from [**HJ99**].

PROBLEM 4.13. 1. For which n is there a Π^0_n n-cohesive set ("yes" for $n \leq 2$ and open for $n \geq 3$)?
2. Characterize the degrees of n-cohesive sets. (For $n = 1$, the *jumps* of such degrees are precisely the degrees $\mathbf{d} >> \mathbf{0}'$ [**JS93**]. For $n \geq 2$, the *jumps* of such degrees are precisely the degrees $\mathbf{d} \geq \mathbf{0^{(n+1)}}$ [**HJ99**].)

5. Index Sets

The notion of an *index* (or Gödel number) e of a computable function φ_e goes back to the beginning of computability theory and the enumeration of the c.e. sets as $W_e = \{n : \varphi_e(n) \text{ converges}\}$ is fundamental to computability theory. An index set for the c.e. sets associated with some property $\mathcal{R}$ of c.e. sets is simply the set of indices e such that W_e has property $\mathcal{R}$. The study of index sets is primarily concerned with determining the complexity of various index sets, in the arithmetical, or sometimes in the difference, hierarchy. For example, $\{e : W_e \text{ is infinite}\}$ is a Π^0_2-complete set, meaning that it is Π^0_2 and that any Π^0_2 set is many-one reducible to it. See [**S87**] for a survey of results on index sets for c.e. sets.

Index sets for computable trees were studied by Lempp [**L87**] in connection with the lattice of c.e. sets. Recently, Gasarch [**G98**] and others studied index sets for various combinatorial problems. Cenzer and Remmel developed the notion of index sets for Π^0_1 classes and gave applications to various mathematical problems in [**CR98a, CR98b, CR99a**].

The enumeration of the Π^0_1 classes is defined as follows. First observe that for any computable tree T, there exists a *primitive recursive* (even polynomial-time) tree U such that $[T] = [U]$, so that every Π^0_1 class is the set of infinite paths through some primitive recursive tree. Let $\pi_0, \pi_1, \ldots$ be an effective enumeration of the primitive recursive functions from ω into $\{0, 1\}$ and define the eth primitive recursive tree to be

$$U_e = \{\sigma \in \omega^{<\omega} : (\forall \tau \preceq \sigma)[\pi_e(\tau) = 1]\}$$

Then $P_e = [U_e]$ defines the eth Π^0_1 class. The use of primitive recursive trees here avoids the problem of totality for computable partial functions, which is Π^0_2-complete.

There are many results in [**CR98a, CR98b, CR99a**] classifying index sets for unbounded, bounded, c.b. and decidable Π^0_1 classes as well as strong Π^0_2 classes. We will just present here a selection of results on index sets for Π^0_1 classes of sets, which can be used for most of the applications from Section two. Thus, for any property $\mathcal{R}$ of sets, let

$$I(\mathcal{R}) = \{e : P_e \cap 2^\omega \text{ has property } \mathcal{R}\}.$$

Of course, all of these results immediately carry over to $P \subseteq k^\omega$ for any finite $k > 2$.

In particular, for any cardinal number κ, let $< \kappa$ denote the property of having $< \kappa$ members and let *cmp.* $< \kappa$ denote the property of having $< \kappa$ *computable* members, and similarly for $= \kappa$, $\geq \kappa$ and so on. A difference of two Σ^0_n sets is said to be a D^0_n set. A real number r in [0,1] may be written as $r_A = \sum_{i \in A} 2^{-i-1}$ and is said to be Σ^m_n if the set A is Σ^m_n. Let s be a Σ^0_1 real and let $p < 1$ be a Π^0_1 real.

THEOREM 5.1. *Let c be a positive integer.*

1. I*(nonempty) is* Π_1^0 *complete.*
2. $I(\leq c)$ *is* Π_2^0 *complete.*
3. $I(=1)$ *is* Π_2^0 *complete and* $I(=c+1)$ *is* D_2^0 *complete.*
4. I*(finite) is* Σ_3^0 *complete.*
5. I*(countable) is* Π_1^1 *complete.*

THEOREM 5.2. *Let c be a positive integer.*

1. I*(nonempty, cmp. empty) and* I*(cmp.* $= 0$*) are* Σ_3^0 *complete.*
2. I*(cmp.* $> c$*) is* Σ_3^0 *complete and* I*(cmp.* $= c$*) is* D_3^0 *complete.*
3. I*(cmp. infinite) is* Π_4^0 *complete.*

THEOREM 5.3. *Let s be a* Σ_1^0 *real and let p be a* Π_1^0 *real.*

1. $I(\textit{measure} \geq s)$ *is* Π_1^0 *complete and* $I(\textit{measure} \leq r)$ *is* Π_2^0 *complete.*
2. I*(perfect) is* Π_3^0 *complete.*
3. I*(thin) and (minimal) are* Π_4^0 *complete sets.*

There are a number of open problems for index sets of Π_1^0 classes. One can consider the problem of having cardinality in a certain set. For example, having finite even cardinality is Σ_3^0 complete by the same methods used for finite cardinality.

PROBLEM 5.4. For a given $C \subseteq \omega$, determine the complexity of the property of having cardinality in C; in particular, for computable, c.e., and Π_1^0 sets C.

The comparison of two Π_1^0 classes can be used to compare given computable models, such as axiomatizable theories or c.e. graphs.

PROBLEM 5.5. For a given cardinality or measure, determine the complexity of the set of pairs (a, b) such that the difference $P_b \setminus P_a$ has the corresponding size.

Here is a new result of this kind which will be needed in Section 6.

THEOREM 5.6. $\{a : P_a \textit{ is clopen}\}$ *is* Σ_2^0 *complete.*

PROOF. It is easily seen that the given index set is Σ_2^0. For the completeness, we define a reduction of the index set $Fin = \{e : W_e \text{ is finite}\}$, which is known to be Σ_2^0 complete. For each e, let $P_{f(e)} = 2^\omega \setminus \bigcup_{n \in W_e} I(0^n * 1)$. It is clear that W_e is finite iff $P_{f(e)}$ is clopen. □

The theorems above have many corollaries [**CR98a**] for the applications discussed in Section 2 where the class of solution sets is a subclass of k^ω for some finite $k \geq 2$. We will sketch the case of logical theories and then give a few results.

One general question is how to assign an index to a given problem and then transfer the index to an index for the solution set to the problem. Given a representation theorem, that any Π_1^0 class of sets can represent the solution set of some computable problem, we need to see how to transfer the index of the class P_e to an index for the mathematical problem.

For axiomatizable theories, let $\gamma_0, \gamma_1, \ldots$ enumerate the sentences of a given effective propositional language $\mathcal{L}$ with infinitely many sentence symbols, or an effective first-order language with equality and another binary predicate symbol but no other nonlogical symbols. Define the eth axiomatizable theory to be

$$\Gamma_e = Cn(\{\gamma_i : i \in W_e\}).$$

where $Cn(\Gamma)$ denotes the set of logical consequence of Γ.

The class of complete extensions of Γ_e is a Π^0_1 class, and this holds uniformly in the sense that it is of the form $P_{g(e)}$ for some primitive recursive function g. The reverse direction, that any Π^0_1 class of sets represents the set of complete extensions of some axiomatizable theory also holds uniformly (by Theorem 47 of [**CR98a**]). That is, there is a primitive recursive function f such that the complete extensions of $\gamma_{f(e)}$ represents P_e.

PROPOSITION 5.7. 1. $\{e : \Gamma_e \text{ is consistent}\}$ is Π^0_1-complete.
2. $\{e : \Gamma_e \text{ is complete}\}$ is Π^0_2-complete.
3. $\{e : \Gamma_e \text{ is essentially undecidable}\}$ is Σ^0_3-complete.
4. $\{e : \Gamma_e \text{ is maximal}\}$ is Π^0_4-complete.

PROOF. First note that *consistent* corresponds to being a nonempty class of complete extensions, *complete* corresponds to having a unique complete extension, *essentially undecidable* corresponds to being consistent and having no decidable complete extension, and *maximal* corresponds to being a thin class of extensions.

Clearly $\{e : \Gamma_e \text{ is consistent}\}$ is a Π^0_1 set. On the other hand, P_e is nonempty iff $\Gamma_{f(e)}$ is consistent, so that

$$e \in I(\text{nonempty}) \iff f(e) \in \{e : \Gamma_e \text{ is consistent}\}.$$

Since $I(\text{nonempty})$ is Π^0_1- complete, it follows that $\{e : \Gamma_e \text{ is consistent}\}$ is also Π^0_1 complete.

The other parts are proved in a similar fashion. □

Results of this type were given in [**CMR99**] for stable models of finite predicate logic programs and in [**CR98a**] for the graph coloring problem.

Gasarch [**G98**] considered the *chromatic number* $\chi(G)$–the least a such that G has an a-coloring, and recursive chromatic number $\chi^r(G)$–the least b such that G has a *computable* b-coloring. It follows from Theorem 3.10 of [**G98**] that the set of indices of computable graphs with chromatic number a and computable chromatic number b, for $a < b$, is Σ^0_3 complete.

PROBLEM 5.8. Given finite $a < b$, determine the complexity of index sets for graphs having a certain number, or measure, of (computable) b-colorings and another number of (computable) a-colorings.

Index sets for Π^0_1 classes and for computably continuous functions on ω^ω as well as on the reals were studied in [**CR99a**], where the index sets for computably continuous functions were primarily related to the set of zeroes or fixed points.

PROBLEM 5.9. Determine the complexity of index sets for computably continous functions with properties such as being differentiable and being integrable.

For the problems only known to represent separating sets, the upper bounds on index sets still follow as above, but the lower bounds now depend on specific results for separating sets. Index sets for just a few properties of classes of separating sets were studied in [**CR98a**]. In particular, the complexity of having $> c$, $< c$ or $= c$ members for finite c, of being finite, and of having no computable member, were all the same as for arbitrary Π^0_1 classes of sets. These results were applied to matching problems in [**CR98a**], and to Dilworth's problem in [**CR98b**].

PROBLEM 5.10. Determine the complexity of index sets for the decomposition of posets problem.

6. The lattice of Π_1^0 classes

In this section, we discuss recent results and open problems on the lattice $\mathcal{E}_\Pi$ of Π_1^0 classes and the lattice $\mathcal{E}_\Pi^*$, of Π_1^0 classes *modulo finite difference.*

The notions of thin and minimal classes are fundamental to the lattice. The thin classes are exactly those infinite Π_1^0 classes which determine complemented initial segments, and the minimal classes are the the atoms in the lattice $\mathcal{E}_\Pi^*$. Since the clopen sets are the complemented elements, the property of being finite or thin is definable in the lattice. The singleton classes are the atoms, so that the property of being a singleton is definable, and similarly the property of having cardinality exactly k, for any finite k. However, it is an open question whether the property of being finite is definable, It follows that finite classes (and hence minimal classes) are preserved under any automorphism of $\mathcal{E}_\Pi$ (that is, finite classes are *invariant*).

PROBLEM 6.1. Determine whether "finite" or "minimal" are definable in $\mathcal{E}_\Pi$.

Since a countable Π_1^0 class P is infinite if and only if the C-B rank of P is greater than 0, we can also ask about the definability and invariance of rank and in particular of rank one for countable classes.

PROBLEM 6.2. Determine for which α it is the case that the family of countable Π_1^0 classes of rank α is definable or invariant in $\mathcal{E}_\Pi$.

Since Π_1^0 classes have a topological and measure-theoretic context, one can ask about the definability of many interesting notions. It is not hard to see, for example, that measure is not definable in the lattice. Cenzer observed at the meeting that there is a computable homeomorphism of 2^ω which maps a (perfect) measure 0 Π_1^0 class to a (perfect) Π_1^0 class of measure $\frac{1}{2}$. Now any computable homeomorphism induces an automorphism of the lattice $\mathcal{E}_\Pi$, so that measure 0 is not preserved under automorphisms and is therefore not definable. Similarly, measure r for any computable $r < 1$ is not definable. However, measure 1 is definable, since the only Π_1^0 class of measure 1 is 2^ω itself.

It was shown in [**CCDH**] that the notions of "finite or thin" and "perfect" are each definable in the lattice. It was also shown there that any two perfect thin classes are automorphic. The question of automorphism between countable rank one classes is raised in [**CCDH**]. Since a countable Π_1^0 class is minimal if and only if it is thin and has exactly one nonisolated point, we pose the following question.

PROBLEM 6.3. Determine whether any two minimal Π_1^0 classes are automorphic.

At the meeting, Cenzer and Remmel obtained partial results. Let P and Q be minimal Π_1^0 classes with unique limit elements A_P and A_Q, respectively, and suppose that for each n, there is at most one path $x \in P$ with $x\lceil n = A_P\lceil n$ but $x \neq A_P$ and similarly for Q and A_Q. Then P and Q are said to be *standard* minimal classes. In this case, A_P and A_Q are computable from $\mathbf{0}'$ and the family of isolated points for each class is uniformly computable from $\mathbf{0}'$, from which it follows that P and Q are automorphic.

The usual notion of the *degree* of a Π_1^0 class P is just the Turing degree of the tree T_P of extendible nodes. The partial results cited above seem to indicate that more information would be useful.

In general, of course, two countable classes of rank one may be homeomorphic without being automorphic. If the class P has a unique limit path A_P which is computable and the class Q has a unique limit path A_Q which is not computable, then P and Q are not automorphic. This is because $\{A_P\}$ would have to map to a subclass of Q under such an automorphism, but it is both an atom and meets every infinite subclass of P, whereas there is no such subclass of Q. Thus we have a more general problem.

PROBLEM 6.4. Determine the orbits of the countable Π^0_1 classes of rank one.

Since rank may not be definable in $\mathcal{E}_\Pi$, we are naturally led to consider the lattice $\mathcal{E}^*_\Pi$, obtained by identifying Π^0_1 classes with finite symmetric difference. Here the finite classes are all 0 and the minimal classes are the atoms, so that finiteness and minimality are definable.

For fixed Π^0_1 classes $P \subseteq Q$, the interval $[P, Q] = \{X \in \mathcal{E}_\Pi : P \subseteq X \subseteq Q\}$. Then $\mathcal{L}(P) = [\emptyset, P]$ is the lattice of subclasses of P and is an *initial segment* of $\mathcal{E}_\Pi$. Assume that P is infinite. Note that P is thin if and only if $\mathcal{L}(P)$ is complemented and that P is minimal if and only if $\mathcal{L}(P)^*$ (i.e. $\mathcal{L}(P)$ modulo finite differences) is isomorphic to the trivial lattice $\{0, 1\}$.

Cenzer and Nies studied initial segments of $\mathcal{E}_\Pi$ in [**CN**]. A distributive lattice L with a least and a greatest element is said to be a *dual reduction lattice* if, for any elements a and b, there exist $a_1 \geq a$ and $b_1 \geq b$ such that $a_1 \vee b_1 = 1$ and $a_1 \wedge b_1 = a \wedge b$. It was observed in [**CN**] that any interval $[P, Q]$ of $\mathcal{E}_\Pi$ is a dual reduction lattice and that, for any *finite* dual reduction lattice, there exists a Π^0_1 class P with $\mathcal{L}(P)^*$ isomorphic to L and, furthermore, with the theory of $\mathcal{L}(P)$ decidable. This is in contrast with the lattice of c.e. sets, where each interval which is not a Boolean algebra must be infinite and has a theory which interprets true arithmetic.

It turns out that decidable Π^0_1 classes here behave more like co-c.e. sets. That is, if P is decidable and countably infinite, and $\mathcal{L}(P)^*$ is finite, then $\mathcal{L}(P)^*$ is a Boolean algebra and, furthermore, if P is decidable and $\mathcal{L}(P)$ is not a Boolean algebra, then the theory of $\mathcal{L}(P)$ interprets true arithmetic and is therefore undecidable ([**CN**]).

PROBLEM 6.5. Determine the possible degree of $Ext(P)$ when $\mathcal{L}(P)^*$ is a finite lattice, but not a Boolean algebra.

We close this section with two interesting problems related to intervals of the lattice.

The first concerns the structure of the lattice $\mathcal{S}(P) = [P, 2^\omega]$ of Π^0_1 supersets of P. It is clear that if P is clopen and $P \neq 2^\omega$, then $\mathcal{S}(P)$ is isomorphic to $\mathcal{E}_\Pi$. It is not hard to see that for any two non-clopen Π^0_1 classes P and Q, $\mathcal{S}(P)$ is isomorphic to $\mathcal{S}(Q)$ ([**CCDH**]). In fact, there is always a computable homeomorphism of $2^\omega - P$ onto $2^\omega - Q$ which induces this isomorphism. (This is a key part of the automorphism result for standard minimal classes discussed above.) Thus there are at most two isomorphism types for $\mathcal{S}(P)$, with $P \neq 2^\omega$.

PROBLEM 6.6. Determine whether $\mathcal{S}(P)$ is isomorphic to $\mathcal{S}(Q)$ for all Π^0_1 classes P, Q other than 2^ω.

Cenzer and Nies showed the following, which was a topic of discussion at the meeting. It is a nice application of index sets and has not been published, so we give the proof here.

THEOREM 6.7. *There is no isomorphism between* $\mathcal{S}(\emptyset)$ *and* $\mathcal{S}(\{0^\omega\})$ *which is computable in* $\mathbf{0}''$.

PROOF. Let P_e be the eth Π^0_1 class, as described above, and let f be a function such that the map taking P_e to $P_{f(e)}$ defines a lattice isomorphism between the family of Π^0_1 classes containing 0^ω and the family $\mathcal{S}(\emptyset) = \mathcal{E}_\Pi$ of all Π^0_1 classes. Let C be the set of indices e such that P_e is complemented in $\mathcal{S}(\{0^\omega\})$. Then $e \in C$ if and only if $P_{f(e)}$ is complemented in $\mathcal{E}_\Pi$, which is to say, if and only if $P_{f(e)}$ is clopen. Since this is a Σ^0_2 property, by Theorem 5.6, it follows that C is computable from f and $\mathbf{0}''$.

Now for c.e. sets W_e, W_e is complemented if and only if W_e is computable, and this is known to be a Σ^0_3-complete problem. We now show that $\{e : W_e \text{ is computable}\}$ may be computed from C. That is, define

$$P_{g(e)} = \{0^\omega\} \cup \bigcup \{I(0^n1) : 1 \notin W_e\}.$$

Observe that the complement of $P_{g(e)}$ in $\mathcal{S}(\{0^\omega\})$ is

$$B_e = \{0^\omega\} \bigcup \{I(0^n1) : 1 \in W_e\},$$

Then it is easy to see that W_e is computable iff $g(e) \in C$. Suppose first that W_e is computable. Then B_e is clearly a Π^0_1 class, so that $P_{g(e)}$ is complemented. On the other hand, suppose that $B_e = P_c$ is a Π^0_1 class. Then $n \in W_e \iff 0^n1 \in Ext(T_c)$, so that W_e is a Π^0_1 set and is therefore computable. □

Finally, suppose that any two minimal classes (with possible extra conditions attached) are automorphic. Then we may hope to extend these results to the more complicated countable classes of rank one studied in [**CN**].

PROBLEM 6.8. Show that if $\mathcal{L}(P)^*$ and $\mathcal{L}(Q)^*$ are isomorphic, finite lattices, then P and Q are automorphic (with possible conditions on P and Q).

References

[B76] D. Bean, *Effective coloration*, J. Symbolic Logic **41** (1976), 469–480.

[C93] D. Cenzer, *Effective dynamics*, in *Logical Methods, in honor of A. Nerode's Sixtieth Birthday*, Birkhauser Progr. Comp. Sci. Appl. Logic **12** (1993), 162–177.

[C] D. Cenzer, *Π^0_1 classes in computability theory*, in *Handbook of Computability*, Stud. Logic Found. Math. **140**, Elsevier, Amsterdam, 1999, 37–85.

[CDJS] D. Cenzer, R. Downey, C. Jockusch and R. Shore, *Countable thin Π^0_1 classes*, Ann. Pure and Appl. Logic **59** (1993), 79–139.

[CR92] D. Cenzer and J. Remmel, *Recursively presented games and strategies*, Math. Social Sciences **24** (1992), 117–139.

[CR95] D. Cenzer and J. Remmel, *Feasible graphs and colorings*, Math. Logic Quarterly **41** (1995), 327–352.

[CR98a] D. Cenzer and J. Remmel, *Index sets for Π^0_1 classes*, Ann. Pure and Appl. Logic **93** (1998), 3–61.

[CR99a] D. Cenzer and J. Remmel, *Index sets in computable analysis*, Theor. Comp. Sci. **219** (1999), 111–150.

[CMR99] D. Cenzer, W. Marek and J. Remmel, *Index sets for finite predicate logic programs*, Proc. Workshop on Complexity-theoretic and Recursion-theoretic methods in Databases, Artificial Intelligence and Finite Model Theory, 1999 Federated Logic Conference (1999), 72–81.

[CN] D. Cenzer and A. Nies, *Initial segments of the lattice of Π^0_1 classes*, J. Symbolic Logic, to appear.

[CR98b] D. Cenzer and J. Remmel, *Π^0_1 Classes in Mathematics*, in *Handbook of Recursive Mathematics*, Vol. 2, Stud. Logic Found. Math. **139**, Elsevier, Amsterdam, 1998, 623–821.

[CRta] D. Cenzer and J. Remmel, *Complexity, Decidability and Completeness*, preprint.

[CCDH] P. Cholak, R. Coles, R. Downey and E. Herrmann, *Automorphisms of the lattice of Π^0_1 classes*, to appear.

[CJS] P. Cholak, C. Jockusch, and T. Slaman, *On the strength of Ramsey's theorem for pairs*, J. Symbolic Logic, to appear.

[D87] R. Downey, *Maximal theories*, Annals of Pure and Applied Logic **33** (1987), 245–282.

[DJS90] R. Downey, C. Jockusch, and M. Stob, *Array nonrecursive sets and multiple permitting arguments* in *Recursion theory week (Oberwolfach, 1989)*, Lecture Notes in Math., **1432**, Springer, Berlin, 1990, 141–173.

[DJS96] R. Downey, C. Jockusch, and M. Stob, *Array nonrecursive degrees and genericity* in *Computability, enumerability, unsolvability*, London Math. Soc. Lecture Note Ser., **224**, Cambridge Univ. Press, Cambridge, 1996, 93–104.

[E61] A. Ehrenfeucht, *Separable theories*, Bull. Acad. Polon. Sci. Ser. Sci. Math. Astronom. Phys. **9** (1961), 17–19.

[FSS] H. Friedman, S. Simpson and R. Smith, *Countable algebra and set existence axioms*, Ann. Pure Appl. Logic **25** (1983), 141–181; addendum, ibid. **27** (1985), 319–320.

[G98] W. Gasarch, *A survey of recursive combinatorics*, in *Handbook of recursive mathematics*, Vol. 2, Stud. Logic Found. Math. **139**, Elsevier, 1998, 1041–1176.

[GS97] M. Groszek and T. Slaman, Π^0_1 *classes and minimal degrees*, Ann. Pure Appl. Logic **87** (1997), 117–144.

[GI] D. Gusfield and R. Irving, *The Stable Marriage Problem*, MIT Press (1989).

[H91] D. Harel, *Hamiltonian paths in infinite graphs*, Israel J. Math. **76** (1991), 317–336.

[H87] J. Hirst, *Combinatorics in Subsystems of Second Order Arithmetic*, Ph.D. thesis, The Pennsylvania State University, 1987.

[H89] J. Hirst, *Marriage theorems and reverse mathematics*, in *Logic and Computation*, ed. W. Sieg, Contemporary Mathematics **106** (1989), 181–196.

[HJ99] T. Hummel and C. Jockusch, *Generalized cohesiveness*, J. Symbolic Logic **64** (1999), 489–516.

[J72] C. Jockusch, *Ramsey's theorem and recursion theory*, J. Symbolic Logic **37** (1972), 268–280.

[J74] C. Jockusch, Π^0_1 *classes and Boolean combinations of recursively enumerable sets*, J. Symbolic Logic **39** (1974), 95-96.

[JLR] C. Jockusch, A. Lewis and J. Remmel, Π^0_1 *classes and Rado's selection principle*, J. Symbolic Logic **56** (1991), 684-693.

[JS72] C. Jockusch and R. Soare, Π^0_1 *classes and degrees of theories*, Trans. Amer. Math. Soc. **173** (1972), 33–56.

[JS93] C. Jockusch and F. Stephan, *A cohesive set which is not high*, Math. Logic Quarterly **39** (1993), 515–530; *Correction* **43** (1997), 569.

[K43] S. Kleene, *Recursive predicates and quantifiers*, Trans. Amer. Math. Soc. **53** (1943), 41-73.

[K55] S. Kleene, *Hierarchies of number-theoretic predicates*, Bull. Amer. Math. Soc. **61** (1955), 193–213.

[Ko98] K. Ko, *On the computability of fractal dimensions and Julia sets*, Ann. Pure App. Logic **93** (1998), 195–216.

[K53] G. Kreisel, *A variant to Hilbert's theory of the foundations of arithmetic*, British J. Philosophy of Science, **4** (1953), 107-129.

[K59] G. Kreisel, *Analysis of the Cantor-Bendixson theorem by means of the analytic hierarchy*, Bull. Acad. Polon. Sci. Ser. Math. Astron. Phys. **7** (1959), 621–626.

[K85] A. Kučera, *Measure,* Π^0_1*-classes and complete extensions of* PA, in *Recursion theory week (Oberwolfach, 1984)*, Lecture Notes in Math., 1141, Springer, Berlin-New York, 1985, 245–259.

[K88] A. Kučera, *On the role of* $0'$ *in recursion theory*, Logic Colloquium '86, F. R. Drake and J. K. Truss, eds., Elsevier (1988), 133–141.

[Kta] M. Kumabe, *A fixed-point free minimal degree*, preprint, 1993.

[L87] S. Lempp, *Hyperarithmetical index sets in recursion theory*, Trans. Amer. Math. Soc. **303**)1987), 559-583.

[MP70] A. Martin and M. Pour-El, *Axiomatizable theories with few axiomatizable extensions*, J. Symbolic Logic **35** (1970), 205–209.

[MR73] A. Manaster and J. Rosenstein, *Effective matchmaking and k-chromatic graphs*, Proc. Amer. Math. Soc. **39** (1973), 371–378.

[M78] D. Misercque, *Problème des mariages et récursivité*, Bull. Math. Soc. Belg. Ser. A **30** (1978), 111-121.

[N97] A. Nies, *Intervals of the lattice of computably enumerable sets*, Bull. London Math. Soc. **29** (1997), 683–692.

[Nta] A. Nies, *Effectively dense boolean algebras and their applications*, to appear.

[O63] V. Orevkov, *A constructive mapping of a square onto itself displacing every constructive point*, Soviet Math.-Doklady **4** (1963), 1253–1256.

[R86] J. Remmel, *Graph colorings and recursively bounded Π_1^0 classes*, Ann. Pure and Appl. Logic **32** (1986), 185–194.

[SS95] D. Seetapun and T. Slaman, *On the strength of Ramsey's Theorem*, Notre Dame J. Formal Logic **36** (1995), 570–582.

[S58] J. Shoenfield, *Degrees of formal systems*, J. Symbolic Logic **23** (1958), 389–392.

[S77] S. Simpson, *Degrees of unsolvability: a survey of results*, in *Handbook of Mathematical Logic*, J. Barwise, ed., North Holland, Amsterdam, New-York, Tokyo, 1977, 631-652.

[S98] S. Simpson, *Subsystems of Second Order Arithmetic*, Perspectives in Math. Logic, Springer-Verlag (1998).

[S62] D. Scott, *Algebras of sets binumerable in complete extensions of arithmetic*, in *Proc. Symp. Pure Math.* vol. 5, Amer. Math. Soc., Providence, R., 1962, 117-121.

[S87] R. Soare, *Recursively enumerable sets and degrees*, Springer, Berlin (1987).

[S71] E. Specker, *Ramsey's Theorem does not hold in recursive set theory*, Logic Colloquium, Manchester 1969, 439–442.

Department of Mathematics, University of Florida, P.O. Box 118105, Gainesville, Florida 32611-8105

E-mail address: cenzer@math.ufl.edu

Department of Mathematics, University of Illinois, 1409 W. Green Street, Urbana, Illinois 61801

E-mail address: jockusch@math.uiuc.edu

Contemporary Mathematics
Volume **257**, 2000

The global structure of computably enumerable sets

Peter A. Cholak

ABSTRACT. We will work in the structure of the computably enumerable sets. The language is just inclusion, $\subseteq$. This structure is called $\mathcal{E}$. Our quest is to partially survey our current understanding of the global structure of $\mathcal{E}$ and the relationship between $\mathcal{E}$ and the computably enumerable degrees and to pose questions whose answers should provide further insight.

1. Introduction

There are four themes that run throughout most work on $\mathcal{E}$: *Definability, orbits/automorphisms/isomorphic substructures, computational complexity*, and *dynamic properties*. Clearly, definability is a general theme throughout all of mathematical logic. The study of orbits and automorphisms (in general) dates from the 1870's and earlier. Mainly we will use Turing reducibility, Turing jump and the related jump classes (low_n and high_n classes) as a measure of computational complexity. This is part of our connection to the computably enumerable degrees. The dynamic properties are more hidden but we will try to highlight them. What is so nice in $\mathcal{E}$ is how all four themes interrelate with each other. At present this is just a bold claim but, in this paper, we will provide elaboration and justification of this claim.

1.1. Preliminaries. Before we start this journey we must deal with some preliminaries. First, this is not the only paper on $\mathcal{E}$ in this collection. Soare [2000] also focuses on $\mathcal{E}$.

All sets will be computably enumerable noncomputable sets and all degrees will be computably enumerable and noncomputable, unless otherwise noted. Our notation and definitions are standard and follow Soare [1987]; however we will try to provide some definitions and notational niceties throughout the text so the reader need not consult Soare [1987].

The sets and relations $0, 1, \cap$ and $\cup$ are definable from $\subseteq$. Hence $\mathcal{E}$ can be considered as a lattice. "X is computable" is definable; "X is computable" iff X is complemented. "X is finite" is definable in $\mathcal{E}$; "X is finite" iff all subsets of X are

1991 *Mathematics Subject Classification.* Primary 03D25.
Key words and phrases Computably enumerable sets, open questions.
Research partially supported by NSF Grant DMS-96-34565.

computable (it takes a little computability theory to show if X is infinite then X has an infinite noncomputable subset).

We will also consider the quotient structure $\mathcal{E}$ modulo the ideal of finite sets, $\mathcal{E}^*$. $\mathcal{E}^*$ is a definable quotient structure of $\mathcal{E}$. We use A^* to denote the equivalence class of A modulo the ideal of finite sets. If there is an automorphism, Φ^*, of $\mathcal{E}^*$ such that $\Phi^*(A^*) = B^*$, then there is an automorphism, Φ, of $\mathcal{E}$ such that $\Phi(A) = B$ (in which case we say A and B are automorphic) [see Soare, 1987, XV.2.7]. Given this, we will blur the difference between $\mathcal{E}$ and $\mathcal{E}^*$.

2. A beautiful example

Perhaps the example we are about to present is overworked but it still highlights our four themes in an excellent fashion.

M is *maximal* if for all W either $W \subseteq^* M$ or $M \cup W =^* \omega$. Friedberg [1958] showed that there is a maximal set [see Soare, 1987, X.3.3].

THEOREM 2.1 (Soare [1974]). *If M_1 and M_2 are maximal sets then there is an automorphism Φ of $\mathcal{E}$ such that $\Phi(M_1) = M_2$. Hence the maximal sets form an orbit.*

THEOREM 2.2 (Martin [1966]). *A degree $\mathbf{h}$ is high iff there is a maximal set M such that $M \in \mathbf{h}$.*

DEFINITION 2.3. A class $\mathcal{D}$ of degrees is *invariant* if there is a class $\mathcal{S}$ of (c.e.) sets such that

i. $\mathbf{d} \in \mathcal{D}$ implies there is a W in $\mathcal{S}$ and $\mathbf{d}$,
ii. $W \in \mathcal{S}$ implies $\deg(W) \in \mathcal{D}$, and
iii. $\mathcal{S}$ is closed under automorphic images.

COROLLARY 2.4. *The high degrees are invariant.*

Let's see how our four themes come into play. Theorem 2.1 implies that the maximal sets are a nonempty elementarily definable orbit. (By elementarily definable we will mean definable by a formula in $\mathcal{L}_{\omega,\omega}$, an elementary formula.) Clearly, Corollary 2.4 involves computational complexity (measured in terms of jump classes). The proofs of Theorems 2.1 and 2.2 rely on dynamic properties.

We say that a function f is *dominant* if f dominates every computable function. The proof of Theorem 2.2 breaks up into two parts: First one shows that the principal function of the complement of a maximal set is a dominant function, and then, one shows that a computably enumerable degree $\mathbf{d}$ is high iff $\mathbf{d}$ contains a dominant function. The idea is that lots of numbers must enter a maximal set.

The proof of Theorem 2.1 relies, in part, on the following: There is a list of computably enumerable sets U_i where $U_i = W_{f(e)}$ and f is computable in $\mathbf{0}''$ such that

- for all e there is an i such that $W_e =^* U_i$,
- for all e there is a σ_e such that σ_e is the e-state of $\overline{M}$ w.r.t. $\{U_i\}_{i\in\omega}$ ($\sigma(i) = 1$ iff $\overline{M} \subseteq^* U_i$) [dymanic isomorphism], and
- if $X_\sigma = \cap\{U_i : \sigma(i) = 1\}$ then $X_\sigma \searrow M = \{x : \exists s, t[x \in (X_{\sigma,s} - M_s) \cap M_t]\}$ is infinite iff $\sigma = \sigma_e$, for some e (or an empty extension of some σ_e) [replete].

These facts can be established using the fact that M is maximal. Several proofs of this can be found, all using some version of Soare's Extension Theorem [see Soare, 1987, XV.4.5]. For example, see Soare [1987, XV.4-6] or Cholak [1994b].

Soare has announced another proof of this using another version of the Extension Theorem which he is calling the New Extension Theorem, see Soare [2000] for more details.

3. Orbits and 1-types

The creative sets (Harrington, see Soare [1987, XV.1.4]); the maximal sets; the quasi-maximal sets of rank k, for any k (Soare, see Soare [1987, XV.4.8]); the Herrmann sets (Herrmann, see Cholak and Downey [n.d.]); the quasi-Herrmann sets of rank k, for any k (Cholak and Downey [n.d.]); the hemi-maximal sets (Downey and Stob [1990]); the hemi-quasi-maximal sets of rank k, for any k (Downey and Stob [1990]); the hemi-Herrmann sets (Cholak and Downey [n.d.]); and the hemi-quasi-Herrmann sets of rank k, for any k, (Cholak and Downey [n.d.]) are all elementarily definable orbits. (The actual definition of these sets is not important, just the fact that they are elementarily definable.) Since each of these orbits is elementarily definable, the defining formula describes a principal type. Hence there are infinitely many 1-types of $\mathcal{E}^*$ and so $\mathcal{E}^*$ is not $\aleph_0$-categorical. Clearly $\mathcal{E}$ has infinitely many 1-types: "X has n elements" is a 1-type in $\mathcal{E}$.

QUESTION 3.1. *i. Show that these (the above list) are the only elementarily definable orbits.*
ii. Show these are the only principal types.
iii. Describe all orbits.

The key to all of the above results is that we have some control over the complement of the set. So to show that the above list is not the complete list of elementarily definable orbits, one should examine the sets where one has some control over the complement.

A set H is hhsimple iff $\mathcal{L}^*(H)$ (the supersets of H modulo the finite sets) is a Boolean algebra. In general, the hhsimple sets will not work to extend the above list. As there are nonautomorphic hhsimple sets whose $\mathcal{L}^*$s are isomorphic. This is due to Lerman, Shore and Soare [1978] and was later extended to all possible $\mathcal{L}^*$s (for all hhsimple sets) by Herrmann [1989]. So given hhsimple sets whose $\mathcal{L}^*$s are isomorphic, one could try to add some additional properties to ensure that the sets are automorphic. Some other good possibilities to extend the above list are r-maximal sets with a maximal superset and the major subsets of a maximal set.

However, I conjecture the Question 3.1 has a positive answer; there are no more such orbits. Even if this conjecture is incorrect, I doubt there is a first-order elementarily definable orbit whose proof is not similar to any of the known results. The next theorem shows that answering Question 3.1 is bound to be hard and supports my conjecture that the answer to Question 3.1 is positive.

THEOREM 3.2 (Cholak, Downey and Harrington [n.d.]). *$\{(A, B) : A \simeq B\}$ is Σ^1_1-complete.*

COROLLARY 3.3. *i. Not all orbits are elementarily definable.*
ii. $\mathcal{E}$ is not a prime model (of its theory).
iii. There is no arithmetical description of all orbits.
iv. The Scott rank of $\mathcal{E}$ is at least ω_1^{CK}.

But can we actually find examples realizing and extending the above corollary? I.e.:

QUESTION 3.4. *i. Find an orbit which is not a principal orbit.*

ii. For all computable ordinal α, find a pair of computably enumerable sets which are Δ^0_α-automorphic but not Δ^0_β-automorphic, for $\beta < \alpha$. (The definition of Δ^0_α-automorphic appears below.)

iii. Can we find an orbit O such that membership in O is Σ^1_1 complete?

Examining types allows us to prove two model-theoretic results: $\mathcal{E}$ is not $\aleph_0$-categorical and is not a prime model. Since, if a model is homogeneous, its Scott rank is at most $\omega + 1$, $\mathcal{E}^*$ is not homogeneous or saturated.

4. Automorphisms

The main question here is :

QUESTION 4.1. *Describe all automorphisms of $\mathcal{E}$.*

However, given the following theorem and Theorem 3.2, this seems like an unanswerable question. But we will argue that it is actually an answerable question.

THEOREM 4.2 (Lachlan, see Soare [1987, XV.2.2]). *There are $2^{\aleph_0}$ automorphisms of $\mathcal{E}$.*

We can sidestep Lachlan's result since all these automorphisms are similar. They just move computably enumerable subsets of almost cohesive sets. There is a tower of computable sets R_i such that for every e, either $W_e \subseteq R_e$ or $W_e \subseteq \overline{R}_e$. Let S_n be an infinite coinfinite computably enumerable subset of $R_{n+1} - R_n$. Given a function f, let Φ_f be the automorphism of $\mathcal{E}$ induced by sending $\Phi_f(S_n)$ to $(R_{n+1} - R_n) - S_n$ and $\Phi_f((R_{n+1} - R_n) - S_n)$ to S_n if $f(n) = 1$; otherwise Φ_f (actually the permutation which induces Φ_f) is the identity on $R_{n+1} - R_n$. But even without Lachlan's result we still would not have an arithmetical description of all automorphisms by Theorem 3.2.

Our understanding of automorphisms of $\mathcal{E}$ is unique to $\mathcal{E}$. In most structures with nontrivial automorphisms we can construct automorphisms via the normal "back and forth" argument. But this is not the case with $\mathcal{E}$. To construct automorphisms of $\mathcal{E}$ we use the properties of being *well-visited* and *well-resided*. Well-visited is Π^0_2 and not being well-resided is Σ^0_3 (we use the negation). Since the complexity of these properties is at most Σ^0_3, the construction of the desired automorphism can be placed on a tree. (We will not discuss the details on this placement nor of the construction of an automorphism of $\mathcal{E}$ but direct the reader to Harrington and Soare [1996b] or Cholak [1995].)

But before we can continue we need some notation: We will need a way to classify the complexity of automorphisms of $\mathcal{E}$. Let Φ be an automorphism of $\mathcal{E}$. We say Φ is a Δ^0_n-automorphism if there is a Δ^0_n function f such that $\Phi(W_e) =^* W_{f(n)}$. Then f is called a presentation of Φ. Two sets A and B are Δ^0_3-automorphic iff there is a Δ^0_3-automorphism Φ such that $\Phi(A) = B$. This method of classification can also be used for isomorphisms between $\mathcal{L}^*(A)$ and $\mathcal{L}^*(B)$. $\mathcal{L}^*(A)$ and $\mathcal{L}^*(B)$ are Δ^0_3-isomorphic (written $\mathcal{L}^*(A) \simeq_{\Delta^0_3} \mathcal{L}^*(B)$) iff there is an isomorphism Φ^* between $\mathcal{L}^*(A)$ and $\mathcal{L}^*(B)$ and a Δ^0_3 function f such that $\Phi^*(W^*_e) = W^*_{f(e)}$.

If an automorphism Φ is constructed on a tree then Φ has a presentation computable in the true path (which is Δ^0_3). Hence all automorphism constructed in this way are Δ^0_3-automorphisms (or even effective automorphisms). All of the known results except Theorems 3.2 and 4.2 involving automorphisms use Δ^0_3-automorphisms.

The structure $\mathcal{E}$ has $2^{\aleph_0}$ automorphisms. Hence, there must be automorphisms which are not Δ^0_3. Even with Lachlan's result there was some slight hope that the relation of whether two sets are automorphic would be Δ^0_3 or at least arithmetical. But Theorem 3.2 shows this is not the case. Therefore we know that there are useful non-Δ^0_3-automorphisms. But the proof of Theorem 3.2 does not allow us to clearly lay our hands on any such automorphisms. The automorphisms constructed for this theorem are based on complex arrangements of Δ^0_3-isomorphisms. (The automorphisms constructed in Lachlan's proof are complex arrangements of effective isomorphisms.) Thus there is some hope that all automorphisms are just complex arrangements of Δ^0_3-isomorphisms. The next collection of results provides us with some factual basis for this hope.

Let $\mathcal{S}(A) = \{B : \exists C(B \sqcup C = A)\}$. $\mathcal{S}(A)$ is the set of splits of A and $\mathcal{S}(A)$ forms a Boolean algebra. $\mathcal{R}(A) = \{R : R \subseteq A$ and R is computable$\}$. $\mathcal{R}(A)$ is the set of computable subsets of A and is an ideal of $\mathcal{S}(A)$. $\mathcal{S}_{\mathcal{R}}(A)$ is the quotient structure $\mathcal{S}(A)$ modulo $\mathcal{R}(A)$. If $W \in S(A)$ then let $W^{R(A)}$ be the equivalence class of W in $\mathcal{S}_{\mathcal{R}}(A)$. $\mathcal{S}_{\mathcal{R}}(A)$ is a Boolean algebra and is definable in $\mathcal{E}$ with a parameter for A. If A and B are automorphic then the structures $\mathcal{S}_{\mathcal{R}}(A)$ and $\mathcal{S}_{\mathcal{R}}(B)$ are isomorphic structures. But something much stronger is true.

THEOREM 4.3 (Cholak and Harrington [n.d.a]). *If A and B are automorphic then the structures $\mathcal{S}_{\mathcal{R}}(A)$ and $\mathcal{S}_{\mathcal{R}}(B)$ are Δ^0_3-isomorphic structures (that is there is an isomorphism Ψ between $\mathcal{S}_{\mathcal{R}}(A)$ and $\mathcal{S}_{\mathcal{R}}(B)$ and a Δ^0_3-function f such that for $W_e \in \mathcal{S}(A)$, $W_{f(e)}$ is in $\Psi(W^{\mathcal{R}(A)})$; and we will write this as $\mathcal{S}_{\mathcal{R}}(A) \simeq_{\Delta^0_3} \mathcal{S}_{\mathcal{R}}(B)$).*

By Theorem 3.2 we cannot go from a Δ^0_3 isomorphism from $\mathcal{S}_{\mathcal{R}}(A)$ to $\mathcal{S}_{\mathcal{R}}(B)$ to an automorphism of $\mathcal{E}$ taking A to B. But there is the hope that we can add extra conditions to this Δ^0_3 isomorphism to help construct the desired automorphism. Theorem 4.3 also provides evidence that we are limited to our current methods in how we can construct automorphisms of $\mathcal{E}$. This is certainly the case for hhsimple sets:

THEOREM 4.4. *Let H_1 and H_2 be hhsimple. H_1 and H_2 are automorphic iff they are Δ^0_3-automorphic iff $\mathcal{L}^*(H_1) \simeq_{\Delta^0_3} \mathcal{L}^*(H_2)$.*

Theorem 4.3 will also have an impact in showing that certain sets cannot be automorphic. Most such results go more or less as follows: First show that the sets A and B cannot be Δ^0_3-automorphic. Then use the failure to be Δ^0_3-automorphic to find a definable property P true of A but not of B. By using Theorem 4.3, perhaps we can formalize this process. If not then at least we can use Theorem 4.3 to diagonalize against all Δ^0_3-isomorphisms and show that certain sets are not automorphic without producing a definable difference.

An announcement of Theorems 4.3 and 4.4 and some other interesting results and examples can be found in Cholak and Harrington [n.d.b].

Perhaps our current understanding of automorphisms of $\mathcal{E}$ may be helpful in other areas, such as the computably enumerable degrees. In our constructions we have a notion of an e-state; enough information, if correctly acted on, to build a partial automorphism between the first e sets. We know that if we act properly, i.e. make moves to copy the well-visited and well-resided states, we can extend our actions to handle all $(e + 1)$-states.

As we mentioned earlier, the normal back and forth argument fails. Back and forth arguments fail because we do not know anything about the n-types of $\mathcal{E}$.

QUESTION 4.5. *Describe the n-types of $\mathcal{E}$.*

The next definition captures α moves of the standard back and forth argument (see Barwise [1973] or Ash and Knight [2000]).

DEFINITION 4.6. *i.* $\vec{A} \leq_1 \vec{B}$ iff every Σ_1 formula realized by $\vec{B}$ is realized by $\vec{A}$.
ii. $\vec{A} \leq_\beta \vec{B}$ iff for all $\vec{D}$, δ if $1 \leq \delta < \beta$ there is a $\vec{C}$ such that $\vec{B}, \vec{D} \leq_\delta \vec{A}, \vec{C}$.

QUESTION 4.7 (Knight). *When is $\vec{A} \leq_\beta \vec{B}$?*

If for all computable α, $A \equiv_\alpha B$ ($A \leq_\alpha B$ and $B \leq_\alpha A$) then A and B are automorphic (see Ash and Knight [2000]). So, by Theorem 3.2, for all computable α, there are automorphic A and B such that $A \not\equiv_\alpha B$.

5. Upward and downward cones

We want to turn to computational complexity issues. First, we will consider the property of being automorphic to a complete set. This issue first arose because of Post's Program. One version of Post's Program is to find a definable property Q in $\mathcal{E}$ such that if $Q(A)$ then Q is incomplete. By Harrington and Soare [1991] this has a positive solution: There is a property $Q(A)$ such that there is an A where $Q(A)$ holds and if $Q(A)$ holds then A is not automorphic to a complete set.

THEOREM 5.1. *All sets listed in the first paragraph of Section 3 (the references are the same as the references listed there) and the almost prompt sets (Harrington and Soare [1996b]) are automorphic to a complete set B.*

This gives rises to the question of which sets are automorphic to complete sets.

QUESTION 5.2. *i. Characterize all sets automorphic to complete sets.*
ii. Is $\{e : W_e$ is automorphic to a complete set$\}$ Σ^1_1-complete?

In terms of the automorphism results, sending A to a complete set B means that we can add numbers to B at any time and continue to build the automorphism. Harrington showed [see Harrington and Soare, 1996b] that for all incomplete $\mathbf{d}$ and all noncomputable A, there is a B such that A and B are automorphic and $B \not\leq_T \mathbf{d}$. But this construction also involves adding things to B. Cholak [1995] and, independently, Harrington and Soare [1996b] showed that every noncomputable set is automorphic a high set B. Hence we can pump up the degree in terms of jump class. Again this construction involved adding numbers to B. Given a random A, we have very little control over the actual degree of B. Controlling the degree of B would involve restraining numbers from B. The question remains: Can we avoid downward cones?

QUESTION 5.3. *Let A be incomplete. Does there exists a B such that $A \simeq B$ and $A \not\leq_T B$?*

One might ask what is the complexity of the index set of those As that are automorphic to a B which cannot Turing compute A? In fact, we can always turn most of our questions into index set questions.

6. Invariant degree classes

Martin [1966] showed that the degrees of the maximal sets are exactly the high degrees. Lachlan [1968a] and Shoenfield [1976] showed that the degrees of sets without maximal supersets are exactly the nonlow$_2$ degrees. So the high degrees and the nonlow$_2$ degrees are invariant. In addition, $\mathbf{L_0} = \mathbf{0}$ and $\overline{\mathbf{L_0}} = \mathbf{R} - \mathbf{0}$ are invariant as witnessed by the computable sets and noncomputable sets, respectively.

Harrington in unpublished work showed that the property of being creative is elementarily definable in $\mathcal{E}$ [see Soare, 1987, XV.1.1]. Hence $\mathbf{H_0}$ is invariant. Harrington and Soare [1996b] showed that every prompt set is automorphic to a complete set and hence $\overline{\mathbf{H_0}}$ is noninvariant. Maass, Shore and Stob [1981] showed that there is a definable class of sets (the promptly simple sets – see below) which splits all jump classes. Also, in as yet unpublished work, Harrington and Soare [see Harrington and Soare, 1996a, Corollary 4.4] have shown that $\overline{\mathbf{L_1}}$ is not invariant. They prove this by showing there is a properly low$_2$ degree $\mathbf{d}$ such that if $A \leq_T \mathbf{d}$ then there is a low B such that A and B are automorphic. (Note the similarities between this and Question 6.5.) The theorem that every noncomputable set is automorphic to a high set implies that no downward closed jump class is invariant.

It is known that the prompt degrees are invariant: Maass [1983] showed that all promptly simple sets with semilow complements are automorphic. Maass et al. [1981] show that every promptly simple set has the splitting property (a definable property, see Maass et al. [1981] for details) and every set which has the splitting property is prompt. Now it is enough to show that there is a promptly simple set with semilow complement in every prompt degree. One such proof can be found in Wald [1999, Theorem 1.4.1]. We cannot help but point out that the prompt degrees have a dynamic definition and are naturally definable in the computably enumerable degrees [for more details see Soare, 1987, XIII].

The most recent result on invariant jump classes is:

THEOREM 6.1 (Cholak and Harrington [n.d.c]). *For all $n \geq 2$, the high$_n$ ($\overline{\text{low}_n}$) computably enumerable degrees are invariant.*

We would like to look slightly closer at this result and some of its corollaries. (See the paper Cholak and Harrington [n.d.c] or the announcement Cholak and Harrington [n.d.b] for more details.)

THEOREM 6.2 (Cholak and Harrington [n.d.c]). *Let $\mathcal{C} = \{\mathbf{a} : \mathbf{a}$ is the Turing degree of a Σ^0_3 set $J \geq_T \mathbf{0}''\}$. Let $\mathcal{D} \subseteq \mathcal{C}$ be such that $\mathcal{D}$ is upward closed. Then there is an $\mathcal{L}(A)$ property $\varphi_{\mathcal{D}}(A)$ such that $D'' \in \mathcal{D}$ iff there is an A such that $A \equiv_T D$ and $\varphi_{\mathcal{D}}(A)$.*

COROLLARY 6.3. *Let $\mathcal{F}$ be a class of computably enumerable degrees such that if $\mathbf{a} \in \mathcal{F}$ and $\mathbf{a}'' \leq_T \mathbf{b}''$ then $\mathbf{b} \in \mathcal{F}$. Then $\mathcal{F}$ is invariant.*

COROLLARY 6.4. *If $\mathbf{a}'' > \mathbf{b}''$ then there is an $A \in \mathbf{a}$ such that for all $B \in \mathbf{b}$, $A \not\approx B$ (in fact, $\mathcal{L}(A)$ and $\mathcal{L}(B)$ are not isomorphic).*

More or less we have shown that the double jump can be encoded into $\mathcal{E}$. What about the single jump? The following question is the strongest possible negation of encoding the single jump in the above fashion (see Corollary 6.4) and is a generalization of the result of Harrington and Soare that $\overline{\mathbf{L_1}}$ is noninvariant.

QUESTION 6.5. *Let J be computably enumerable in and above $\mathbf{0}''$. There are degrees $\mathbf{a}$ and $\mathbf{b}$ such that $\mathbf{a}' \neq \mathbf{b}'$, $\mathbf{a}'' \equiv_T \mathbf{b}'' \equiv_T J$, and for all $A \leq_T \mathbf{a}$ there is a B such that $B \leq_T \mathbf{b}$ and $A \simeq B$.*

The "no fat orbit" result of Downey and Harrington [1996] (this result will be discussed later) says that the degrees in the above conjecture most likely are both prompt or both tardy (not prompt). So the fact that the prompt degrees are invariant provides some evidence that the conjecture should be true. In addition, we have some partial results in this direction: For example, in Cholak [1995], it is shown, for all A and for all $\mathbf{H_1}$ degrees $\mathbf{h}$ there is a $B \in \mathbf{h}$ such that $\mathcal{L}^*(A) \simeq \mathcal{L}^*(B)$. Recently Harrington has announced a generalization of the above theorem:

THEOREM 6.6 (Harrington). *For all A and degrees $\mathbf{d}$ if $A' \leq_T \mathbf{d}'$ there is a $B \in \mathbf{d}$ such that $\mathcal{L}^*(A) \simeq \mathcal{L}^*(B)$.*

One might also want to try to combine a positive answer to Question 6.5 with Shore's noninversion theorem for the jump [Shore, 1988] or the result of Cooper [1989] that there is a degree computably enumerable in and above $\mathbf{0}'$ which is not the jump of a tardy degree. In addition, one could add other degree theoretic conditions on $\mathbf{a}$ and $\mathbf{b}$.

By Corollary 6.3, $\mathbf{H}_\omega = \cup_{n\in\omega}\mathbf{H_n}$ is invariant. Solovay has shown $\mathbf{H}_\omega$ is $\Sigma^0_{\omega+1}$-complete and hence nonarithmetical [see Soare, 1987, XII.4.14]. Let $\mathcal{C}$ be the class of computably enumerable sets witnessing that $\mathbf{H}_\omega$ is invariant. So a computably enumerable degree $\mathbf{d}$ is in $\mathbf{H}_\omega$ if there is a computably enumerable set A in both $\mathbf{d}$ and $\mathcal{C}$. Whether A is in $\mathbf{d}$ is arithmetical. So whether A is in $\mathcal{C}$ cannot be arithmetical. Hence $\mathcal{C}$ is not definable in $\mathcal{E}$ by a $\mathcal{L}_{\omega,\omega}$ formula. (Otherwise it would be arithmetical.)

The above class $\mathcal{C}$ provides another[1] negative answer to Shoenfield's question if every class $\mathcal{C}$ of computably enumerable sets closed under automorphic images is definable in $\mathcal{E}$ by a $\mathcal{L}_{\omega,\omega}$ formula. We note that such a $\mathcal{C}$ is definable in $\mathcal{E}$ by a $\mathcal{L}_{\omega_1,\omega}$ formula: $\mathcal{C}$ is a countable union of orbits. Each orbit is definable by a $\mathcal{L}_{\omega_1,\omega}$ formula. So $\mathcal{C}$ is definable by the countable disjunction the formulas defining the orbits in $\mathcal{C}$.

The formula $\varphi_D(A)$ in Theorem 6.2 is a $\mathcal{L}_{\omega_1,\omega}$ formula. One might ask if this formula can be improved to an elementary formula. The above work concerning $\mathbf{H}_\omega$ shows that in general this is not possible. But it leaves open the possibility that the complexity of the defining formula can be decreased in some cases. Call a class $\mathcal{D}$ of computably enumerable degrees an *elementarily definable invariant degree class* if $\mathcal{D}$ is invariant and the class $\mathcal{C}$ of computably enumerable sets witnessing $\mathcal{D}$ is invariant is elementarily definable. The high degrees and the nonlow$_2$ degrees are both elementarily definable invariant degree classes. Are the high$_2$ degrees an elementarily definable invariant degree class?

Corollary 6.3 shows that any class of degrees $\mathcal{F}$ such that if $\mathbf{a} \in \mathcal{F}$ and $\mathbf{a}'' = \mathbf{b}''$ then $\mathbf{b} \in \mathcal{F}$ is invariant. This is not true for the single jump, for example as the nonlow degrees are not invariant. But this says nothing about classes which are unrelated to the jump operator.

QUESTION 6.7. *Characterize the (elementarily definable) invariant degree classes.*

[1] The work in Lempp [1987] and Nies [1997] also provides negative answers.

In particular:

QUESTION 6.8. *Are the array noncomputable degrees invariant?*

As evidence that the above question might have a positive answer, we note that in the collection of all Π^0_1 classes, Cholak, Coles, Downey and Herrmann [n.d.] have recently shown that the anc degrees are invariant.

But how should we approach this question? Downey, Jockusch and Stob [1990] showed that every nonlow$_2$ degree is anc but there are low$_2$ anc degrees. The nonlow$_2$ degrees are invariant. There are two proofs of this: First Lachlan [1968a] and Shoenfield [1976] showed that the degrees of atomless sets are exactly the nonlow$_2$ degrees and second, via Theorem 6.1. One could try to find another definable property which exactly describes the anc degrees. Or one could try to push the second proof through to work for the anc degrees. Both of these plans seem unlikely. The most likely outcome is there is an anc degree $\mathbf{d}$ such that every $A \in \mathbf{d}$ is automorphic to a nonanc set. This is similar to Question 6.5 and Harrington and Soare's result that there is properly low$_2$ degree $\mathbf{d}$ such that every $A \in \mathbf{d}$ is automorphic to a low set.

Downey and Harrington [1996] have shown that there is no fat orbit. That is, they showed there is a definable property $S(A)$, a prompt $\mathbf{L_1}$ degree $\mathbf{d_1}$, a prompt $\mathbf{H_2}$ degree $\mathbf{d_2}$ greater than $\mathbf{d_1}$, and a tardy $\mathbf{H_2}$ degree $\mathbf{e}$ such that for all $E \leq_T \mathbf{e}$, $\neg S(E)$ and if $\mathbf{d_1} \leq_T D \leq_T \mathbf{d_2}$ then $S(D)$. This result points out the dichotomy between the prompt and tardy degrees. A corollary is that, except for the high degrees, no single orbit can witness that a member of the $\mathbf{H}_n$, $\mathbf{L}_n$ hierarchy is invariant. Downey and Harrington ask whether their result can be extended to include the case when $\mathbf{e}$ is high. Harrington and Cholak conjecture that this is not the case:

CONJECTURE 6.9 (Cholak-Harrington). *For every noncomputable degree* $\mathbf{a}$ *there is a set A whose orbit contains every high degree.*

All of the above theorems and questions explore invariant degree classes and orbits containing only degrees in these classes. The "no fat orbit" theorem says that we can separate certain degrees via orbits: we can separate some prompt degree from a tardy degree. A positive answer to the above conjecture implies that we cannot separate a random degree from any high degree via orbits. The following question generalizes this idea.

QUESTION 6.10. *Is there a tardy ($\mathbf{L_n}$, $\overline{\mathbf{H_n}}$) set A whose orbit contains a set of every prompt ($\overline{\mathbf{L_n}}$, $\mathbf{H_n}$) degree?*

7. Decidability and coding issues

Harrington and Herrmann developed various coding methods in $\mathcal{E}^*$. They both were able to use their methods to show that the theory of $\mathcal{E}^*$ is undecidable [Herrmann, 1984; Harrington, 1983]. Harrington was able to extend his methods to show that the theory of $\mathcal{E}^*$ has degree $\mathbf{0}^{(\omega)}$ [see Cholak, 1994a; Harrington and Nies, 1998]. The coding method used for this result has been further developed in Harrington and Nies [1998], where is it is shown that the structure $(\omega, +, \times, 0, 1, \leq)$ can be coded in $\mathcal{E}^*$ with parameters but one cannot code a linear order into $\mathcal{E}^*$ without parameters. Nies [1998] later sharpened this result to show that the standard model of arithmetic can be coded in $\mathcal{E}^*$ with 5 alternations of quantifiers and the Π_6-theory of $\mathcal{E}^*$ is undecidable.

QUESTION 7.1. *How many alternations of quantifiers does it take to define a model of arithmetic? 4 or less?*

On the decidable side, Lachlan [1968b] showed that the $\exists\forall$-theory of $\mathcal{E}^*$ is decidable (he included $0, 1, \cup$ and $\cap$ in the language). This was later slightly expanded by Lerman and Soare [1980]. See Soare [1987, XVI.2] for an overview of these results.

QUESTION 7.2. *Is the $\forall\exists\forall$-theory of $\mathcal{E}^*$ decidable? At what exact level does the theory of $\mathcal{E}^*$ fail to be decidable?*

We should mention that the proofs of Theorems 4.3 and 6.2 involve a new definable encoding. The actual details of this coding are complex. These details and some theorems concerning this coding will appear in Cholak and Harrington [n.d.c]. We feel that this coding has great potential. This coding is simpler to decode than the previous coding. More or less, it can be decoded at the Σ_3^A level. Perhaps it can be used to answer the above two questions.

8. Connections with the computably enumerable degrees?

Even given Cooper's claim that there is an automorphism of the computably enumerable degrees, the biinterpretability conjecture *with parameters* remains a strong possibility to globally understand the computably enumerable degrees ($\mathcal{R}$) [see Cooper, 1999; Shore, 1999; Slaman, 1999].

One of the corollaries of the biinterpretability conjecture with parameters is that the underlying structure has at most countably many automorphisms. We know that $\mathcal{E}$ has $2^{\aleph_0}$ automorphisms; hence the biinterpretability conjecture with parameters for $\mathcal{E}$ fails. It would be nice to suggest a global paradigm for $\mathcal{E}$.

It is still open if $\mathcal{R}$ has $2^{\aleph_0}$ automorphisms. If so, we suggest that any paradigm to describe the computably enumerable degrees can be used to describe the computably enumerable sets and vice-versa.

We will close by pointing out one of the many similarities between the computably enumerable sets and degrees. By Corollary 6.3, any class of computably enumerable degrees closed under the double jump is invariant in $\mathcal{E}$. Nies, Shore and Slaman [1998] have shown that any relation on the computably enumerable degrees which is invariant (used in a different sense) under the double jump is definable in $\mathcal{R}$ iff it is definable in first order arithmetic. See Shore [2000] for more details.

References

Ash, C. J. and Knight, J. F. [2000]. *Computable structures and the hyperarithmetical hierarchy.* In prepartion.

Barwise, J. [1973]. Back and forth through infinitary logic, pp. 5–34. MAA Studies in Math., Vol. 8.

Cholak, P. [1994a]. Notes on 3 theorems by Leo Harrington. Handwritten Notes.

Cholak, P. [1994b]. The translation theorem, *Arch. Math. Logic* **33**: 87–108.

Cholak, P. [1995]. Automorphisms of the lattice of recursively enumerable sets, *Mem. Amer. Math. Soc.* **113**(541): viii+151.

Cholak, P., Coles, R., Downey, R. and Herrmann, E. [n.d.]. Automorphisms of the lattice of Π_1^0 classes; perfect thin classes and anc degrees. Submitted, Draft available.

Cholak, P. and Downey, R. [n.d.]. Some orbits for $\mathcal{E}^*$. Preprint Available.

Cholak, P., Downey, R. and Harrington, L. A. [n.d.]. Automorphisms of the computably enumerable sets: Σ_1^1-completeness. In preparation.

Cholak, P. and Harrington, L. A. [n.d.a]. Δ_3^0-automorphisms of the computably enumerable sets. In preparation.

Cholak, P. and Harrington, L. A. [n.d.b]. Definable encodings in the computably enumerable sets. Submitted, Draft available.

Cholak, P. and Harrington, L. A. [n.d.c]. On the definability of the double jump in the computably enumerable sets. Submitted, Draft available.

Cooper, S. [1999]. Local degree theory, *in* E. R. Griffor (ed.), *Handbook of computability theory*, Vol. 140 of *Studies in Logic*, North–Holland Publishing Co., chapter 4, pp. 121–154.

Cooper, S. B. [1989]. A jump class of noncappable degrees, *J. Symbolic Logic* **54**: 324–353.

Downey, R. G., Jockusch, Jr., C. G. and Stob, M. [1990]. Array nonrecursive sets and multiple permitting arguments, *in* K. Ambos-Spies, G. H. Muller and G. E. Sacks (eds), *Recursion Theory Week, Oberwolfach 1989*, Vol. 1432 of *Lecture Notes in Mathematics*, Springer–Verlag, Heidelberg, pp. 141–174.

Downey, R. G. and Stob, M. [1990]. Automorphisms and splittings of recursively enumerable sets, *Proceedings of the Fourth Asian Logic Conference*, CSSK Centre, pp. 75–87.

Downey, R. and Harrington, L. [1996]. There is no fat orbit, *Ann. Pure Appl. Logic* **80**(3): 277–289.

Friedberg, R. M. [1958]. Three theorems on recursive enumeration. I. decomposition. II. maximal set. III. enumeration without duplication., *J. Symbolic Logic* **23**: 309–316.

Harrington, L. A. [1983]. The undecidability of the lattice of recursively enumerable sets. Handwritten Notes.

Harrington, L. A. and Nies, A. [1998]. Coding in the partial order of enumerable sets, *Adv. Math.* **133**(1): 133–162.

Harrington, L. A. and Soare, R. I. [1991]. Post's program and incomplete recursively enumerable sets, *Proc. Nat. Acad. Sci. U.S.A.* **88**: 10242–10246.

Harrington, L. A. and Soare, R. I. [1996a]. Definability, automorphisms, and dynamic properties of computably enumerable sets, *Bull. Symbolic Logic* **2**(2): 199–213.

Harrington, L. A. and Soare, R. I. [1996b]. The Δ_3^0-automorphism method and noninvariant classes of degrees, *J. Amer. Math. Soc.* **9**(3): 617–666.

Herrmann, E. [1984]. The undecidability of the elementary theory of the lattice of recursively enumerable sets, *Frege conference, 1984 (Schwerin, 1984)*, Akademie-Verlag, Berlin, pp. 66–72.

Herrmann, E. [1989]. Automorphisms of the lattice of recursively enumerable sets and hyperhypersimple sets, *Logic, methodology and philosophy of science, VIII (Moscow, 1987)*, North-Holland, Amsterdam, pp. 179–190.

Lachlan, A. H. [1968a]. Degrees of recursively enumerable sets which have no maximal supersets., *J. Symbolic Logic* **33**: 431–443.

Lachlan, A. H. [1968b]. The elementary theory of the lattice of recursively enumerable sets, *Duke Math. J.* **35**: 123–146.

Lempp, S. [1987]. Hyperarithmetical index sets in recursion theory, *Trans. Amer. Math. Soc.* **303**: 559–583.

Lerman, M., Shore, R. A. and Soare, R. I. [1978]. r-maximal major subsets, *Israel J. Math.* **31**(1): 1–18.

Lerman, M. and Soare, R. I. [1980]. A decidable fragment of the elementary theory of the lattice of recursively enumerable sets, *Trans. Amer. Math. Soc.* **257**(1): 1–37.

Maass, W. [1983]. Characterization of recursively enumerable sets with supersets effectively isomorphic to all recursively enumerable sets, *Trans. Amer. Math. Soc.* **279**: 311–336.

Maass, W., Shore, R. A. and Stob, M. [1981]. Splitting properties and jump classes, *Israel J. Math.* **39**: 210–224.

Martin, D. A. [1966]. Classes of recursively enumerable sets and degrees of unsolvability, *Z. Math. Logik Grundlag. Math.* **12**: 295–310.

Nies, A. [1997]. Intervals of the lattice of computably enumerable sets and effective boolean algebras, *Bull. Lond. Math. Soc.* **29**: 683–92.

Nies, A. [1998]. Coding methods in computability theory and complexity theory. Habilitationsschrift, Universität Heidelberg.

Nies, A., Shore, R. A. and Slaman, T. A. [1998]. Interpretability and definability in the recursively enumerable degrees, *Proc. London Math. Soc. (3)* **77**(2): 241–291.

Shoenfield, J. R. [1976]. Degrees of classes of recursively enumerable sets, *J. Symbolic Logic* **41**: 695–696.

Shore, R. [1999]. The recursively enumerable degrees, *in* E. R. Griffor (ed.), *Handbook of computability theory*, Vol. 140 of *Studies in Logic*, North–Holland Publishing Co., chapter 6, pp. 169–198.

Shore, R. [2000]. Natural definability in degree structures, *in* P. Cholak, S. Lemmp, M. Lerman and R. Shore (eds), *Computability Theory and Its Applications: Current Trends and Open Problems*, American Mathematical Society.

Shore, R. A. [1988]. A noninversion theorem for the jump operator, *Ann. Pure Appl. Logic* **40**: 277–303.

Slaman, T. [1999]. The global structure of the Turing degrees, *in* E. R. Griffor (ed.), *Handbook of computability theory*, Vol. 140 of *Studies in Logic*, North–Holland Publishing Co., chapter 5, pp. 155–168.

Soare, R. I. [1974]. Automorphisms of the lattice of recursively enumerable sets I: maximal sets, *Ann. of Math. (2)* **100**: 80–120.

Soare, R. I. [1987]. *Recursively Enumerable Sets and Degrees*, Perspectives in Mathematical Logic, Omega Series, Springer–Verlag, Heidelberg.

Soare, R. I. [2000]. Extensions, automorphisms, and definability, *in* P. Cholak, S. Lemmp, M. Lerman and R. Shore (eds), *Computability Theory and Its Applications: Current Trends and Open Problems*, American Mathematical Society.

Wald, K. [1999]. *Automorphism and noninvariant properites of the computably enumerable sets*, PhD thesis, University of Chicago.

Department of Mathematics, University of Notre Dame, Notre Dame, IN 46556-5659
E-mail address: Peter.Cholak.1@nd.edu

Contemporary Mathematics
Volume **257**, 2000

Computability Theory in Arithmetic: Provability, Structure and Techniques

C. T. Chong and Yue Yang

ABSTRACT. We list and discuss open questions in computability theory in arithmetic, classified broadly under three main categories: Reverse computability theory, structure theory of degrees, as well as techniques and methods peculiar to models of the theory.

1. Introduction

By *computability (recursion) theory in arithmetic* we mean computability theory on fragments of Peano arithmetic. More precisely, starting with a judicious choice of subsets of the Peano axioms, one studies computability-theoretic problems for models of these subsets. This area began life as a branch of reverse mathematics called reverse computability theory. The general question which guided (and continues to exert influence on) its development was:

Given a theorem in computability theory, what axioms are required to prove the theorem?

However, instead of using subsystems of second order arithmetic as the underlying framework for discussion—as is normally done in reverse mathematics—reverse computability theory is based on fragments of first order Peano arithmetic. In models of such theories, second order objects are often either referred to by their codes or through first order definitions. In this setting, the universe of models of such theories are no longer necessarily ω (the set of standard natural numbers). Indeed it is in the domain of nonstandard models that distinct new ideas and techniques, which have no parallel in the classical theory, are introduced. The variety of models makes the subject both very interesting and rich in problems. Recent discoveries show that computability theory on nonstandard models goes beyond the study of 'reverse problems' and is a subject in its own right. This is amplified in the list of open questions we give below. The term 'computability theory in arithmetic' is used to underscore this development, as well as to emphasize the inseparable link between this study and the arithmetical hierarchy.

Our list of problems is classified roughly into three main categories:

1991 *Mathematics Subject Classification.* 03D20, 03F30, 03H15.

(a) Foundational—dealing with issues concerning reverse mathematics;
(b) Structure theory—on the interaction between models and their associated degree-theoretic structures;
(c) Techniques and methods—especially where those developed for classical computability theory do not apply in the new setting.

It should be mentioned that all three are very much intertwined with one another, as problems in one category are very often either inspired by, or lead to, problems in one or both of the other categories. Questions are listed respectively in Sections 2 and 3 for (a), 4, 5 and 8 for (b), and 6 and 7 for (c).

All formulas in this paper are first order formulas in the language of Peano arithmetic (with parameters).

We begin with some basic definitions. Let PA^- be Peano axioms without the induction scheme, but equipped with the exponential function. Let $I\Sigma_n$ denote the induction scheme for Σ_n formulas. The Σ_n *bounding (collection) scheme* $B\Sigma_n$ states that for any Σ_n formula $\varphi(x,y)$,

$$(\forall x < a)(\exists y)\varphi(x,y) \to (\exists t)(\forall x < a)(\exists y < t)\varphi(x,y).$$

Intuitively, this says that in every model of $B\Sigma_n$, every Σ_n function on a closed interval has a bounded range.

The following theorem indicates that the hierarchy of theories is nondegenerate.

Theorem 1.1 (Paris and Kirby [**12**]). *Let $n \geq 1$. Over $PA^- + I\Sigma_0$,*

$$\cdots \Longrightarrow I\Sigma_{n+1} \Longrightarrow B\Sigma_{n+1} \Longrightarrow I\Sigma_n \Longrightarrow B\Sigma_n \Longrightarrow \dots,$$

but not conversely.

Central to the study of nonstandard models is the notion of a *cut*. This, by definition, is a nonempty proper subset closed downwards and under the successor function. A cut is Σ_n if it can be so defined.

Definition 1.1. *$\mathcal{M}$ is a $B\Sigma_n$* model *if $\mathcal{M}$ satisfies $B\Sigma_n$ but not $I\Sigma_n$. Similarly, $\mathcal{M}$ is an $I\Sigma_n$* model *if $\mathcal{M}$ satisfies $I\Sigma_n$ but not $B\Sigma_{n+1}$.*

Observe that $\mathcal{M}$ is a $B\Sigma_n$ model if and only if there is a Σ_n cut, since a $B\Sigma_n$ model is also a model of $I\Sigma_{n-1}$, therefore, has no Σ_{n-1} cuts. Using $PA^- + B\Sigma_1$ as the base theory, the fundamental notions of computability theory such as computable functions, computably enumerable (c.e., otherwise known as recursively enumerable) sets, etc. may be defined. From this one may develop a meaningful computability theory. The usual equivalence of c.e. sets with Σ_1-definable sets (with parameters) applies, and there is a version of Church's thesis for models of $PA^- + B\Sigma_1$ as well. Together these are sufficient for the development of 'predegree' computability theory in the general setting.

To go beyond that, the notion of a 'finite' set is crucial. If $\mathcal{M}$ is a model of $PA^- + B\Sigma_1$, we say that $X \subset \mathcal{M}$ is $\mathcal{M}$-finite if it is *coded*, i.e. there is a number $a \in \mathcal{M}$ such that $x \in X$ if and only if x divides a. With this one may define Turing reducibility. Caution needs to be taken, however, as Turing reducibility is in general not transitive. Groszek and Slaman [**11**] have exhibited an $I\Sigma_1$ model in which Turing reducibility is not transitive on c.e. sets. This may be extended to show that for $n \geq 1$, Turing reducibility is not, in general, transitive on Σ_n sets under $PA^- + I\Sigma_n$. Since transitivity is necessary for the definition of a Turing degree, one introduces the following relation: In a $B\Sigma_n$ or $I\Sigma_n$ model, we say that a set A in $\mathcal{M}$ (that is, A is a subset of $\mathcal{M}$) is *strongly reducible* to a set B if there is an algorithm which, given any $\mathcal{M}$-finite set X, it decides if $X \subseteq A$ or $X \subseteq \mathcal{M} \setminus A$ using $\mathcal{M}$-finite information about B. It is straightforward to verify that this is a transitive relation. Two sets are said to be of the same Turing degree if each is strongly reducible to the other.

We now rephrase the main theme of reverse computability theory in a more precise form:

Given a theorem of computability theory, how much induction (or bounding) is required to prove the theorem?

The following results illustrate some of the progress made thus far:

Theorem 1.2. *Over $PA^- + B\Sigma_n$, $I\Sigma_n$ is equivalent to*

1. (Chong and Yang [**6**]) *The existence of a high$_n$ degree.*
2. (Chong and Yang [**5**]) *The existence of a low$_{n+1}$ degree.*

Theorem 1.3. *Over $PA^- + B\Sigma_2$, $I\Sigma_2$ is equivalent to*

1. (Chong, Qian, Slaman and Yang [**3**]) *The existence of a c.e. minimal pair.*
2. (Chong and Yang [**7**]) *The existence of a c.e. maximal set.*

Theorem 1.4 (Chong and Mourad [**2**]). *$PA^- + B\Sigma_1$ proves the Friedberg-Muchnik Theorem.*

Theorem 1.5 (Groszek, Mytilinaios and Slaman [**9**]). *$PA^- + B\Sigma_2$ proves the Sacks Density Theorem.*

The reader will notice that whereas the first two theorems pinpoint precisely the proof-theoretic strength of the computability-theoretic theorem in question (over a base theory), the last two provide only a sufficient condition for the theorem to hold. We take this up in the next section.

2. $\mathbf{0}^{(n)}$ Priority Constructions

As shown by Groszek and Slaman [**10**] and Yang [**21**], there are natural connections between $\mathbf{0}^{(n)}$ priority constructions and $PA^- + I\Sigma_n$. The general heuristic is that to carry out a $\mathbf{0}^{(n)}$ priority construction, $I\Sigma_n$ is sufficient. However, except for finite injury arguments, i.e., $\mathbf{0}'$ priority constructions, there is no satisfactory abstract framework available, not even for the simplest infinite injury arguments, i.e., $\mathbf{0}''$ priority constructions. This leads us to the first in our list of questions.

Open Question 1. *Find an abstract framework for infinite injury arguments such that most (if not all) standard $\mathbf{0}''$ injury constructions fit into it, and $I\Sigma_2$ is* sufficient *to carry out the abstract construction.*

The matter of *necessity* in the above question is more complex. Theorem 1.2 and Theorem 1.5 together show that $\mathbf{0}''$ priority constructions do not always require the same power of induction for successful execution. Indeed in the case of Theorem 1.2 the main tool used in establishing the necessity of $I\Sigma_2$ for the existence of a high c.e. set was Lemma 3.1 (the Coding Lemma) below. It imposes a restriction on the range of the jump operator in $B\Sigma_2$ models, leading to the conclusion that high degrees do not exist (and therefore eliminating any possibility of a priority construction for this purpose).

Open Question 2. *Classify degree-theoretic theorems using* $\mathbf{0}''$ *priority constructions which are provable under* $PA^- + B\Sigma_2$, *hence not equivalent over* $B\Sigma_2$ *to* $I\Sigma_2$ *(for example, the Density Theorem).*

Thus broadly speaking there exist at least two types of $\mathbf{0}''$ constructions: Those requiring full Σ_2 induction and those requiring only Σ_2 bounding. This provides a way to more finely measure the relative complexity of such infinite injury constructions.

If one goes beyond $\mathbf{0}''$ priority, then besides the result in Theorem 1.2 on jump hierarchies, whose proof is a relativised version of that for Σ_2 level, little is known about $\mathbf{0}^{(n)}$ priority constructions, where $n \geq 3$.

Open Question 3. *Investigate* $\mathbf{0}^{(n)}$ *constructions and classify the relevant theorems in the hierarchy of theories. More specifically, are there natural degree-theoretic theorems which are equivalent to* $I\Sigma_3$ *over* $B\Sigma_3$*?*

3. $I\Sigma_n$ Models

As seen in the previous section, progress has been made on using $PA^- + B\Sigma_n$ as the base theory, and proving the equivalence of theorems with a stronger theory such as $I\Sigma_n$. Little is known, however, concerning the complementary problem: Starting with the base theory $PA^- + I\Sigma_{n-1}$ and proving the equivalence of computability-theoretic results with stronger theories such as $B\Sigma_n$.

An advantage of working with $B\Sigma_n$ models is that there is an abundance of codes in these models. The following coding lemma plays an essential role in the study of sets and their degrees. A set X is $\Delta_{n,A}$ if both $X \cap A$ and $A \setminus X$ are Δ_n.

Lemma 3.1 ((Coding Lemma) Chong and Mourad [**2**]). *Let* $\mathcal{M}$ *be a* $B\Sigma_n$ *model. Then any bounded* $\Delta_{n,A}$ *subset* X *of* A *is coded on* A*. In other words, there is an* $\mathcal{M}$*-finite set whose intersection with* A *is equal to* X*.*

The Coding Lemma allows one to use as parameters bounded sets with Δ_n definitions. When $n = 2$ this is useful for controlling the jump of a Δ_2 set. There are $B\Sigma_n$ models, which have ω as a Σ_n cut, such that every real is coded on ω. These are called *countably saturated* models. The solutions of many problems assume much simpler forms in these models. For example, Mytilinaios and Slaman [**14**] showed that the existence of a high degree is not provable in $PA^- + B\Sigma_2$ by noting that there is no high degree in any countably saturated model. The proof that $I\Sigma_2$ is necessary (Chong and Yang [**6**]) requires delicate applications of the Coding Lemma for $B\Sigma_2$ models.

When working in $I\Sigma_{n-1}$ models, Lemma 3.1 no longer holds. At the level of $n = 2$, there is a special difficulty arising from a model exhibited by Groszek and Slaman [**11**]:

Lemma 3.2 (Groszek and Slaman). *There is an* $I\Sigma_1$ *model* $\mathcal{M}$ *in which* ω *is a* Σ_2 *cut and there is a* Δ_2 *bijection* g *from* ω *onto* $\mathcal{M}$*.*

In this model, we can arrange requirements in $\mathcal{M}$ to be of order type ω via a computable approximation of g. In this way one can mimic many classical constructions on a universe which is essentially ω. For example,

Theorem 3.1. *Let* $\mathcal{M}$ *be the model above. Then*

1. (Yang [**22**]) *There exists an incomplete high c.e. set in* $\mathcal{M}$*.*

2. (Chong [**1**]) *There exists a maximal c.e. set in* $\mathcal{M}$.

Consequently, in Theorems 1.2 and 1.3, the base theory $PA^- + B\Sigma_2$ may not be weakened to $PA^- + I\Sigma_1$. This prompts one to pose the following questions:

Open Question 4. *Study computability theory in* $I\Sigma_n$ *models. In particular, is there a theorem about c.e. degrees which is equivalent to* $B\Sigma_2$ *over the base theory* $I\Sigma_1$*?*

Open Question 5. *Is there an* $I\Sigma_1$ *model in which there is no maximal c.e. set (or incomplete high c.e. set)?*

While it has not been done, we believe that the Density Theorem also holds in the Groszek-Slaman model. The following question is therefore appropriate:

Open Question 6. *Is there an* $I\Sigma_1$ *model in which the Sacks Density Theorem fails?*

It is worth noting that studying $I\Sigma_n$ models may shed light on some problems in model theory. For example, the following problem is taken from [**8**].

Open Question 7 (Paris). *Is* $B\Sigma_n$ *equivalent to* $I\Delta_n$ *(i.e. induction for* Δ_n *formulas)?*

An answer to the following model-theoretic question will also be useful for computability theory.

Open Question 8. *Is there an* $I\Sigma_1$ *model whose universe is not the* Δ_2 *projection of a cut?*

4. The Minimal Degree Problem

A noncomputable degree is minimal if there is no nontrivial degree below it. Improving on the result of Spector [**20**], Sacks [**17**] showed that there is a minimal degree below $\mathbf{0}'$. In α-recursion theory, Shore [**18**] showed the existence of a minimal α-degree below $\mathbf{0}'$ for all Σ_2-admissible ordinals α. The case for non-Σ_2-admissible ordinals, in particular $\aleph_\omega^L$, remains open after more than 25 years, despite attempts by many (Sacks had commented to us recently that he did not expect to see a solution in his lifetime). This makes it one of the most outstanding open problems in computability theory.

It can be shown that Sacks' theorem is provable in $PA^- + I\Sigma_2$. On the other hand, we have a parallel version of the minimal degree problem in the absence of Σ_2-induction.

Open Question 9. *Does* $PA^- + B\Sigma_2$ *prove the existence of a minimal degree, especially one below* $\mathbf{0}'$*?*

The difficulty encountered in a priority construction of a minimal $\aleph_\omega^L$-degree appears again as a major obstacle when one attempts a similar approach in a $B\Sigma_2$ model: The existence of a Σ_2-cofinal function on ω as an ordinal (for $\aleph_\omega^L$) and as a Σ_2-cut (in a $B\Sigma_2$-model for example) makes it impossible for the Sacks construction to satisfy more than the first ω requirements. This is due to the fact that the stages where the first ω requirements are met are computed in a $\Sigma_2(\mathbf{0}')$ manner, so that a uniform bound on these stages is not assured.

It is likely that the above problem, if solved, will point to a direction for solving the minimal α-degree problem.

On the other hand, if one is interested only in the existence of a minimal degree, and not those below $\mathbf{0}'$, then some positive results are known through an analysis of definable cuts.

Theorem 4.1. 1. (Chong and Mourad [**2**]) *The degree of a* Σ_n *cut is minimal in a countably saturated* $B\Sigma_n$ *model.*
2. (Mourad [**13**]) *In any* $B\Sigma_1$ *model, the degree of a* Σ_1 *cut is minimal among c.e. degrees.*

The discovery of cuts as minimal degrees leads to several related questions.

Open Question 10. *In a (not necessarily countably saturated)* $B\Sigma_n$ *model, what is the degree of a* Σ_n *cut? Is it minimal?*

In a countably saturated model of $B\Sigma_2$, it is known that every Σ_4 cut is actually Σ_2. The proof involves essential use of coding, see Qian [**16**]. Without saturation, it is not clear if there is a genuine Δ_3 cut.

Open Question 11. *Let* $\mathcal{M}$ *be a* $B\Sigma_2$ *model. Are there any genuine* Δ_3 *cuts?*

5. Global Degree Structure

The previous section indicates that cuts in nonstandard models sometimes serve as candidates for natural solutions of problems in computability theory. The next few results show that these objects play a major role in determining the global degree structure in $B\Sigma_n$ models.

Theorem 5.1. 1. (Chong and Mourad [**2**]) *The degree of a* Σ_n *cut and* $\mathbf{0}^{(n-1)}$ *form a minimal pair.*
2. *If the model is countably saturated, then every* Σ_n *degree is either below* $\mathbf{0}^{(n-1)}$ *(hence* Δ_n*) or lies above the degree of a* Σ_n *cut.*
3. (Qian [**16**]) *There is a jump-embedding of degrees from those below* $\mathbf{0}^{(n-1)}$ *into those above the degree of a* Σ_n *cut.*
4. (Chong and Yang [**6**]) *(*$n = 2$*) Every incomplete c.e. degree is low if and only if there is sufficient coding in the model. Furthermore, if* A *is c.e. and is not low, then its jump is the degree of* $I \oplus \emptyset'$.

In a $B\Sigma_n$ model, let $\mathcal{D}$ denote the collection of degrees and let $\mathcal{D}(\geq \deg(I))$ denote the collection of degrees above the degree of a Σ_n cut I.

Open Question 12. *In a countably saturated* $B\Sigma_n$ *model, is the theory of* $\mathcal{D}$ *elementarily equivalent to the theory of* $\mathcal{D}(\geq deg(I))$*? What about a general* $B\Sigma_n$ *model? Does the answer depend on* n*?*

Open Question 13. *Classify all* $B\Sigma_n$ *models up to elementary equivalence in the theory of Turing degrees.*

In connection with the above problem, the recent result of Nies, Shore and Slaman [**15**] is relevant. They have shown:

(*) There is a copy $\mathcal{N}_0$ of the standard model of arithmetic in the c.e. degrees $\mathcal{R}$ and a definable map $f : \mathcal{R} \to \mathcal{N}_0$ such that for each c.e. degree $\boldsymbol{a}$, $f(\boldsymbol{a})$ is the (code for) the least index of a c.e. set whose double jump has degree $\boldsymbol{a}''$.

This result provides a great amount of information about the theory of $\mathcal{R}$ beyond interpreting arithmetic in $\mathcal{R}$. For example it implies that the class of high c.e. degrees is definable. In the context of computability theory in arithmetic, the copy $\mathcal{N}_0$ has to be appropriately modified. Shore has suggested that the problem related to (*) is an area worth exploring.

Open Question 14. *Investigate (*) in models of fragments of Peano arithmetic.*

Returning to questions on the definability of degrees, some partial results are known:

Theorem 5.2 (Qian [**16**]). *Let $\mathcal{M}$ be a countably saturated $B\Sigma_n$ model and let I be a Σ_n cut. Then the degree of I is definable in the degrees below $\mathbf{0}^{(n)}$.*

Open Question 15. *Is the degree of a Σ_n cut definable in $\mathcal{D}$ in a countably saturated, $B\Sigma_n$ model? How about in any $B\Sigma_n$ model?*

Not much is known about the structure of degrees with respect to other notions of reducibility in computability theory in arithmetic.

Open Question 16. *Study the structure of the class of enumeration degrees, many-one degrees, truth-table degrees etc. for models of fragments of PA.*

6. Uniform Solutions for the Friedberg-Muchnik Theorem

It remains open in classical computability theory whether there is a degree-invariant solution to Post's Problem: Does there exist an index e such that for all X, $X <_T W_e^X <_T X'$? There is a model-theoretic analog in our setting, for the unrelativized Post's problem.

Open Question 17. *Is there a standard number $e \in \omega$ such that for all models of $PA^- + B\Sigma_1$, $\emptyset <_T W_e <_T \emptyset'$? Or more generally, are there $d, e \in \omega$ such that for any model of $PA^- + B\Sigma_1$, the degree of W_d is incomparable with the degree of W_e?*

The relevant results relating to this topic include Theorem 1.4 and the following:

Theorem 6.1 (Slaman and Woodin [**19**]). *If $\mathcal{M}$ is a model of $PA^- + B\Sigma_1$, then there exists a c.e. set W such that $\emptyset <_T W <_T \emptyset'$.*

However both proofs are nonuniform. One considers two cases: if $I\Sigma_1$ holds then the normal finite injury proof applies; if $I\Sigma_1$ fails then any Σ_1 cut is a solution to Post's problem (Theorem 6.1) while a construction involving the notion of 'union of cuts' gives an incomparable pair of c.e. sets (Theorem 1.4).

Observe that these two cases cannot be combined since Σ_1 cuts exist only in the absence of Σ_1 induction. This brings up the notion of *regularity*. We say a set A is regular if for any $m \in \mathcal{M}$, $A \cap \{x : x \leq m\}$ is $\mathcal{M}$-finite. It is clear that finite injury arguments always produce regular sets because of restraints imposed on priority constructions. However, it is known that in any $B\Sigma_1$ model, all regular c.e. sets are computable from I and, according to Theorem 4.1 (2), I is the least c.e. noncomputable degree. Thus finite injury priority method fails to produce an incomparable pair in $B\Sigma_1$ models. Indeed for countably saturated models, every regular c.e. set is computable, so that the structure of the c.e. degrees becomes interesting only when one looks at nonregular sets and their degrees.

7. Nonregular Sets

The simplest example of a nonregular set is a cut. Working with nonregular sets is always a challenge since priority constructions are no longer applicable. However, they figure prominently in the models we study in this paper. Thus in $B\Sigma_1$ models, every c.e. degree above the degree of a Σ_1 cut is nonregular, i. e. contains a nonregular set. On the other hand, since the degree of a Σ_1 cut is a minimal c.e. degree which lies below every noncomputable c.e. degree, an incomparable pair of c.e. degrees (solving the Friedberg-Muchnik problem) exists only in the region of c.e. degrees between the degree of a Σ_1 cut and $\mathbf{0}'$. Such degrees were exhibited in the proof of Theorem 1.4. We call this the *special region.* This leads us to the next two questions.

Open Question 18. *Study the structure of Turing degrees in the* special region *for $B\Sigma_1$ models. More specifically, does the Sacks splitting Theorem or Density Theorem hold above deg(I)? Is there a minimal cover of I? What kind of initial segments can be embedded into the c.e. degrees in $B\Sigma_1$ models?*

Open Question 19. *Investigate the structure of degrees above nonregular degrees.*

8. Full Peano Arithmetic

For our final list of questions, we consider models of PA. In such models, all cuts are undefinable and every Σ_n set, for $n < \omega$, is regular. It is not difficult to see that there is only one theory of arithmetical degrees (i.e. degrees of Σ_n sets) for these models. Hence interesting questions are found only when one goes beyond the arithmetical hierarchy. The following is a sample which is a counterexample to the Friedberg Jump Inversion Theorem for full Peano arithmetic (the corresponding problem for fragments of PA is studied in Chong and Yang [**5**]).

Theorem 8.1 (Chong, Qian and Yang [**4**]). *Let $\mathcal{M}$ be a model of PA which is countably saturated. Then the set $Y = \{(x, n) : n < \omega \ \& \ x \in \emptyset^{(n)}\}$ is not the jump of any set in $\mathcal{M}$.*

Again the source for the above counterexample to jump inversion comes from the failure of the original Friedberg construction in handling nonregular sets.

Open Question 20. *Investigate the Friedberg Jump Inversion Theorem in models of full Peano arithmetic.*

It is also known that in such a nonstandard model, ω is a set of minimal degree, and that it is definable in the degrees below the degree of Y.

Open Question 21. *Is there a cone of minimal covers for the collection of degrees in a countably saturated model $\mathcal{M}$ of PA?*

Open Question 22. *Is the degree of ω (for the model $\mathcal{M}$) definable in $\mathcal{D}$?*

References

[1] C. T. Chong, *Maximal sets and fragments of Peano arithmetic*, Nagoya Math. J. **115** (1989), 165–183.

[2] C. T. Chong and K. J. Mourad, *Σ_n definable sets without Σ_n induction*, Trans. Amer. Math. Soc. **334** (1992), no. 1, 349–363.

[3] C. T. Chong, Lei Qian, Theodore A. Slaman, and Yue Yang, Σ_2 *induction and infinite injury priority arguments, part III: Prompt sets, minimal pairs and Shoenfield's conjecture*, To appear.

[4] C. T. Chong, Lei Qian, and Yue Yang, *The Friedberg jump inversion theorem revisited: A study of undefinable cuts*, To appear in Proceeding of Logic Colloquium '98, Prague.

[5] C. T. Chong and Yue Yang, *Local degree theory in fragments of Peano arithmetic*, in preparation.

[6] ———, Σ_2 *induction and infinite injury priority arguments, part II: Tame* Σ_2 *coding and the jump operator*, Ann. Pure Appl. Logic **87** (1997), no. 2, 103–116, Logic Colloquium '95 Haifa.

[7] ———, Σ_2 *induction and infinite injury priority arguments, part I: Maximal sets and the jump operator*, J. Symbolic Logic **63** (1998), no. 3, 797–814.

[8] Peter Clote and Jan Krajíček, *Open problems*, Arithmetic, Proof Theory, and Computational Complexity, (Prague, 1991) (Peter Clote and Jan Krajíček eds.), Oxford University Press, 1993, pp. 1–19.

[9] Marcia J. Groszek, M. E. Mytilinaios, and Theodore A. Slaman, *The Sacks density theorem and* Σ_2*-bounding*, J. Symbolic Logic **61** (1996), no. 2, 450–467.

[10] Marcia J. Groszek and Theodore A. Slaman, *Foundations of the priority method I: Finite and infinite injury*, Preprint.

[11] ———, *On Turing reducibility*, Preprint, 1994.

[12] L. A. Kirby and J. B. Paris, *Initial segments of models of Peano's axioms*, Set Theory and Hierarchy Theory V (Bierutowice, Poland, 1976) (Heidelberg), Lecture Notes in Mathematics, vol. 619, Springer–Verlag, 1977, pp. 211–226.

[13] Karim Joseph Mourad, *Recursion theoretic statements equivalent to induction axioms for arithmetic*, Ph.D. thesis, University of Chicago, 1988.

[14] Michael E. Mytilinaios and Theodore A. Slaman, Σ_2*-collection and the infinite injury priority method*, J. Symbolic Logic **53** (1988), no. 1, 212–221.

[15] André Nies, Richard A. Shore and Theodore A. Slaman, *Interpretability and definability in the recursively enumerable degrees*, Proc. London Math. Society **77** (1998), no. 3, 241–291.

[16] Lei Qian, *Degree structure in Skolem models of Peano arithmetic*, Ph.D. thesis, National University of Singapore, in preparation.

[17] Gerald E. Sacks, *A minimal degree below* $\mathbf{o}'$, Bull. Amer. Math. Soc. **67** (1961), 416–419.

[18] Richard A. Shore, *Minimal* α*-degrees*, Ann. Math. Logic **4** (1972), 393–414.

[19] Theodore A. Slaman and W. Hugh Woodin, Σ_1*-collection and the finite injury method*, Mathematical Logic and Applications (Heidelberg) (Juichi Shinoda, Theodore A. Slaman, and T. Tugué, eds.), Springer–Verlag, 1989, pp. 178–188.

[20] C. Spector, *On the degrees of recursive unsolvability*, Ann. of Math. **64** (1956), 581–592.

[21] Yue Yang, *Iterated trees and fragments of arithmetic*, Arch. Math. Logic **34** (1995), no. 2, 97–112.

[22] ———, *The thickness lemma from* $P^- + I\Sigma_1 + \neg B\Sigma_2$, J. Symbolic Logic **60** (1995), no. 2, 505–511.

Department of Mathematics, Faculty of Science, National University of Singapore, Lower Kent Ridge Road, Singapore 119260.

E-mail address: `chongct@math.nus.edu.sg`

Department of Mathematics, Faculty of Science, National University of Singapore, Lower Kent Ridge Road, Singapore 119260.

E-mail address: `matyangy@math.nus.edu.sg`

Contemporary Mathematics
Volume **257**, 2000

How Many Turing Degrees are There?

Randall Dougherty and Alexander S. Kechris

Abstract. A Borel equivalence relation on a Polish space is *countable* if all of its equivalence classes are countable. Standard examples of countable Borel equivalence relations (on the space of subsets of the integers) that occur in recursion theory are: recursive isomorphism, Turing equivalence, arithmetic equivalence, etc. There is a canonical hierarchy of complexity of countable Borel equivalence relations imposed by the notion of Borel reducibility. We will survey results and conjectures concerning the problem of identifying the place in this hierarchy of these equivalence relations from recursion theory and also discuss some of their implications.

The obvious answer to the question of the title is: *continuum many*. There is however a different way of looking at this question, which leads to some very interesting open problems in the interface of recursion theory and descriptive set theory. Our goal in this paper is to explain the context in which this and related problems can be formulated, i.e., the theory of Borel equivalence relations, and survey some of the progress to date.

1. Formulation of the problem

We denote by $\equiv_T$ the **Turing equivalence relation** on $\mathcal{P}(\mathbb{N}) = \{X : X \subseteq \mathbb{N}\}$, which we identify with $2^{\mathbb{N}}$, viewing sets as characteristic functions. (We use the standard set-theoretic convention that $n = \{0, 1, \ldots, n-1\}$ for all natural numbers n.) Then $\equiv_T$ is a Borel (in fact Σ^0_3) equivalence relation on $2^{\mathbb{N}}$. We denote by $\mathcal{D}$ the quotient space $2^{\mathbb{N}}/(\equiv_T)$, i.e., the set of **Turing degrees**.

Now consider general Borel equivalence relations on $2^{\mathbb{N}}$ or even arbitrary **Polish** (separable completely metrizable) spaces. We measure their complexity by studying the following partial (pre)order of **Borel reducibility**: if E, F are Borel equivalence relations on X, Y respectively, then a **Borel reduction** of E into F is a Borel map $f : X \to Y$ such that

$$xEy \iff f(x)Ff(y).$$

If such an f exists we say that E is **Borel reducible** to F and denote this by

$$E \leq_B F.$$

Let also

$$E \sim_B F \iff E \leq_B F \;\&\; F \leq_B E$$

1991 *Mathematics Subject Classification.* Primary 03D30, 03E15; Secondary 04A15, 54H05.

The first author was partially supported by NSF Grant DMS 9158092.
The second author was partially supported by NSF Grant DMS 9619880.

(this defines the concept of **bi-reducibility**) and

$$E <_B F \iff E \leq_B F \;\&\; F \not\leq_B E.$$

Let us say that a function $f_* : X/E \to Y/F$ is **Borel** if it has a Borel lifting, i.e., there is a Borel function $f : X \to Y$ such that $f_*([x]_E) = [f(x)]_F$ for all $x \in X$. Then it is clear that $E \leq_B F$ is equivalent to the assertion that there is a **Borel injection** from X/E into Y/F, which we express by saying that the **Borel cardinality**, $|E|_B$, of E is less than or equal to to that of F; in symbols,

$$|E|_B \leq |F|_B \iff E \leq_B F.$$

Then define

$$|E|_B = |F|_B \iff E \sim_B F,$$

i.e., X/E, Y/F have the same Borel cardinality, and

$$|E|_B < |F|_B \iff E <_B F,$$

i.e., X/E has (strictly) smaller Borel cardinality then Y/F.

We are now ready to formulate our problem as follows, where, by abusing notation, we write below $|\mathcal{D}|_B$ instead of $|\equiv_T|_B$ and call this the Borel cardinality of $\mathcal{D}$, instead of $\equiv_T$:

Question: What is the Borel cardinality, $|\mathcal{D}|_B$, of the set of Turing degrees $\mathcal{D}$?

If we denote the classical (Cantor) cardinality of $\mathcal{D}$ by $|\mathcal{D}|$, then we have $|\mathcal{D}| = |\mathbb{R}|$. However, it is not hard to see that the Borel cardinality of $\mathcal{D}$ is bigger than that of the continuum. Let $=_X$ be the identity relation on the Polish space X. So $|{=_\mathbb{R}}|_B$ is the Borel cardinality which naturally represents the classical cardinality of the continuum.

Fact. $(\equiv_T) >_B (=_\mathbb{R})$.

PROOF. It is standard that there is a perfect set of pairwise Turing incomparable subsets of $\mathbb{N}$, so $(=_\mathbb{R}) \leq_B (\equiv_T)$. If on the other hand $f : 2^\mathbb{N} \to \mathbb{R}$ is Borel and Turing-invariant, i.e., $x \equiv_T y \implies f(x) = f(y)$, then for each Borel set $A \subseteq \mathbb{R}$, $f^{-1}(A)$ is a Turing-invariant Borel subset of $2^\mathbb{N}$, so it has measure 0 or 1. It follows that, for each n, the nth digit in the decimal expansion of $f(x)$ is fixed on a set of measure 1. So there is a Turing-invariant Borel set of measure 1 on which f is constant, therefore f cannot be a reduction of $\equiv_T$ into $=_\mathbb{R}$. Thus $(\equiv_T) \not\leq_B (=_\mathbb{R})$. $\square$

We now have our question but it is not clear yet what kind of answer we should expect. In what sense can we hope to compute $|\mathcal{D}|_B$? To understand this, we have to dig a little deeper into the theory of Borel equivalence relations.

For our purposes, a crucial property of the Turing equivalence relation is that it has countable equivalence classes. In general, we call a Borel equivalence relation **countable** if every one of its classes is countable. We will next review some basic facts of the theory of countable Borel equivalence relations, for which we refer the reader to the papers Kechris [**K2**], Dougherty-Jackson-Kechris [**DJK**], Jackson-Kechris-Louveau [**JKL**], Kechris [**K1**], and Adams-Kechris [**AK**].

(i) (Feldman-Moore [**FM**]) Every countable Borel equivalence relation is generated by a Borel action of a countable group.

More precisely, given a countable Borel equivalence E on a Polish space X, there is a countable group G and a Borel action $(g, x) \mapsto g \cdot x$ of G on X such that, if E_G^X is defined by

$$x E_G^X y \iff \exists g{\in}G\ (g \cdot x = y),$$

then $E = E_G^X$.

In particular, $\equiv_T$ is given by a Borel action of a countable group on $2^{\mathbb{N}}$. It seems like an interesting, but somewhat vague, question to find out whether one can obtain such a representation that has some recursion theoretic significance.

REMARK 1.1. Using the Feldman-Moore theorem and related facts, within a Schröder-Bernstein argument, one can show that, for countable Borel equivalence relations E and F, $E \sim_B F$ is equivalent to the existence of a Borel bijection of X/E with Y/F.

(ii) There is a **universal** countable Borel equivalence relation, in the sense of $\leq_B$.

That is, there is a countable Borel equivalence relation E such that, for any countable Borel equivalence relation F, we have $F \leq_B E$. This E is clearly unique, up to $\sim_B$, and denoted by E_∞.

An example of a universal countable Borel equivalence is given by the orbit equivalence relation of the shift action of F_2, the free group on two generators, on 2^{F_2} given by

$$g \cdot x(h) = x(g^{-1}h), \qquad g, h \in F_2, \quad x \in 2^{F_2}.$$

(iii) There is a smallest, in the sense of $\leq_B$, countable Borel equivalence relation on uncountable Polish spaces, namely $=_{\mathbb{R}}$.

So for every countable Borel equivalence relation E on an uncountable Polish space, we have $(=_{\mathbb{R}}) \leq_B E$. If $(=_{\mathbb{R}}) \sim_B E$, we say that E is **smooth**. For example, $\equiv_T$ is not smooth. Another example of a non-smooth countable Borel equivalence is the following one, defined on $2^{\mathbb{N}}$:

$$x E_0 y \iff \exists n\, \forall m{\geq}n\ (x(m) = y(m)).$$

This turns out to be the smallest, in the sense of $\leq_B$, non-smooth countable Borel equivalence relation. This is a particular instance of the general Glimm-Effros Dichotomy proved in Harrington-Kechris-Louveau [**HKL**], but this special case can already be derived from Effros [**E**].

(iv) (Glimm-Effros Dichotomy) If E is a countable Borel equivalence relation which is not smooth, then $E_0 \leq_B E$.

(v) $E_0 <_B E_\infty$.

Thus we have

$$(=_{\mathbb{R}}) <_B E_0 <_B E_\infty$$

and every other countable Borel equivalence relation on an uncountable space is in the interval (E_0, E_∞).

(vi) (Adams-Kechris [**AK**]) There are continuum many pairwise incomparable, under $\leq_B$, countable Borel equivalence relations.

We now have all the ingredients to formulate a precise conjecture, in response to the question about the Borel cardinality of $\mathcal{D}$. This was originally formulated (as a question) in Kechris [**K2**] and listed (as a conjecture) in Slaman's list of Questions in Recursion Theory, item 2.3, posted in http://math.berkeley.edu/~slaman/.

Conjecture: $\equiv_T$ is a universal countable Borel equivalence relation, i.e., $(\equiv_T) \sim_B E_\infty$.

2. Known results and implications

There is some information already available about the complexity of $\equiv_T$.

THEOREM 2.1. *(Slaman-Steel* [**SS**]*)* $E_0 <_B (\equiv_T)$.

This has been strengthened in Kechris [**K1**] to show that $\equiv_T$ is not amenable and in Jackson-Kechris-Louveau [**JKL**] to show that $\equiv_T$ is not treeable, all indications that $\equiv_T$ is quite complex.

One of the intriguing implications of the conjecture that $\equiv_T$ is universal concerns the existence of unusual functions on the Turing degrees. Recall that we call a function $f : \mathcal{D}^n \to \mathcal{D}$ **Borel** if there is a Borel function $F : (2^{\mathbb{N}})^n \to 2^{\mathbb{N}}$ such that

$$f([x_1]_T, \ldots, [x_n]_T) = [F(x_1, \ldots, x_n)]_T$$

for all $x_1, \ldots, x_n \in 2^{\mathbb{N}}$, where $[x]_T$ is the Turing degree of $x \in 2^{\mathbb{N}}$. A **pairing function** on $\mathcal{D}$ is a bijection $\langle , \rangle : \mathcal{D}^2 \to \mathcal{D}$.

Fact. If $\equiv_T$ is universal, then there is a Borel pairing function on $\mathcal{D}$.

PROOF. If E, F are Borel equivalence relations on X, Y respectively, let $E \times F$ be the Borel equivalence relation on $X \times Y$ given by

$$(x, y)(E \times F)(x', y') \iff xEx' \;\&\; yFy'.$$

Clearly $E_\infty \times E_\infty \geq_B E_\infty$, so, since E_∞ is universal, $E_\infty \times E_\infty \sim_B E_\infty$. Hence, if $(\equiv_T) \sim_B E_\infty$, we have

$$(\equiv_T) \times (\equiv_T) \sim_B (\equiv_T),$$

which shows that there is a Borel pairing function on $\mathcal{D}$. □

The well-known Martin Conjecture (or the 5th Victoria Delfino problem), see Kechris-Moschovakis, Eds. [**KM**] or Slaman's list, item 2.2, seeks to classify definable functions on $\mathcal{D}$, asymptotically, i.e., up to identification on a cone of degrees. One part of the conjecture asserts, in particular, that if a Borel $f : \mathcal{D} \to \mathcal{D}$ is not constant on a cone, then $f(d) \geq d$ on a cone. We can now easily see the following:

Fact. If $\equiv_T$ is universal, then Martin's Conjecture fails.

PROOF. Fix $d_0 \neq d_1$ in $\mathcal{D}$ and let $\langle , \rangle$ be a Borel pairing function on $\mathcal{D}$. Let $f_0(d) = \langle d_0, d \rangle$ and $f_1(d) = \langle d_1, d \rangle$. Then $f_i : \mathcal{D} \to \mathcal{D}$ is Borel for $i = 0, 1$ and, if $A_i = \mathrm{rng}(f_i)$, then $A_0 \cap A_1 = \varnothing$. Since $\equiv_T$ is countable, one can show that the inverse of the pairing function $\langle , \rangle$ is also Borel, so the sets A_i are Borel.

Clearly f_0 and f_1 are injective, so they are not constant on a cone. Thus, if Martin's Conjecture were true, we would have that $f_i(d) \geq d$ on a cone for $i = 0, 1$. Then A_0 and A_1 would be cofinal in the Turing degrees, so, by Borel Determinacy, each would contain a cone, contradiction. □

3. Some more questions and answers

There are of course several other notions of equivalence and degree studied in recursion theory, and similar questions and conjecture can be considered for them too. We will concentrate here on one of the finest, *recursive isomorphism*, and one of the coarsest, *arithmetic equivalence*.

Let S_∞ be the group of permutations of $\mathbb{N}$, and let S_r be the subgroup consisting of all recursive permutations. We let $\equiv_r$ denote **recursive isomorphism** for subsets of $\mathbb{N}$. Via our identification of $\mathcal{P}(\mathbb{N})$ with $2^{\mathbb{N}}$, we have for $x, y \in 2^{\mathbb{N}}$:

$$x \equiv_r y \iff \exists \pi \in S_r \ (x \circ \pi = y).$$

For any $n \in \{2, 3, 4, \dots\} \cup \{\mathbb{N}\}$ we also define recursive isomorphism on $n^{\mathbb{N}}$ by

$$x \equiv_r^n y \iff \exists \pi \in S_r \ (x \circ \pi = y),$$

so that $(\equiv_r^2) = (\equiv_r)$.

It is well-known that $(\equiv_T) \leq_B (\equiv_r)$, because $x \equiv_T y \iff x' \equiv_r y'$, where x' is the Turing jump of x. Hence, if $\equiv_T$ is universal, then $\equiv_r$ is universal; and proving that $\equiv_r$ is universal could be viewed as providing additional evidence that $\equiv_T$ is universal.

Finally, we denote by $\equiv_A$ the notion of **arithmetic equivalence** on $2^{\mathbb{N}}$. So $(\equiv_r) \subseteq (\equiv_T) \subseteq (\equiv_A)$.

Again, one can conjecture that $\equiv_r$ and $\equiv_A$ are universal. Here, though, we have some answers.

THEOREM 3.1. *(Slaman-Steel, unpublished). Arithmetic equivalence, $\equiv_A$, is universal, i.e., $(\equiv_A) \sim_B E_\infty$.*

So arithmetical equivalence has a Borel pairing function, and the arithmetical analogue of Martin's Conjecture fails.

The problem for recursive equivalence is still open, but there has been a lot of progress.

THEOREM 3.2. *(Dougherty-Kechris* [**DK**]*). Recursive isomorphism on $\mathbb{N}^{\mathbb{N}}$ is universal, i.e., $(\equiv_r^{\mathbb{N}}) \sim_B E_\infty$.*

This was very recently improved to

THEOREM 3.3. *(Andretta-Camerlo-Hjorth* [**ACH**]*). Recursive isomorphism on $5^{\mathbb{N}}$ is universal, i.e., $(\equiv_r^5) \sim_B E_\infty$.*

However, it is not yet clear how to reduce 5 to 2.

Actually, Theorems 3.2 and 3.3 are much more general. In each case, one actually shows that there is a fixed subgroup S_0 consisting of primitive recursive (in fact much simpler) permutations such that the result is true if S_r is replaced by any countable group S with $S_0 \subseteq S \subseteq S_\infty$.

There is one last problem related to Theorem 3.2, that has further interesting implications.

First recall that an action of a group G on a set X is called **free** if $g \cdot x \neq x$ for any $x \in X$ and $g \neq 1_G$. Also recall from §2 that every countable Borel equivalence relation is induced by a Borel action of a countable group G. From considerations in ergodic theory, it turns out that it is not always possible to find a **free** such action that induces it; see Adams [**A**]. It has been observed though that every known example of a countable Borel equivalence relation E, which cannot be induced by a free Borel action of a countable group, admits an invariant Borel probability measure (**measure** for short). (A measure is **invariant** for E if it is invariant for any Borel action of a countable group that generates it.) It has in fact been conjectured that this is always the case. In other words, a countable Borel equivalence relation which does not admit an invariant measure can be induced by a free Borel action of a countable group.

By using the arguments in §2 of Dougherty-Jackson-Kechris [**DJK**] and a theorem of Nadkarni [**N**], it can be seen that this last assertion is equivalent to the following:

(†) There is a universal countable Borel equivalence relation, which is induced by a free Borel action of a countable group.

We return now to Theorem 3.2. We have that $\equiv_r^{\mathbb{N}}$ is induced by the following Borel action of S_r on $\mathbb{N}^{\mathbb{N}}$:

$$\pi \cdot x = x \circ \pi^{-1}.$$

This action is not free, but its restriction to

$$[\mathbb{N}]^{\mathbb{N}} = \{x \in \mathbb{N}^{\mathbb{N}} : x \text{ is one-to-one}\}$$

is. It is natural to conjecture that Theorem 3.2 can be strengthened to the statement that $(\equiv_r)\restriction[\mathbb{N}]^{\mathbb{N}}$ is universal. If this turns out to be the case, this will also prove (†).

4. Some proofs

We will give here our proof of Theorem 3.2 (and a related result). This comes from the unpublished Dougherty-Kechris [**DK**]. Although Theorem 3.2 has now been superseded by Theorem 3.3, our proof uses different methods and may find other applications in the future.

As we indicated in §3, one has in fact a stronger result. For any subgroup S of S_∞, and any X, let for $x, y \in X^{\mathbb{N}}$:

$$x \equiv_S^X y \iff \exists \pi \in S\ (x \circ \pi = y).$$

So $(\equiv_r^{\mathbb{N}}) = (\equiv_{S_r}^{\mathbb{N}})$. We call S **primitive recursive** if $S = \{g_n : n \in \mathbb{N}\}$, with $g(n, m) = g_n(m)$ primitive recursive. We now have:

THEOREM 4.1. *There is a primitive recursive countable group $S_0 \subseteq S_\infty$ such that for any countable group S with $S_0 \subseteq S \subseteq S_\infty$, we have that $\equiv_S^{\mathbb{N}}$ is a universal countable Borel equivalence relation. In particular this is true for $\equiv_r^{\mathbb{N}}$.*

PROOF. To explain the basic idea, consider a countable infinite group H and fix a one-to-one enumeration $H = \{h_n : n \in \mathbb{N}\}$ of it. Then any $h_a \in H$ corresponds to a permutation $\tilde{a} \in S_\infty$ given by $h_{\tilde{a}(n)} = h_n h_a$ (the right regular representation). Fix also a bijection $\langle , \rangle : \mathbb{N}^2 \to \mathbb{N}$ and let $\pi_a \in S_\infty$ be defined by

$$\pi_a(\langle n, m \rangle) = \langle \tilde{a}(n), m \rangle.$$

Now given an action $(h, x) \mapsto h \cdot x$ of H into a space of the form $X^{\mathbb{N}}$ and the corresponding equivalence relation E_H, define the function $f : X^{\mathbb{N}} \to X^{\mathbb{N}}$ by

$$f(x)(\langle n, m \rangle) = (h_n \cdot x)(m).$$

Then we have

$$\begin{aligned} f(h_a \cdot x)(\langle n, m\rangle) &= (h_n \cdot (h_a \cdot x))(m) \\ &= (h_{\tilde{a}(n)} \cdot x)(m) \\ &= f(x)(\langle \tilde{a}(n), m\rangle) \\ &= (f(x) \circ \pi_a)(\langle n, m\rangle); \end{aligned}$$

hence, $f(h_a \cdot x) = f(x) \circ \pi_a$. It follows that if $H_0 = \{\pi_a : a \in \mathbb{N}\}$ (a countable subgroup of S_∞), then

$$xE_H y \iff f(x) \equiv^X_{H_0} f(y). \tag{*}$$

Unfortunately, if $S_\infty \supseteq H' \supseteq H_0$, H' a countable group, then we cannot, in general, replace H_0 by H' in (*) since it could be that $f(x) \equiv^X_{H'} f(y)$ via some $\pi \in H' \setminus H_0$. After appropriately choosing H, X, and the action of H on $X^{\mathbb{N}}$ (so that at least E_H is universal), we will modify $f(x)$ to $f^*(x) \in (X^*)^{\mathbb{N}}$, for some X^*, by encoding in it some further information, so that even if $f(x) \equiv^{X^*}_{H'} f(y)$ via some $\pi \in H' \setminus H_0$ we can still conclude that $xE_H y$. In particular, although the X we will start with will be finite, this encoding will require X^* to be infinite. Moreover, we will be forced to restrict the x's to some subset of $X^{\mathbb{N}}$, say $Y \subseteq X^{\mathbb{N}}$, so we will also need to make sure that $E_H \restriction Y$ is universal.

We will now implement this idea. We fix some notation first:

For any X and countable group G, we have the shift action of G on X^G given by

$$g \cdot x(h) = x(g^{-1}h).$$

This induces for any subgroup $H \subseteq G$ an action of H on X^G and we denote the corresponding equivalence relation by $E(H, X^G)$. If G is infinite, fixing a one-to-one enumeration of G, we can view this as an action of H on $X^{\mathbb{N}}$.

Now fix a one-to-one enumeration $\{g_n : n \in \mathbb{N}\}$ of the free group F_2 on two generators, with $g_0 = 1$ where 1 is the identity element of F_2. Define $\tilde{a}$ and π_a as above by the formulas $g_{\tilde{a}(n)} = g_n g_a$ and $\pi_a(\langle n, m\rangle) = \langle \tilde{a}(n), m\rangle$, and let

$$S_0 = \{\pi_a : a \in \mathbb{N}\}.$$

If $\{g_n : n \in \mathbb{N}\}$ and $\langle , \rangle$ are chosen appropriately, then S_0 is primitive recursive. Fix also any countable group S such that $S_\infty \supseteq S \supseteq S_0$; we will show that $\equiv^{\mathbb{N}}_S$ is universal. Say $S = \{\rho_i : i \in \mathbb{N}\}$.

We call $i \in \mathbb{N}$ **bad** if

(i) $\forall n \forall m \exists n' (\rho_i(\langle n, m\rangle) = \langle n', m\rangle)$; and

(ii) if $\rho_i(\langle 0, m\rangle) = \langle n_m, m\rangle$ for all m, then $n_m \to \infty$ as $m \to \infty$.

We can now easily define $n^{(i)}_j, m^{(i)}_j \in \mathbb{N}$ for $i, j \in \mathbb{N}$ such that:

(a) $0 < n^{(i)}_j < n^{(i)}_{j+1}$ and $0 < m^{(i)}_j < m^{(i)}_{j+1}$;

(b) $(i, j) \neq (i', j') \implies m^{(i)}_j \neq m^{(i')}_{j'}$;

(c) if i is bad, then $n_{m^{(i)}_j} = n^{(i)}_j$.

Also, for the free group F_k with k generators and $g \in F_k$, $m \in \mathbb{N}$, let $B_k(g, m)$ be the ball of radius m around g in the tree of F_k; i.e., $B_k(g, m)$ is the set of all products gh where h is a word in F_k of length at most m.

Now consider the shift action of F_2 on 9^{F_3} (9 is a large enough number here) and the Borel set $A \subseteq 9^{F_3}$ defined by

$$y \in A \iff \forall i \forall j \left[[(g_{n_j^{(i)}} \cdot y) \restriction B_3(1, m_j^{(i)}) = (g_{n_{j+1}^{(i)}} \cdot y) \restriction B_3(1, m_j^{(i)})] \implies \right.$$
$$\left. g_{n_j^{(i)}} \cdot y = g_{n_{j+1}^{(i)}} \cdot y \right],$$

where 1 is the identity element of F_3.

LEMMA 4.2. $E(F_2, 9^{F_3}) \restriction A \leq_B (\equiv_S^{\mathbb{N}})$.

PROOF. Fix an injection c from the countable set $\bigcup_m 9^{B_3(1,m)}$ to $\mathbb{N}$. Now define $f^* : A \to \mathbb{N}^{\mathbb{N}}$ by $f^*(x) = x^*$, where $x^*(\langle n, m \rangle) = c((g_n \cdot x) \restriction B_3(1, m))$. Thus $x^*(\langle n, m \rangle)$ encodes the values of $g_n \cdot x$ at the ball of radius m around $1 \in F_3$. In particular, $x^*(\langle n, m \rangle)$ encodes (i.e., uniquely determines) m as well. (If we were to take $f(x)$ as in the intuitive explanation in the beginning of this proof, then $f(x)(\langle n, m \rangle)$ would be just $g_n \cdot x(p_m)$, where $\{p_m : m \in \mathbb{N}\}$ is a one-to-one enumeration of F_3.)

We claim that

$$xE(F_2, 9^{F_3})y \iff x^* \equiv_S^{\mathbb{N}} y^*,$$

which completes the proof.

$\Rightarrow$: Clearly $y = g_a \cdot x \implies y^* = x^* \circ \pi_a$.

$\Leftarrow$: Say now $\pi \in S$ is such that $y^* = x^* \circ \pi$, i.e., $y^*(\langle n, m \rangle) = x^*(\pi(\langle n, m \rangle))$. Since $x^*(\langle n, m \rangle)$ encodes m, it follows that there is a function $\pi' : \mathbb{N} \to \mathbb{N}$ such that $\pi(\langle n, m \rangle) = \langle \pi'(\langle n, m \rangle), m \rangle$ for all n and m; that is, the second coordinate is left fixed by π. (Note that all π_a have this property, of course. By our encoding we have forced any π as above to have it as well.)

We now have two cases:

(I) $\pi'(0, m)$ does not tend to ∞ as $m \to \infty$. So there must exist a number ℓ such that, for infinitely many m, $\pi'(0, m) = \ell$. For any such m, we have $y^*(\langle 0, m \rangle) = x^*(\langle \ell, m \rangle)$, i.e., $y \restriction B_3(1, m) = (g_\ell \cdot x) \restriction B_3(1, m)$; since there are arbitrarily large such m, it follows that $y = g_\ell \cdot x$, so $xE(F_2, 9^{F_3})y$.

(II) $\pi'(0, m) \to \infty$ as $m \to \infty$. So if $\pi = \rho_i$, then i is bad. For any j, we have $y^*(\langle 0, m_j^{(i)} \rangle) = x^*(\langle n_j^{(i)}, m_j^{(i)} \rangle)$, i.e., $y \restriction B_3(1, m_j^{(i)}) = (g_{n_j^{(i)}} \cdot x) \restriction B_3(1, m_j^{(i)})$; but we also have $y \restriction B_3(1, m_{j+1}^{(i)}) = (g_{n_{j+1}^{(i)}} \cdot x) \restriction B_3(1, m_{j+1}^{(i)})$, and $m_j^{(i)} < m_{j+1}^{(i)}$, so we get $(g_{n_j^{(i)}} \cdot x) \restriction B_3(1, m_j^{(i)}) = (g_{n_{j+1}^{(i)}} \cdot x) \restriction B_3(1, m_j^{(i)})$. So, since $x \in A$, we have $g_{n_j^{(i)}} \cdot x = g_{n_{j+1}^{(i)}} \cdot x$ for all j, i.e., $g_{n_0^{(i)}} \cdot x = g_{n_1^{(i)}} \cdot x = g_{n_2^{(i)}} \cdot x = \cdots$. It follows that $y \restriction B_3(1, m_j^{(i)}) = (g_{n_0^{(i)}} \cdot x) \restriction B_3(1, m_j^{(i)})$ for all j; since $m_j^{(i)} \to \infty$ as $j \to \infty$, we have $y = g_{n_0^{(i)}} \cdot x$, so $xE(F_2, 9^{F_3})y$ again. □

It remains to show that $E(F_2, 9^{F_3}) \restriction A$ is universal. For that we will show that

$$E(F_2, 2^{F_2}) \leq_B E(F_2, 9^{F_3}) \restriction A,$$

which is enough, since $E(F_2, 2^{F_2})$ is universal (see, e.g., Dougherty-Jackson-Kechris [**DJK**]).

LEMMA 4.3. *There is a Borel injection* $f : 2^{F_2} \to 9^{F_3}$ *with* $f(2^{F_2}) \subseteq A$ *which preserves the group action of* F_2 *(i.e., for all* $g \in F_2$ *and* $x \in 2^{F_2}$, $f(g \cdot x) = g \cdot f(x)$*). So in particular*

$$E(F_2, 2^{F_2}) \leq E(F_2, 9^{F_3}) \restriction A.$$

To prove this lemma, we will need the following technical sublemma.

SUBLEMMA. *For each $w \in F_2 \setminus \{1\}$, there is a Borel injection $f_w : 2^{F_2} \to 6^{F_2}$ which preserves the group action of F_2 and satisfies*

$$f_w(x)(g) = f_w(x)(gw) \implies g^{-1} \cdot x = w^{-1}g^{-1} \cdot x$$

for all $g \in F_2$ and $x \in 2^{F_2}$.

We will assume this and complete the proof.

PROOF OF LEMMA 4.3. Let $\{\alpha_1, \alpha_2\}$ be the generators of F_2 and $\{\alpha_1, \alpha_2, \alpha_3\}$ the generators of F_3. Define $f(x)$ for $x \in 2^{F_2}$ as follows:

(i) If $g \in F_2$, then $f(x)(g) = x(g)$.

(ii) If $g = h\alpha_3^{-p}g'$, with $h \in F_2$, $p > 0$, and g' not starting with $\alpha_3^{\pm 1}$, then $f(x)(g) = 2$.

(iii) If $g = h\alpha_3^p g'$, with h, g' as in (ii) and $p > 0$, $p \neq m_j^{(i)}$ for all i, j, then $f(x)(g) = 2$.

(iv) If $g = h\alpha_3^{m_j^{(i)}} g'$, with h, g' as in (ii), then $f(x)(g) = f_{w_j^{(i)}}(x)(h) + 3$, where

$$w_j^{(i)} = g_{n_j^{(i)}} g_{n_{j+1}^{(i)}}^{-1}.$$

It is easy to check that f is one-to-one and preserves the action of F_2. So it remains to verify that $f(x) \in A$.

So fix i, j with

$$(g_{n_j^{(i)}} \cdot f(x)) \upharpoonright B_3(1, m_j^{(i)}) = (g_{n_{j+1}^{(i)}} \cdot f(x)) \upharpoonright B_3(1, m_j^{(i)}).$$

If $d = \alpha_3^{m_j^{(i)}}$, then $d \in B_3(1, m_j^{(i)})$, so

$$f(x)(g_{n_j^{(i)}}^{-1} d) = f(x)(g_{n_{j+1}^{(i)}}^{-1} d),$$

thus

$$\begin{aligned} f_{w_j^{(i)}}(x)(g_{n_j^{(i)}}^{-1}) &= f_{w_j^{(i)}}(x)(g_{n_{j+1}^{(i)}}^{-1}) \\ &= f_{w_j^{(i)}}(x)(g_{n_j^{(i)}}^{-1} w_j^{(i)}). \end{aligned}$$

By the sublemma, $g_{n_j^{(i)}} \cdot x = (w_j^{(i)})^{-1} g_{n_j^{(i)}} \cdot x = g_{n_{j+1}^{(i)}} \cdot x$, so

$$\begin{aligned} g_{n_j^{(i)}} \cdot f(x) &= f(g_{n_j^{(i)}} \cdot x) \\ &= f(g_{n_{j+1}^{(i)}} \cdot x) \\ &= g_{n_{j+1}^{(i)}} \cdot f(x); \end{aligned}$$

since i, j were arbitrary, $f(x) \in A$. □

It remains to prove the sublemma.

PROOF OF SUBLEMMA. View F_2 as a rooted tree in the usual way (1 is the root of this tree, and there is an edge between g and $g\alpha_i$ for any group element g and generator α_i). Thus $x \in 2^{F_2}$ is a labeling of this tree using labels 0,1. Similarly for 6^{F_2}. Then $g^{-1} \cdot x$ is the same labeling except that the root of the tree is at g instead of 1. So the condition

$$\forall g\ [g^{-1} \cdot x \neq w^{-1}g^{-1} \cdot x \implies f_w(x)(g) \neq f_w(x)(gw)]$$

just means that if x, viewed from root g, is different from x viewed from gw, then the label of $f_w(x)$ at g is different from the label of $f_w(x)$ at gw. Moreover, to guarantee that $f_w(g' \cdot x) = g' \cdot f_w(x)$ for each $g' \in F_2$, we will make sure that the value of $f_w(x)$ at any g depends only on the labeling x viewed from root g (and not on g itself).

Given $x \in 2^{F_2}$ and $g \in F_2$, we have two cases:

(I) $g^{-1} \cdot x = w^{-1}g^{-1} \cdot x$, i.e., x looks the same from root g and root gw (note that this only depends on how x looks from root g).

Then put $f_w(x)(g) = \langle x(g), 0\rangle$, where $\langle , \rangle$ is a bijection of 2×3 with 6.

(II) $g^{-1} \cdot x \neq w^{-1}g^{-1} \cdot x$. So x looks different from roots g, gw. In particular there is a least $n = n_g(x)$ so that for some $i, j \in \mathbb{Z}$ and $h \in F_2$ of length n we have $x(gw^i h) \neq x(gw^j h)$. Clearly $n_{gw^i}(x) = n_g(x)$ for any integer i (note that $(gw^i)^{-1} \cdot x \neq w^{-1}(gw^i)^{-1} \cdot x$ as well).

The functions $p_j : B_2(1, n_g(x)) \to 2$ given by

$$p_j(h) = x(gw^j h)$$

are thus not all equal. So fix $p \in 2^{B_2(1,n_g(x))}$ with $Z = \{j \in \mathbb{Z} : p_j = p) \neq \varnothing$ and p least such (in some ordering of $2^{B_2(1,n_g(x))}$ fixed in advance). The value of p would be the same if we started with gw^i instead of g; the set $\tilde{Z}$ we would get from gw^i is a translate of Z ($j \in \tilde{Z}$ iff $j + i \in Z$).

Also $\{j \in \mathbb{Z} : p_j \neq p\} \neq \varnothing$. If Z has a largest element i_0, let $f_w(x)(g) = \langle x(g), 0\rangle$, if i_0 is even, and $f_w(x)(g) = \langle x(g), 1\rangle$, if i_0 is odd. If Z has no largest element but has a least element i_0, define $f_w(x)(g)$ the same way. Proceed similarly if $\mathbb{Z} \setminus Z$ has a least or largest element. So assume both Z and $\mathbb{Z} \setminus Z$ are unbounded in both directions. Put

$$Z' = \{j \in Z : j + 1 \notin Z\}.$$

Let finally $f_w(x)(g) = \langle x(g), 0\rangle$ if $0 \in Z'$, $f_w(x)(g) = \langle x(g), 1\rangle$ if $0 \notin Z'$, but the least positive element of Z' if odd, and $f_w(x)(g) = \langle x(g), 2\rangle$ if this least positive element is even.

This completes the definition of f; it is straightforward to verify that it has the desired properties. □

This completes the proof of Theorem 4.1. □

We conclude with another application of these ideas.

For a countable group G consider the shift action of G on X^G. We call $x \in X^G$ a **left-free** point if for all distinct $g, g' \in G$ there exists $h \in G$ such that $x(hg) \neq x(hg')$. We call $x \in X^G$ a **right-free** or just **free** point, if for all distinct $g, g' \in G$ there exists $h \in G$ such that $x(gh) \neq x(g'h)$; equivalently, $g \cdot x \neq g' \cdot x$ for $g \neq g'$, or simply $g \cdot x \neq x$ for all $g \neq 1_G$. Denote by LF the set of left-free points and F the set of free points. Note that LF and F are Borel G-invariant subsets of X^G. If G is abelian, clearly $LF = F$. But LF and F are very different for free groups in the following sense.

THEOREM 4.4. *The equivalence relation $E(F_3, 4^{F_3})\restriction LF$ is universal for countable Borel equivalence relations but $E(F_3, 4^{F_3})\restriction F$ is not.*

PROOF. The equivalence relation $E(F_3, 4^{F_3})\restriction F$ is not universal because it is treeable; see Kechris [**K2**]. For the first assertion we will show that $E(F_2, 2^{F_2}) \leq_B E(F_3, 4^{F_3})\restriction LF$.

Fix a left-free point z_0 in $\{2,3\}^{F_2}$. Define then $f : 2^{F_2} \to 4^{F_3}$ by:

(i) If $h \in F_2, f(x)(h) = x(h)$.

(ii) If $h \notin F_2$, express the reduced word for h in the form $h = h_1\alpha_3^{\pm 1}h'$ with $h' \in F_2$, and put $f(x)(h) = z_0(h')$.

It is easy to check that $xE(F_2, 2^{F_2})y \iff f(x)E(F_3, 4^{F_3})f(y)$. It remains to verify that $f(x) \in LF$. Let g and g' be distinct elements of F_3; we must find $h \in F_3$ such that $f(x)(hg) \neq f(x)(hg')$.

Consider two cases:

(1) $g^{-1}g' \in F_2$. Then let $p \in F_2$ be such that $z_0(p) \neq z_0(pg^{-1}g')$, and let h be such that $hg = \alpha_3 p$. Then $f(x)(hg) = f(x)(\alpha_3 p) = z_0(p) \neq z_0(pg^{-1}g') = f(x)(hgg^{-1}g') = f(x)(hg')$.

(2) $g^{-1}g' \notin F_2$. Let $h = g^{-1}$. Then

$$f(x)(hg) = f(x)(1) = x(1) \in \{0,1\}$$

but

$$f(x)(hg') = f(x)(g^{-1}g') = z_0(h') \in \{2,3\}$$

for some $h' \in F_2$. □

References

[A] S. Adams, *An equivalence relation that is not freely generated*, Proc. Amer. Math. Soc., **102** (1988), 565–566.

[AK] S. Adams and A. S. Kechris, *Linear algebraic groups and descriptive set theory*, preprint, 1999.

[ACH] A. Andretta, R. Camerlo, and G. Hjorth, *Conjugacy equivalence relation on subgroups*, preprint, 1999.

[DJK] R. Dougherty, S. Jackson, and A. S. Kechris, *The structure of hyperfinite Borel equivalence relations*, Trans. Amer. Math. Soc., **341** (1994), 193–225.

[DK] R. Dougherty and A. S. Kechris, *The universality of recursive isomorphism on* ω^ω, preprint, 1991.

[E] E. G. Effros, *Transformation groups and C**-*algebras*, Ann. of Math., **81(2)** (1965), 38–55.

[FM] J. Feldman and C. C. Moore, *Ergodic equivalence relations, cohomology and von Neumann algebras, I*, Trans. Amer. Math. Soc., **234** (1977), 289–324.

[HKL] L. Harrington, A. S. Kechris, and A. Louveau, *A Glimm-Effros dichotomy for Borel equivalence relations*, J. Amer. Math. Soc., **3** (1990), 903–928.

[JKL] S. Jackson, A. S. Kechris, and A. Louveau, *Countable Borel equivalence relations*, in preparation.

[K1] A. S. Kechris, *Amenable equivalence relations and Turing degrees*, J. Symb. Logic, **56** (1991), 182–194.

[K2] A. S. Kechris, *The structure of Borel equivalence relations in Polish spaces*, Set Theory of the Continuum, H. Judah, W. Just, H. Woodin, Eds., MSRI Publications, Vol. 26, Springer-Verlag, New York, 1992, 89–102.

[K3] A. S. Kechris, *New directions in descriptive set theory*, Bull. Symb. Logic, **5** (1999), 161–174.

[KM] A. S. Kechris and Y.N. Moschovakis (Eds.), *Cabal Seminar 76–77*, Lecture Notes in Math., vol. 689, Springer-Verlag, 1978.

[N] M. G. Nadkarni, *On the existence of a finite invariant measure*, Proc. Indian Acad. Sci. Math. Sci., **100** (1990), 203–220.

[SS] T. Slaman and J. Steel, *Definable functions on degrees*, Cabal Seminar, 81–85, Lecture Notes in Math., vol. 1333, Springer-Verlag, 1988, 37–55.

Department of Mathematics, Ohio State University, Columbus, Ohio 43210
E-mail address: `rld@math.ohio-state.edu`

Department of Mathematics, California Institute of Technology, Pasadena, California 91125
E-mail address: `kechris@caltech.edu`

Contemporary Mathematics
Volume **257**, 2000

Questions in Computable Algebra and Combinatorics

Rod Downey and J. B. Remmel

1. Introduction

Computable mathematics studies the effective content of theorems and constructions in mathematics. There have been extensive studies of the effective content of theorems and constructions of a wide variety of results from algebra, analysis, model theory and combinatorics. A good general introduction to the study of computable mathematics can be found in the two volume set "Handbook of Recursive Mathematics: vol. 1 and 2" edited by Yu. L. Ershov, S. S. Goncharov, A. Nerode, and J. B. Remmel [**42**].

In computable mathematics, one most often assumes that the underlying structures are computable, i.e. the universe of the structure is a computable set of natural numbers and the relations and functions of the structure are uniformly computable. One then uses the tools of modern computability theory to study the effective content of theorems and constructions on these computable structures. However, as we shall see, there are a number of interesting results in computable mathematics which are concerned with more general structures such as Σ_n^0 structures where the underlying universe, relations, and functions are Σ_n^0 as well as a number of interesting results concerning more restricted structures such as polynomial-time structures where the underlying universe is a polynomial-time subset of $\{0,1\}^*$ and the relations and functions are restrictions of polynomial-time relations and functions.

In this article, we will focus on two areas of computable mathematics, namely computable algebra and combinatorics. The goal of this article is to present a number of open questions in both computable algebra and computable combinatorics and to give the reader a sense of the research activity in these fields. Our philosophy is to try to highlight questions, whose solutions we feel will *either* give insight into algebra or combinatorics, *or* will require new technology in the computability-theoretical techniques needed. A good historical example of the first phenomenon is the word problem for finitely presented groups which needed the development of a great deal of group theoretical machinery for its solution by Novikov [**110**] and Boone [**10**]. A good example of the latter phenomenon is the recent solution by Coles, Downey and Slaman [**17**] of the question of whether all rank one torsion free

1991 *Mathematics Subject Classification.* Primary 03D45; Secondary 03D25.
Downey's research is supported by the Marsden Fund of New Zealand.

Abelian groups of finite type have first jump degree which relied on the discovery that every set has a least jump enumeration[1].

In view of this, in some instances we will try to comment on why we feel that the relevant question is interesting. Also, where possible we will try to give an attribution of the question. Any question without such an attribution is either ours or folklore. While we have done our best to get views of others, our article is necessarily a reflection of our own view of the mathematical universe, and hence we apologise in advance to those working in areas to which we have not given due justice.

In a short article such as this we cannot hope to provide all the theoretical background which leads to our open questions, but we will give enough to frame the questions.

Our notation is standard and follows Soare [**144**].

2. Preliminaries

2.1. Presentations. Since we are concerned with questions of effectivity in algebraic and combinatorial structures, we need to be concerned with how we present the relevant structures.

Consider a model, or, more precisely, a structure $\mathcal{A} = (A, \{R_i^{\mathcal{A}}\}_{i\in S}, \{f_i^{\mathcal{A}}\}_{i\in T}, \{c_i^{\mathcal{A}}\}_{i\in U})$ where S, T, and U are initial segments of the natural numbers $\omega = \{0, 1, \dots\}$. We let $\phi_{e,n}$ denote the partial computable function of n-variables computed by the e-th Turing machine in some standard enumeration of Turing machines.

DEFINITION 2.1. $\mathcal{A}$ is *computable* if

(i) A is a computable subset of ω.
(ii) The set of relations $\{R_i^{\mathcal{A}}\}$ is uniformly computable, i.e. there exists a computable function G such that $G(i) = [e_i, n_i]$ where $R_i^{\mathcal{A}}$ is the n_i-ary relation whose characteristic function is computed by ϕ_{e_i,n_i}.
(iii) The set of functions $\{f_i^{\mathcal{A}}\}$ is uniformly computable, i.e. there exists a computable function F such that $F(i) = [e_i, n_i]$ where $f_i^{\mathcal{A}} = \phi_{e_i,n_i}|A^{n_i}$.
(iv) There is a computable function H such that $H(i) = c_i^{\mathcal{A}}$.

Thus $\mathcal{A}$ is computable iff its atomic diagram is computable. We say that a computable structure $\mathcal{A}$ is *decidable* if its complete diagram is computable and we say that $\mathcal{A}$ is Σ_n*-decidable* if its Σ_n-diagram is computable. We say that two computable structures $\mathcal{A} = (A, \{R_i^{\mathcal{A}}\}_{i\in S}, \{f_i^{\mathcal{A}}\}_{i\in T}, \{c_i^{\mathcal{A}}\}_{i\in U})$ and $\mathcal{B} \doteq (B, \{R_i^{\mathcal{B}}\}_{i\in S}, \{f_i^{\mathcal{B}}\}_{i\in T}, \{c_i^{\mathcal{B}}\}_{i\in U})$ are *computably isomorphic* if there is an isomorphism $f : A \to B$ which is a partial computable function. It is quite easy to see that any infinite computable structure $\mathcal{A}$ is computably isomorphic to a computable structure $\mathcal{B}$ whose universe $B = \omega$. Since computable mathematics is only interested in results which are invariant under computable isomorphisms, there is no loss in generality in making the convention that all computable structures have universe ω. Thus, for example, a linear ordering is computable precisely if its ordering relation is a computable relation on ω.

As mentioned in the introduction, one often works with more general structures in computable algebra and combinatorics. For any structure $\mathcal{A}$ whose universe A

[1]That is, for all sets A, the collection $\{\deg(B)' : A \text{ is computably enumerable in } B\}$ always has a least element.

is a subset of ω, one can assign a degree to $\mathcal{A}$, $deg(\mathcal{A})$, as the join of all the degrees of $\{R_i^{\mathcal{A}} : i \in S\}, \{f_i^{\mathcal{A}} : i \in T\}, \{c_i^{\mathcal{A}} : i \in U\}$. One can also look at this classification of structures "descriptively". That is, we say that $\mathcal{A}$ is Σ_n^0 *presented* or, simply, Σ_n if its atomic diagram is Σ_n. Again, there is no loss in generality in assuming that a Σ_n structure has universe ω. Thus a linear ordering L with domain ω is Σ_n iff its ordering relation $\preceq_L$ is a Σ_n^0 relation. We should note that a special case of Σ_1 structures arises naturally in many situations in algebra. Namely, one often considers the quotient structures, e.g. groups modulo a normal subgroup, rings or Boolean algebras modulo an ideal, vector spaces modulo a subspace. If the underlying structure is computable, but the normal subgroup, ideal, subspace etc. is only computably enumerable (c.e.), then one is naturally led to consider structures $\mathcal{A} = (A, =^{\mathcal{A}}, \{R_i^{\mathcal{A}}\}_{i\in S}, \{f_i^{\mathcal{A}}\}_{i\in T}, \{c_i^{\mathcal{A}}\}_{i\in U})$ where A, $R_i^{\mathcal{A}}$, $f_i^{\mathcal{A}}$ and the underlying equality relation $=^{\mathcal{A}}$ are c.e. In such cases, we shall often refer to these structures as *c.e. presented* structures.

We should note that there is rich area of computable algebra and combinatorics where one considers more restricted resource bounded versions of computation such as polynomial-time, exponential time, polynomial-space, etc. For example, we say that a structure $\mathcal{A} = (A, \{R_i^{\mathcal{A}}\}_{i\in S}, \{f_i^{\mathcal{A}}\}_{i\in T},\{c_i^{\mathcal{A}}\}_{i\in U})$ is **polynomial-time** if

1. A is a polynomial-time computable subset of $\{0,1\}^*$.
2. The set of relations $\{R_i^{\mathcal{A}}\}$ is uniformly polynomial-time computable, i.e. there exists a computable function G such that $G(i) = [n_i, e_i]$ where $R_i^{\mathcal{A}}$ is the n_i-ary relation whose characteristic function is computed by ϕ_{n_i,e_i} and there exists a computable function G' such that $G'(i) = m_i$ where for all $(x_1, \dots, x_{n_i}) \in \{0,1\}^{n_i}$, it takes at most $(max\{2, |x_1|, \dots, |x_{n_i}|\})^{m_i}$ steps to compute $\phi_{n_i,e_i}(x_1, \dots, x_{n_i})$.
3. The set of functions $\{f_i^{\mathcal{A}}\}$ is uniformly polynomial-time computable, i.e. there exists a computable function F such that $F(i) = [n_i, e_i]$ where $f_i^{\mathcal{A}}$ is the n_i-ary function computed by ϕ_{n_i,e_i} and there exists a computable function F' such that $F'(i) = m_i$ where for all $(x_1, \dots, x_{n_i}) \in \{0,1\}^{n_i}$, it takes at most $(\max\{2, |x_1|, \dots, |x_{n_i}|\})^{m_i}$ steps to compute $\phi_{n_i,e_i}(x_1, \dots, x_{n_i})$.
4. There is a polynomial-time computable function H such that $H(i) = c_i^{\mathcal{A}}$.

Exponential time, polynomial-space, and exponential space structures can be defined in a similar manner.

We note that if $\mathcal{A}$ is polynomial-time computable, it does not automatically follow that its atomic diagram is polynomial-time computable. Thus we say that $\mathcal{A}$ is *uniformly polynomial-time computable* iff its atomic diagram is polynomial-time computable. We say that two polynomial-time structures $\mathcal{A} = (A, \{R_i^{\mathcal{A}}\}_{i\in S}$, $\{f_i^{\mathcal{A}}\}_{i\in T}$, $\{c_i^{\mathcal{A}}\}_{i\in U})$ and $\mathcal{B} = (B, \{R_i^{\mathcal{B}}\}_{i\in S}$, $\{f_i^{\mathcal{B}}\}_{i\in T}$, $\{c_i^{\mathcal{B}}\}_{i\in U})$ are *polynomial-time isomorphic*, if there is an isomorphism $f : A \to B$ such that both f and its inverse f^{-1} are the restrictions of polynomial-time functions. We note that it is not the case that any two polynomial-time sets are polynomial-time isomorphic. For example, if we let $Bin(\omega)$ denote the set of binary representations of the natural numbers and $Tal(\omega) = \{0\}^*$ denote the tally representation of the natural numbers, then $Bin(\omega)$ and $Tal(\omega)$ are not polynomial-time isomorphic. Thus unlike the case of computable structures, we cannot just restrict ourselves to the consideration of structures whose underlying universe is $\{0,1\}^*$ or $Bin(\omega)$ or $Tal(\omega)$. We will look further at polynomial-time structures in Section 8.

Given any notion of effective structures, there is often a natural notion of effective categoricity associated with it. Below we give the definitions of computable and polynomial-time categoricity. We note that by our remarks above, the notion of polynomial-time categoricity naturally depends on the underlying universe.

DEFINITION 2.2. (i) We say a computable structure $\mathcal{A}$ is *computably categorical* if all computable structures $\mathcal{B}$ which are isomorphic to $\mathcal{A}$ are computably isomorphic to $\mathcal{A}$.
(ii) We say a p-time structure $\mathcal{A}$ with universe A is *p-time categorical over A* if all p-time structures $\mathcal{B}$ with universe A which are isomorphic to $\mathcal{A}$ are p-time isomorphic to $\mathcal{A}$.

Finally, we should note that the Russian school of computable mathematics developed slightly different notions of computable structures. That is, in the Russian literature, one distinguishes the underlying abstract structure $\mathcal{A}$ from its various presentations. This point of view leads to the following definitions. Given a model,

$$\mathcal{A} = (A, \{R_i^{\mathcal{A}}\}_{i\in S}, \{f_i^{\mathcal{A}}\}_{i\in T}, \{c_i^{\mathcal{A}}\}_{i\in U})$$

where S, T, and U are initial segments of ω, a **constructivization (strong constructivization)** of $\mathcal{A}$ is a surjective mapping $\alpha : \omega \to A$ such that for any atomic formula (resp. *any* formula) ϕ, the relations

$$\{\overline{a} : \mathcal{A} \models \phi(\alpha(\overline{a}))\}$$

are computable uniformly in ϕ. It is easy to see that $\alpha(\mathcal{A})$ is a computable model if α is a constructivization of $\mathcal{A}$ and is a decidable model if α is a strong constructivization. Thus it is easy to translate results about computable and decidable models into the language of constructivizations and strong constructivizations and vice versa.

2.2. Basic Questions. In this paper, we will be concerned mainly with well-studied classical structures and theorems in algebra and combinatorics. Thus for computable algebra, we will be concerned with groups, rings, fields, orderings, etc. and for computable combinatorics we will consider classical results about graphs, colorings, matchings, etc. Our general concern will be loosely based on interpreting the following basic types of questions in the various arenas.

(i) *Study the presentations of a structure.*
What is the set of degrees of presentations of a structure? Compare and contrast c.e. presented vs. computable structures, computable vs. n-decidable structures, etc. Thus we try to understand the relationship between classical isomorphism-type and effective isomorphism-type. Richter's Ph.D. Thesis [**125**] was ground-breaking here.

An important category here consists of the *spectrum* problems, which looks at the set of degrees within an isomorphism-type, or the degrees of images of a relation on a structure. Ash-Nerode [**5**] and Goncharov [**55**] are good examples of these types of considerations.

A related issue is the question of the *number* of computable, decidable, etc. models. For instance, is a structure, like the rationals as an ordering, computably categorical in the sense that it has only one computable model up to computable isomorphism? This is an example of a *computable dimension* problem. How many degrees can a relation on a computable model have? For instance, Downey and Moses [**35**] constructed a computably presented linear ordering for which the adjacency relation is *intrinsically complete*, whereas Hirschfeldt [**63**] has proven that a computable relation on a computably presented linear ordering with a spectrum with at least two elements has infinitely many elements.

Clearly, since we are considering *effective* structures, it is entirely natural to wish to understand how our classification tool, *effective* isomorphism, relates to the historical one. Aside from this intrinsic interest, *another* good reason for such analyses is that many of the remaining questions will need insight into the model theory of the relevant structures to solve them. For instance, it is unknown whether there is a low_n Boolean algebra with no computable copy, as we will see. This question would seem purely technical, but seems to need a complete understanding of a reasonable set of invariants for countable Boolean algebras. A related question is whether there is a Boolean algebra of rank 1 which is c.e. presented and not isomorphic to a computably presented one. This question was solved by Downey and Jockusch [**30**], by demonstrating a new algebraic construction for folding essentially any countable Boolean algebra into a rank 1 one in such a way that the original one can be recovered by two quantifiers.

(ii) *Study the effective content of classical theorems.* Two nice classical theorems are that every field has an algebraic closure, and that any two algebraic closures of the same field are isomorphic. Rabin [**117**] proved that any computably presented field has a computably presented algebraic closure, but Frölich and Shepherdson [**50**] proved that they are not necessarily computably unique. What's going on here?

In some sense the explanation was provided by Metakides and Nerode [**99, 100**] who showed that uniqueness happened iff there was an additional effectivity condition on the field, namely a *separable splitting algorithm.* If there was such an algorithm, then one could construct the algebraic closure by the usual sequential addition of roots, and this construction gives computable uniqueness. Therefore the Rabin construction actually constructs the algebraic closure *using a different method.*

Another nice example of this phenomenon stems from Kalantari's (and others (e.g. [**20**])) investigations of Stone's separation theorem in [**72**]. There one looked at a computable vector space over a computable ordered field with *both* a computable dependence algorithm and a computable algorithm to decide if a vector lay in the convex closure of a finite set of vectors. The question naturally arose as to whether both of these conditions were necessary. This question was answered by Nevins [**109**], who demonstrated that either implies the other.

Notice that this program has several subprograms.

(iia) *If a theorem is not effectively true, how can we modify the theorem to make it effectively true?* One example is the result mentioned in the preceding paragraph. Another example would be the classical observation that every infinite linear ordering has either an infinite ascending chain or an infinite descending chain. Tennenbaum proved that this can fail if we ask that everything be computable. There are several analogues of this theorem which don't fail. For instance, Manaster (see [**23**]) proved that there is either an infinite Π_1 ascending or descending sequence. Rosenstein [**131**] proved that there must be a computable subsequence of type ω, ω^*, $\omega + \omega^*$ or $\omega + \zeta\eta + \omega^*$, and Lerman [**91**] proved that none from this list could be deleted.

(iib) *If a theorem is effectively false, can we measure how badly it fails?* A good instance is the result that every formally real field admits an ordering. Metakides and Nerode [**100**] proved that the cones of orderings of computable formally real fields are precisely the Π^0_1 classes.

(iic) *Study the set of "solutions" to a computable instance of a combinatorial problem*, e.g.: Can the set of k-colorings of a computable graph represent an arbitrary Π^1_0-class?

We remark that yet another good reason for this type of investigation is that there are usually reverse mathematical spinoffs. Here one uses proof theory in the form of subsystems of second-order arithmetic to analyse and calibrate the theorems of classical mathematics. An immediate corollary to the proof of the quoted result of Metakides and Nerode is that the statement that every countable formally real field has an ordering is equivalent over RCA_0 to WKL_0 (See Friedman, Simpson, Smith [**47**]). Generally most results in reverse mathematics have, at their hearts, a computability theory construction. In many ways, the two subjects are simply reflections of the same basic philosophy, and enrich each other. In fact, many good questions in computable algebra and combinatorics are motivated by questions from reverse mathematics.

Ramsey's theorem for pairs and two colours, $RT(2,2)$, provides an excellent illustration of this phenomenon. It is unknown exactly what its proof theoretical strength is. In his Ph.D. Thesis, Hirst [**65**] used a computability-theoretical argument to show that it is not provable in WKL_0. Does it imply WKL_0? It was not known till the work of Seetapun (see Hummel [**67**] or Seetapun and Slaman [**136**]) that it was not as strong as ACA_0. Seetapun used a complicated forcing argument, later modified by Jockusch to a simpler forcing argument, to construct a computable coloring of pairs that avoided $\mathbf{0}'$, and concluded that $RT(2,2)$ did not imply ACA_0. We refer the reader to Simpson's article in this volume (Simpson, [**139**]) for background and an excellent source of questions of this ilk.

(iii) *Study the lattice of c.e. substructures of a computable structure, e.g. subspaces of a vector space, subfields of a field, ideals or filters of a Boolean algebra, subalgebras of a Boolean algebra, order ideals of a poset, etc.* There is a persistent intuition going back to Post that structural properties of the poset of c.e. sets have reflections in computability properties. For instance, the high degrees are exactly the degrees of maximal sets, and every double-jump class is an invariant class for the c.e. sets.

It seems reasonable to try to similarly understand computation via other lattices, to compare with the c.e. sets and to clarify their relationship with the degrees. Furthermore, since classically the lattices of (e.g.) subgroups of groups give insight into groups, one would expect the same from the effective versions.

For instance, Metakides and Nerode [**99**] constructed a computable coinfinite dimensional subspace V of V_∞ such that every c.e. set independent of V is finite. The quotient space V_∞/V is then an infinite dimensional computable vector space, the only c.e. independent subsets of which are finite. Friedman, Simpson and Smith [**47**] constructed a computable infinite dimensional vector space, such that every infinite independent set computes the halting problem. Shore [**137**] proved that given any effective sequence of c.e. degrees $\mathbf{a}_1, \mathbf{a}_2, \ldots.$ and a degree $\mathbf{a}$ above all the $\mathbf{a}_i$, there is a computable vector space V such that computing dependence for k-tuples (k-fixed) has degree $\mathbf{a}_k$, and computing dependence has degree $\mathbf{a}$. Thus, for instance, it is possible to have a vector space where for any fixed k we can compute dependence for k-tuples, but the complexity of computing dependence if k is allowed to vary is that of the Halting problem.

Another illustration is provided by the recent work of Cholak, Coles, Downey, and Herrmann [**15**], who proved that for the lattice of ideals of the free Boolean algebra, the analogous notion to maximal set provides an invariant class in the c.e. degrees consisting of the *array noncomputable* degrees.

3. Linear Orderings and Boolean algebras

We begin with the spectrum problem. Recall the definition originally due to Harizanov for relations.

DEFINITION 3.1. (i) Given a structure $\mathcal{A}$,

$$Spec(\mathcal{A}) = \{deg(B) : B \text{ is isomorphic to } \mathcal{A}\}.$$

(ii) Given a relation R on a computable structure $\mathcal{A}$, the *spectrum* of R is the set

$$\{\deg(\widehat{R}) : \text{there is a computable } \widehat{\mathcal{A}} \cong \mathcal{A} \text{ with } \widehat{R} \text{ the image of } R\}.$$

The basic question is:

QUESTION 3.2. (i) What spectra can linear orderings have?
(ii) Can we classify the orderings that have computable copies? decidable copies? etc. Here we mean a classification by order-type.

We know that spectra of linear orderings are closed upwards (Knight [**86**]). We know that if $\mathbf{a}$ is in spec(L) then there is an ordering $\widehat{L} \cong L$ with degree forming a minimal pair with $\mathbf{a}$ and hence if L does not have a computable copy then there is no minimum in the spectrum. What else can be said? An important test case is the following question.

QUESTION 3.3 (Downey). Is there a linear ordering with no computable copy whose spectrum contains all nonzero degrees?

Of relevance here is the following theorem of Miller.

THEOREM 3.4 (Miller [**102**]). *There is a linear order $\mathcal{L}$ such that $Spec(\mathcal{L})$ contains all nonzero Δ_2^0 degrees but does not contain* $\mathbf{0}$.

The method used by Miller is a modification of an earlier technique of Jockusch and Soare [**69**]. It builds a certain ordering which only relies on "permissions" to work, and always gives the same order-type provided that we get such permisssions. This technique might be able to be pushed to hyperimmune degrees, but otherwise we seem to need a genuinely new idea.

Slaman [**140**] and Wehner [**151**] independently produced a structure $\mathcal{A}$ such that $Spec(\mathcal{A})$ consists of all nonzero degrees. Work of Hirschfeldt, Khoussainov,

Shore and Slinko [**64**] shows that $\mathcal{A}$ can be a graph, a lattice, an integral domain, or an Abelian group. Downey and Jockusch [**29**] show that $\mathcal{A}$ cannot be a Boolean algebra. Miller also asked the following question.

QUESTION 3.5 (Miller). Can a spectrum of a linear ordering separate the high degrees H_1 from the low degrees L_1?

Notice that the refined notions of a spectrum considered by Miller gives rise to differing versions. For a class C of degrees the *C-spectrum* is $Spec(L) \cap C$. If C is the class of c.e. degrees (say) then the *strong* c.e. spectrum is the class of c.e. degrees realized by c.e. presented models isomorphic to $\mathcal{L}$. It is not too difficult to show that the strong c.e. spectrum and the c.e. spectrum coincide for *linear orderings*.

QUESTION 3.6. (i) What is the relationship between the c.e. spectrum and the strong c.e. spectrum of a c.e. structure?
(ii) What are the possible spectra of the adjacency relation for a computable linear ordering $\mathcal{L}$? We have seen it can have one element, and it can have all c.e. degrees except $\mathbf{0}$ (Downey [**24**]). But must it have either one or infinitely many degrees, etc?
(iii) (Hirschfeldt) In general, if a relation R on a linear ordering has a spectrum with more than one element, must it have an infinite spectrum? (In his thesis, Hirschfeldt proved that the answer is yes if the spectrum contains a computable copy of L with R computable.)

Related here is the question of whether there can be relations on orderings with one element spectra where that element is not computable from $\mathbf{0}^{\alpha}$ for some α. Can, for instance, the adjacency relation have a one element spectrum and it not be a member of $\{\mathbf{0}, \mathbf{0}'\}$? An even more interesting question is whether one can have a nontrivial single element spectrum for a relation on a computable Boolean algebra.

Boolean algebras and linear orderings are related by the Stone construction. That is, each Boolean algebra is effectively isomorphic to $Intal(L)$, the algebra of left open right closed intervals of a linear ordering L.

QUESTION 3.7. (i) What is the relationship between the degree of B and the degree of L with $B \cong_{eff} Intal(L)$?
(ii) Given a computable Boolean algebra $A = Intal(L)$ for a computable linear ordering L and a computable $\widehat{B} \cong B$, is there a computable copy $\widehat{L}$ of L such that $\widehat{B} \cong_{comp} Intal(\widehat{L})$. The answer here is probably not. But what can be said?

Exact knowledge of the above would give insight into how to transfer some results from linear orderings to Boolean algebras. There is definite complexity here. For instance, Jockusch and Soare [**69**] constructed a low linear ordering not isomorphic to a computable one. However, Downey and Jockusch [**29**] proved that every low Boolean algebra is isomorphic to a computable one. Later Thurber [**149**] improved this to low$_2$ and then Knight and Stob [**88**] to low$_4$.

QUESTION 3.8. Is every low$_n$ Boolean algebra isomorphic to a computable one?

This question would seem purely technical, but its solution either way would seem to need much greater understanding of countable Boolean algebras and their invariants than we currently possess.

Of course, there are analogous questions for Boolean algebras concerning spectra.

QUESTION 3.9. (i) What are the possible spectra of a Boolean algebra? By the work of Feiner [**43, 44**] we know that a Boolean algebra can be intrinsically nonlow$_n$ for all $n \in \omega$ (that is, not isomorphic to any low$_n$ Boolean algebra). What else can be said?
(ii) What is the possible value for the spectrum of the atomicity relation for a computable Boolean algebra. In the Stone construction, adjacencies and atoms coincide, so the spectrum of the atomicity relation must be built as a union of spectra for adjacencies. Downey [**21**] proved that each spectrum must contain an incomplete degree, and Remmel [**120**], Theorem 2.12(ii), proved that such spectra are closed upwards.
(iii) **(Hirschfeldt)** If a relation on a Boolean algebra has a spectrum with at least 2 elements does it have to have infinitely many? What if the relation has a computable copy?

Before we turn to questions intrinsic to Boolean algebras, we consider some questions more specific to orderings. Recall the following result. A self-embedding of an ordering $\mathcal{L}$ is an isotone injective map $f : \mathcal{L} \mapsto \mathcal{L}$.

THEOREM 3.10 (Dushnik and Miller [**41**]). *1) Every countable infinite linear order has a nontrivial self-embedding.*
2) There is a linear order which is isomorphic to a dense subset of the reals which has no nontrivial self-embedding.

The effective content of this theorem has been examined by Hay and Rosenstein (see Rosenstein [**131**]), Downey [**23**], and Downey and Lempp [**34**]. Let $f : \mathcal{Q} \mapsto \omega$. We say that an ordering L is *η-like* if $L = \Sigma_{q \in \mathcal{Q}} f(q)$. That is, L is obtained from the rationals by replacing points with finite blocks. We say that the ordering is *strongly* η-like if the range of f is bounded.

CONJECTURE 3.11 (Downey and Moses [**36**]). Every computable copy of a computable linear order $\mathcal{L}$ has a nontrivial computable self-embedding iff $\mathcal{L}$ contains a closed interval which is strongly η-like.

We remark that by Downey [**23**] there is no *uniform* proof of this conjecture. There are a number of related questions concerning nontrivial automorphisms and categoricity. An automorphism $f : L \mapsto L$ is called strongly nontrivial if for all x, $[x, f(x)]$ has infinitely many points.

CONJECTURE 3.12 (Kierstead [**83**]). Every computable copy of an ordering L has a strongly nontrivial Π_1 automorphism iff L has an interval of type η.

There are a number of related more technical questions. There are basic questions also, which seem fairly unapproachable.

QUESTION 3.13. Classify, by order-type those computable orderings $\mathcal{L}$ such that
(i) Every computable copy of $\mathcal{L}$ has a Π_n nontrivial automorphism.
(ii) Every computable copy of $\mathcal{L}$ has a nontrivial Π_n self-embedding.
(iii) $\mathcal{L}$ is Δ^0_n categorical. (For $n = 1$ the answer is that $\mathcal{L}$ has a finite number of adjacencies (Remmel [**118**], Goncharov [**54**]), and the answer is known for $n = 2$ by McCoy [**97**].)

In a slightly differing direction, Moses [**105**] proved the following.

THEOREM 3.14 (Moses [**105**]). *A relation R is intrinsically computable*[2] *in a computable linear ordering iff it can be defined by a quantifier free formula with finitely many parameters.*

This naturally leads to the question below.

QUESTION 3.15 (Moses). Can one characterize via some definability criterion the intrinsically computable relations in decidable (Σ_1 decidable, etc.) linear orderings?

Moses also asked some specific question related to spectra of adjacencies and decidable models.

QUESTION 3.16 (Moses). (i) Can one characterize the computable linear orderings with intrinsically complete successivities?
(ii) Can one characterize the computable linear orderings which have no computable copy with a computable set of successivities?
(iii) Can one characterize the computable linear orderings which have no decidable copy?

Naturally there are the related questions for atoms in Boolean algebras.

We finish this section with some questions which seem more specific to Boolean algebras. We recall the following definitions.
1) Let B be a Boolean algebra.
(i) $At(B)$ = set of atoms of B,
(ii) $Atl(B)$ = set of atomless elements of B,
(iii) $IAt(B)$ = ideal generated by the set of atoms of B,
(iv) $F_0(B) = \{0_B\}$
$F_1(B) = IAt(B)$
$F_{\alpha+1}(B) = \{x \in B : x/F_\alpha(B) \in IAt(B/F_\alpha(B))\}$
$F_\beta(B) = \bigcup_{\alpha<\beta} F_\beta(B)$ if β a limit ordinal.
(v) B is *atomic* if every element of B has an atom below it.
(vi) B is α-atomic if $B/F_\beta(B)$ is atomic for all $\beta < \alpha$.

QUESTION 3.17. What can be said about the theory of Boolean algebra in extended logics?

For instance, we know:
a) **Goncharov** [**53**] Any ω-atomic computable Boolean algebra is isomorphic to a decidable Boolean algebra.

b) **Pinus** [**115**] Any ω-atomic computable Boolean algebra is isomorphic to an $L(Q)$-decidable Boolean algebra where $L(Q)$ is the extension of the first order language by the quantifier Q = there exist infinitely many.

c) **Ershov (1973)** Any superatomic[3] computable Boolean algebra is isomorphic to an $L(I)$-decidable Boolean algebra where $L(I)$ is the restriction of the second order language whose unary predicate variables range over ideals.

While we are at it, we mention one long-standing question of Rosenstein.

---

[2]Recall that a relation on a structure $\mathcal{A}$ is *intrinsically computable* if it is computable in each computable copy of $\mathcal{A}$.

[3]That is, every sublgebra is atomic.

QUESTION 3.18 (Rosenstein). Suppose that T is a complete theory of linear orderings with a computable model and a prime model. Must T have a computable prime model?

4. Extensions of Partial Orderings to Linear Ones.

One of the basic theorems about partial orderings is the fact that they all have linear extensions. That is, given a poset $(P, \preceq)$ there is a linear ordering $(L, \leq)$ with the same field P such that for all x, y, $x \prec y$ implies $x < y$. This result is easily seen to be effectively true. This short section will look at several questions emanating from refinements of this theorem.

It is known that any well-founded partial ordering has a well-founded linear extension. The best way to think of this is that if we cannot embed ω^* into $(P, \preceq)$, then there is a linear extension also free of ω^*. Bonnet [**9**] has completely classified the linear order-types τ such that if $(P, \preceq)$ is τ-free, then there is an extension $(L, \leq)$ which is also τ-free. (Actually, this classifications uses higher cardinalities. Hence, one should look at Jullien's thesis [**71**] for a similar classification for countable orderings.)

QUESTION 4.1 (Downey). What is the effective content of the Bonnet-Jullien result? Is there an effective analogue?

Of particular interest is the result for $(P, \preceq)$ scattered. (That is, we cannot embed η into $(P, \preceq)$.) Slaman and Woodin [**141**] proved that those posets with dense extensions are essentially unclassifiable since the set is not Borel. Are there similar results for other constraints on extensions? We remark that the classical proofs of Bonnet's theorem that η can be avoided, part of the Bonnet-Jullien classification, are proof-theoretically the same as Π^1_2-CA. It would be remarkable if the result was equivalent. Downey, Hirschfeldt, Lempp and Solomon [**27**] have results in this area.

One can also look at computable versions. Thus, a poset can be called *computably well-founded* if one cannot computably embed an ω^* sequence into it. Although every well-founded computable poset has a computable well-founded extension (Kierstead and Rosenstein, see Rosenstein [**132**]), there exist computable computably well-founded posets with no computably well-founded computable linear extension.

QUESTION 4.2 (Rosenstein). (i) What is the situation for other properties, e.g. it is known by Downey, Hirschfeldt, Lempp, and Solomon [**27**] that there is a $(P, \preceq)$ which is computably scattered yet no computable linear extension is computably scattered.
(ii) If $(P, \preceq)$ is computably well-founded (resp. computably scattered), then it has a Δ_2 computably well-founded (resp. computably scattered) linear extension (see [**132**]). Can this be sharpened to, for instance, Π_1?

QUESTION 4.3 (Downey). Classify the computable partial orderings with precisely one computable linear extension.

There is a whole host of other questions of similar ilk we could examine and generalize up the arithmetical hierarchy, but we will confine ourselves to the above and return to posets when we look at computable combinatorics in Section 5.

5. Computable Combinatorics

We begin with some basic definitions. Let $G = (V, E)$ be a graph[4] and $x \in V$.
1) $N(x) = \{y \in V : (x, y) \in E\}$,
$\delta(x) = |N(x)|$,
$\Delta(G) = max\{\delta(x) : x \in V\}$.
2) G is *computable* if V and E are computable.
3) G is *highly computable* if V, E, and $N(x)$ are computable.
4) G is A-*computable* if V and E are computable and $N(x)$ is A-computable (Gasarch and Lee [**51**]).
5) $\omega(G) = max\{n : G$ contains an n-clique$\}$.
6) $\chi(G) = min\{n : G$ is n-colorable$\}$,
$\chi^c(G) = min\{n : G$ is computably n-colorable$\}$,
$\chi^p(G) = min\{n : G$ is polynomial-time n-colorable$\}$, etc.
7) A graph is *perfect* if for every induced subgraph H, $\omega(H) = \chi(H)$.

There have been a lot of theorems about computable analogues of theorems for graphs (and posets). To frame our question we recall some of those results.

THEOREM 5.1 (Bean [**7**]). (i) *Let a, b be such that $2 \leq a < b \leq \infty$. Let X be an infinite computable set. There exists a computable graph $G = (V, E)$ such that $\chi(G) = a$, $\chi^c(G) = b$, and $V \subseteq X$. If $a \leq 4$ then G can be taken to be planar.*
(ii) *There exists a computable graph G such that $\chi(G) = 2$, $\chi^c(G) = \infty$, and G is planar.*

THEOREM 5.2 (Gasarch and Lee [**51**]). *For any noncomputable c.e. set A, there exists an A-computable graph G such that $\chi(G) = 2$, $\chi^c(G) = \infty$, and G is planar.*

QUESTION 5.3. Gasarch asked if the theorem is true for all $A \leq_T 0'$. Jockusch observed that any such A has a noncomputable c.e. set $B \leq_T A$. For graphs which are computable relative to A, Theorem 5.2 is true for degrees which contain a hyperimmune set.

What can be said about general $A \leq_T \emptyset'$?

THEOREM 5.4 (Remmel [**122**]). *For every computable tree T, there is a computable k-colorable graph G such that there is a degree-preserving 1:1 correspondence between the infinite paths through T and the set of k-colorings of G (up to a permutation of the colors).*

QUESTION 5.5. Is the theorem true for restricted classes of graphs? E.g. planar graphs, comparability graphs, co-comparability graphs, subgraphs of a given graph, etc.

THEOREM 5.6 (Schmerl [**133**]). *1) If G is highly computable and n-colorable, then G is computably $(2n-1)$-colorable. Moreover, given an index for G (as a highly computable graph), one can computably find an index for a $(2n-1)$-coloring of G.*
2) For every $n \geq 2$, there is a highly computable n-colorable graph H which is not computably $(2n-2)$-colorable.

[4]Here G is undirected, although similar questions and notions can be considered in the directed case.

THEOREM 5.7 (Kierstead [**80**]). *Every perfect, k-colorable, highly computable graph is computably $(k+1)$-colorable. Moreover there exist perfect, k-colorable, highly computable graphs which cannot be computably k-colored.*

THEOREM 5.8 (Brooks' Theorem). *For any graph G, $\chi(G) \leq \Delta(G)$ unless either $\Delta(G) \leq \omega(G)$, or both $\Delta(G) = 2$ and G contains an odd cycle.*

THEOREM 5.9 (Schmerl [**134**]). *For any highly computable graph G, $\chi^c(G) \leq \Delta(G)$ unless $\Delta(G) \leq \omega(G)$, or both $\Delta(G) = 2$ and G contains an odd cycle.*

QUESTION 5.10. (i) **(Gasarch)** What happens to such results when we replace highly computable graphs by decidable graphs?
(ii) What happens to such results when we replace highly computable graphs by $L(Q)$-decidable graphs?
(iii) What is the complexity of such colorings if we assume the underlying graph is polynomial-time?

Remark: Bean [**7**] showed that all negative results that he obtained for highly computable graphs also hold for decidable graphs, e.g. for every $k \geq 2$ there exists a decidable graph G such that $\chi(G) = k$ and $\chi^c(G) = 2k - 1$.

A graph is called *k-edge colorable* if there is a coloring of the edges of G with k-colors such that any two edges of G which share a common vertex have different colors.

e-$\chi(G)$ = the minimum k such that G is k-edge colorable.

THEOREM 5.11 (Vizing's Theorem). *For any graph G, e-$\chi(G) \leq \Delta(G)$.*

THEOREM 5.12 (Kierstead [**80**]). *For any highly computable graph G, e-$\chi^c(G) \leq e$-$\chi(G) + 1 \leq \Delta(G) + 2$.*

THEOREM 5.13 (Manaster and Rosenstein [**96**]). *There exists a highly computable graph G such that e-$\chi(G) = \Delta(G)$, but e-$\chi(G) < e$-$\chi^c(G)$.*

QUESTION 5.14 (Kierstead). Is Vizing's Theorem effectively true?

THEOREM 5.15 (Kierstead). *If G is a computable directed graph that does not have an induced subgraph of the form*
(1) directed 3-cycle, or
(2) $\circ \longrightarrow \circ \longrightarrow \circ \longleftarrow \circ$, *or*
(3) $\circ \longleftarrow \circ \longrightarrow \circ \longrightarrow \circ$,
then $\chi^c(G) \leq 2^{\omega(G)}$ where $\omega(G)$ is the size of the largest clique.

We finish the section on graphs with one question which is apparently wide open.

QUESTION 5.16 (Kierstead). Lovasz proved that if $n_1, n_2 \in \omega$ and $n_1 + n_2 - 1 = \Delta(G)$, then G can be partitioned into V_1, V_2 with the maximum degree of the subgraph induced by $V_i \leq n_i$ for $i = 1, 2$. Kierstead asked if this result is effectively true for highly computable G.

5.1. Dilworth's Theorem and Dimension. We turn now to combinatorial aspects of the theory of partial orderings. We recall some definitions.

Let $\mathcal{P} = (P, \leq_P)$ be a partially ordered set (poset).
1) *width*$(\mathcal{P})$ = maximal size of an antichain in $\mathcal{P}$.

2) *height*$(\mathcal{P})$ = maximal size of a chain in $\mathcal{P}$.
3) *dim*$(\mathcal{P})$ = the minimum number of linear orders $L_1 = (P, \leq_1), \ldots, L_n = (P, \leq_n)$ such that

$$x \leq_P y \iff x \leq_i y \text{ for } i = 1, \ldots, n$$

= the minimum n such that $\mathcal{P}$ can be embedded in Q^n (under the product ordering).

If $w \in N$, then a *w-cover* of $\mathcal{P} = (P, \leq_P)$ is a set of w disjoint chains such that every element of P is in some chain. The fundamental theorem of partial orderings is the following.

THEOREM 5.17 (Dilworth's Theorem). *Every finite poset of width w has a w-cover.*

While we will see that Dilworth's theorem is *not* effectively true, amazingly, it has an effective counterpart.

THEOREM 5.18 (Kierstead [**79**]). *Every computable poset of width w can be partitioned into $\frac{5^w - 1}{4}$ computable chains.*

We let $w^c(\mathcal{P})$ denote the *computable* chain cover number. That is the minimum number of computable chains needed to cover the poset.

THEOREM 5.19 (Szemerédi, Trotter (see Kierstead [**82**])). *For every $w \geq 2$, there exists a computable partial order $\mathcal{P}$ of width w such that*

$$w^c(\mathcal{P}) = \binom{w+1}{2}.$$

THEOREM 5.20 (Kierstead [**82, 79**]). *There is a computable partial order of width 2 which cannot be partitioned into 4 computable chains.*

QUESTION 5.21 (Kierstead). Is the constant $\frac{5^w - 1}{4}$ the best we can do? The difference between the the quadratic lower bound and the exponential upper one is quite large.

Let $\mathcal{P} = (P, \leq_P)$ be a computable poset.

1) The *comparability graph* of $\mathcal{P}$, $C(\mathcal{P}) = (P, E_C)$ is the graph where $(x, y) \in E_C$ iff x and y are comparable in $\mathcal{P}$.
2) The *co-comparability graph* of $\mathcal{P}$, $I(\mathcal{P}) = (P, E_I)$ is the graph where $(x, y) \in E_I$ iff x and y are not comparable in $\mathcal{P}$.
3) Note that a chain in $\mathcal{P}$ is a clique in $C(\mathcal{P})$.
4) Note that an antichain in $\mathcal{P}$ is a clique in $I(\mathcal{P})$.
5) Dilworth's Theorem says that the co-comparability graph $C(\mathcal{P})$ of a poset of width w can be colored with w colors, i.e. $\chi(C(\mathcal{P})) = \omega(C(\mathcal{P}))$.

QUESTION 5.22 (Trotter, Schmerl). Does there exist a function f such that, for any computable co-comparability graph G, $\chi^c(G) \leq f(\omega(G))$? (Trotter in [**127**] asked the question also for Chvatal graphs.)

THEOREM 5.23 (Schmerl (see Kierstead [**82**]). *Every computable poset of height h can be covered by $\binom{h}{2}$ computable antichains. Moreover, for all positive integers h, there exists a computable poset of height h which cannot be covered by $\binom{h}{2} - 1$ computable antichains.*

THEOREM 5.24. *Every acyclic graph is the comparability graph of an ordered set of height two.*

Thus Bean's result shows that there exists a computable comparability graph G such that $\omega(G) = 2$ but $\chi^c(G) = \infty$.

1) An *interval order* is an ordered set whose points can be represented by intervals of the real line so that a point $x = (a, b)$ is less than a point $y = (c, d)$ iff $b \leq c$.
2) An *interval graph* is the comparability graph for an interval order.

THEOREM 5.25 (Kierstead and Trotter [**84**]). *For every computable interval graph G, $\chi^c(G) \leq 3\omega(G) - 2$. Moreover, for every w, there is a computable interval graph G such that $\omega(G) = w$ and $\chi^c(G) = 3w - 2$.*

Fact: The dimension of a partial order is $\leq$ its width.

THEOREM 5.26 (Kierstead, McNulty and Trotter [**85**]). *Let $\mathcal{P}$ be a computable poset. Then:*
(1) If $w^c(\mathcal{P}) \leq 2$, the computable dimension of $\mathcal{P}$ is ≤ 5.
(2) If $w^c(\mathcal{P}) \leq 3$, the computable dimension of $\mathcal{P}$ is ≤ 6 and moreover, there exists a computable partial order $\mathcal{Q}$ with $w^c(\mathcal{Q}) = 3$ and computable dimension 6).
(3) There is a computable partial order $\mathcal{Q}$ with $w^c(\mathcal{Q}) = 4$ which has no finite computable dimension ($\mathcal{Q}$ also has width 3).
(4) If the computable width of $\mathcal{P}$ is $\leq w$, then the computable dimension of $\mathcal{P}$ is $\leq 2w$.

THEOREM 5.27 (Hopkins [**66**]). *Let $\mathcal{P}$ be a computable interval order of width w. Then:*
(1) If $w = 2$, $\mathcal{P}$ has computable dimension ≤ 3.
(2) $\mathcal{P}$ has computable dimension $\leq 4w - 4$.
(3) For all $w \geq 2$, there exists a computable interval order $\mathcal{Q}$ of width w such that $\mathcal{Q}$ has computable dimension $\lceil \frac{4}{3}w \rceil$ [5].

A *crown* is a partial order on $\{a_1, \ldots, a_n, b_1, \ldots, b_n\}$ $(n \geq 3)$ such that (1) for all $i \leq n$, $a_i < b_i$, (2) for all $i \leq n - 1$, $a_{i+1} < b_i$, (3) $a_1 < b_n$, and (4) no other relation exists between the elements.
A partial order is *crown-free* if none of its induced suborders are crowns.

THEOREM 5.28 (Kierstead, McNulty, Trotter [**85**]). *(1) Every crown-free computable poset with computable width w has computable dimension $\leq w!$, and:*
(2) For $w \geq 3$ there is a computable crown-free poset with computable width w, but computable dimension at least $w\binom{w}{t}$ where $t = \lfloor \frac{(w-1)}{2} \rfloor$ [6].

COROLLARY 5.29 (Kierstead, McNulty, Trotter [**85**]). Every crown-free computable poset of width w has computable dimension $\leq (\frac{5^w - 1}{4})!$.

QUESTION 5.30 (Kierstead et. al.). What are the best bounds in the theorems above?

[5]Here, as usual, $\lceil z \rceil$ denotes the least integer $\geq z$.
[6]Here, as usual, $\lfloor z \rfloor$ denotes the greatest integer $\leq z$.

5.2. Ramsey-Type Theorems. Ramsey-type theorems assert that in large enough samples there is some well-behaved piece. Of course Ramsey's theorem itself asserts that if we colour the n element subsets of ω with k colours then there is an infinite homogeneous set: a subset A such that $[A]^n$ is monochromatic. But there are a host of Ramsey-type theorems such as Van der Waerden's theorem on arithmetical progressions, Szemerédi's theorem, the Hales-Jewett theorem, etc. Most of these have a constellation of computability-theoretic and associated "reverse mathematics" questions related to them. In fact, for most we know neither their proof theoretical strengthnor most recursion theoretical aspects.

Below are some assorted questions garnered from the literature. The reader is also directed to the articles by Simpson's [**139**] and Cenzer and Jockusch [**12**] in this volume, Hummel [**67**], and to the recent paper by Cholak, Jockusch and Slaman [**16**] where the connections between the computability-theoretical aspects and the reverse mathematical ones (in the sense of what is needed to get a reversal) are made quite explicit.

QUESTION 5.31 (Cholak, Jockusch, Slaman [**16**]). (i) For each noncomputable set C and each computable 2-colouring of $[\omega]^2$, is there an infinite low_2 homogeneous set X with $C \not\leq_T X$?
(ii) What degrees $\mathbf{d}$ have the property that every 2-colouring of $[\omega]^2$ has an infinite homogeneous set of degree at most $\mathbf{d}$?
(iii) For every 2-colouring C of $[\omega]^2$ not of PA degree, is there an infinite homogeneous H such that $C \oplus H$ does not have PA degree?

We refer the reader to Cholak, Jockusch and Slaman [**16**] for more on these questions. We remark that one of the questions there (For every infinite set $A \leq_T \emptyset'$, does one of A or $\overline{A}$ have an infinite low subset?) was solved negatively by Downey, Hirschfeldt, Lempp and Solomon [**25**].

Similar questions can be asked for the many well-quasi-ordering results in the literature. Kruskal's theorem (that finite trees are well-quasi-ordered under topological embedding as posets) has been investigated by Harvey Friedman, as have some variations. But what about the WQO results for graphs of bounded treewidth, posets under the chain minor ordering, etc.? Moreover, how fast do obstruction sets grow as a function of the genus for Robertson and Seymour's Kuratowski Theorem for surfaces of genus $\leq g$?

5.3. Some Other Combinatorial Problems Which Have Not Been Studied. The following is just a sample of other problems in combinatorics whose effective content has not been analysed.
1) Computable Genus.
2) Spherical Partial Orders (see Felsner, Fishburn, and Trotter [**46**]).
Every partial order $\mathcal{P}$ has a representation of the form

$$x \leq_P y \iff F(x) \subseteq F(y)$$

where F is some mapping of $\mathcal{P}$ to $\mathcal{R}^d$, for some d. Idea: restrict the range of F to be a sphere in R^d.
3) Computable Crossing Numbers.
4) Stable Marriage Problems (but see Cenzer and Jockusch [**12**]).

6. Groups

6.1. Groups in General. We begin with a question whose solution would seem to need real insight into group theory.

QUESTION 6.1 (Downey). (i) Is there a finitely presented group which is Σ_n-decidable but not Σ_{n+1}-decidable?
(ii) Is there an undecidable finitely presented group that is n-decidable for each n?

Actually the question is of interest in many cases, say for Abelian groups, even if the finitely presented hypothesis is removed.

There are also several questions that can be gleaned from Miller's survey paper [**101**].

Recall that a property P of groups G is called a *Markov Property* if:

(i) There is some finitely presented group H such that for all finitely presented G, if H embeds in G then $G \notin P$.
(ii) There is a finitely presented X such that $X \in P$.

The Adian-Rabin theorem states that *for any Markov property P, there is no algorithm which, when given a finite presentation of a group G determines if G satisfies P.* This is sort of a Rice's theorem for groups. We remark that for many Markov properties the complexity of the index set is known. For instance, if the property is c.e. then the proof of the Adian-Rabin theorem (see Miller [**101**], Rabin [**116**]) shows that it is Σ_1-complete. Boone and Rogers [**11**] showed that having a solvable word problem is Σ_3-complete and Lempp [**90**] proved that being torsion free is Π_2-complete.

QUESTION 6.2 (Miller [**101**]). (i) Is the index set of finitely presented residually finite groups Π_2-complete?
(ii) Is the index set of finitely presented simple groups Π_2-complete?
(iii) Is the index set of solvable finitely presented groups Σ_3-complete?

It would be interesting to develop analogues of the Adian-Rabin theorem for higher index sets, such as an analogue of Markov properties that always gives $\mathbf{0}''$-hard problems.

6.2. Abelian Groups. To state some of the questions we will need some definitions leading to the classical invariants for Abelian groups.

Let G be a group, $G[p^n] = \{x \in G \mid p^n x = 0\}$. We let $\mathbf{Z}_{p^n}$ denote the cyclic group of order p^n and let $\mathbf{Z}_{p^\infty} = \overset{lim}{\leftarrow} \mathbf{Z}_{p^n}$ (The set of all rational numbers whose denominator is a power of p under addition mod 1.)

DEFINITION 6.3. 1) A group G is *divisible* iff for all $g \in G$ and $n \in \omega - \{0\}$, $n|g$ iff $nG = G$ for all $n > 0$.
2) A group G is *p-divisible* iff $p^n G = G$ for all $n > 0$ iff $pG = G$.
3) A group G is *reduced* if it has no divisible subgroups other than $\{0_G\}$.

It is a well-known classical theorem that: *Every Abelian group A is the direct sum of a divisible group D and a reduced group C, $A = D \oplus C$.* The effective content of this theorem has been investigated first by Khisamiev and Khisamiev [**76**] and later by Friedman-Simpson-Smith [**47**] who proved that its proof theoretical strength is Π^1_1-CA_0.

A subgroup G of an Abelian group A is called **pure** if for all $g \in G$, the equation $nx = g$ is solvable in G whenever it is solvable in A. A subgroup G of an Abelian group A is p-**pure** if for all $k > 0$,

$$p^k G = G \cap p^k A,$$

which happens iff $pG = G \cap pA$. We can now state the relevant invariants.

DEFINITION 6.4 (Szmielew invariants of G). 1) $G \models A_{pnk}$ iff G contains $\mathbf{Z}^{\mathbf{k}}_{\mathbf{p}^{\mathbf{n}}}$ as a pure subgroup.
2) $G \models B_{pnk}$ iff $\mathbf{Z}^{\mathbf{k}}_{\mathbf{p}^{\mathbf{n}}} \leq \mathbf{G}$.
3) $G \models C'_{pnk}(g_1, \ldots, g_k)$ iff the images $g'_1, \ldots, g'_k$ of $g_1, \ldots, g_k$ in the factor group $\overline{G} = G/G[p^n]$ are such that $g'_1 + p\overline{G}, \ldots, g'_k + p\overline{G}$ are linearly independent in $\overline{G}/p\overline{G}$.
4) $C_{pnk} = \exists x_1, \ldots, x_k C'_{pnk}(x_1, \ldots, x_k)$.

The following numbers are also called the Szmielew invariants of G.

$$\begin{aligned}
\alpha_{pn}(G) &= \sup\{k \in \omega \mid G \models A_{pnk}\},\\
\beta_p(G) &= \inf\{\sup\{k \in \omega \mid G \models B_{pnk}\} \mid n \in \omega\},\\
\gamma_p(G) &= \inf\{\sup\{k \in \omega \mid G \models C_{pnk}\} \mid n \in \omega\}.
\end{aligned}$$

THEOREM 6.5 (Szmielew [**147**]). *Two Abelian groups G and H are elementarily equivalent if and only if these invariants of G and H are the same.*

Let T be a complete theory of groups. We let

$$\begin{aligned}
A(T) &= \{(p^{n+1}, k) \mid A_{pnk} \in T\},\\
B(T) &= \{(p^{n+1}, k) \mid B_{pnk} \in T\},\\
C(T) &= \{(p, k) \mid \exists N_p[\forall n \geq N_p(\neg A_{pn1} \in T) \wedge C_{pN_pk} \in T]\},\\
P(T) &= \{p \mid \exists N_p[\forall n \geq N_p(\neg A_{pn1} \in T)]\}.
\end{aligned}$$

A set C' of pairs (p, k), $k > 0$, is called a T*-extension* of the set $C(T)$ if
(i) $C(T) \subseteq C'$,
(ii) if $(p, k) \in C' \setminus C(T)$ then $p \notin P(T)$, and
(iii) if $(p, k) \in C'$ and $0 < s < k$ then $(p, s) \in C'$.

Khisamiev [**74**] was the first to really examine the computability aspects of Abelian groups. He proved the following.

THEOREM 6.6 (Khisamiev [**74**]). *A complete theory T of Abelian groups has an X-computable model if and only if the following hold.*
(i) $A(T) \in \Sigma^X_2$,
(ii) $B(T) \in \Sigma^X_1$,
(iii) The set $C(T)$ has a T-extension $C' \in \Sigma^X_2$.

This leads to several open questions.

QUESTION 6.7. 1) **(Khisamiev)** Find necessary and sufficient conditions, similar to the theorem above, for a theory of Abelian groups with a c.e. set of axioms to have a computable model.
2) **(Knight, Harizanov)** Find an algebraic characterization of the countable reduced Abelian p-groups which have computable copies.

There are several questions related to computable dimension.

THEOREM 6.8 (Goncharov [**55**], Smith [**142**]). *An Abelian p-group A is computably categorical iff either*
1) $A \cong \mathbf{Z}^{\omega}_{p^\infty} \oplus F$ *or*
2) $A \cong \mathbf{Z}^{m}_{p^\infty} \oplus \mathbf{Z}^{\omega}_{p^n} \oplus F$
where F *is a finite p-group and* $m, n < \omega$.

THEOREM 6.9 (Nurtazin). *A torsion-free Abelian group is computably categorical iff it has finite rank.*

The most basic question here is the following.

QUESTION 6.10 (Goncharov). Find algebraic criteria for when a computable Abelian group is computably categorical.

Naturally there are questions such as giving an algebraic characterization of when an Abelian group, p-group, or torsion free Abelian group is Δ^0_n-categorical.

Recently, the jump degrees of Abelian p-groups (Oates [**111**]) and torsion free Abelian groups were examined by Downey and Jockusch (in [**22**]). It was found that every rank 1 group has a 2nd jump degree, this being a kind of explanation of Baer's theorem about type sequences (see Coles, Downey, Slaman [**17**]). Is there a version for groups of finite rank not equal to 1?

6.3. Open Questions on the Groups of Computable Automorphisms of a Computable Model.

QUESTION 6.11. (i) **(Goncharov)** Find a description of the class of all groups isomorphic to the group of all computable automorphisms of a computable model.

(ii) **(Goncharov)** Is the class of all groups of computable automorphisms of a computable model closed under free products?

(iii) **(Nies)** Let G_δ be the group of all permutations computable in a Turing degree δ. Define G^*_δ as the quotient of this group modulo the group of all finitary permutations. Is it true that G^*_δ is embeddable into G^*_γ iff $\delta' \leq_T \gamma'$?

Morozov [**104**] proved that
1) G_δ is embeddable into G_γ iff $\delta \leq_T \gamma$.
2) G^*_δ is isomorphic to G^*_γ iff $\delta = \gamma$.
3) Similar results hold for G_δ modulo the set of all even permutations, i.e. permutations which are the product of an even number of two cycles.

6.4. Ordered Groups. The primary sources of work on computability aspects of the theory of ordered groups is Downey and Kurtz [**32**], Hatzikiriakou-Simpson [**61**], and above all, Solomon [**145**]. There, for instance, it is shown that the basic result that torsion free Abelian groups are orderable fails to hold effectively, and is equivalent to WKL_0. There are many open questions. Here is a selection.

QUESTION 6.12. (i) **(Downey and Kurtz)** Given a Π^0_1 class C, is there an orderable computable group whose cone of orderings (up to reversal) is in correspondence with the class C?

(ii) **(Downey and Kurtz)** Is every orderable computable group isomorphic to a computably orderable group?

(iii) **(Solomon)** Characterize the computably categorical ordered groups. Solomon remarks that since the groups are torsion free in the Abelian case, one would guess that the answer is "precisely when they have finite rank." In unpublished work, Goncharov, Lempp, and Solomon have shown that Archimedian ordered groups are computably categorical iff they have finite rank. But the case with infinitely many equivalence classes remains open. It is also not known if they can have finite computable dimension which is not 1.

(iv) (Downey) Does every low ordered computable group have a computable copy? That is, we fix the low order and ask for a copy where both the ordering and the group have are computable. Solomon remarks that the answer seems open even in the Abelian case.

(v) (Downey) Same as (iv) for low ordered fields.

(vi) (Downey) Does every computable formally real computable field have a presentation which admits a computable ordering?

(vii) (Solomon) Does every computable ordered Abelian group have a presentation which admits a computable basis. Dobritsa [**19**] proved this for computable Abelian groups, but maybe in the ordered case, it is not possible to make *both* the basis and the ordering computable at the same time.

We remark that (i) above has a negative answer if we ask that the group be Abelian (Solomon [**145**]).

7. Fields, Rings and the Like

A longstanding question of Friedman, Simpson and Smith [**48**] is the following.

QUESTION 7.1. Let P be a Π^0_1 class. Is there a computable commutative ring R with 1 such that the ideals of R are in 1-1 correspondence with the members of P?

For many types of algebraic structures the possible computable dimension is known.

QUESTION 7.2 (Goncharov). Is there a field with finite computable dimension $\neq 1$?

If the answer is no then it will be interesting to look at skew fields (Shore). For integral domains, Kudinov (personal communication) has shown the answer to be yes.

There are other more general questions one might ask about rings. Many classical theorems were proven using noneffective means and await analysis. For instance, while the effective content of Stone duality is fairly well-understood, the more general Zariski topology is not. Also properties involving radicals are a central topic in the general structure of rings, but have not been analysed effectively.

8. Polynomial-Time and Other Special Structures

There is a lot of work on various time and space bounded models. We look at some of this material and the particular sorts of questions one can ask below. Let $\mathcal{A}$ be a structure as before. We recall from the introductory section the basic definitions for polynomial-time structures.

DEFINITION 8.1. $\mathcal{A}$ is **polynomial-time** (abbrev. **p-time**) if
1) A is a polynomial-time subset of $\{0,1\}^*$.
2) The set of relations $\{R_i^{\mathcal{A}}\}$ is uniformly polynomial-time computable, i.e. there exists a computable function G such that $G(i) = [n_i, e_i]$ where $R_i^{\mathcal{A}}$ is the n_i-ary relation whose characteristic function is computed by ϕ_{n_i,e_i} and there exists a computable function G' such that $G'(i) = m_i$ where for all $(x_1, \ldots, x_{n_i}) \in \{0,1\}^{n_i}$, it takes at most $(max\{2, |x_1|, \ldots, |x_{n_i}|\})^{m_i}$ steps to compute $\phi_{n_i,e_i}(x_1, \ldots, x_{n_i})$.
3) The set of functions $\{f_i^{\mathcal{A}}\}$ are uniformly polynomial-time computable.
4) There is a polynomial-time computable function H such that $H(i) = c_i^{\mathcal{A}}$.

Note: The fact that $\mathcal{A}$ is polynomial-time computable does not necessarily imply that its atomic diagram is polynomial-time computable. Thus we say that $\mathcal{A}$ is **uniformly polynomial-time computable** iff its atomic diagram is polynomial-time computable.

Similarly we can define $\mathcal{A}$ being **p-time decidable**, **p-time Σ_n-decidable**, etc. A related idea is that of automatic structures. Let $\mathcal{A} = (A, \{R_i^{\mathcal{A}}\}_{i \in S}, \{f_i^{\mathcal{A}}\}_{i \in T}, \{c_i^{\mathcal{A}}\}_{i \in U})$ where $A \subseteq \Sigma^*$ for some finite alphabet Σ.

DEFINITION 8.2. Let $\nu : D \to A$ be a surjective mapping where D is an automaton recognizable subset of Σ^*. The mapping ν is an **automatic (strongly automatic, asynchronous automatic)** presentation of $\mathcal{A}$ if ν satisfies:
1) There exists a 2-variable automaton (strong automaton, asynchronous automaton) which for any two words $\alpha, \beta \in D$ decides whether $\nu(\alpha) = \nu(\beta)$.
2) For each $j \in S$, there exists an n_j-variable automaton (strong automaton, asynchronous automaton) which, for all $(\alpha_1, \ldots, \alpha_{n_j})$, decides whether $\mathcal{A} \models R_j^{\mathcal{A}}(\nu(\alpha_1), \ldots, \nu(\alpha_{n_j}))$.
3) For each $j \in t$, there exists an $n_j + 1$-variable automaton (strong automaton, asynchronous automaton) which, for all $(\alpha_1, \ldots, \alpha_{n_j}, \alpha_{n_j+1})$, decides whether

$$\mathcal{A} \models f_j^{\mathcal{A}}(\nu(\alpha_1), \ldots, \nu(\alpha_{n_j})) = \nu(\alpha_{n_j+1}).$$

There are several variations depending upon how we let the automata act on n-tuples:

$\alpha_1 = \sigma_{01} \ldots \sigma_{k_1 1}$

$\vdots$

$\alpha_n = \sigma_{0n} \ldots \sigma_{k_n n}$

n-automata: Let an ordinary automaton act on the words individually but have the final or accepting states be n-tuples.

strong n-automata: Pad the words with a new symbol so that all words have the same length and then have the underlying letters of the automaton alphabet be n-tuples.

asynchronous automata: Allow transitions where we make a move on some subset of the n-tuples.

8.1. Basic Questions for Polynomial-Time Models. Let $\mathcal{C}$ be a class of computable models, e.g. linear orderings, groups, etc.
1) Is every $\mathcal{A}$ in $\mathcal{C}$ isomorphic to a p-time model?

2) Is every $\mathcal{A}$ in $\mathcal{C}$ computably isomorphic to a p-time model?
3) Is every $\mathcal{A}$ in $\mathcal{C}$ isomorphic to a p-time model over a standard universe such as $Bin(\omega)$ or $Tal(\omega)$?
4) Is every $\mathcal{A}$ in $\mathcal{C}$ computably isomorphic to a p-time model over a standard universe such as $Bin(\omega)$ or $Tal(\omega)$?
5) Uniqueness? $A \cong_p B$ iff there exist p-time functions f and g such that $f \mid A : A \to B$ and $g \mid B : B \to A$.

Analogously to the computable version we have the following definition.

DEFINITION 8.3. (i) We say a p-time structure $\mathcal{A}$ is **strongly p-time categorical** if all p-time structures $\mathcal{B}$ which are isomorphic to $\mathcal{A}$ are p-time isomorphic to $\mathcal{A}$.
(ii) We say a p-time structure $\mathcal{A}$ with universe A is **p-time categorical over** A if all p-time structures $\mathcal{B}$ with universe A which are isomorphic to $\mathcal{A}$ are p-time isomorphic to $\mathcal{A}$.

To clarify the situation we give some examples below.

Example 1. $\mathcal{A} = (Tal(N), S, A)$ where $S(x) = x + 1$ and $A \subseteq N$. $\mathcal{A}$ is isomorphic to a p-time model $\mathcal{B}$ implies that A is double exponential time.

Example 2. $\mathcal{A} = (Tal(N), f)$ where $f : Tal(N) \to Tal(N)$. The complexity of possible presentations of $\mathcal{A}$ is characterized by the orbits of f. Consider the extremes given by $f(x) = 2x$ versus $f(x)$ equals the Ackermann function.

Example 3. (Cenzer-Remmel [**13**]) $(N, S, +, -, \times, 2^x, \leq)$ has a polynomial-time model.

QUESTION 8.4. Does $(N, S, +, -, \times, 2^x, 3^x \leq)$ have a polynomial-time model?

THEOREM 8.5 (Grigorieff [**56**]). *Any computable relational structure*

$$\mathcal{A} = (A, \{R_i^A\}_{i \in S}, \{c_i^A\}_{i \in U})$$

is computably isomorphic to a p-time structure.

THEOREM 8.6 (Cenzer-Remmel [**13**]). *Let $A \subseteq \{0,1\}^*$ be any p-time set. Then there exists a computable linear ordering $\mathcal{L}$ isomorphic to $\omega + \omega^*$ which is not computably isomorphic to any p-time linear ordering whose universe is A.*

THEOREM 8.7 (Grigorieff [**56**]). *Every computable linear ordering $\mathcal{L}$ is isomorphic to a real time linear ordering $\mathcal{L}' = (Bin(\omega), <_{\mathcal{L}'})$.*

The proof of Theorem 8.7 boils down to two cases as follows.

Case A: $\mathcal{L}$ has a computable increasing sequence S

$$S = (s_0 <_{\mathcal{L}} s_1 <_{\mathcal{L}} s_2 <_{\mathcal{L}} \dots)$$

such that S is cofinal, or S has a limit, or $\mathcal{L}$ has a computable decreasing sequence D

$$D = (d_0 >_{\mathcal{L}} d_1 >_{\mathcal{L}} d_2 >_{\mathcal{L}} \dots)$$

such that D is cofinal or D has a limit.

Case B: Not case A.

Then $\mathcal{L} \cong \omega + Z \cdot \lambda + \omega^*$.

THEOREM 8.8 (Remmel). *Let $A \subseteq \{0,1\}^*$ be any infinite p-time set and let $\mathcal{L}$ be a computable linear ordering which is isomorphic to $\omega + Z \cdot \lambda + \omega^*$ for some linear ordering λ. Then there exists a computable linear ordering $\mathcal{K}$ which is isomorphic to $\mathcal{L}$ but which is not computably isomorphic to any p-time linear ordering whose universe is A.*

Similar results hold for Boolean algebras and graphs.

QUESTION 8.9. Are there similar results for other structures?

COROLLARY 8.10 (Cenzer-Remmel [**13**]). Every computable Boolean algebra is computably isomorphic to a computable Boolean algebra over B where $B = Tal(\omega)$ or $B = Bin(\omega)$.

8.2. Questions for Automatic Models.

THEOREM 8.11 (Khoussainov-Nerode [**77**]). *1) For each n, ω^n has an automatic presentation. The rationals have an automatic presentation. The ordinal ω^ω has an asynchronous automatic presentation.*
2) For each n, the Boolean algebra $\mathcal{B}_{n\times\omega}$ has an asynchronous automatic presentation.
3) Automatic (asynchronous) orderings are closed under $+$ and $\times$.

QUESTION 8.12 (Khoussainov-Nerode). Find the first ordinal α which does not have an automatic (asynchronous) presentation?

9. Lattices of C.E. Substructures

9.1. Subspaces and the Like. Let V_∞ be an infinite dimensional computable vector space over a computable field F with a dependence algorithm. Let $L(V_\infty)$ denote the lattice of c.e. subspaces of V_∞ and $L^*(V_\infty)$ denote the lattice of c.e. subspaces of V_∞ modulo finite dimensional subspaces.

QUESTION 9.1. 1) Does every r-maximal subset of a computable basis for V generate an r-maximal subspace?
2) Does every atomless subset of a computable basis for V generate an atomless subspace?
3) What are the principal filters of $L^*(V_\infty)$? (Conjecture: **(Downey)** Every Σ^0_3 bounded modular lattice is a filter in $L^*(V_\infty)$.)
4) Is the $\forall\exists$ theory of $L^*(V_\infty)$-decidable?

We remark that this is all related to the classical question below. For $X \subset V_\infty$ let $sp(X)$ denote the subspace spanned by X.

QUESTION 9.2 (Downey). Suppose that $V_1 \oplus V_2 = R$ for a vector space R, and B is a basis of R. Are there $B_1 \sqcup B_2 = B$ such that $V_1 \oplus sp(B_1) = V_2 \oplus sp(B_2) = R$? (Note that this means that we require $V_i \cap sp(B_i) = \{0\}$.) What if $R = V_\infty$? If both V_i are decidable, do there exist computable B_i and is there an algorithm for effectively computing them?

We remark that a positive solution to this question for $R = V_\infty$ implies that r-maximal subsets of bases generate r-maximal spaces.

We remark that V_∞ is unlikely to have any interesting orbits (although this is an interesting question) since every automorphism is induced by a computable semilinear invertible transformation (Guichard [**57**]). Hence all automorphic spaces have, in particular the same 1-degree.

9.2. Lattice of C.E. Subalgebras and Ideals (Filters) of a Computable Boolean Algebra. Let $\mathcal{N}$ denote a decidable presentation of the Boolean algebra of finite and cofinite subsets of N, $\mathcal{Q}$ denote a decidable presentation of the countable atomless Boolean algebra, $\mathcal{C}$ denote a decidable presentation of the Boolean algebra generated by the closed intervals of the rationals Q.

QUESTION 9.3. Is there an analogue of supermaximal subspaces in $\mathcal{N}$? (Recall that c.e. $V \subseteq V_\infty$ is called supermaximal iff for all superspaces W of V, either $W = V_\infty$ or $W =^* V$.)

We can look at the lattice of ideals of $\mathcal{Q}$. This is dually the lattice of Π^0_1 classes. It is distributive and has infinitely many orbits. A c.e. ideal I with infinitely many extensions is called *maximal* if for all c.e. extensions J of I, there is an x such that $J = \langle I \cup \{x\} \rangle$. I is called perfect if $\mathcal{Q}/I \cong \mathcal{Q}$. There is a general programme of examining the lattice of ideals of $\mathcal{Q}$.

QUESTION 9.4. What degrees contain maximal ideals?

It is known by the work of Cholak, Coles, Downey and Herrmann [**15**] that the degrees containing perfect maximal ideals are exactly the array noncomputable degrees of Downey, Jockusch and Stob [**31**].

QUESTION 9.5. What is the relationship between the Boolean algebra $\mathcal{Q}/I$ and the possible degrees for I, where I is maximal?

QUESTION 9.6 (Cholak, Coles, Downey, Herrmann). Cholak, Coles, Downey, and Herrmann [**15**], proved that any two perfect maximal ideals are automorphic. Is this true for any other $\mathcal{Q}/I$? For instance, does the result still hold if we ask that $\mathcal{Q}/I$ be a copy of the Boolean algebra of finite and cofinite sets with the ideal generated by the atoms immune?

There are clearly a large number of interesting questions concerning the automorphism group of the lattice of ideals of $\mathcal{Q}$. We state only one more.

QUESTION 9.7 (Herrmann). Given any two nontrivial ideals I and J, are the lattices of c.e. subideals isomorphic? It is known that the number of isomorphism-types is at most 2.

10. Subalgebras

QUESTION 10.1 (Remmel). What is the relationship between the degrees of splittings and c.e. splittings of c.e. subalgebras of a computable Boolean algebra?

The background to the question is provided by the following set of results.

THEOREM 10.2 (Downey and Remmel [**39**]). *Let $V \in L(V_\infty)$. Suppose that A is a c.e. set. Then V has a basis B such that $B \equiv_{wtt} A$ iff $A \leq_{wtt} V$.*

That is, the correct reduction here is exactly weak truth-table reduction.

THEOREM 10.3 (Downey, Remmel, Welch [**40**]). *1) Let V be a c.e. subspace of V_∞. Then δ is the degree of a c.e. basis of V iff there exist c.e. subspaces W_1 and W_2 such such that $W_1 \oplus W_2 = V$ and $deg(W_1) = depdeg(W_1) = \delta$, where $depdeg(Z)$ denotes the dependence degree of Z which is the degree of computing independence over Z.*
2) There exist $Q, V, W \in L(V_\infty)$ such that $Q \oplus V = W$ and $\emptyset <_T Q <_T W$, but such that for all $V', Q' \in L(V_\infty)$ if

$$V' \oplus Q' = W \text{ and } Q \equiv_T Q',$$

then $deg(Q') \vee deg(V') = deg(W)$. (And same for depdeg.)

For Boolean algebras we have the following.

THEOREM 10.4 (Remmel, J. Yang (1997) see [**152**]). *Let δ_1 and δ_2 be any two c.e. degrees.*
1) There exist c.e. subalgebras W_1, W_2 of $\mathcal{N}$ such that $W_1 \oplus W_2 = \mathcal{N}$ and $deg(W_i) = \delta_i$ for $i = 1, 2$.
2) Similar results hold for $\mathcal{Q} \times B$, $\tilde{N} \times B$, and $\mathcal{C}$ for any computable Boolean algebra B.
3) Every noncomputable c.e. subalgebra of $\mathcal{N}^k$ has a noncomputable c.e. generating sequence.

10.1. NP and Similar Subspaces. Let $st(V_\infty)$ denote the standard representation of V_∞ (i.e. over $\{0,1\}^*$) and $tal(V_\infty)$ the tally representation.

A c.e. subspace S of V_∞ is *simple* if $dim(V_\infty/S) = \infty$ and for any infinite dimensional c.e. subspace W of V_∞, $W \cap S \neq \{\vec{0}\}$. By analogy, we define the following.

Let A be an oracle. An NP^A subspace S of $st(V_\infty)$ $(tal(V_\infty))$ is *NP^A-simple* if the dimension of $st(V_\infty)/S$ $(tal(V_\infty)/S)$ is infinite and for any infinite NP^A subspace W of $st(V_\infty)$ $(tal(V_\infty))$, $W \cap S \neq \{bin(\vec{0})\}$ $(W \cap S \neq \{tal(\vec{0})\})$. An NP^A subspace M is *NP^A-maximal* if the dimension of $st(V_\infty)/M$ $(tal(V_\infty)/M)$ is infinite and for any NP^A subspace W of $st(V_\infty)$ $(tal(V_\infty))$, either the dimension of $st(V_\infty)/W$ $(tal(V_\infty)/W)$ or the dimension of W/M is finite. An NP^A subspace M is *NP^A-supermaximal* if the dimension of $st(V_\infty)/M$ $(tal(V_\infty)/M)$ is infinite and for any NP^A subspace W of $st(V_\infty)$ $(tal(V_\infty))$, either $st(V_\infty) = W$ $(tal(V_\infty) = W)$ or the dimension of W/M is finite. A NP^A subspace S of $st(V_\infty)$ $(tal(V_\infty))$ is *P^A-simple* if the dimension of $st(V_\infty)/S$ $(tal(V_\infty)/S)$ is infinite and for any infinite dimensional P^A subspace W of $st(V_\infty)$ $(tal(V_\infty))$, $W \cap S \neq \{bin(\vec{0})\}$ $(W \cap S \neq \{tal(\vec{0})\})$.

It follows by Nerode and Remmel [**108**] that there exist oracles A such that there exists an infinite dimensional NP^A subspace V of $tal(V_\infty)$ such that V has no infinite dimensional subspace $W \in P^A$. Thus while a subspace W which is NP^A-simple is certainly P^A-simple, it is not clear that every P^A-simple subspace of $tal(V_\infty)$ is NP^A-simple.

There are a lot of variations to explore. The main theme is to understand the possible structure of NP^A spaces for various oracles A. Recently Downey and Nies [**37**] proved that, for subsets of ω (that is the lattice of NP^A *sets*), the structure can be very complicated indeed.

Assume the underlying field F of V_∞ is finite.

QUESTION 10.5 (Remmel). Does there exist a c.e. oracle D such that there exists an NP^D-supermaximal NP^D-simple subspace in $tal(V_\infty)$?

References

[1] Ash, C. J., Categoricity in the hyperarithmetical degrees, Ann. Pure and Appl. Logic, **34**(1987), 1–34.

[2] Ash, C. J., Labelling systems and r.e. structures, Ann. Pure and Appl. Logic, **47**(1990), 99–120.

[3] Ash, C., P. Cholak, and J. Knight, Permitting, forcing, and copying of a given recursive relation, Ann. Pure and Appl. Logic, **87**(1997), 219–236.

[4] Ash, C. J., C. Jockusch and J. F. Knight, Jumps of orderings, Trans. Amer. Math. Soc., **319**(1990), 573–599.

[5] Ash, C. J. and A. Nerode, Intrinsically recursive relations, In: Crossley [**18**], 26–41.

[6] Baumslag, G., E. Dyer, and C. Miller III, On the integral homology of finitely presented groups, Topology, **22**(1983), 27–46.

[7] Bean, D., Effective coloration, J. Symb. Logic, **1**(1976), 469–480.

[8] Bean, D., Recursive Euler and Hamilton paths, Proc. Amer. Math. Soc., **55**(1976), 385–394.

[9] Bonnet, R., Stratification et extension des genres de chaines denombrables, CRAS (Paris), **269**(1969), 880–882.

[10] Boone, W. W., Certain simple unsolvable problems in group theory, I–VI Nederlakad. Wentenschappen, Proc. Ser. A, **57**(1954), 231–237, 492–497, **58**(1955), 252–256, 571–577, **60**(1957), 22–27, 227–232.

[11] Boone, W. and H. Rogers Jr., On a problem of J. H. C. Whitehead and a problem of Alonzo Church, Math. Scand., **19**(1966), 185–192.

[12] Cenzer, D. and C. Jockusch, Π^0_1 classes - structure and applications, this volume.

[13] Cenzer, D. and J. B. Remmel, Polynomial-time vs recursive models, Ann. Pure and Appl. Logic, **54**(1991), 17–58.

[14] Chisholm, J. and M. Moses, Undecidable linear orderings that are n-recursive for each n, to appear, Notre Dame J. Formal Logic.

[15] Cholak, P., R. Coles, R. Downey and E. Herrmann Automorphisms of the lattice of Π^1_0 classes: perfect thin classes and anr degrees, submitted.

[16] Cholak, P., C. Jockusch, and T. Slaman, On the strength of Ramsey's Theorem for pairs, J. Symb. Logic, to appear.

[17] Coles, R., R. Downey, and T. Slaman, Every set has a minimal jump enumeration, Bull. London Math. Soc., to appear.

[18] Crossley, J. N. (ed), *Aspects of Effective Algebra*, Upside Down A Book Co., Varra Glen, Vic. Australia, 1981.

[19] Dobritsa, V., Some constructivizations of Abelian groups, Siberian Math Journal, **24**(2)(1983), 167–173.

[20] Downey R. G., Some remarks on a theorem of Iraj Kalantari concerning convexity and recursion theory, Z. Math. Logik Grund. Math., **30**(1984), 295–302.

[21] Downey, R., Every recursive Boolean algebra is isomorphic to one with incomplete atoms, Ann. Pure and Appl. Logic, **60**(1990), 193–206.

[22] Downey, R. On presentations of algebraic structures, In: *Complexity, Logic and Recursion Theory*, A. Sorbi, ed., Marcel Dekker, Lecture Notes in Pure and Applied Mathematics, Vol. 197, 1997, 157–206.

[23] Downey, R., Computability theory and linear orderings, In: *Handbook of Recursive Mathematics, Vol. 2,* Y. Ershov, S. Goncharov, A. Nerode, J. Remmel, and W. Marek eds., Studies in Logic Vol. 139, North-Holland, 1998, 823–976.

[24] Downey, R. G., Computability, Definability and algebraic structures, In: *Proceedings 7th Asian Logic Conference, 1999*, World Scientific, to appear.

[25] Downey, R., D. Hirschfeldt, S. Lempp and R. Solomon, A Δ^0_2 set such that neither it not its complement has a low subset, in preparation.

[26] Downey, R., D. Hirschfeldt, S. Lempp and R. Solomon, Group presentations and reverse mathematics, in preparation.

[27] Downey, R., D. Hirschfeldt, S. Lempp and R. Solomon, Extensions of partial orderings, in preparation.

[28] Downey, R., and D. Hirschfeldt, Spectra of relations on orderings and Boolean algebras, in preparation.

[29] Downey, R. and C. Jockusch, Every low Boolean algebra is isomorphic to a recursive one, Proc. Amer. Math. Soc., **122**(1994), 871–880.

[30] Downey, R. and C. Jockusch, Effective presentability of Boolean algebras of Cantor-Bendixson rank 1, J. Symb. Logic, **64**(1999), 45–52.

[31] Downey, R., C. Jockusch and M. Stob, Array recursive sets and multiple permitting arguments In: *Proceedings Oberwolfach 1989*, Springer-Verlag, *Lecture Notes in Mathematics 1990*, 141–174.

[32] Downey, R. G. and J. F. Knight, Orderings with α-th jump degree $\mathbf{0}^{(\alpha)}$, Proc. Amer. Math. Soc., **14**(1992), 545–552.

[33] Downey, R. G. and S. A. Kurtz, Recursion theory and ordered groups, Ann. Pure and Appl. Logic, **52**(1986), 137–451.

[34] Downey, R. G. and S. Lempp, On the proof theoretical strength of the Dushnik-Miller theorem for countable linear orderings, In: *Recursion Theory and Complexity,*, M. Arslanov and S. Lempp, eds., de Gruyter, 1999, 55–58.

[35] Downey, R. G. and M. F. Moses, Recursive linear orderings with incomplete successivities, Trans. Amer. Math. Soc., **320**(1991), 653–668.

[36] Downey, R. G. and M. F. Moses, On choice sets and strongly nontrivial self-embeddings of recursive linear orders. Z. Math. Logik Grund. Math., **35**(1989), 237–246.

[37] Downey, R. and A. Nies, Undecidability results for low complexity degree structures, In: *Proceedings Complexity Theory, 12th Annual Conference*, 1997, 128–132.

[38] Downey, R. and A. Nies, Undecidability results for low complexity time classes, J. Computing and Sys. Sci., to appear.

[39] Downey, R. G. and J. B. Remmel, Classification of degree classes associated with r.e. subspaces, Ann. Pure and Appl. Logic, **42**(1989), 105–125.

[40] Downey, R., J. Remmel and L. Welch, Degrees of splittings and bases of an r.e. vector space, Trans. Amer. Math. Soc., **302**(1987), 683–714.

[41] Dushnik, B. and Miller, E.W., Concerning similarity transformations of linearly ordered sets, Bull. Amer. Math. Soc., **46**(1940), 322–326.

[42] Ershov, Y., S. Goncharov, A. Nerode, and J. Remmel eds., *Handbook of Recursive Mathematics,* (V. Marek, associate editor), Studies in Logic Vol. 139, North-Holland, 1998.

[43] Feiner, L. J., *Orderings and Boolean algebras not isomoprhic to recursive ones*, Ph.D. Thesis, MIT, 1967.

[44] Feiner, L. J., Hierarchies of Boolean algebras, J. Symb. Logic, **35**(1970), 365–373.

[45] Feiner, L. J., Degrees of nonrecursive presentability, Proc. Amer. Math. Soc., **38**(1973), 621–624.

[46] Felsner, L., P. Fishburn, and W. Trotter, Finite three-dimensional partial orderings which are not sphere orderings, Discrete Math., **201**(1999), 101–132.

[47] Friedman, H., S. Simpson and R. Smith, Countable algebra and set existence axioms, Ann. Pure and Appl. Logic, **25**(1983), 141–181.

[48] Friedman, H., S. Simpson and R. Smith, Addendum to "Countable algebra and set existence axioms," Ann. Pure and Appl. Logic, **28**(1985), 319–320.

[49] Friedman, H. and L. Stanley, A Borel reducibility for countable models, J. Symb. Logic, **54**(1989), 894–914.

[50] Frölich, A, and J. Shepherdson, Effective procedures in field theory, Trans. Roy. Soc. London, Ser. A, **248**(1956), 407–432.

[51] Gasarch, W. I. and Lee, Andrew C.Y., On the finiteness of the recursive chromatic number, Ann. Pure and Appl. Logic, **93**(1998), 73–81.

[52] Goncharov, S. S., Constructivizability of superatomic Boolean algebras, Algebra and Logic, **12**(1972), 17–22.

[53] Goncharov, S. S., Some properties of the constructivizaton of Boolean algebras, Sibirski Math. Zh., **16**(1975), 264–278.

[54] Goncharov, S. S., On the number of nonautoequivalent constructivizations, Algebra and Logic, **16**(1977), 169–185.

[55] Goncharov, S. S., Autostability of models and Abelian groups, Algebra and Logic, **19**(1980), 13–27.

[56] Grigorieff, S., Every recursive linear ordering has a copy in DTIME–SPACE $(n, \log(n))$, J. Symb. Logic, **55**(1990), 260–276.

[57] Guichard D., Automorphisms of substructure lattices in recursive algebra, Ann. Pure and Appl. Logic, **25**(1983), 47–58.

[58] Harizanov, V., Some effects of Ash-Nerode and other decidability conditions on degree spectra, Ann. Pure and Appl. Logic., **54**(1991), 51–65.

[59] Harizanov, V., Uncountable degree spectra, Ann. Pure and Appl. Logic., **54**(1991), 255–263.

[60] Harizanov, V., Turing degree of the nonzero member in a two element degree spectrum, Ann. Pure and Appl. Logic, **60**(1993), 1–30.

[61] Hatzikiriakou K. and S. G. Simpson, WKL_0 and orderings of countable Abelian groups, In: *Logic and Computation, Contemporary Math.* Vol. 106, 1999, 177–180.

[62] Higman, G., Subgroups of finitely presented groups, Proc. Royal Soc. London, **262**(1961), 455–475.

[63] Hirschfeldt, D., *Degree Spectra of Relations on Computable Structures*, Ph.D. Thesis, Cornell University, 1999.

[64] Hirschfeldt, D., B. Khoussainov, R. Shore and A. Slinko, Degree spectra and computable dimension in algebraic structures, in preparation.

[65] Hirst, J., *Combinatorics in Subsystems of Second Order Arithmetic,* Ph.D. Thesis, Penn. State University, 1987.

[66] Hopkins. L., *Some Problems Involving Combinatorial Structures Determined by Intersections of Intervals and Arcs,* Ph.D. Thesis, Univ. of South Carolina, 1981.

[67] Hummel, T., Effective versions of Ramsey's theorem: Avoiding the cone above $\mathbf{0}'$, J. Symb. Logic, **59**(1994), 1301–1325.

[68] Jockusch, C. G., Ramsey's theorem and recursion theory, J. Symb. Logic, **37**(1972), 268–279.

[69] Jockusch, C. G. and R. I. Soare, Degrees of orderings not isomorphic to recursive linear orderings, Ann. Pure and Appl. Logic, **52**(1991), 39–64.

[70] Jockusch, C. and R. Soare, Boolean algebras, Stone spaces, and the iterated Turing jump, J. Symb. Logic, **59**(1994), 1121–1138.

[71] Jullien, P., *Contribution à l'étude des types d'ordre dispersés,* Ph.D. Thesis, Marseille, 1969.

[72] Kalantari, I, Effective content of a theorem of M. H. Stone, In: Crossley [**18**], 128–146.

[73] Kechris, A. S., New directions in descriptive set theory, Bull. Symb. Logic, **5** (1999), 161–174.

[74] Khisamiev, N., Connections between constructivizability and nonconstructivizability for different classes of Abelian groups, Algebra and Logic, **23**(1984), 220–233.

[75] Khisamiev, N., Hierarchies of torsion free Abelian groups, Algebra and Logic, **25**(1986), 128–142.

[76] Khisamiev, N., and Z. Khisamiev, Nonconstructavizability of the reduced part of a strongly constructive torsion-free Abelian group, Algebra and Logic, **24**(1985), 69–76.

[77] Khoussainov, B. and A. Nerode, On automata representable structures, in preparation.

[78] Khoussainov, B. and A. Nerode, *Automatic Model Theory*, monograph, in preparation.

[79] Kierstead, H., An effective version of Dilworth's theorem, Trans. Amer. Math. Soc., **268**(1981), 63–77.

[80] Kierstead, H., Recursive colorings of highly recursive graphs, Can. J. Math., **33**(1981), 1279–1290.

[81] Kierstead, H., An effective version of Hall's theorem, Proc. Amer. Math. Soc., **88**(1983), 124–128.

[82] Kierstead, H., Recursive ordered sets, In: *Combinatorics and Ordered Sets* I. Rival ed., Contemporary Math., vol. 57, Amer. Math. Soc., Providence, 1986.

[83] Kierstead, H., On Π_1-automorphisms of recursive linear orderings, J. Symb. Logic, **52**(1987), 681–688.

[84] Kierstead, H. and Trotter, W., An extrema problem in recursive combinatorics, Congress Numerantium, **33**(1981), 143–153.

[85] Kierstead, H., G. McNulty and W. Trotter, A theory of recursive dimension of ordered sets, Order, **1**(1984), 67–82.

[86] Knight, J. F., Effective constructions of models, In: *Logic Colloquium*, J. Paris, A. Wilkie and G. Wilmers eds., North-Holland, Amsterdam, 1986.

[87] Knight, J. F., Degrees coded into jumps of orderings, J. Symb. Logic, **51**(1986), 1034–1042.

[88] Knight, J., and M. Stob, Computable boolean Algebras, J. Symb. Logic, to appear.

[89] Kopytov, A., and V. Kokorin, *Fully Ordered Groups,* John Wiley and Sons, 1974.

[90] Lempp, S., The computational complexity of torsion-freeness of finitely presented groups, Bull. Aust. Math. Soc., **56**(2)(1997), 273–277.

[91] Lerman, M., On recursive linear orderings, In: *Logic Year 1979–80: The University of Connecticut*, M. Lerman, J. Schmerl and R. Soare eds., *Lecture Notes In Mathematics* vol. 859, Springer-Verlag, Heidelberg, New York, 1981, 132–142.

[92] Lerman, M., *Degrees of Unsolvability*, Springer-Verlag, Heidelberg, New York, 1983.

[93] Lerman, M. and J. Rosenstein, Recursive linear orderings, In: *Patras Logic Symposium*, G. Metakides ed., Stud. Log. Found. Math. vol. 109, 1982, North-Holland, Amsterdam, 123–136.

[94] Lin, C., Recursion theory on countable Abelian groups, Ph.D. Thesis, Cornell University, 1977.

[95] Lin, C., The effective content of Ulm's theorem, In: Crossley [**18**], 147-160.

[96] Manaster, A. and J. Rosenstein, Effective matchmaking (recursion theoretical aspects of a theorem of P. Hall), Proc. London Math. Soc., **25**(1971), 615–654.

[97] McCoy, C., Ph.D. Thesis, University of Notre Dame, 1999.

[98] McNulty, G., Infinite ordered sets, a recursive perspective, In: *Proc. of Symp. on Ordered Sets*, I. Rival ed., D. Reidel, Dortrecht, 1982.

[99] Metakides, G. and A. Nerode, Recursion theory and algebra, In: *Algebra and Logic*, J. N. Crossley ed., Lecture Notes in Math., vol. 450, Springer-Verlag, New York, 1975, 209–219.

[100] Metakides, G. and A. Nerode, Effective content of field theory, Ann. Math. Logic, **17**(1979), 289–320.

[101] Miller, C. F. III, Decision problems for groups-survey and reflections, In: *Algorithms and Classification in Combinatorial Group Theory*, G. Baumslag and C. Miller, eds., Springer-Verlag, New York (1992), 1-59.

[102] Miller, R., The Δ_2^0 spectrum of a linear ordering, to appear.

[103] Morozov A., Recursive automorphisms of atomic Boolean algebras, Algebra and Logic, **29**(1990), 310–330.

[104] Morozov A., Turing reducibility as algebraic embeddability, Siberian J. Math., **38**(1997), 312–313.

[105] Moses, M. F., *Recursive Properties of Isomorphism Types*, Ph.D. Thesis, Monash University, Melb. Australia, 1983.

[106] Moses, M. F., Recursive linear orders with recursive successivities, Ann. Pure and Appl. Logic, **27**(1984), 253–264.

[107] Moses, M. F., Relations intrinsically recursive in linear orderings, Z. Math. Logik Grund. Math., **32**(5) (1986), 467–472.

[108] Nerode, A. and Remmel, J.B., Complexity-theoretic algebra: Vector space bases, Feasible mathematics, In: *Proc. Math. Sci. Inst. Workshop, Ithaca/NY (USA) 1989*, Prog. Comput. Sci. Appl. Log. Vol. 9, 1990, 293–319.

[109] Nevins T., Degrees of convex dependence in recursively enumerable vector spaces, Ann. Pure and Appl. Logic, **60**(1993), 31–47.

[110] Novikov, P. S., On the algorithmic unsolvability of the word problem in group theory, Trudy Mat. Inst. Steklov, **44**(1955).

[111] Oates, S., *Jump Degrees of Groups*, Ph.D. Thesis, University of Notre Dame, 1989.

[112] Odifreddi, P., *Classical Recursion Theory, Vol. I*, North-Holland, Amsterdam, 1989.

[113] Odifreddi, P., *Classical Recursion Theory, Vol. II*, North-Holland, Amsterdam, 1999.

[114] Peretyat'kin, M., Every recursively enumerable extension of a theory of linear orderings has a constructive model, Algebra i Logik, **12**(1973), 211–219.

[115] Pinus, A., Effective linear orderings, Siberian Math. J., **16**(1975), 956–962.

[116] Rabin, M., Recursive unsolvability of group theoretic problems, Annals of Math., **67**(1958), 172–194.

[117] Rabin, M., Computable algebra, general theory and theory of computable fields, Trans. Amer. Math. Soc., **5**(1960), 341–360.

[118] Remmel, J. B., Recursively categorical linear orderings, Proc. Amer. Math. Soc., **83**(1981), 379–386.

[119] Remmel, J. B., Recursive Boolean algebras with recursive atoms, J. Symb. Logic, **46**(1981), 595–615.

[120] Remmel, J. B., Recursive isomorphism types of recursive Boolean algebras, J. Symb. Logic, **46**(1981), 572–594.

[121] Remmel, J. B., On the effectiveness of the Schröder-Bernstein theorem, Proc. Amer. Math. Soc., **83**(1981) 379–386.

[122] Remmel, J. B., Graph coloring and recursively bounded Π_1^0 classes, Ann. Pure and Appl. Logic, **32**(1986), 185–194.

[123] Remmel, J. B., Recursively rigid Boolean algebras, Ann. Pure and Appl. Logic, **36**(1987), 39–52.

[124] Remmel, J. B., Recursive Boolean algebras, In: *Handbook of Boolean algebras, Vol. 3*, D. Monk ed., North-Holland, Amsterdam, 1990, 1097-1166.

[125] Richter, L, *Degrees of Structures*, Ph.D. Thesis, University of Illinois at Urbana-Champaign, 1979.

[126] Richter, L. J., Degrees of structures, J. Symb. Logic, **46**(1981), 723–731.

[127] Rival, I. (ed.), *Ordered Sets*, Nato Advanced Study Inst. Series 83, D. Reidel Co., Dordrecht, 1981.

[128] Rival, I. (ed.), *Graphs and Order*, Nato Advanced Study Inst. Series, vol. 147, D. Reidel Co., Dordrecht, 1984.

[129] Rival, I. (ed.), *Combinatorics and Ordered Sets*, Contemporary Math., vol. 57, Amer. Math. Soc., Providence, 1986.

[130] Rogers, H. J., *Theory of Recursive Functions and Effective Computability*, McGraw-Hill, New York, 1967.

[131] Rosenstein, J., *Linear Orderings*, Academic Press, New York, 1982.

[132] Rosenstein, J., Recursive linear orderings, In: *Orders: Descriptions and Roles*, M. Pouzet and D. Richard, eds., Ann. Discrete Math., vol. 23, North-Holland, Amsterdam, 1982, 465–476.

[133] Schmerl, J., Recursive colorings of graphs, Can. J. Math., **32**(1980), 821–883.

[134] Schmerl, J., The effective version of Brooks' theorem, Can. J. Math., **34**(1982), 1036–1046.

[135] Schmerl, J., Recursion theoretical aspects of graphs and orders, In: I. Rival [1984], 467–486.

[136] Seetapun, D., and T. Slaman, On the strength of Ramsey's theorem, Notre Dame J. Formal Logic, **3**(1995), 570–582.

[137] Shore, R., Controlling the dependence degree of a recursively enumerable vector space, J. Symb. Logic, **4**(1978), 13–22.

[138] Shore, R., On the strength of Fraisse's conjecture, In: *Logical Methods*, J. Crossley, J. Remmel, R. Shore and M. Sweedler, eds., Birkhauser, Boston, 1993, 782–813.

[139] Simpson, S., Questions in reverse mathematics, this volume.

[140] Slaman, T., Relative to any nonrecursive set, Proc. Amer. Math. Soc, **126**(1998), 2117–2122.

[141] Slaman, T. and H. Woodin, Extending partial orders to dense linear orders, Ann. Pure and Appl. Logic, **94**(1998), 253–261.

[142] Smith, R., Two theorems on autostability in p-groups, In: *Logic Year 1979-80: The University of Connecticut*, M. Lerman, J. Schmerl, R. Soare eds., *Lecture Notes in Mathematics*, Vol. 859, Springer-Verlag, Heidelberg, New York, 1981, 302–311.

[143] Soare, R. I., Recursion theory and Dedekind cuts, Trans. Amer. Math. Soc., **140**(1969), 271–294.

[144] Soare, R. I., *Recursively Enumerable Sets and Degrees*, Springer-Verlag, Heidelberg, New York, 1987.

[145] Solomon, R., *Reverse Mathematics and Ordered Groups*, Ph.D. Thesis, Cornell University, 1998.

[146] Solomon, R., Ordered groups: a case study in reverse mathematics, Bull. Symb. Logic, **5**(1999), 45–58.

[147] Szmielew, W., Elementary properties of Abelian groups, Fund. Math., **41**(1955), 203–271.

[148] Szpilrajn, E., Sur l'extension de l'ordre partiel, Fund. Math., **16**(1930), 386–389.

[149] Thurber, J., *Degrees of Boolean Algebras*, Ph.D. Thesis, University of Notre Dame, 1994.

[150] Watnick, R., *Recursive and constructive linear orderings*, Ph.D. Thesis, Rutgers University, 1980.
[151] Wehner, S., Enumerations, countable structures and Turing degrees, Proc. Amer. Math. Soc., **126**(1998), 2131–2139.
[152] Yang J., *On recursive Boolean algebras*, Ph.D. Thesis, Univ. California, San Diego, La Jolla, CA, (1996).

School of Mathematical and Computing Sciences, Victoria University, P.O. Box 600, Wellington, New Zealand
E-mail address: rod.downey@vuw.ac.nz

Mathematics Department, University of California at San Diego, La Jolla, CA 92093-0012, U.S.A.
E-mail address: jremmel@ucsd.edu

Contemporary Mathematics
Volume **257**, 2000

Issues and Problems in Reverse Mathematics

Harvey Friedman and Stephen G. Simpson

Abstract. The basic reference for Reverse Mathematics is Simpson's recently published book *Subsystems of Second Order Arithmetic* [**52**]. The web site for the book is `http://www.math.psu.edu/simpson/sosoa/`. This article includes a general discussion of the nature and goals of Reverse Mathematics, as well as a write-up of some representative open problems in Reverse Mathematics.

Contents

1. The Emergence of Reverse Mathematics

The comprehensive reference for Reverse Mathematics is Simpson's self-contained treatise [**52**]. A brief account of the emergence of Reverse Mathematics in the 1970's from Simpson's point of view has already appeared in the notes at the end of Simpson [**52**, §I.9]. The rest of this section is a brief account from H. Friedman's point of view.

1991 *Mathematics Subject Classification.* Primary 03B30, 03F35, Secondary 05C55, 06C99, 26E40, 46B99.

Key words and phrases. Reverse Mathematics, Tietze Extension Theorem, DNR, Ramsey, WQO, BQO.

Friedman is supported by NSF grant DMS-9970459.

Gödel's Second Incompleteness Theorem showed that Hilbert's envisioned consistency proof for mathematics is impossible. But this is only the beginning of the story.

What has emerged is a broad and deep investigation into what mathematics can be proved in what formal systems.

In the late 60's and early 70's, H. Friedman investigated the logical principles involved in the proofs of several classical mathematical theorems. He discovered that each of these theorems are provably equivalent to certain natural logical principles — assuming some comparatively weak underlying principles in order to get started.

Some of this early work that precedes the emergence of Reverse Mathematics is documented in the unpublished series of three papers [**13**]. In particular, the first paper refers to a 1969 result:

> "In 1969, I discovered that a certain subsystem of second order arithmetic based on a *mathematical* statement (that every perfect tree which does not have at most countably many paths has a perfect subtree), was provably equivalent to a *logical principle* (the weak Π^1_1 axiom of choice) modulo a weak base theory (comprehension for arithmetic formulae)."
>
> (Emphasis from the original text.)

Friedman discussed a number of these results in his invited address at the 1974 International Congress of Mathematicians, and in the subsequent publication based on that talk, [**16**]. Friedman explicitly articulated (what was later called) the general program of Reverse Mathematics as "principal theme I" on page 235 of [**16**]:

> "I. When the theorem is proved from the right axioms, the axioms can be proved from the theorem."

In [**16**], a number of formal systems were introduced for this program, in connection with a number of classical mathematical theorems.

Specifically, the systems codifying logical principles used in [**16**] were RCA, WKL, ACA, ATR, Π^1_1-CA along with HCA, HAC, HDC, TI, RFN, weak Π^1_1-AC. All of these eleven systems are in the (first order two-sorted) language of second order arithmetic and incorporate the full scheme of induction in the language of second order arithmetic.

The first five of these systems are equivalent to the five principal formal systems of current Reverse Mathematics — augmented with the full scheme of induction in the language of second order arithmetic. The next four are equivalent to what are normally referred to as Δ^1_1-CA, Σ^1_1-AC, Σ^1_1-DC, BI (bar induction), again with full induction. The equivalence of RFN and TI, as well as of weak Π^1_1-AC and ATR, is claimed in [**16**].

In [**16**], seven very well known theorems from classical analysis are adjoined to RCA in order to form seven formal systems which are claimed to be provably equivalent to various of WKL, ACA, ATR, Π^1_1-CA.

Four additional theorems from classical analysis, normally stated with third order objects (sets of reals), are instead stated in terms of arithmetic predicates; and are also adjoined to RCA (in one case, to ACA). Three of these are claimed to be logically equivalent to HAC, ATR, Π^1_1-CA.

In summary, in [**16**], the principal theme I is forcefully illustrated with standard theorems from classical analysis, stated in conventional mathematical terms, via the systems WKL, ACA, ATR, Π^1_1-CA; RCA serves as the base theory. It was also less forcefully illustrated in [**16**] through some standard theorems from classical analysis stated "logically" in terms of arithmetic predicates, via one additional system HAC (Σ^1_1-AC).

The subsequent development of Reverse Mathematics has focused on a wide range of mathematics, stated mathematically in the language of second order arithmetic, mostly (but not exclusively) involving these same five formal systems (but with restricted induction), with RCA (with restricted induction) serving as the base theory.

The five formal systems with restricted induction — which are the mainstay of Reverse Mathematics — were published as JSL abstracts [**17**], and are already mentioned in [**16**]. These are now commonly written as RCA_0, WKL_0, ACA_0, ATR_0, Π^1_1-CA_0.

These abstracts discuss a reworking of results from [**16**] in terms of these systems with restricted induction. These results with restricted induction are sharper than the corresponding results in [**16**]. The proofs of most of the results claimed in [**17**] can be found in [**52**].

Success in the 70's with equivalents with these five systems over RCA_0 suggested a far-reaching systematic development. The novelty is that mathematical theorems are used to derive logical principles, not just vice versa. Friedman promoted the name "Reverse Mathematics" for the ensuing systematic investigation and classification.

By the late 70's, Simpson was intensively involved in the development of Reverse Mathematics. Simpson emphasized that in the vast preponderance of cases, the equivalences are with the four systems WKL_0, ACA_0, ATR_0, Π^1_1-CA_0 over the base theory RCA_0. The additional formal system HAC used by Friedman for an equivalence, as well as other systems under discussion by Friedman in the early days, have played little subsequent role.

This situation still largely holds today, with some important exceptions. Most notably, there is Simpson's WWKL_0 (see §5), and Simpson's RCA_0 + "ω^ω is well ordered," which are both developed by Simpson (see [**52**]).

There can be no question that Friedman was influenced by prior work of S. Feferman, G. Kreisel, and others on subsystems of second arithmetic, to which Friedman is indebted. Friedman's 1967 Ph.D. thesis [**15**] contains a number of results that answer questions or are relevant to prior work of Feferman and Kreisel.

We repeat that the major comprehensive reference for Reverse Mathematics as it stands today is [**52**].

2. The Reverse Mathematics Enterprise

A rather ambitious and comprehensive statement of the Reverse Mathematics enterprise can be stated informally as follows.

> A. For every theorem of mathematics, determine which sets of logical principles are sufficient to prove it.

We do not merely mean

> * For every theorem of mathematics, find some set of logical principles that are sufficient to prove it.

We want to know which sets of logical principles are and are not sufficient to prove it.

The present incarnation of Reverse Mathematics is based on these three crucial elements:

i. The choice of formal language for expressing mathematical statements. This is the two-sorted first order language of second order arithmetic given in [**52**], page 2, and written as L_2.
ii. The way in which mathematical statements are presented as sentences within L_2. I.e., coding, as discussed throughout [**52**].
iii. The choice of a formal system based on L_2 that is to serve as the base theory. I.e., the system RCA_0 given in [**52**], page 63.

Thus the present incarnation of Reverse Mathematics is limited to those mathematical statements which can be reasonably presented as sentences within L_2. For many mathematical statements, such a presentation is obvious and canonical. For others, there are reasonable presentations, but some issues arise as to alternative presentations. For still others, no (known) presentation captures the intended mathematical content.

We will use "L_2-appropriate" to indicate that there is a reasonable presentation as a sentence within L_2.

Given an L_2-appropriate theorem of mathematics, how does Reverse Mathematics "determine which sets of logical principles are sufficient to prove it"?

At first blush, this might simply mean: make a list of all finite sets of logical principles, and mark which ones suffice and which ones do not suffice to prove the theorem.

But such a crude plan taken in the most literal sense would be not be practical, communicable, or informative. It is not clear what a logical principle is in general, and also a comprehensive list would be so large that the number of subsets listed would be unworkable.

Instead, one considers the equivalence relation on L_2-appropriate mathematical statements Φ,Ψ, given by:

B. The sets of logical principles sufficient to prove Φ are the same as the sets of logical principles sufficient to prove Ψ.

We are all familiar with lots of logical principles, but it is not clear what a logical principle is in general. So how can such an equivalence be established?

Clearly, if

C. Φ,Ψ are themselves logically equivalent.

then by a principle of transitivity, B follows.

If logical equivalence is taken as provability in predicate calculus based on L_2, then it turns out that this particular equivalence relation is (believed to be) grossly unmanageable and uninformative, with an enormous number of equivalence classes.

This is where the choice of base theory, RCA_0, comes in. So we instead consider the equivalence relation

B′. The sets of logical principles containing RCA_0 sufficient to prove Φ are exactly the sets of logical principles containing RCA_0 sufficient to prove Ψ.

And likewise, if

C′. Φ,Ψ are themselves logically equivalent over RCA_0.

then B′ holds.

There is the restricted equivalence relation

D′. Φ,Ψ are themselves logically equivalent to a single "natural" set of logical principles over RCA_0.

The main discovery of the present incarnation of Reverse Mathematics is that C′ and D′ are manageable and informative. In particular,

I. The number of equivalence classes under C′ appears to be very small relative to the number of L_2-appropriate mathematical statements. In particular, Reverse Mathematicians have been very successful in determining the equivalence relation C′ among the L_2-appropriate "core curriculum" theorems; i.e., L_2-appropriate mathematical theorems from the standard, generally required undergraduate/graduate mathematics curriculum.

II. The three relations B′, C′, and D′ appear to be identical for L_2-appropriate core curriculum theorems. No counterexample is known.

III. In fact, as far as we know, every L_2-appropriate core curriculum theorem is logically equivalent to one of the seven systems RCA_0, WWKL_0, WKL_0, $\mathsf{RCA}_0 + \omega^\omega$ is well ordered, ACA_0, ATR_0, Π^1_1-CA_0, over RCA_0. The vast majority of L_2-appropriate core curriculum theorems that have been considered have been so classified. An interesting exception is the Lebesgue Monotone Differentiation Theorem, stated, for example, in the particularly clean formulation "every monotone continuous function is somewhere differentiable." Its equivalence with ACA_0 over RCA_0 is not known.

IV. If appropriate "extended curriculum" theorems are considered, then additional equivalence classes arise; e.g., the usual infinite Ramsey's Theorem is equivalent to a natural weak extension ACA_0' of ACA_0 over RCA_0; see §7. Also, there is a higher rate of open problems, where we do not even know the logical implications between the theorem and the seven formal systems from III, over RCA_0. For example, Ramsey's Theorem for pairs is in this category, although Ramsey's Theorem for triples is known to be equivalent to ACA_0 over RCA_0. See §7. (An extended curriculum theorem is one which is a standard part of any first level graduate course in a major area of mathematics, even if that area is not generally required).

V. If we consider the general mathematical literature, then the phenomenon in IV is accentuated, with more equivalence classes and more open problems. E.g., Kruskal's Theorem and its natural restrictions and refinements represent more equivalence classes and logical principles. But, e.g., the Nash Williams Infinite Tree Theorem or Laver's Embedding Theorem are not intelligibly classified. E.g., it is not known whether the former is provable in Π^1_1-CA_0, and it is not known whether the latter is provable in ATR_0, or even in Π^1_1-CA_0. See §8.

VI. The evidence is that every L_2-appropriate theorem from the general mathematical literature is logically equivalent to a "sensible" formal system of logical principles, over RCA_0. A consequence of this is that the equivalence relations B′, C′, and D′ are equivalent. In the largely comprehensive development of Reverse Mathematics in [**52**], the following sensible formal systems of logical principles arise: RCA_0, WWKL_0, WKL_0, RCA_0 + "ω^ω is well ordered," ACA_0, ACA_0', ACA_0^+, ATR_0, Π^1_1-CA_0, Π^1_1-TR_0.

VII. Some additional systems of the form $\mathsf{RCA}_0 +$ "λ is well ordered", for various proof-theoretic ordinals λ, appear in connection with WQO theory. See §8. Some stronger systems arise in the classification of theorems first proved by mathematical logicians. E.g., Borel determinacy and its fragments. See [**52**, pages 239–240].

In summary, the Reverse Mathematics classification of mathematical statements is informative and adequately robust, with relatively few equivalence classes, each with a natural corresponding formal system of logical principles.

We now turn to some aspects of Reverse Mathematics as a research enterprise.

a. The problems range essentially continuously from easy to (apparently) extremely difficult. This range is apparent from examination of [**52**], including the open problems there, as well as those discussed in this paper.
b. The problems arise in connection with a huge range of mathematical topics, including those in the extended mathematics curriculum (see IV above). One is dealing directly with great mathematics that has stood the test of time.
c. Two factors restrict the already considerable size and scope of Reverse Mathematics. One is the requirement that the mathematical theorem to be classified is L_2-appropriate. For a very large portion of mathematics, this is not a major restriction. More importantly, the preponderance of L_2-appropriate theorems are already provable in the base theory, RCA_0, and thus are logically equivalent to RCA_0 over RCA_0. The fact that a theorem is provable in RCA_0 may be interesting and nontrivial, but does not illustrate the essence of Reverse Mathematics.
d. The program of expanding the Reverse Mathematics enterprise by enlarging the language and weakening the base theory is discussed below in §10. This program is already under way as illustrated by the use of RCA_0^* as a base theory (see [**52**], pages 410–412, and this §10), as well as Simpson's investigations into the use of third order concepts to directly formalize Borel sets (Simpson [**51**]).
e. It is hoped that such an extended Reverse Mathematics enterprise will prove to be as robust as present Reverse Mathematics, but will touch virtually all areas of mathematics. In particular, we hope that most mathematical theorems will be subject to informative analysis by one or more versions of extended Reverse Mathematics, in the sense that the theorem will be classified at a level that is higher than the base theory on which that version is based.
f. The projected size and scope of even present Reverse Mathematics, is not to be underestimated. For example, take the two volume set on Abelian Group Theory by Fuchs, [**25**], and restrict everything to countable groups, where the essence of most everything there lives anyways. A significant number of the Propositions and Theorems in those books lend themselves to Reverse Mathematics treatment. See §9.

Obviously, Reverse Mathematics has significant overlap with recursive analysis, recursive algebra, constructive analysis, constructive algebra, descriptive set theory, and recursion theory.

The development borrows heavily from many methods and techniques in these areas. None of these areas — including Reverse Mathematics — reflect all known interesting logical aspects of mathematical theorems.

Reverse Mathematics does simultaneously reflect recursion-theoretic aspects of mathematics such as primitive recursion, the jump, arithmetic sets, hyperarithmetic sets, the hyperjump; descriptive set-theoretic aspects such as continuity, Borel measurability, analyticity; and proof-theoretic aspects such as ω^ω, ε_0, Γ_0, and higher proof-theoretic ordinals.

There are important metamathematical phenomena for which Reverse Mathematics is particularly well suited. For example, consider Simpson's result [**49**] that the celebrated Hilbert Basis Theorem (for multidimensional polynomial rings over fields) is equivalent to "ω^ω is well ordered" over RCA_0. The Hilbert Basis Theorem does not appear to be readily analyzable in terms of these other subjects, recursive analysis, recursive algebra, constructive analysis, constructive algebra, descriptive set theory, and recursion theory. The same can be said to the equivalents with induction spoken about by Chong at this Boulder conference.

The program of classifying mathematical theorems in the sense of Reverse Mathematics is well on its way, though at an early stage even for the standard math curriculum. This curriculum has a permanent place in history, even if somewhat more abstract than most current research.

It is the view of the authors that the systematic classification of mathematics as represented, say, by the mathematics curriculum, according to the scheme of Reverse Mathematics is a fundamental classification project that is as interesting and important to foundations of mathematics as other fundamental classification projects are for their subjects.

You understand more about a mathematical theorem when you have classified it in terms of Reverse Mathematics. And you have done something particularly significant when the mathematical theorem itself is a great classic — even if the Reverse Mathematics result is not unexpected. Sometimes the Reverse Mathematics results lead to unexpected developments. E.g., Simpson's work on Hilbert's Basis Theorem in Reverse Mathematics [**49**] led to ongoing work of Friedman relating Ackermann's function to algebraic sets that is becoming of interest to algebraic geometers. See [**12**].

Reverse Mathematics is a great source of problems for recursion theorists. We have seen this in impressive work of Cenzer, Cholak, Chong, Downey, Groszek, Harrington, Jockusch, Lempp, Remmel, Shore, Slaman, etc.

The essence of Reverse Mathematics can be stated to scholars outside the mathematical logic community with some effectiveness in the following way:

> "There is an accepted classification scheme of mathematical theorems according to their inherent logical strength, and I have just classified the Fundamental Theorem of Rally Dah Dah; it is at the same level as the Fundamental Theorems of Rally Dah Dee, and Rally Dah Doo."

3. The Interpretability Conjecture

Although the original five systems form a proper chain under inclusion of theorems, some of the newer systems that have arisen such as $\mathsf{RCA}_0 +$ "ω^ω is well ordered" break that chain.

However, we still have the following chain under relative interpretability: RCA_0 $= \mathsf{WWKL}_0 = \mathsf{WKL}_0 < \mathsf{RCA}_0 +$ "ω^ω is well ordered" $< \mathsf{ACA}_0 < \mathsf{ACA}_0' < \mathsf{ACA}_0^+ <$ $\mathsf{ATR}_0 < \Pi^1_1\text{-}\mathsf{CA}_0 < \Pi^1_1\text{-}\mathsf{TR}_0$.

The interpretability of WKL_0 in RCA_0 is not obvious and is due to Friedman (unpublished), by the following argument. The Harrington forcing interpretation of WKL_0 into RCA_0 is not an interpretation in the required sense; see [**52**], 369–372. However, it does provide a reduction within EFA (exponential function arithmetic, or $\mathrm{I}\Sigma_0(\exp)$) of any inconsistency in WKL_0 to an inconsistency in RCA_0 in which the cut degree is raised by at most a fixed constant. This is the hypothesis needed for interpretability in a theorem of Friedman; see [**55**], page 219.

INTERPRETABILITY CONJECTURE. Let X, Y be any finite sets of actual mathematical theorems in the published literature, which are L_2-appropriate. Then $\mathsf{RCA}_0 + X$ is interpretable in $\mathsf{RCA}_0 + Y$ or $\mathsf{RCA}_0 + Y$ is interpretable in $\mathsf{RCA}_0 + X$.

It is well known that there are examples of X, Y such that $\mathsf{RCA}_0 + X$ is not interpretable in $\mathsf{RCA}_0 + Y$, and $\mathsf{RCA}_0 + Y$ is not interpretable in $\mathsf{RCA}_0 + X$, but they are not natural in the sense of actual mathematical theorems in the published literature.

4. The Coding Issue

Reverse Mathematics requires substantial coding. This is a byproduct of the fact that the primitives are so restricted.

The coding involved is essentially routine and robust for discrete mathematics. Some interesting issues arise with real and complex analysis, measure theory, and functional analysis.

The coding issues are particularly delicate for the two weakest systems RCA_0 and WKL_0. ACA_0 is sufficiently power packed that a wide variety of coding schemes are provably equivalent. But one wants a coding that is appropriate for the base theory RCA_0, once and for all.

Analysis in Reverse Mathematics uses Cauchy sequences with 2^{-n} convergence, with the associated equality relation. Dedekind cuts have the advantage that one can use identity. These can be seen to be the "same" in RCA_0. But the third approach via Cauchy sequences without a rate of convergence, again with the associated equality relation, is not the "same" in RCA_0.

Adhering to the first two approaches creates no real problem until infinite sequences of real numbers. There is a natural coding of infinite sequences of infinite sequences of rationals and of sets of rationals. In [**52**], Simpson uses the former to code infinite sequences of reals and infinite sequences of Cauchy sequences with 2^{-n} convergence. But infinite sequences of Dedekind cuts are bad.

There is a key criterion for choice of coding that is implicitly used: that RCA_0 should prove as much as possible. E.g., if real numbers are Cauchy sequences of rationals, we can't prove in RCA_0 that every real number is the limit of a sequence of rationals with arbitrarily fast convergence.

PROBLEM. Make formal sense of this key criterion for coding: being able to "prove as much as possible". Show that 2^{-n} convergent Cauchy sequences and Dedekind cuts are "optimal" over RCA_0. Similarly, show that Simpson's coding of infinite sequences of reals is also "optimal."

The problem becomes most delicate when it comes to continuous functions.

Simpson [**52**] treats complete separable metric spaces and continuous functions between them, the real line being an important special case.

Most mathematics naturally lies within the realm of complete separable metric spaces and continuous functions between them defined on open, closed, compact or G_δ subsets. This is the central coding issue.

> PROBLEM. Continuation of the previous problem: Show that Simpson's neighborhood condition coding of partial continuous functions between complete separable metric spaces is "optimal". (It amounts to a coding of continuous functions on a G_δ.)

We emphasize our view that the handling of the critical coding in RCA_0 in [**52**] is canonical, but we are asking for theorems supporting this view.

5. Real Analysis and Topology

Much is known concerning Reverse Mathematics for real analysis and the topology of complete separable metric spaces. Some of the inspiration for this comes from recursive analysis [**43**] and Bishop-style constructivism [**3**]. We shall not discuss those connections here, but see Simpson's book [**52**] for more information.

Giusto/Simpson [**26**] includes a rather thorough Reverse Mathematics discussion of various notions of closed set, and of various forms of the Tietze Extension Theorem for real-valued continuous functions on closed sets, in compact metric spaces. The purpose of this section is to call attention to one open problem left over from that paper.

Let X be a compact metric space. For concreteness we may take $X = [0,1]$, the unit interval. In RCA_0 we define $K \subseteq X$ to be *closed* if it is the complement of a sequence of open balls; *separably closed* if it is the closure of a sequence of points; *located* if the distance function $d(x, K)$ exists as a continuous real-valued function on X; *weakly located* if the predicate $d(x, K) > r$ is Σ^0_1 (allowing parameters, of course). $\mathrm{C}(X)$ denotes the separable Banach space of continuous real-valued functions on X which have a modulus of uniform continuity. The *Strong Tietze Theorem* for K is the statement that every $\phi \in \mathrm{C}(K)$ extends to some $\widetilde{\phi} \in \mathrm{C}(X)$. See [**26**] for details.

Known results from [**26**] are:

(1) The Strong Tietze Theorem for closed, weakly located sets is provable in RCA_0.
(2) The Strong Tietze Theorem for separably closed sets is equivalent to WKL_0 over RCA_0.

There remain open questions concerning the status of

(3) the Strong Tietze Theorem for closed sets, and
(4) the Strong Tietze Theorem for closed, separably closed sets.

It is known from [**26**] that (3) and (4) are provable in WKL_0 and not provable in RCA_0. There is a partial reversal: (3) or (4) implies the DNR Axiom over RCA_0. We shall outline the proof of this below. But first we discuss the DNR Axiom.

The DNR Axiom says: For every $A \subseteq \mathbb{N}$ there exists $f : \mathbb{N} \to \mathbb{N}$ which is *diagonally nonrecursive* relative to A, i.e., $f(n) \neq \{n\}^A(n)$ for all $n \in \mathbb{N}$. Here $\mathbb{N}$ is the set of natural numbers. It would be possible to restate the DNR Axiom in a combinatorial way, not involving recursion theory, but we shall not do so here.

The DNR Axiom is known to be weaker than WKL_0 (= RCA_0 + Weak König's Lemma). Indeed, the DNR Axiom is provable in the strictly weaker system WWKL_0 (= RCA_0 + Weak Weak König's Lemma) which arises in connection with Reverse Mathematics for measure theory. (See [**52**, §X.1], [**26**], [**5**].) Because of Kumabe's result [**39**], it seems likely that the DNR Axiom is strictly weaker than WWKL_0.

Recursion theorists can understand these variants of Weak König's Lemma in terms of separating sets, recursively bounded Π^0_1 classes, etc. Thus there is a close connection with the talks of Jockusch and Cenzer at this conference (AMS-IMS-SIAM conference, Computability Theory and Applications, June 13–17, 1999, Boulder, Colo.). See also the abstracts at `http://www.math.psu.edu/simpson/cta/`. In descending order we have:

1. WKL_0 is just RCA_0 plus any of the following, relativized to arbitrary $A \subseteq \mathbb{N}$:
   (a) for every infinite recursive tree $T \subseteq \{0,1\}^{<\mathbb{N}}$, there exists a path through T.
   (b) for every disjoint pair of r.e. sets, there exists a separating set.
   (c) there exists a $\{0,1\}$-valued DNR function, i.e., a function $f : \mathbb{N} \to \{0,1\}$ such that $f(n) \neq \{n\}(n)$ for all $n \in \mathbb{N}$.
2. WWKL_0 is just RCA_0 plus either of the following, relativized to arbitrary $A \subseteq \mathbb{N}$:
   (a) for every recursive tree $T \subseteq \{0,1\}^{<\mathbb{N}}$ such that
   $$\lim_n \frac{|\{\sigma \in T : \mathrm{lh}(\sigma) = n\}|}{2^n} \neq 0$$
   there exists a path through T.
   (b) there exists a 1-random real (see Kučera [**36, 37, 38**]).
3. The DNR Axiom is equivalent over RCA_0 to the following, relativized to arbitrary $A \subseteq \mathbb{N}$:
   (a) there exists a DNR function, i.e., a function $f : \mathbb{N} \to \mathbb{N}$ such that $f(n) \neq \{n\}(n)$ for all $n \in \mathbb{N}$.

Unfortunately, we don't know much about how to use the DNR Axiom in mathematical arguments. Unlike WKL_0 and WWKL_0, the DNR Axiom seems weak and therefore difficult to apply.

We shall now end this section with an outline of the proof that the Strong Tietze Theorem for closed, separably closed subsets of $[0,1]$ implies the DNR Axiom.

We may as well assume that Weak König's Lemma fails. For each n let I_n be the closed interval $[1/2^{2n+1}, 1/2^{2n}]$. Since Weak König's Lemma fails, the Heine/Borel Covering Lemma fails, so let (a_{nk}, b_{nk}), $k \in \mathbb{N}$, be a covering of I_n by open intervals with no finite subcovering. We may assume that these coverings are disjoint from one another.

If $\{n\}(n)$ is defined, let s_n be the least s such that $\{n\}_s(n)$ is defined, and put

$$J_n = I_n \setminus \bigcup_{k=0}^{s_n} (a_{nk}, b_{nk}).$$

Let

$$K = \{0\} \cup \bigcup \{J_n : \{n\}(n) \text{ is defined}\}.$$

It can be shown that $K \subseteq [0,1]$ is closed, separably closed, and not weakly located.

Define a real-valued continuous function $\phi(x) = \pm x$ on K, as follows. First let $p_i(x)$, $i \in \mathbb{N}$, be a fixed, one-to-one, recursive enumeration of $\mathbb{Q}[x]$, the ring of polynomials with rational coefficients in one indeterminate, x. Using this, define $\phi(0) = 0$ and, for $\{n\}(n) = i$ and $x \in J_n$, $\phi(x) = x$ if $|p_i(z) - z| \geq 1/2^{2n+2}$ for some $z \in J_n$, $\phi(x) = -x$ otherwise. It can be shown that $\phi \in \mathrm{C}(K)$.

By the Strong Tietze Theorem for K, let $\widetilde{\phi} \in \mathrm{C}([0,1])$ be an extension of ϕ from K to all of $[0,1]$. By Weierstrass polynomial approximation in RCA_0, let $p_{i_n}(x)$, $n \in \mathbb{N}$, be a sequence of polynomials such that $\sup\{|\widetilde{\phi}(x) - p_{i_n}(x)| : 0 \leq x \leq 1\} < 1/2^{2n+2}$ for all n. It is not difficult to show that the function $f : \mathbb{N} \to \mathbb{N}$ given by $f(n) = i_n$ is DNR.

By relativizing the above to an arbitrary $A \subseteq \mathbb{N}$, we get a function that is DNR relative to A. This completes the proof.

Note: Recursion theorists may want to view the above as a standard diagonal construction leading to a recursive counterexample to (4). However, from the viewpoint of Reverse Mathematics, there seems to be something unusual going on here. Usually, a recursive counterexample leads to a reversal to ACA_0 or WKL_0, but in this instance all we seem to get is a reversal to the DNR Axiom.

6. Banach Space Theory

Regarding the Reverse Mathematics status of well-known theorems of functional analysis, much is known, but many questions remain.

For example, the Open Mapping Theorem for separable Banach spaces is known to be provable in a system called RCA_0^+ which is of the same proof-theoretic strength as RCA_0 and indeed conservative over RCA_0 for Π^1_1 sentences. But whether it is provable in RCA_0 or even WKL_0 remains unknown. See Brown/Simpson [**6**], Mytilinaios/Slaman [**42**], Simpson [**50**].

As another example, consider the Krein/Šmulian Theorem for separable Banach spaces. This is a somewhat lesser known but still basic theorem of functional analysis. It says that a convex set in the dual of a separable Banach space is weak-$*$-closed if and only if it is bounded-weak-$*$-closed. It is known from Humphreys/Simpson [**33**] that this statement is provable in ACA_0, but the exact strength is unknown.

7. Ramsey Theory

An important contribution to Reverse Mathematics is the recent paper [**8**] by Cholak/Jockusch/Slaman on Ramsey's Theorem for pairs, and the last section of that paper mentions some open problems.

However, *Ramsey theory* is a large subject with many interesting results in addition to the familiar Ramsey Theorem. Moreover, there are important connections between Ramsey theory and dynamical systems theory, especially topological dynamics and ergodic theory. For a survey of the area, see the monograph of Graham/Rothschild/Spencer [**28**].

The purpose of this section is to mention some open problems regarding Reverse Mathematics and Ramsey theory in this broader sense. Such problems may be especially attractive, because Ramsey theory is a branch of mathematics where one could expect to find statements of high logical strength.

Hindman's Theorem. A famous and important Ramsey-type result is *Hindman's Theorem*:

> For any coloring of $\mathbb{N}$ with finitely many colors, there exists an infinite set $H \subseteq \mathbb{N}$ such that all sums of finite subsets of H have the same color.

Hindman's Theorem is well known to be closely related to the *Auslander/Ellis Theorem* in topological dynamics:

> For every state x in a compact dynamical system, there exists a state y which is proximal to x and uniformly recurrent.

(A *compact dynamical system* consists of a compact metric space X and a continuous function $T : X \to X$. A state $x \in X$ is said to be *uniformly recurrent* if for all $\epsilon > 0$ there exists m such that for all n there exists $k < m$ such that $d(T^{n+k}x, x) < \epsilon$. Two states $x, y \in X$ are said to be *proximal* if for all $\epsilon > 0$ there exist infinitely many n such that $d(T^n x, T^n y) < \epsilon$.)

There has been a great deal of interest in the constructive or effective aspect of Hindman's Theorem and the Auslander/Ellis Theorem. Some of the known proofs are highly set-theoretical and cannot even be formalized in second-order arithmetic. For an extensive discussion, including several proofs of Hindman's Theorem, see [**28**].

Simpson conjectures that Hindman's Theorem and the Auslander/Ellis Theorem are equivalent to ACA_0 over RCA_0. The known partial results in this direction are in Blass/Hirst/Simpson [**4**]. There it was shown that Hindman's Theorem and the Auslander/Ellis Theorem are provable in ACA_0^+, which consists of ACA_0 plus "for all $A \subseteq \mathbb{N}$, the ωth Turing jump $A^{(\omega)}$ of A exists". The proof of Hindman's Theorem in ACA_0^+ involves a delicate effectivization of Hindman's original proof. A reversal was obtained by showing that Hindman's Theorem implies ACA_0 over RCA_0. The problem here is to close the gap between ACA_0 and ACA_0^+.

Szemerédi's Theorem. Another well known result of Ramsey theory is *Van der Waerden's Theorem*:

> If $\mathbb{N}$ is colored with finitely many colors, then for each k there is a color containing an arithmetic progression of length k.

Using a method of Shelah, Van der Waerden's Theorem is provable in RCA_0. The so-called "density version" of Van der Waerden's Theorem is due to Szemerédi:

> If $A \subseteq \mathbb{N}$ is such that
>
> $$\limsup_{n\to\infty} \frac{|A \cap \{1, \ldots, n\}|}{n} \neq 0$$
>
> then A contains arithmetic progressions of arbitrary finite length.

For a long time the strength of Szemerédi's Theorem was an open problem. In particular, it was unknown whether Szemerédi's Theorem is provable in ACA_0. Recent results of Gowers [**27**] appear to indicate that Szemerédi's Theorem is provable in RCA_0 or even RCA_0^*.

The Dual Ramsey Theorem. Let $(\mathbb{N})^k$ denote the set of partitions of $\mathbb{N}$ into exactly k nonempty pieces. Let $(\mathbb{N})^\infty$ denote the set of partitions of $\mathbb{N}$ into infinitely many nonempty pieces. For $X \in (\mathbb{N})^\infty$, $(X)^k$ is the set of all $Y \in (\mathbb{N})^k$ such that Y is coarser than X. The Dual Ramsey Theorem of Carlson/Simpson [**7**] reads as follows:

> If $(\mathbb{N})^k$ is colored with finitely many Borel colors, then there exists $X \in (\mathbb{N})^\infty$ such that $(X)^k$ is monochromatic.

This also holds with k replaced by ∞. There are some important generalizations of this due to Carlson. This kind of result has been used by Furstenberg and Katznelson to obtain a "density version" of the Hales/Jewett Theorem. For references, see Simpson's book [**52**, remark X.3.6].

There are many open problems concerning the strength of the Dual Ramsey Theorem and related theorems. Slaman [**54**] has shown that the Dual Ramsey Theorem is provable in Π^1_1-CA_0. No interesting reversal is known.

Let A be a fixed finite alphabet. A^* denotes the set of *words*, i.e., finite strings of elements of A. An *infinite variable word* is an infinite string W of elements of $A \cup \{x_n : n \in \mathbb{N}\}$ such that each x_n occurs at least once, and all occurrences of x_n precede all occurrences of x_{n+1}, for all n. If $s = a_0 \cdots a_{n-1} \in A^*$, we denote by $W(s)$ the word which results from W upon replacing all occurrences of x_m by a_m for each $m < n$, then truncating just before the first occurrence of x_n. A key lemma of Carlson/Simpson [**7**] reads as follows:

> If A^* is colored with finitely many colors, then there exists an infinite variable word W such that $W(A^*) = \{W(s) : s \in A^*\}$ is monochromatic.

The strength of this lemma is unknown. In particular, it is unknown whether this lemma is true recursively, *i.e.*, whether W can be taken to be recursive in the given coloring. For more background on this problem, see Simpson [**48**].

Consequences of Ramsey's Theorem. In [**19**], the following consequence of Ramsey's Theorem is presented.

> THEOREM 1. Let $F : \mathbb{N}^k \to \mathbb{N}$. There is an infinite $A \subset \mathbb{N}$ such that for all $x_1, \ldots, x_k \in A$, if $F(x_1, \ldots, x_k)$ is in A then $F(x_1, \ldots, x_k)$ is among $x_1, \ldots, x_k$.

We can think of A as being kind of "free" of F to the extent that we can demand.

Here is an even more basic consequence of Ramsey's Theorem which trivially follows from Theorem 1.

> THEOREM 2. Let $F : \mathbb{N}^k \to \mathbb{N}$. There is an infinite $A \subset \mathbb{N}$ such that $F[A^k]$ is not $\mathbb{N}$.

For $k \geq 1$, write T1(k) and T2(k) for Theorems 1 and 2, respectively, in dimension k. Write T1 and T2 for the full theorems.

What is the Reverse Mathematics of the T1(k), T2(k), and T1, T2?

By way of background, Ramsey's Theorem (RT) for arbitrary tuples is equivalent to the system $\mathsf{ACA}_0' = \mathsf{ACA}_0 + \forall n \, \forall X$ (the n-th Turing jump of X exists), over RCA_0. And Ramsey's Theorem for 3-tuples (RT(3)) or any specific arity ≥ 3 is equivalent to ACA_0 over RCA_0. See [**52**], 122–125, 404.

The case of 2-tuples (RT(2)) is not fully understood and is discussed in detail in [**8**], with plenty of open questions.

It is easy to see that RCA_0 proves T1(1) and also that RT implies T1. Friedman has shown T2 is not provable in ACA_0 using the following fact about ACA_0. (This fact can be proved using the method of [**52**, Lemma IX.4.3, Theorem IX.4.4].)

> Suppose ACA_0 proves $\forall X \, \exists Y \, \varphi(X, Y)$, where φ is an arithmetical formula whose free variables are among those shown. (Note that T1

and T2 can be put in this form.) Then there exists an arithmetical operator H, defined without parameters, such that ACA_0 proves $\forall X\, \varphi(X, H(X))$.

The existence of such an arithmetical operator is refuted in the case of T2 by the following.

LEMMA. For all $k > 1$ there exists a (primitive) recursive $F : \mathbb{N}^k \to \mathbb{N}$ such that for all infinite A recursive in $0^{(k-2)}$, $F[A^k] = \mathbb{N}$.

We now sketch a proof of the Lemma. The idea is to use a set S recursive in $0^{(k-1)}$ such that for any infinite A recursive in $0^{(k-2)}$, there is at most one element of S in the closed interval from any sufficiently large element of A to the next element of A.

It is well known that for any Σ^0_k formula $A(n,r)$ there is a Σ_0 formula B such that for all $n, r << m_1 << \cdots << m_{k-1}$, $A(n,r)$ iff $B(n,r,m_1,\ldots,m_{k-1})$.

In particular, choose a Σ_0 formula B such that for all $n, r << m_1 << \cdots << m_{k-1}$, $B(n,r,m_1,\ldots,m_{k-1})$ if and only if there are exactly r elements of S that are $\leq n$. Define $F(n,m_1,\ldots,m_{k-1})$ to be the least r such that $B(n,r,m_1,\ldots,m_{k-1})$ if it exists; 0 otherwise.

Let A be infinite and recursive in $0^{(k-2)}$. For any $n \in A$, there exist $m_1, \ldots, m_{k-1} \in A$ such that $F(n,m_1,\ldots,m_{k-1})$ is the number of elements of S that are $\leq n$. Now as n goes through the elements of A, the number of elements of S that are $\leq n$ must go through the natural numbers such that, eventually, no natural number is skipped. Hence $F[A^k]$ is cofinite.

This establishes the Lemma with "$F[A^k]$ is cofinite". To complete the sketch, replace F with the function $G(x_1,\ldots,x_k)$ = the numerical difference between $F(x_1,\ldots,x_k)$ and the greatest square $\leq F(x_1,\ldots,x_k)$.

What is the logical relationship between RT, T1, T2, RT(2), T1(2), T2(2), and the T1(k), T2(k), over RCA_0? It is obvious that RT(2) implies T2(2).

8. WQO Theory

We now turn from Ramsey theory to another important branch of combinatorics: WQO theory. Like Ramsey theory, WQO theory is of special interest from the viewpoint of Reverse Mathematics, because many of the proofs seem to need unusually strong set existence axioms.

A *quasiordering* is a set Q together with a reflexive, transitive relation $\leq$ on Q. A *well quasiordering* (abbreviated WQO) is a quasiordering such that for every function $f : \mathbb{N} \to Q$ there exist $m, n \in \mathbb{N}$ such that $m < n$ and $f(m) \leq f(n)$. Let $[\mathbb{N}]^\infty$ be the space of infinite subsets of $\mathbb{N}$. A *better quasiordering* (abbreviated BQO) is a quasiordering such that for every Borel function $f : [\mathbb{N}]^\infty \to Q$ there exists $X \in [\mathbb{N}]^\infty$ such that $f(X) \leq f(X \setminus \{\min(X)\})$. It can be shown that every BQO is a WQO but not conversely.

Generally speaking, WQO theory is an appropriate tool when considering quasiorderings of finite structures, but BQO theory is better adapted to infinite structures. For example, a famous theorem of WQO theory is *Kruskal's Theorem*:

Finite trees are WQO under embeddability.

(Here a *tree* is a connected acyclic graph, and *embeddings* are required to take vertices to vertices, and edges to paths.) Kruskal's Theorem has been generalized to infinite trees, but the proof is much more difficult and involves BQO theory. Detailed references are in [**52**, §X.3].

There are some important results about the strength of various theorems of WQO theory. For instance, Friedman (see Simpson [**47**]) showed that Kruskal's Theorem is not provable in ATR_0, and he characterized exactly the strength of Kruskal's Theorem, in proof-theoretic terms. This had remarkable consequences for Friedman's foundational program of finding mathematically natural, finite combinatorial statements which are proof-theoretically strong.

Consider now the following generalization of Kruskal's Theorem, due to Kriz [**35**]. Let T_1 and T_2 be finite trees where each edge is labeled with a positive integer. Write $T_1 \leq T_2$ to mean that there exists an embedding of T_1 into T_2 such that the label of each edge of T_1 is less than or equal to the minimum of the labels of the corresponding edges of T_2. *Kriz's Theorem* says that this quasiordering is a WQO.

What is the strength of Kriz's Theorem? By results of Friedman (see Simpson [**47**]), Kriz's Theorem is at least as strong as Π^1_1-CA_0. It may be much stronger, but little is known. This is an open problem which may have a big payoff.

We now consider a famous theorem of BQO theory. If Q is a countable quasiordering, let $\widetilde{Q}$ be the set of countable transfinite sequences of elements of Q. Quasiorder $\widetilde{Q}$ by putting $s \leq t$ if and only if there exists a one-to-one order-preserving map $f : \mathrm{lh}(s) \to \mathrm{lh}(t)$ such that $s(i) \leq t(f(i))$ for all $i < \mathrm{lh}(s)$. The *Nash-Williams Transfinite Sequence Theorem* [**46, 56**] says that if Q is BQO then $\widetilde{Q}$ is BQO.

Marcone [**41**] has shown that the Nash-Williams Theorem is provable in Π^1_1-CA_0 but not equivalent to Π^1_1-CA_0. Shore [**44**] has shown that the Nash-Williams Theorem implies ATR_0 over RCA_0. There remains the problem of closing the gap. We conjecture that the Nash-Williams Theorem is provable in ATR_0, hence equivalent to ATR_0 over RCA_0.

Another famous theorem of BQO theory is *Laver's Theorem* [**46, 56**]:

> The set of all countable linear orderings is WQO (in fact BQO) under embeddability.

The strength of Laver's Theorem is an open problem. Shore [**44**] has shown that Laver's Theorem implies ATR_0 over RCA_0, and we conjecture that Laver's Theorem is provable in ATR_0.

9. Countable Abelian Group Theory

The theory of countable Abelian groups is a rich area of investigation for Reverse Mathematics. The classic references for this branch of algebra include [**25**], [**29**], and [**34**]. The two-volume book [**25**] is the most comprehensive.

There are standard theorems in countable Abelian group theory representing each of the four standard systems of Reverse Mathematics above RCA_0. Here are some examples.

1. ACA_0 is equivalent over RCA_0 to the assertion that every countable Abelian group has a subgroup consisting of the torsion elements. See [**52**], 118.
2. ATR_0 is equivalent over RCA_0 to the assertion that every countable reduced Abelian p-group has an Ulm resolution. See [**52**], 201.
3. Π^1_1-CA_0 is equivalent over RCA_0 to the assertion that every countable Abelian group is a direct sum of a divisible group and a reduced group. See [**52**], 230.

4. Π^1_1-CA_0 is equivalent over RCA_0 to the assertion that every countable Abelian p-group has an Ulm resolution. See [**14**].

Ulm theory is crucial in the theory of torsion Abelian groups (every element has finite order). It involves transfinite recursion. [**34**] discusses two test problems for Ulm theory:

a) Let G, H be countable torsion Abelian groups, where $G+G$ and $H+H$ are isomorphic. Then G, H are isomorphic.
b) Let G, H be countable torsion Abelian groups, where G is a direct summand of H and H is a direct summand of G. Then G, H are isomorphic.

[**34**] emphasizes that these test problems immediately follow from Ulm theory. The relevant Ulm theory for countable Abelian p-groups, through 4 above, is easily formalized in Π^1_1-CA_0. And the relevant Ulm theory for reduced countable Abelian p-groups, through 2 above, is easily formalized in ATR_0. From these considerations, test problems a) and b) are provable in Π^1_1-CA_0, and test problems a) and b) in the reduced case are provable in ATR_0. See [**14**].

PROBLEMS. Are test problems a) and b) equivalent to Π^1_1-CA_0 over RCA_0? Are test problems a) and b) in the reduced case equivalent to ATR_0 over RCA_0?

A number of equivalences of natural consequences of Ulm theory are shown to be equivalent to ATR_0 over RCA_0 in [**14**]. See also [**18**]. For example, both of the following are shown in [**14**] to be equivalent to ATR_0 over RCA_0.

I. Let p be prime and G, H be countable Abelian p-groups. Either G is embeddable into H^∞ or H is embeddable into G^∞.
II. Let p be prime and G, H be countable reduced Abelian p-groups. Either G is embeddable into H^∞ or H is embeddable into G^∞.

10. Replacing RCA_0 by a Weaker Base Theory

In all but §X.4 of Simpson's book [**52**], RCA_0 is taken as the base theory for Reverse Mathematics. That is to say, reversals are stated as theorems of RCA_0. An important research direction for the future is to replace RCA_0 by weaker base theories. In this way we can hope to substantially broaden the scope of Reverse Mathematics, by obtaining reversals for many ordinary mathematical theorems which are provable in RCA_0.

A start on this has already been made. Simpson/Smith [**53**] defined RCA_0^* to be the same as RCA_0 except that Σ^0_1 induction is weakened to Σ^0_0 induction, and exponentiation of natural numbers is assumed. Thus RCA_0 is equivalent to RCA_0^* plus Σ^0_1 induction. It turns out that RCA_0^* is conservative over EFA (elementary function arithmetic) for Π^0_2 sentences, just as RCA_0 is conservative over PRA (primitive recursive arithmetic) for Π^0_2 sentences.

One project for the future is to redo all of the known results in Reverse Mathematics using RCA_0^* as the base theory. The groundwork for this has already been laid, but there are some difficulties. For example, we know that Ramsey's Theorem for exponent 3 is equivalent to ACA_0 over RCA_0, but it unclear whether RCA_0 can be replaced by RCA_0^*. Other problems of this nature are listed in Simpson's book [**52**, remark X.4.3].

Another project is to find ordinary mathematical theorems that are equivalent to Σ^0_1 induction over RCA_0^*. Several results of this kind are already known and are mentioned in Simpson's book [**52**, §X.4]. For example, Hatzikiriakou [**31**] has shown that the well-known structure theorem for finitely generated Abelian groups is equivalent to Σ^0_1-induction over RCA_0^*.

A more visionary project would be to replace RCA_0^* by even weaker base theories, dropping exponentiation and Δ^0_1 comprehension. One could even consider base theories that are conservative over the theory of discrete ordered rings. Initial steps in this direction are discussed in [**20**], [**21**], [**22**], [**23**], [**24**].

References

[1] *FOM e-mail list*, `http://www.math.psu.edu/simpson/fom/`, September 1997 to the present.

[2] K. Ambos-Spies, G. H. Müller, and G. E. Sacks (eds.), *Recursion Theory Week*, Lecture Notes in Mathematics, no. 1432, Springer-Verlag, 1990, ix + 393 pages.

[3] Errett Bishop and Douglas Bridges, *Constructive Analysis*, Grundlehren der mathematischen Wißenschaften, no. 279, Springer-Verlag, 1985, xii + 477 pages.

[4] Andreas R. Blass, Jeffry L. Hirst, and Stephen G. Simpson, *Logical analysis of some theorems of combinatorics and topological dynamics*, [**45**], 1987, pp. 125–156.

[5] Douglas K. Brown, Mariagnese Giusto, and Stephen G. Simpson, *Vitali's theorem and WWKL*, Archive for Mathematical Logic (1999), 18 pages, accepted April 1998, to appear.

[6] Douglas K. Brown and Stephen G. Simpson, *The Baire category theorem in weak subsystems of second order arithmetic*, Journal of Symbolic Logic **58** (1993), 557–578.

[7] Timothy J. Carlson and Stephen G. Simpson, *A dual form of Ramsey's theorem*, Advances in Mathematics **53** (1984), 265–290.

[8] Peter Cholak, Carl G. Jockusch, Jr., and Theodore A. Slaman, *On the strength of Ramsey's theorem for pairs*, Journal of Symbolic Logic (1999), 71 pages, to appear.

[9] S. B. Cooper, T. A. Slaman, and S. S. Wainer (eds.), *Computability, Enumerability, Unsolvability: Directions in Recursion Theory*, London Mathematical Society Lecture Note Series, no. 224, Cambridge University Press, 1996, vii + 347 pages.

[10] J. N. Crossley, J. B. Remmel, R. A. Shore, and M. E. Sweedler (eds.), *Logical Methods*, Birkhäuser, 1993, 813 pages.

[11] H.-D. Ebbinghaus, G.H. Müller, and G.E. Sacks (eds.), *Recursion Theory Week*, Lecture Notes in Mathematics, no. 1141, Springer-Verlag, 1985, ix + 418 pages.

[12] Harvey Friedman, *The Ackerman function in elementary algebraic geometry*, December 10, 1999, unpublished manuscript, 20 pages.

[13] ———, *The analysis of mathematical text, and their calibration in terms of intrinsic stength, I – III*, State University of New York at Buffalo, April 3, April 8, May 19, 1975.

[14] ———, *Metamathematics of Ulm theory*, June 24, 1999, unpublished, 34 pages.

[15] ———, *Subsystems of Set Theory and Analysis*, Ph.D. thesis, Massachusetts Institute of Technology, 1967, 83 pages.

[16] ———, *Some systems of second order arithmetic and their use*, Proceedings of the International Congress of Mathematicians, Vancouver 1974, vol. 1, Canadian Mathematical Congress, 1975, pp. 235–242.

[17] ———, *Systems of second order arithmetic with restricted induction, I, II* (abstracts), Journal of Symbolic Logic **41** (1976), 557–559.

[18] ———, *#49:Ulm theory/Reverse Mathematics*, FOM e-mail list [**1**], July 17, 1999.

[19] ———, *#53:Free sets/Reverse Math*, FOM e-mail list [**1**], July 19, 1999.

[20] ———, *#74:Reverse arithmetic beginnings*, FOM e-mail list [**1**], December 22, 1999.

[21] ———, *#75:Finite Reverse Mathematics*, FOM e-mail list [**1**], December 28, 1999.

[22] ———, *#76:Finite set theories*, FOM e-mail list [**1**], December 28, 1999.

[23] ———, *#77:Missing axiom/atonement*, FOM e-mail list [**1**], January 4, 2000.

[24] ———, *#78:Quadratic axioms/literature conjectures*, FOM e-mail list [**1**], January 7, 2000.

[25] Laszlo Fuchs, *Infinite Abelian Groups*, Pure and Applied Mathematics, Academic Press, 1970–1973, Volume I, xi + 290 pages, Volume II, ix + 360 pages.

[26] Mariagnese Giusto and Stephen G. Simpson, *Located sets and reverse mathematics*, Journal of Symbolic Logic (2000), 37 pages, accepted August 1998, to appear.

[27] W. Timothy Gowers, *A new proof of Szemerédi's theorem*, preprint, 126 pages, January 2000.

[28] Ronald L. Graham, Bruce L. Rothschild, and Joel H. Spencer, *Ramsey Theory*, 2nd ed., Wiley, New York, 1990, xi + 196 pages.
[29] Phillip A. Griffith, *Infinite Abelian Group Theory*, University of Chicago Press, 1970, 152 pages.
[30] L. A. Harrington, M. Morley, A. Scedrov, and S. G. Simpson (eds.), *Harvey Friedman's Research on the Foundations of Mathematics*, Studies in Logic and the Foundations of Mathematics, North-Holland, 1985, xvi + 408 pages.
[31] Kostas Hatzikiriakou, *Algebraic disguises of Σ^0_1 induction*, Archive for Mathematical Logic **29** (1989), 47–51.
[32] W. Hodges, M. Hyland, C. Steinhorn, and J. Truss (eds.), *Logic: From Foundations to Applications, Keele 1993*, Oxford Science Publications, Oxford University Press, 1996, xiii + 536 pages.
[33] A. James Humphreys and Stephen G. Simpson, *Separable Banach space theory needs strong set existence axioms*, Transactions of the American Mathematical Society **348** (1996), 4231–4255.
[34] Irving Kaplansky, *Infinite Abelian Groups*, revised ed., University of Michigan Press, 1969, vii + 95 pages.
[35] Igor Kriz, *Well-quasiordering finite trees with gap-condition*, Annals of Mathematics **130** (1989), 215–226.
[36] Antonín Kučera, *Measure, Π^0_1 classes and complete extensions of PA*, [**11**], 1985, pp. 245–259.
[37] ———, *Randomness and generalizations of fixed point free functions*, [**2**], 1990, pp. 245–254.
[38] ———, *On relative randomness*, Annals of Pure and Applied Logic **63** (1993), 61–67.
[39] Masahiro Kumabe, *A fixed point free minimal degree*, preprint, 48 pages, 1997.
[40] Richard Mansfield and Galen Weitkamp, *Recursive Aspects of Descriptive Set Theory*, Oxford Logic Guides, Oxford University Press, 1985, vi + 144 pages.
[41] Alberto Marcone, *On the logical strength of Nash-Williams' theorem on transfinite sequences*, [**32**], 1996, pp. 327–351.
[42] Michael E. Mytilinaios and Theodore A. Slaman, *On a question of Brown and Simpson*, [**9**], 1996, pp. 205–218.
[43] Marian B. Pour-El and J. Ian Richards, *Computability in Analysis and Physics*, Perspectives in Mathematical Logic, Springer-Verlag, 1988, xi + 206 pages.
[44] Richard A. Shore, *On the strength of Fraïssé's conjecture*, [**10**], 1993, pp. 782–813.
[45] S. G. Simpson (ed.), *Logic and Combinatorics*, Contemporary Mathematics, American Mathematical Society, 1987, xi + 394 pages.
[46] Stephen G. Simpson, *BQO theory and Fraïssé's conjecture*, [**40**], 1985, pp. 124–138.
[47] ———, *Nonprovability of certain combinatorial properties of finite trees*, [**30**], 1985, pp. 87–117.
[48] ———, *Recursion-theoretic aspects of the dual Ramsey theorem*, [**11**], 1986, pp. 356–371.
[49] ———, *Ordinal numbers and the Hilbert basis theorem*, Journal of Symbolic Logic **53** (1988), 961–974.
[50] ———, *Partial realizations of Hilbert's program*, Journal of Symbolic Logic **53** (1988), 349–363.
[51] ———, *The Borel universe (a positive posting)*, FOM e-mail list [**1**], December 6, 1997.
[52] ———, *Subsystems of Second Order Arithmetic*, Perspectives in Mathematical Logic, Springer-Verlag, 1999, xiv + 445 pages.
[53] Stephen G. Simpson and Rick L. Smith, *Factorization of polynomials and Σ^0_1 induction*, Annals of Pure and Applied Logic **31** (1986), 289–306.
[54] Theodore A. Slaman, *A note on the dual Ramsey theorem*, 4 pages, unpublished, January 1997.
[55] Craig Smorynski, *Nonstandard models and related developments*, [**30**], 1985, pp. 179–229.
[56] Fons van Engelen, Arnold W. Miller, and John Steel, *Rigid Borel sets and better quasi-order theory*, [**45**], 1987, pp. 199–222.

Department of Mathematics, Ohio State University
E-mail address: `friedman@math.ohio-state.edu`

Department of Mathematics, Pennsylvania State University
E-mail address: `simpson@math.psu.edu`

Contemporary Mathematics
Volume **257**, 2000

Open Problems in the Theory of Constructive Algebraic Systems

Sergey Goncharov and Bakhadyr Khoussainov

Abstract. In this paper we concentrate on open problems in two directions in the development of the theory of constructive algebraic systems. The first direction deals with universal algebras whose positive open diagrams can be computably enumerated. These algebras are called positive algebras. Here we emphasize the interplay between universal algebra and computability theory. We propose a systematic study of positive algebras as a new direction in the development of the theory of constructive algebraic systems. The second direction concerns the traditional topics in constructive model theory. First we propose the study of constructive models of theories with few models such as countably categorical theories, uncountably categorical theories, and Ehrenfeucht theories. Next, we propose the study of computable isomorphisms and computable dimensions of such models. We also discuss issues related to the computability-theoretic complexity of relations in constructive algebraic systems.

1. Introduction

The use of constructive objects (e.g. numbers, symbols) has played a significant role in the development of mathematics. Greeks and Persians were fascinated with what can be constructed and studied using symbols. For example, they wanted to find explicit formulas for solutions of algebraic equations. This approach continued into the middle of the nineteenth century, when Kronecker used explicit formulas and algorithms in the study of algebra and geometry. Newton and Leibniz solved geometric and physical problems by translating them into symbolic form. In the 1930s the work of Church, Kleene and Turing formalized the notion of computable functions, that is, the functions that can be computed algorithmically. In the 1930s Church and Kleene applied the notion of a computable function to study the effective content of the theory of ordinals. In the late 1930s the algebraist van der Waerden considered and studied fields *given explicitly*, where "a field Δ is given

---

1991 *Mathematics Subject Classification.* Primary: 03C05, 03D45, 03D80.

Key words and phrases. algebraic systems, models, first order theories, computable functions and predicates, computable isomorphisms, constructive/recursive/computable algebraic systems, numberings.

Goncharov's work has been supported by grant RFBR N99-01-00485.

Khoussainov's work has been partially supported by the University of Auckland research committee.

explicitly if its elements are uniquely represented by distinguishable symbols with which addition, subtraction, multiplication and division can be performed by a finite number of operations". By the end of the 1940s the definition of a computable function and the Church-Turing thesis were widely accepted. In the 1950s Frölich and Shepherdson introduced and studied recursive fields by formalizing the notion of "fields given explicitly". Later Rabin continued research on recursive fields and provided examples of recursive fields that fail to have factorization algorithms. In the 1960s Malcev began a systematic study of interactions between algebra, model theory and computability theory. In the early 1970s fundamental work of Ershov and Nerode led to a vast amount of research in the former Soviet Union and the United States in the area. The area has now become known as the theory of constructive (effective, computable) algebraic systems. A goal of this theory is to study the effective content of the techniques, concepts and theorems in the theory of algebraic systems, in particular in model theory and universal algebra. The area is also devoted to the study of interactions between notions and concepts of computability theory and the theory of algebraic systems.

In this paper we discuss some open problems in the theory of constructive algebraic systems and suggest possible directions for further research in the area. We emphasize two directions. The first direction of research is quite general and proposes a systematic development of the theory of positive (or equivalently, computably enumerable) algebras. Generally speaking these are universal algebras whose positive atomic diagrams can be computably enumerated. For example, recursively presented algebras (e.g. finitely presented groups, rings, etc.) have computably enumerable (c.e.) positive diagrams. The Lindenbaum Boolean algebras of computably enumerable theories (e.g. Peano arithmetic) also have c.e. positive diagrams. We suggest the systematic development of the area of positive algebras and think that results and methods of computability theory can fruitfully be applied here. The second direction is more specific and concentrates on traditional topics in the theory of constructive algebraic systems. In particular, the paper is devoted to the study of constructive models of countably categorical theories, uncountably categorical theories, and Ehrenfeucht theories. In this direction we specify three areas of research and open problems. The first area is related to constructing models, the second area is related to the study of computable isomorphisms of constructive models, and finally the third area is related to the study of computability-theoretic properties of relations in constructive algebraic systems.

The paper consists of five sections including the introduction, conclusion, and references. In the next section we provide the basic notions about numbered algebraic systems and models. The section introduces positive universal algebras, contains some results about these algebras, and suggests a systematic development of the theory of positive algebras. The next section, Section 3, considers the questions related to finding constructive presentations (also known as computable presentations or constructivizations) of models of countably categorical, uncountably categorical, and Ehrenfeucht theories. Section 4 is devoted to the study of computable isomorphisms of models. As in the previous section, an emphasis is given to problems related to computable isomorphisms of models of countably categorical, uncountably categorical, and Ehrenfeucht theories. The section also discusses topics related to the dependency of computability-theoretic properties of relations on constructive presentations. Finally, the last section is a conclusion.

We assume that the reader knows some basic facts and concepts from model theory, universal algebra, and computability theory. Some knowledge of the first several chapters of the classic textbooks by Chang and Keisler on model theory [**4**], Grätzer on universal algebra [**15**], Soare on computability theory [**39**], and Malcev on general theory of algebraic systems [**31**] will suffice to follow the paper. We will assume that all algebraic systems considered in this paper are countable unless otherwise stated.

2. Numbered Algebraic Systems

The goal of this section is twofold. On the one hand, we give definitions of constructive and positive algebraic systems, the central notions of this paper. On the other hand, we propose a development of the theory of positive universal algebras, a relatively new area in the field of constructive algebraic systems. This section of the paper consists of five parts. The first part gives basic terminology about numberings. The second and third parts define constructive algebraic systems and positive universal algebras. The last two parts study some general properties of positive algebras.

2.1. Basics of Numeration Theory. In the theory of constructive algebraic systems the notion of numbering plays a central role. A **numbering** of a set A is a map $\nu : \omega \to A$ from the set ω of natural numbers onto the set A. The pair (A, ν) is then called a **numbered set**. For an element $a \in A$, if $\nu(n) = a$ then n is called a **ν-name of the element**. The basic idea in the definition of a numbering is to give constructive names to the elements of the set A, and thus, in some sense, to coordinate or represent elements of the set A by constructive means. Note that the same element $a \in A$ can have several ν-names. A natural set associated with the numbering ν is the equivalence relation E_ν on the set ω that identifies those numbers x, y that name the same element. Formally, we give the following definition:

DEFINITION 2.1. Let (A, ν) be a numbered set. The **equivalence relation induced by** ν is the set $E_\nu = \{(x, y) | \nu(x) = \nu(y)\}$.

Given two numberings ν and μ of the set A, from a computational point of view, one is naturally interested in how these two numbering are related to each other. This leads one to the following natural notion of a reducibility between numberings of the set A. A numbering ν is **reducible** to a numbering μ, written $\nu \leq \mu$, if there exists a computable function from ω into ω such that $\nu(n) = \mu(f(n))$ for all $n \in \omega$. Informally, this means that there exists an effective procedure that applied to any ν-name of an element produces a μ-name of the same element. Two numberings ν and μ are **equivalent** if $\nu \leq \mu$ and $\mu \leq \nu$. Let $N(A)$ be the set of all equivalence classes of all numberings of A. The reducibility relation naturally induces a partial order, also denoted by $\leq$, on $N(A)$. Thus, we have a partially ordered set $(N(A), \leq)$ which constitutes one of the main objects of study in the theory of numerations [**5**]. Note that this partially ordered set is an uppersemilattice, where the join operation $\nu_1 \oplus \nu_2$ is defined as follows. For all $n \in \omega$ if $n = 2k$ then $\nu_1 \oplus \nu_2(n) = \nu_1(k)$; if $n = 2k+1$ then $\nu_1 \oplus \nu_2(n) = \nu_2(k)$. An interesting note is that when A consists of exactly two elements then the semilattice $(N(A), \leq)$ is isomorphic to the semilattice of all many-one reducibility degrees.

Depending on the set A, one considers certain types of numberings which can be of some interest. For example, A can be a set of some c.e. sets (or computable functions) in which case numberings known as computable numberings are of particular interest. A numbering ν of A is called **computable** if the set $\{(n,m)|n \in \nu(m)\}$ is a c.e. set. For example, the standard Kleene enumeration of all c.e. sets is a computable numbering. Informally, if A has a computable numbering then one can uniformly list all sets in A possibly with repetitions. This leads to considering the set $N_C(A)$ of all equivalence classes of computable numberings and studying the partially ordered set $(N_C(A), \leq)$. Computable numberings have been extensively studied, especially by the Novosibirsk school of logic. The paper by Badaev and Goncharov in this volume discusses issues related to the theory of numberings. We also refer the reader interested in the subject of the theory of numberings to the book by Ershov[**5**].

2.2. Numbered Algebraic Systems. We fix a language $L =< f_0^{n_0}, f_1^{n_1}, \dots, P_0^{m_0}, P_1^{m_1}, \dots, c_0, c_1, \dots >$ for which the functions $i \to n_i$ and $j \to m_j$ are computable. Such languages are called **computable languages**. The symbols $f_i^{n_i}$ and $P_j^{m_j}$ are operation and predicate symbols, respectively. If the language contains no predicate symbols, then the algebraic systems of the language are called **universal algebras**, or for short **algebras**. We denote algebraic systems by letters $\mathcal{A}$, $\mathcal{B}$, etc. The domains are, respectively, denoted by A, B, etc. A **numbered algebraic system** is a pair $(\mathcal{A}, \nu)$, where ν is a numbering of the domain of $\mathcal{A}$.

In order to motivate our next definitions we recall several notions from model theory. Let $\mathcal{A}$ be an algebraic system. When one is interested in properties of $\mathcal{A}$ from the predicate calculus point of view, e.g. the space of types of $\mathcal{A}$, elementary embeddings of $\mathcal{A}$, first order definable relations on $\mathcal{A}$, then the full diagram of $\mathcal{A}$, that is the set $FD(\mathcal{A}) = \{\phi(a_1, \dots, a_n) \mid \phi(x_1, \dots, x_n)$ is a formula, $\mathcal{A} \models \phi(a_1, \dots, a_n)$, $a_1, \dots, a_n \in A\}$, gives the necessary information about the properties. On the other hand, when one is interested in $\mathcal{A}$ from an algebraic point of view, e.g. subsystems of $\mathcal{A}$, embeddings of $\mathcal{A}$, homomorphisms of $\mathcal{A}$, then it is natural to consider the atomic diagram of $\mathcal{A}$, that is the set $AD(\mathcal{A}) = \{\phi(a_1, \dots, a_n) \mid \phi(x_1, \dots, x_n)$ is an atomic formula or a negation of an atomic formula, $\mathcal{A} \models \phi(a_1, \dots, a_n), a_1, \dots, a_n \in A\}$.

As we introduce numberings of the algebraic systems, one can consider the full and atomic diagrams of these systems under the numberings. Formally, let $(\mathcal{A}, \nu)$ be a numbered algebraic system. We expand the system by adding new constants a_i to the language L so that the value of each a_i is the element $\nu(i)$ for all $i \in \omega$. Let L_1 be the expanded language. The **full diagram of $\mathcal{A}$ under the numbering** ν is $FD_\nu(\mathcal{A}) = \{\phi(\nu(i_1), \dots, \nu(i_n)) \mid \mathcal{A} \models \phi(\nu(i_1), \dots, \nu(i_n))$, and $\phi(x_1, \dots, x_n)$ is a formula of L_1, $i_1, \dots, i_n, n \in \omega\}$. Similarly, the **atomic diagram of $\mathcal{A}$ under the numbering** ν is $AD_\nu(\mathcal{A}) = \{\phi(\nu(i_1), \dots, \nu(i_n)) \mid \phi(x_1, \dots, x_n)$ is an atomic formula or the negation of an atomic formula of L_1, $i_1, \dots, i_n \in \omega$, and $\mathcal{A} \models \phi(\nu(i_1), \dots, \nu(i_n))\}$. Here is a central definition of this paper.

DEFINITION 2.2. A pair $(\mathcal{A}, \nu)$ is a **strongly constructive algebraic system** if the set $FD_\nu(\mathcal{A})$ is a computable set. In this case, ν is called a **strong constructivization** of $\mathcal{A}$. Similarly, the system $(\mathcal{A}, \nu)$ is **constructive** if the set $AD_\nu(\mathcal{A})$ is a computable set. In this case, ν is called a **constructivization**, or equivalently a **constructive presentation** of $\mathcal{A}$.

Clearly every strongly constructive algebraic system is a constructive one. The converse does not always hold as $(\omega, +, \times, \leq)$ is an example of a constructive but not strongly constructive system whose constructivization is the identity mapping.

We say that two constructive models $(\mathcal{A}, \nu)$ and $(\mathcal{A}, \mu)$ are **computably isomorphic** if there exists an automorphism $\alpha : \mathcal{A} \to \mathcal{A}$ and a computable function $f : \omega \to \omega$ such that $\alpha(\nu(n)) = \mu(f(n))$ for all n. In this case the constructivizations ν and μ are called **autoequivalent** constructivizations. In other words, autoequivalent constructivizations are those that are equivalent up to an automorphism of the system.

LEMMA 2.3. *Any constructive system $(\mathcal{A}, \nu)$ is computably isomorphic to a constructive system $(\mathcal{A}, \mu)$ such that μ is a one to one mapping.*

Proof. Since ν is a constructivization, the equivalence relation $E_\nu = \{(n, m) | \nu(n) = \nu(m)\}$ is a computable set. Let $k_0 < k_1 < k_2 < \ldots$ be an effective list of all minimal elements in the E_ν-equivalence classes. For every $n \in \omega$, set $\mu(n) = \nu(k_n)$. Clearly, μ is a one to one constructivization that is equivalent to ν. This proves the lemma.

In the literature there is an equivalent terminology for constructive models and constructivizations that does not refer to numberings. These are computable algebraic systems and computable presentations. An algebraic system is called **computable** if the domain of the system is ω and the atomic diagram is a computable set. Clearly, every computable system is a constructive system whose constructivization is the identity mapping from ω onto ω. An algebraic system is called **decidable** if the domain of the system is ω and the full diagram is a computable set. Clearly, every decidable system is a strongly constructive system. The lemma above shows the opposite, that is, every constructive (strongly constructive) algebraic system can be considered as a computable (decidable) algebraic system. Indeed, assume that $(\mathcal{A}, \nu)$ is a constructive algebraic system. Then by the lemma above, we can assume that ν is a one to one mapping. The numbering ν naturally induces an algebraic system with domain ω isomorphic to $\mathcal{A}$. If $(\mathcal{A}, \nu)$ is constructive then the system induced is computable. If $(\mathcal{A}, \nu)$ is strongly constructive then the system induced is decidable.

2.3. Numbered Algebras. Fix a computable language $L = < f_0^{n_0}, f_1^{n_1}, \ldots >$ with no predicate symbols. The algebraic systems of the language are now algebras. The study of numberings of algebras is of particular interest from a computability point of view because any algebra can be numbered in such a way that all the basic operations of the algebra become computable under the numbering. In other words, any algebra can, in some sense, be effectivized if it is numbered properly. To formalize this we give the following definition.

DEFINITION 2.4. A **numbered algebra** is a pair $(\mathcal{A}, \nu)$ for which there exists a computable sequence of computable functions $\psi_0^{n_0}, \psi_1^{n_1}, \ldots$ such that for all $i, t_1, \ldots, t_{n_i} \in \omega$ we have $f_i^{n_i}(\nu(t_1), \ldots, \nu(t_{n_i})) = \nu(\psi_i^{n_i}(t_1, \ldots, t_{n_i}))$. In this case ν is called a **numbering of the algebra**.

Thus, informally if ν is a numbering of an algebra $\mathcal{A}$ then all the basic operations of $\mathcal{A}$ can be carried out effectively under ν.

THEOREM 2.5. *Any algebra possesses a numbering.*

Proof. Indeed, let $\mathcal{A}$ be an infinite algebra. Consider the absolutely free algebra $\mathcal{F}$ of the language L whose set of generators is the domain A. Since A is a countable set, we can assume that $A = \omega$. Let $\nu : \omega \to F$ be a one to one numbering of the algebra so that the set $\{(t, n) \mid n \in \omega, t \in F, t = \nu(n)\}$ is computable. Then clearly $(\mathcal{F}, \nu)$ is a constructive algebra. Let h be a homomorphism from $\mathcal{F}$ onto $\mathcal{A}$ such that $h(i) = i$ for all $i \in \omega$. We note that such a homomorphism exists because $\mathcal{F}$ is absolutely free. Thus, the pair $(\mathcal{A}, \mu)$, where μ is defined by $\mu(n) = h(\nu(n))$, is a numbered algebra. This proves the theorem.

The theorem suggests that the complexity of a numbered algebra $(\mathcal{A}, \nu)$ can be identified with the complexity of the relation E_ν. We give the following definition.

DEFINITION 2.6. A numbered algebra $(\mathcal{A}, \nu)$ is a **Σ_n-algebra** (**Π_n-algebra**) if the relation E_ν is a Σ_n-set (Π_n-set). If $(\mathcal{A}, \nu)$ is both a Σ_n-algebra and a Π_n-algebra then we call it a **Δ_n-algebra**.

Thus, Δ_1-algebras are exactly the class of all constructive algebras. Of course, there are natural examples of nonconstructive numbered algebras that have been intensively studied in computability theory. For example, the lattice $\mathcal{E} = (\{W_i\}_{i\in\omega}, \bigcup, \bigcap)$ of all computably enumerable sets is a Π_2-algebra, where $i \to W_i$ is a standard enumeration of all c.e. sets. Similarly, the algebra $\mathcal{E}^\star$, obtained from $\mathcal{E}$ by factoring it modulo finite sets is an example of a Σ_3-algebra.

For any Σ_1-algebra $(\mathcal{A}, \nu)$ the **positive diagram** of this algebra, that is, the set $\{\phi(\nu(t_1), \ldots, \nu(t_n)) \mid \mathcal{A} \models \phi(\nu(t_1), \ldots, \nu(t_n)),\ \phi(x_1, \ldots, x_n)$ is an atomic formula$\}$ is a c.e. set. Similarly, for a Π_1-algebra $(\mathcal{A}, \nu)$ the **negative diagram** of the algebra, that is, the set $\{\phi(\nu(t_1), \ldots, \nu(t_n)) \mid \mathcal{A} \models \phi(\nu(t_1), \ldots, \nu(t_n))$ $\phi(x_1, \ldots, x_n)$ is the negation of atomic formula$\}$ is a c.e. set. This observation suggests the following definition.

DEFINITION 2.7. Any Σ_1-algebra is called a **positive algebra**. Any Π_1-algebra is called a **negative algebra**.

One of the general programs in the theory of constructive algebraic systems is the study of numbered algebras, in particular the study of Σ_n-algebras and/or Π_n-algebras. This is an open area of research where many results of universal algebra can be studied from a computability theory point of view. We pose this as an open problem for research:

PROBLEM 1. Develop the general theory of Σ_n-algebras.

The following two sections give some interesting examples of results about positive algebras.

2.4. Positive Algebras. In this section we are interested in finding algebraic conditions for positive algebras to possess constructivizations. Recall that a **congruence** on an algebra $\mathcal{A}$ is an equivalence relation η on A such that, for every basic n-ary operation f and all $(x_1, \ldots, x_n), (y_1, \ldots, y_n) \in A^n$, the condition $(x_1, y_1), \ldots, (x_n, y_n) \in \eta$ implies that the pair $(f(x_1, \ldots, x_n), f(y_1, \ldots, y_n)) \in \eta$.

THEOREM 2.8. *A positive algebra $(\mathcal{A}, \nu)$ is constructive if and only if there is a c.e. set $S \subset \omega^2$ such that $\nu(x) \neq \nu(y)$ for all $(x, y) \in S$, and for any nonzero congruence relation η we have $(\nu(x), \nu(y)) \in \eta$ for some $(x, y) \in S$.*

Proof. We first show that there exists an effective procedure that given an index of a c.e. set $X \subset \omega^2$ produces an index of a c.e. set $Y \subset \omega^2$ such that the relation $\nu(Y) = \{(\nu(x), \nu(y)) | (x, y) \in Y\}$ is the smallest congruence relation containing $\nu(X)$. To prove this, note that any congruence relation η that contains $\nu(X)$ must satisfy the following three conditions: 1) $\{(a, a) | a \in A\} \subset \eta$; 2) $\nu(X) \subset \eta$; 3) For every basic operation f of arity n and $(x_1, \dots, x_n), (y_1, \dots, y_n) \in A^n$ the condition $(x_1, y_1), \dots, (x_n, y_n) \in \eta$ implies that $(f(x_1, \dots, x_n), f(y_1, \dots, y_n) \in \eta$. Thus, we can computably enumerate the set Y required in the lemma by satisfying the following three properties that correspond to the conditions 1), 2) and 3) above: 1) $E_\nu \subset Y$; 2) $X \subset Y$; 3) For every basic operation f of arity n, for all $s, t \in \omega$, $(x_1, \dots, x_n), (y_1, \dots, y_n) \in \omega^n$ the conditions $(x_1, y_1), \dots, (x_n, y_n) \in Y$, $\nu(t) = f(\nu(x_1), \dots, \nu(x_n))$, and $\nu(s) = f(\nu(y_1), \dots, \nu(y_n))$ imply that $(s, t) \in Y$.

Now in order to prove the theorem, we need to show that the equivalence relation E_ν is computable. Take $x, y \in \omega$. Consider the set Y such that $\nu(Y)$ is the minimal congruence relation that contains the pair $(\nu(x), \nu(y))$. Note that if $\nu(y) = \nu(x)$ then $Y = E_\nu$. Otherwise, there exists a pair $(n, m) \in S$ such that $(n, m) \in Y$. Thus, using the properties of S and that S, Y, and E_ν are c.e. sets, we can decide if $\nu(x) = \nu(y)$. Hence ν is a constructivization. The theorem is proved.

A corollary of this theorem is the following fact that gives an example of how algebraic properties may influence effective properties of numberings. An algebra is **quasisimple** if it has only a finite number of congruence relations.

COROLLARY 2.9. *Any positive and quasisimple algebra $(\mathcal{A}, \nu)$ is constructive.*

Proof. Let $\eta_0, \dots, \eta_k$ be all nonzero congruences of $\mathcal{A}$. For each $i \leq k$ take (x_i, y_i) such that $\nu(x_i) \neq \nu(y_i)$ and $(\nu(x_i), \nu(y_i)) \in \eta_i$. The the set $S = \{(x_1, y_1), \dots, (x_k, y_k)\}$ satisfies the assumptions of Theorem 2.8.

We recall, before we state the next corollary, that the index of an equivalence relation is the number of its equivalence classes.

COROLLARY 2.10. *If $(\mathcal{A}, \nu)$ is a positive finitely generated algebra with nonzero congruences of finite index only and the language of $\mathcal{A}$ is finite then $(\mathcal{A}, \nu)$ is constructive.*

Proof. Let $c_1, \dots, c_n$ be generators of $\mathcal{A}$. Consider the **ground terms** of the language expanded by constants $c_1, \dots, c_n$, that is the terms that contain no variables. We inductively define the **height** of terms as follows. The height of each constant is 0. Let $t = f(t_1, \dots, t_n)$ be a term. Let m be the maximum among all the height of $t_1, \dots, t_n$. Then the height of t is $m + 1$. Note that for any given m we can effectively compute the number of terms of height m. For any finite algebra $\mathcal{B}$ whose generators are $c_1, \dots, c_n$ there exists a number m with the following property. For any term t of height $m + 1$ there exists a term t' such that the equality $t = t'$ holds in $\mathcal{B}$ and the height of t' does not exceed m. Now for the positive algebra $(\mathcal{A}, \nu)$ we define the set S as follows. A pair (x, y) belongs to S if and only if the minimum congruence relation that contains the pair $(\nu(x), \nu(y))$ is of finite index. The set S is c.e. and satisfies the assumptions of the theorem above.

The next corollary is a result of McKinzey [**32**]. We need some preliminary notions. An algebra $\mathcal{A}$ is **residually finite in a class K of algebras** if, for any two distinct elements a, b of the algebra, there exists a homomorphism h of the algebra onto a finite algebra from K such that $h(a) \neq h(b)$. If K is the class of all finite algebras then $\mathcal{A}$ is called **residually finite**. A **conditional equation** is a universally quantified formula of the type $t_1 = q_1 \& \ldots \& t_n = q_n \to t = q$, where $t, q, t_1, q_1, \ldots, t_n, q_n$ are terms of the signature expanded by finitely many constants $c_0, \ldots, c_n$. Let C be a set of conditional equations. Then there exists the **free algebra**, denoted $\mathcal{F}_C$, that satisfies the following properties: 1) $\mathcal{F}_C$ is finitely generated with generators $c_0, c_1, \ldots, c_n$. 2) $\mathcal{F}_C$ satisfies C. 3) Any algebra with properties 1) and 2) is a homomorphic image of $\mathcal{F}_C$. 4) The set $\{t_1 = t_2 | \mathcal{A}$ satisfies the equality $t_1 = t_2$ and t_1, t_2 contain no variables $\}$ is c.e. in C. The first three properties define $\mathcal{F}_C$ up to isomorphism. If C is finite then the algebra $\mathcal{F}_C$ is called a **finitely presented algebra**.

Let ν be a one to one numbering of all ground terms of the language $L \bigcup \{c_0, \ldots, c_n\}$ such that the set $\{t = \nu(n) | n \in \omega, t$ is a ground term $\}$ is computable. The numbering ν induces a natural numbering ν_s of the algebra $\mathcal{F}_C$ (see Theorem 2.5). Then if C is a c.e. set then the numbered algebra $(\mathcal{F}_C, \nu_s)$ is a positive algebra.

COROLLARY 2.11. *If $\mathcal{F}_C$ is finitely presented and residually finite in the class of all algebras satisfying C then $(\mathcal{A}, \nu_s)$ is a constructive algebra.*

Proof. We can assume that C is finite. Let $\mathcal{A}_0, \mathcal{A}_1, \ldots$ be an effective sequence of all finite algebras that satisfy C. Note that there exists a natural homomorphism h_i from $\mathcal{F}_C$ onto $\mathcal{A}_i$. Moreover the set $\{(i, x, y) | h_i(\nu_s(x)) = h_i(\nu_s(y))\}$ is computable. Consider the set $S \subset \omega^2$ such that $(x, y) \in S$ if and only if there exists an algebra $\mathcal{A}_i$ such that the images of $\nu_s(x)$ and $\nu_s(y)$ are distinct in $\mathcal{A}_i$ under the natural homomorphism h_i. The set S satisfies the assumptions of Theorem 2.8.

2.5. Positive Algebras with Countably Many Congruences. In this section we provide some interesting computability-theoretic and algebraic properties of positive algebras with a countable number of congruences. Positive quasisimple algebras and algebras with congruences of finite index only are examples of algebras with countably many congruences. In this section we always assume that the language L is finite. All the results in this section were first obtained by N. Kasymov [**22**] [**23**]. All the algebras considered in this section are infinite.

For a positive algebra $(\mathcal{A}, \nu)$, the **characteristic transversal** $tr(\nu)$ of ν is the set $\{x | \forall y (y < x \to \nu(x) \neq \nu(y))\}$. For $x, y \in \omega$, let $\eta_\nu(x, y)$ be the smallest congruence containing the pair $(\nu(x), \nu(y))$. Note, $E_\nu \subset \nu^{-1}(\eta_\nu(x, y))$ and an index of $\nu^{-1}(\eta_\nu(x, y))$ can be obtained effectively given x and y.

THEOREM 2.12. *For any positive algebra $(\mathcal{A}, \nu)$ with congruences of finite index only the set $tr(\nu)$ is either computable or hyperimmune.*

Proof. Assume that $tr(\nu)$ is not hyperimmune. We need to show that $tr(\nu)$ is computable. It suffices to show that the equivalence relation E_ν is a computable set. Since $tr(\nu)$ is not hyperimmune there must exist a strong array $S_0, S_1, \ldots$ of finite disjoint sets such that $S_i \bigcap tr(\nu) \neq \emptyset$ for all i. Hence for every S_m there exists a $z \in S_m$ such that for all $t < z$ we have $(t, z) \notin E_\nu$. Consider $\eta_\nu(x, y)$. If $\nu(x) \neq \nu(y)$ then $\eta_\nu(x, y)$ is of finite index. Hence there must exist an S_m such that for every $z \in S_m$ there exists a $t < z$ for which $(\nu(z), \nu(t)) \in \eta_\nu(x, y)$. Also, if $\nu(x) = \nu(y)$ then $E_\nu = \eta_\nu(x, y)$. These two are mutually exclusive cases which can be effectively checked given x and y. Hence E_ν is computable. The theorem is proved.

The theorem above implies the next result. The result shows that the assumption that $(\mathcal{A}, \nu)$ is a positive algebra, with congruences of finite index only, has not only computability-theoretic implications but also purely algebraic implications for the algebra $\mathcal{A}$.

THEOREM 2.13. *If $(\mathcal{A}, \nu)$ is a positive but not constructive algebra with congruences of finite index only, then the algebra $\mathcal{A}$ is locally finite, residually finite and the language of $\mathcal{A}$ contains at least one binary function symbol.*

Proof. Recall that an algebra is **locally finite** if every finitely generated subalgebra of the algebra is finite. Assume that $(\mathcal{A}, \nu)$ is not locally finite. Hence there exist finitely many elements $a_1, \ldots, a_n \in A$ such that the subalgebra $\mathcal{B}$ generated by the elements is infinite. Note that $\nu^{-1}(B)$ is a c.e. set. Since $\mathcal{B}$ is infinite and finitely generated there exists a computable sequence $B_0 \subset B_1 \subset \ldots$ of subsets of ω such that $B = \bigcup_i B_i$, the function $i \to |B_i|$ is computable, and $\nu(B_{i+1}) \setminus \nu(B_i) \neq \emptyset$. Hence the function $i \to max(B_i)$ majorizes the characteristic transversal $tr(\nu)$. But by the previous theorem $tr(\nu)$ is hyperimmune since, by assumption, $(\mathcal{A}, \nu)$ is not a constructive algebra. This is a contradiction. Hence $\mathcal{A}$ is locally finite.

Now we prove that $\mathcal{A}$ is residually finite. Let $\nu(x)$ and $\nu(y)$ be counterexamples to the fact that $\mathcal{A}$ is residually finite. Hence for all $n, m \in \omega$ if $\nu(n) \neq \nu(m)$ then, since every congruence is of finite index, we have $(\nu(x), \nu(y)) \in \eta_\nu(n, m)$. We conclude that the equality $\nu(n) = \nu(m)$ can be checked effectively which implies that $(\mathcal{A}, \nu)$ is a constructive algebra. This contradicts with the assumption.

Now we show that the language of $\mathcal{A}$ contains at least one binary operation symbol. Assume that all basic operations of $\mathcal{A}$ are unary. Take an element $b \in A$. Consider the subalgebra $\mathcal{B}$ generated by b. The subalgebra $\mathcal{B}$ is finite. Consider the equivalence relation $\{(a, a) \mid a \in A\} \bigcup B^2$. The equivalence relation is a congruence relation on $\mathcal{A}$ because all the basic operations are unary. However the index of the congruence relation is not finite. This is a contradiction with the assumption. The theorem is proved.

By Corollary 2.10 every positive finitely generated algebra with congruences of finite index only is a constructive algebra. On the other hand, Kassimov provided examples of positive algebras without constructivizations every congruence of which is of finite index [**22**].

The last example of results about positive algebras is the next theorem that has an implication on the cardinality of the congruence lattice of a given algebra. It gives a sufficient condition for a positive algebra not to have countably many congruence relations. We need one notion. We say that a numbered algebra $(\mathcal{A}, \nu)$ is **effectively infinite** if there exists an infinite c.e. set X such that $\nu(x) \neq \nu(y)$ for all distinct $x, y \in X$.

THEOREM 2.14. *Every positive finitely generated noneffectively infinite algebra $(\mathcal{A}, \nu)$ has a continuum number of congruence relations.*

Proof. Let $a_0, \ldots, a_n$ be the generators of the algebra $\mathcal{A}$. Consider the ground terms of the language of $\mathcal{A}$ expanded by the constants $a_0, \ldots, a_n$. Each ground term t naturally defines an element of $\mathcal{A}$ which we also denote by t. Let $t_1, q_1, \ldots, t_s, q_s$ be a set of ground terms such that in the algebra $\mathcal{A}$ the set of inequalities $I = \{t_i \neq q_i \mid i = 1, \ldots, s\}$ holds.

CLAIM 1. There exist ground terms p, q and s, t such that $\eta_\nu(p,q)$ and $\eta_\nu(s,t)$ are each of infinite index, not comparable, and inequalities in I are true in any quotient algebra with respect to these congruence relations.

Proof. Let F be the set of all pairs of ground terms (t_1, t_2) such that $\eta_\nu(t_1,t_2)$ is of finite index. Let V be the set of all pairs (t_1,t_2) of ground terms such that the quotient algebra with respect to $\eta_\nu(t_1,t_2)$ does not satisfy I. Now note that F and V are computably enumerable sets. Hence if for all (t_1,t_2) either $(t_1,t_2) \in E_\nu$ or $(t_1,t_2) \in F$ or $(t_1,t_2) \in V$ then ν would be a constructivization. Therefore, there exists a pair p, q such that $p \neq q$, $\eta_\nu(p,q)$ has an infinite index, and the quotient algebra with respect to $\eta_\nu(p,q)$ satisfies I. We now need to show that there exists a pair s,t that satisfies the claim. Assume that such s and t do not exist. Then for all ground terms s,t for which $(s,t) \notin \eta_\nu(p,q)$ either $\eta_\nu(s,t)$ is of finite index or the quotient algebra with respect to $\eta_\nu(s,t)$ does not satisfy I or $(p,q) \in \eta_\nu(s,t)$. Using this fact, we now show that the algebra $(\mathcal{A},\nu)$ is effectively infinite. We construct a c.e. infinite sequence $r_0, r_1, \ldots$ of ground terms as follows.

Stage 0. Let r_0 be any ground term.

Stage n+1. Assume that the sequence $r_0, \ldots, r_n$ has been constructed and $r_i \neq r_j$ holds true in the quotient $\mathcal{A}$ with respect to $\eta_\nu(p,q)$ for all $i \neq j$. Note that there exists an r such that $r \neq r_i$ holds in the quotient $\mathcal{A}$ with respect to $\eta_\nu(p,q)$. Find an r such that for all $i \leq n$ either $\eta_\nu(r,r_i)$ is of finite index or the quotient algebra with respect to $\eta_\nu(r,r_i)$ does not satisfy I or $(p,q) \in \eta_\nu(r,r_i)$. Note that such an r exists and can be found effectively. Let r_{n+1} be the first such r found.

Thus, we have contradicted with the assumption that $(\mathcal{A},\nu)$ is not effectively infinite. Hence the claimed pairs $(p,q),(s,t)$ exist. The claim is proved.

Now it is easy to embed the set of paths through the binary tree into the lattice of congruences of $\mathcal{A}$. We send the root of the tree into $\eta_\emptyset = \{(a,a) \mid a \in A\}$. We set $I_\emptyset = \emptyset$. Assume that the congruence relation $\eta_{i_0 \ldots i_n}$ and the finite set $I_{i_0 \ldots i_n}$ have been defined, where $i_0, \ldots, i_n \in \{0,1\}$. Consider the algebra $\mathcal{B}$ obtained by factoring $\mathcal{A}$ with respect to $\eta_{i_0 \ldots i_n}$. The algebra is positive and is not effectively infinite by the induction hypothesis. By the claim there exists $(p,q),(s,t)$ such that $\eta_\nu(p,q)$ and $\eta_\nu(s,t)$ are each of infinite index, not comparable, and inequalities in $I_{i_0 \ldots i_n}$ hold in any quotient algebra with respect to these congruence relations. Map $i_0 \ldots i_n 0$ into $\eta'_\nu(p,q)$, $i_0 \ldots i_n 1$ onto $\eta'_\nu(s,t)$, where $\eta'_\nu(r_1,r_2)$ denotes the smallest congruence of $\mathcal{B}$ that contains r_1 and r_2. Set $I_{i_0 \ldots i_n 0} = I_{i_0 \ldots i_n} \bigcup \{s \neq t\}$ and $I_{i_0 \ldots i_n 1} = I_{i_0 \ldots i_n} \bigcup \{p \neq q\}$. Clearly each path corresponds to a congruence relation and the correspondence is one to one. The theorem is proved.

COROLLARY 2.15. *Every finitely generated infinite positive algebra with countably many congruence relations is effectively infinite.*

We end this section by proposing the following research program.

PROBLEM 2. Develop the theory of positive algebras.

3. Constructive Models of Theories

In contrast to the previous section, this section and the next are devoted to specific open problems related to constructive models of theories and computable isomorphisms. Many of the problems have been open and known for many years and, perhaps, new ideas, constructions, and concepts will be needed to solve these problems. This section consists of three parts. The first part is devoted to constructive models of countably categorical theories. The second part discusses the issues related to constructivizations of models of uncountably categorical theories. The last part will deal with constructive models of Ehrenfeucht theories.

3.1. Constructive Countably Categorical Models. Before we discuss the issues related to constructive models of countably categorical theories, we recall some basic general facts about constructive models of theories. By a theory we always mean a set of sentences closed under deduction. So, let T be a consistent theory. The Completeness Theorem states that T has a model. A proof of this result can be based on a Henkin type construction. If, in addition, T is a decidable theory then the construction can be carried out effectively. Hence the full diagram of the model constructed is decidable. Thus, we have the following fundamental theorem, known as The Effective Completeness Theorem:

THEOREM 3.1. *If T is decidable and consistent theory then T has a strongly constructive model.*

The Effective Completeness Theorem suggests several fundamental questions about models of theories: Which models of T have strong constructivizations? If T has a prime model then does the prime model have a strong constructivization? If T has a saturated model then does the saturated model have a strong constructivization? When does a given homogeneous model of T have a strong constructivization? How many strongly constructive models can T have? There has been extensive research in the study of these question. Basically all these questions have been answered. The following theorem proved in [**14**] and [**18**] and other places characterizes the theories with strongly constructive prime models:

THEOREM 3.2. *A decidable complete theory has a strongly constructive prime model if and only if there exists an algorithm that, for any formula consistent with T, produces a principal type of the theory that contains the formula.*

The following theorem proved in [**14**] and [**34**] characterizes all the theories whose saturated models have strong constructivizations.

THEOREM 3.3. *A decidable complete theory T has a strongly constructive saturated model if and only if the set of all types of T has a computable numbering.*

For a survey of results related to strongly constructive models of theories we refer the reader to papers in [**6**], especially the paper by Harizanov. Now we concentrate our attention on constructivizations of **countably categorical models**, that is the models of countably categorical theories. Recall the following definition.

DEFINITION 3.4. A theory T is **countably categorical** if T has exactly one countable model up to isomorphism.

From a model theory point of view, countably categorical theories and their models have been very well-studied and understood. Clearly, if T is countably categorical then we have the following result that follows from The Effective Completeness Theorem:

THEOREM 3.5. *A countably categorical theory T is decidable if and only if all models of T have strong constructivizations if and only if T has a strongly constructive model.*

Thus, the question of the existence of strongly constructive models for countably categorical categorical theories is answered via their decidability.

The Ryll-Nardzewski Theorem characterizes countably categorical theories in terms of types. The theorem states that a theory T is countably categorical if and only if for every n the number of n-types of T is finite. This theorem leads us to define the type function of the theory T, denoted by $type_T$, that associates with each $n \geq 1$ the number of n-types of the theory. Note that if T is decidable then $type_T$ is a Δ_2^0-function. A natural question arises as to what can be said about the computability-theoretic complexity of the type function $type_T$ when T is decidable. The following theorem, proved independently by Herrmann, Schmerl and Venning, answers the question.

THEOREM 3.6. [**38**] *For any c.e. degree* $\mathbf{x}$ *there exists a decidable countably categorical theory* T *such that the type function* $type_T$ *is of degree* $\mathbf{x}$.

These two theorems basically answer the questions related to the existence of strongly constructive models for countably categorical decidable theories. However, the theorems and their proofs do not give a clear picture about the existence of constructive models of countably categorical theories when one omits the assumption of decidability. We pose the following problem:

PROBLEM 3. Characterize the countably categorical theories that have constructive models.

We note that there has been some research on this problem. Lerman and Schmerl in [**29**] give some sufficient conditions for countably categorical *arithmetic* theories to have a constructive model. More precisely, they have shown that if T is a countably categorical arithmetical theory such that the set of all sentences beginning with an existential quantifier and having $n+1$ alternations of quantifiers is Σ_{n+1}^0 for each n, then T has a constructive model. However, this result cannot be considered as a solution to the problem. We do not even know of any example that satisfies the conditions of this results for sufficiently large n. Hence one of the ways to approach the problem above is to actually build such theories:

CONJECTURE 1. For every $n \geq 1$, there exists a countably categorical theory of Turing degree $\mathbf{0}^{(n)}$ that has a constructive model.

It is not hard to see that if a theory T has a constructive model then T is computable in $\mathbf{0}^{(\omega)}$, the degree of the ω-jump of a computable set. In fact, this bound is sharp as there exist theories (the theory of $(\omega, +, \times, \leq)$ for example) with computable models which are Turing equivalent to $\mathbf{0}^{(\omega)}$. We now end this section with a formulation of the next conjecture that states that a nonarithmetical countably categorical theory with constructive models exists.

CONJECTURE 2. There exists a countably categorical theory of Turing degree $\mathbf{0}^{(\omega)}$ that has a constructive model.

3.2. Constructive Uncountably Categorical Models. Another class of theories well-studied in model theory is the class of uncountably categorical theories. In this section we deal with models of uncountably categorical theories. Throughout this section, we assume that the theories considered are *not* countably categorical. We recall the definition.

DEFINITION 3.7. A theory T is **uncountably categorical** if any two models of T of cardinality ω_1 are isomorphic. Models of uncountably categorical theories are called **uncountably categorical models.**

Typical examples of uncountably categorical theories are the following: the theory of algebraically closed fields of fixed characteristic, the theory of vector spaces over a fixed countable field, the theory of the structure (ω, S), where S is the successor function on ω. Roughly speaking, all the countable models of each of these theories can be listed in an $\omega + 1$ chain: the first element of the chain is the prime model, the last element of the chain is the saturated model; moreover, for any two models of the theory one can be elementarily embedded into the other. It turns out these are one of the basic structural properties of the class of models of an uncountably categorical theory. More precisely, in [**3**] Baldwin and Lachlan showed that all models of any uncountably categorical theory T can be listed into the following chain, denoted by $chain(T)$, of elementary embeddings:

$$\mathcal{A}_0 \preceq \mathcal{A}_1 \preceq \mathcal{A}_2 \preceq \ldots \mathcal{A}_\omega,$$

where $\mathcal{A}_0$ is the prime model of T, $\mathcal{A}_\omega$ is the saturated model of T, and each $\mathcal{A}_{i+1}$ is a minimal proper elementary extension of $\mathcal{A}_i$. As we are interested in constructive models of T the following definition is central.

DEFINITION 3.8. Let T be an uncountably categorical theory. **The spectrum of constructive models of** T, denoted by $SCM(T)$, is the set

$$\{i \mid \text{the model } \mathcal{A}_i \text{ of } T \text{ has a constructivization}\}.$$

Thus a natural open problem about constructive models of uncountably categorical theories is the following:

PROBLEM 4. Characterize all the subsets X of $\omega \bigcup \{\omega\}$ for which there exist uncountably categorical theories T such that $SCM(T) = X$.

An initial step in the study of this problem is to assume that the theory T is decidable. Clearly, in general, the decidability of a theory T (independently of whether or not T is uncountably categorical) does not imply that all the models of T have strong constructivizations. However, one of the important results in the study of (strongly) constructive models of a given theory T is the following result of Harrington [**18**] and Khisamiev [**27**]:

THEOREM 3.9. *Let T be an uncountably categorical theory. Then T is decidable if and only if T has a decidable model if and only if all models of T have decidable presentations.*

The proof of this theorem consists of two steps. The first step uses a purely model theoretic fact that states that every model $\mathcal{A}$ of an uncountably categorical theory T can be considered as the prime model of a uncountably categorical theory T' with a strongly minimal formula. Moreover, if T is decidable then so is T'. The second step of the proof shows that for any decidable uncountably categorical theory there exists an algorithm that, given a formula consistent with the theory, produces a principal type that contains the formula. Then Theorem 3.2 is applied to show that the model $\mathcal{A}$ has a strong constructivization.

Again, we have a situation similar to the one for countably categorical theories. The theorem above basically answers the question related to the existence of strongly constructive models for uncountably categorical decidable theories. However, neither the theorem nor the proof of the theorem gives a clear picture for building constructive models of uncountably categorical theories when one omits the assumption of decidability. We pose the following problem:

PROBLEM 5. Characterize uncountably categorical theories that have constructive models.

The situation for uncountably categorical theories is also complicated by the following fact. In general, the existence of a constructive model for an uncountably categorical theory T does not imply that all models of T have constructivizations. Indeed, in [**7**] Goncharov showed that there exists an uncountably categorical theory T for which $SCM(T) = \{0\}$, that is, the only model of T which has a constructivization is the prime model of T. Kudaibergenov extended this result by showing that for every $n \geq 0$ there exists an uncountably categorical T such that $SCM(T) = \{0, 1 \ldots, n\}$ [**28**]. This theory T basically codes a noncomputable, computably enumerable set X in such a way that from the open diagram of the model $\mathcal{A}_{n+1}$ in $chain(T)$ one can decode the set X. More complicated codings are needed to prove the following theorem:

THEOREM 3.10. [**26**] *There exist uncountably categorical theories T_1 and T_2 such that $SCM(T_1) = \omega$, and $SCM(T_2) = \omega \bigcup \{\omega\} \setminus \{0\}$.*

The constructions of the theory T_1 is based on a coding of a Σ_2-set, while the construction of the theory T_2 is based on a coding of a Π_2-set. This theorem together with the theorem by Khissamiev and Harrington are the only known results so far about the spectra of constructive models of uncountably categorical theories. We also note that the theories T_1 and T_2 of the theorem above have infinite languages. Recently Herwig, Lempp and Ziegler [**19**] have constructed an uncountably categorical theory T of a finite language such that $SCM(T) = \{0\}$.

It is interesting to note that all the known uncountably categorical theories that have constructive models are computable in $\mathbf{0}''$. Khoussainov, Lempp, and Solomon have recently constructed examples (the paper is in preparation) of uncountably categorical theories that compute $\mathbf{0}^{(n)}$, $n > 2$, and all of whose models have constructivizations. In connection with this result Lempp has asked the following question:

QUESTION 6. If an uncountably categorical theory has a constructive model then must the theory be arithmetical?

We note that the theorem of Harrington and Khissamiev can be relativized to show that if T is uncountably categorical and arithmetical then all models of T have arithmetical numberings. Hence a positive answer to the question of Lempp implies that all models of an uncountably categorical theory with constructive models have arithmetical numberings. However, we state the following hypotheses which, if correct, would negatively answer the question of Lempp:

CONJECTURE 3. There exists a nonarithmetical uncountably categorical theory with a constructive model.

One of the ways to approach this conjecture is, for example, to answer to the following question:

QUESTION 7. Does there exist an uncountably categorical theory T whose models are $\mathcal{A}_0 \preceq \mathcal{A}_1 \preceq \ldots \preceq \mathcal{A}_\omega$ such that $\mathcal{A}_0$ has a constructivization, and each $\mathcal{A}_{(i+1)}$, $i \in \omega$, has a constructivization computable in $\mathbf{0}^{(i+1)}$ but does not have constructivizations computable in $\mathbf{0}^{(i)}$?

As we have already stated, there are examples of uncountably categorical theories T with models which have no constructivizations but with $SCM(T) \neq \emptyset$. On the other hand, we also know that in some cases all models of T have constructivizations. We single out such theories in the following definition.

DEFINITION 3.11. A theory T has **constructively complete** if all countable models of T have constructivizations.

Decidable countably categorical and decidable uncountably categorical theories are examples of theories which are constructively complete. There are also examples of undecidable constructively complete theories. In general, constructively complete theories do not need to be complete. For example, the theory of all dense linearly ordered sets is constructively complete. We end this section by posing the following question:

QUESTION 8. When is an uncountably categorical theory T constructively complete?

3.3. Constructive Models of Ehrenfeucht Theories. Another class of theories that has been well studied and has attracted considerable attention is the class of Ehrenfeucht theories. Here is an exact definition.

DEFINITION 3.12. A theory is an **Ehrenfeucht theory** if it has finitely many models.

Vaught proved that no complete theory has exactly two models. On the other hand, for each $n \geq 3$ there exists a theory with exactly n models. For example, the theory of the model $(Q, \leq, c_0, c_1, \ldots)$, where $(Q, \leq)$ is the natural ordering of rationals and $c_0 < c_1 < c_2 < \ldots$, has exactly three models. This example can be generalized to give theories with exactly $n \geq 4$ models. Inspired by Theorem 3.9 Nerode posed the following question: If an Ehrenfeucht theory T is decidable then do all models of T have strong constructivizations? It turns out that models of decidable Ehrenfeucht theories are not as well behaved as decidable uncountably or decidable countably categorical theories. For instance, the following theorem is true.

THEOREM 3.13. [**34**] [**36**] *For each $n \geq 3$ there exists a decidable Ehrenfeucht theory T_0 that admits elimination of quantifiers, has exactly n models and exactly one model with a strong constructivization.*

In relation to this theorem we make the following comments. First of all we note that the prime model of any decidable Ehrenfeucht theory must have a strong constructivization. This follows from an effective version of the Omitting Types Theorem for decidable theories [**33**] which is not discussed in this paper. Hence the strongly constructive model of the theory T_0 in the theorem above is a prime model. Secondly, the reason that not all models of T_0 have strong constructivizations is that T_0 has a noncomputable type. Based on this Morley asked the following question that has become known as Morley's problem:

QUESTION 9. If all types of an Ehrenfeucht theory T are computable then do all models of T have strong constructivizations?

This is an open problem which has been attempted by many with no success. Ash and Millar obtained several interesting results in the study of this question. One of the results is the following. We say that an Ehrenfeucht theory T is **persistently Ehrenfeucht** if any complete extension of T with finitely many new constants is also an Ehrenfeucht theory. Here is the theorem:

THEOREM 3.14. [**2**] *If T is persistently Ehrenfeucht all of whose types are arithmetical then all models of T have arithmetic presentations.*

In relation to this theorem and Morley's problem, it is interesting to note that the following question, asked by Goncharov and Millar, is still open:

QUESTION 10. If T is an arithmetic Ehrenfeucht theory whose types are arithmetical, do then all models of T have arithmetic presentations?

We now briefly discuss the problem of existence of constructive models of Ehrenfeucht theories. As for categorical theories, there has not been much research about finding constructive models for (undecidable) Ehrenfeucht theories. We note that the results in finding constructive models for undecidable Ehrenfeucht theories can be quite different from those about decidable Ehrenfeucht theories. We give an example. If all types of an Ehrenfeucht theory T are computable then T must have at least three strongly constructive models (a proof of this can, for example, be found in [**25**]). Therefore for any decidable Ehrenfeucht theory T with exactly three models, the saturated model of T has a strong constructivization if and only if all models of T have strong constructivizations. We also recall that the prime model of every decidable Ehrenfeucht theory has a strong constructivization. In contrast to this, in [**26**] the following theorem is proved:

THEOREM 3.15. *There exists an Ehrenfeucht theory with exactly three models of which only the saturated one has a constructivization.*

We conclude this section with the following research proposal.

PROBLEM 11. Work towards characterizing strongly constructive or/and constructive models of Ehrenfeucht theories.

4. Computable Isomorphisms

There has been extensive research on computable isomorphisms of constructive algebraic systems. Many researchers have worked on problems and research directions discussed in this section. These are still in the center of research interest and play a significant role in the creation of new ideas, theorems and concepts. This section consists of four parts. The first part introduces the basic concepts about computable isomorphisms between constructive algebraic systems. In the second, we discuss computable isomorphisms of countably categorical models. The third part is devoted to computable isomorphisms of uncountably categorical models. In the final part we deal with the problem of the dependency of computability-theoretic properties of relations on constructivizations.

4.1. Basic Notions. A fundamental concept in the theory of constructive models is the notion of a computable isomorphism. Informally, the notion of a computable isomorphism allows one to compare constructivizations of a given model, and to tell whether or not two constructivizations have the same computability-theoretic properties. We recall the definition.

DEFINITION 4.1. Two constructive algebraic systems $(\mathcal{A}, \nu)$ and $(\mathcal{A}, \mu)$ are **computably isomorphic** if there exists an automorphism α of $\mathcal{A}$ and a computable function f such that $\alpha\nu(n) = \mu(f(n))$ for all $n \in \omega$. In this case we also say that ν and μ are **autoequivalent**.

One of the fundamental properties of computably isomorphic structures is that they cannot be distinguished in terms of computability-theoretic properties of definable relations. This means that for any relation R invariant under the automorphisms of $\mathcal{A}$, the Turing degrees of R under the constructivizations ν and μ are equivalent, that is, $\nu^{-1}(R)$ and $\mu^{-1}(R)$ have the same Turing degree. In addition, if ν and μ are one to one, then $\nu^{-1}(R)$ and $\mu^{-1}(R)$ are computably isomorphic. One of the important concepts introduced in the study of computable isomorphisms is Goncharov's notion of dimension. Here is the definition.

DEFINITION 4.2. The **computable dimension** of an algebraic system $\mathcal{A}$, denoted $dim(\mathcal{A})$, is the maximal number of its nonautoequivalent constructivizations.

Informally, the computable dimension tells us as how many effective realizations the algebraic system $\mathcal{A}$ possesses. In computability-theoretic terms the computable dimension of a given algebraic system can be thought of as the number of its computable isomorphism types. Thus, if the dimension of $\mathcal{A}$ is 1 then $\mathcal{A}$ has exactly one effective realization. We single out the algebraic systems of dimension 1, and give the following definition first introduced by Malcev.

DEFINITION 4.3. An algebraic system $\mathcal{A}$ is **autostable** if $dim(\mathcal{A}) = 1$. The system $\mathcal{A}$ is **strongly autostable** if all its strong constructivizations are autoequivalent.

An important notion introduced by Goncharov in the study of computable isomorphisms is the notion of an effectively infinite algebraic system. We say that a sequence $(\mathcal{A}_0, \nu_0), (\mathcal{A}_1, \nu_1), \ldots$ of constructive models is **effective** if the set $\{(i, \phi) | \phi \in AD_{\nu_i}(\mathcal{A}_i)\}$ is computably enumerable. Informally, an effective sequence of constructive models is one which can be constructed in a uniform manner.

DEFINITION 4.4. An algebraic system $\mathcal{A}$ is **effectively infinite** if there exists an algorithm that applied to any index of an effective sequence of constructive systems $(\mathcal{A}, \nu_0), (\mathcal{A}, \nu_1), \ldots$ produces a constructive algebraic system $(\mathcal{A}, \nu)$ such that $(\mathcal{A}, \nu)$ is not computably isomorphic to $(\mathcal{A}, \nu_i)$ for $i \in \omega$.

Thus, if an algebraic system $\mathcal{A}$ is effectively infinite then it has infinite computable dimension. One of the first important results in the study of autostable models is the following result of Nurtazin [**35**] that basically characterizes strongly autostable algebraic systems:

THEOREM 4.5. *A strongly constructive algebraic system $(\mathcal{A}, \nu)$ is strongly autostable if and only if there exists a finite number $a_0, \ldots, a_n \in A$ of elements such that the following properties hold:*

1. *The set of all complete formulas of the theory T of the algebraic system $(\mathcal{A}, a_0, \ldots, a_n)$ is computable.*
2. *The system $(\mathcal{A}, a_0, \ldots, a_n)$ is the prime model of its own theory T.*

Moreover, if $(\mathcal{A}, \nu)$ is not strongly autostable then there exists an algorithm that applied to any index of an effective sequence of strongly constructive systems $(\mathcal{A}, \nu_0), (A, \nu_1), \ldots$ produces a strongly constructive algebraic system $(\mathcal{A}, \nu)$ such that $(\mathcal{A}, \nu)$ is not computably isomorphic to $(\mathcal{A}, \nu_i)$ for any $i \in \omega$.

Thus, by this theorem, the computable dimension of any non-strongly autostable strongly constructive algebraic system is infinite.

4.2. Isomorphisms of Countably Categorical Models. Let $\mathcal{A}$ be an algebraic system. Natural questions that arise about the computable isomorphisms of $\mathcal{A}$ are the following. Is $\mathcal{A}$ autostable? If $\mathcal{A}$ is autostable then why is it so? If $\mathcal{A}$ is not autostable then why is it not? What is the dimension of $\mathcal{A}$? Can $\mathcal{A}$ have infinite dimension? Can $\mathcal{A}$ have finite dimension?, etc.

These and related questions have been extensively studied with respect to known classes of algebraic systems such as linearly ordered sets, Boolean algebras, Abelian groups, rings, groups, partially ordered sets, fields, vector spaces, etc. One of the first results obtained in the study of these questions is the following theorem proved independently by Goncharov and Remmel.

THEOREM 4.6. *A linearly ordered set is autostable if and only if the set of adjacent pairs of the linearly ordered set is finite. Similarly, a Boolean algebra is autostable if and only if the set of all atoms of the algebra is finite. Moreover, nonautostable linearly ordered sets and Boolean algebras are effectively infinite.*

An immediate corollary of this theorem is the following result which has not been explicitly stated in the literature.

COROLLARY 4.7. *A linearly ordered set is autostable if and only if the linearly ordered set is countably categorical. Similarly, a Boolean algebra is autostable if and only if the algebra is countably categorical.*

This corollary suggests the study of computable dimensions of those algebraic systems whose theories and/or algebraic, model-theoretic properties are well-understood. Thus, one can study the computable dimensions of countably categorical models.

Using the result of Nurtazin mentioned at the end of the previous section, the following theorem characterizes all strongly autostable countably categorical models.

THEOREM 4.8. *A strongly constructive model $(\mathcal{A}, \nu)$ of a countably categorical theory T is strongly autostable if and only if the type function $type_T$ of the theory T is computable.*

Proof. Assume that the type function $type_T$ is a computable function. Let $(\mathcal{A}, \mu)$ be a strongly constructive model. We want to show that ν and μ are autoequivalent. We claim that for any m-tuple $b_1, \ldots, b_m \in A$ with $\mu(n_1) = b_1, \ldots, \mu(n_m) = b_m$ we can effectively find a complete formula of the type determined by $(b_1, \ldots, b_m)$. Indeed, to do this we compute $type_T(m)$ and then find consistent with T formulas

$$\phi_1(x_1, \ldots, x_m), \ldots, \phi_{t(m)}(x_1, \ldots, x_m)$$

with exactly m number of variables such that

$$\phi_i(x_1, \dots, x_m) \to \phi_j(x_1, \dots, x_m) \notin T$$

for all $1 \leq i \neq j \leq t(m) = type_T(m)$. These formulas can be found effectively because $type_T$ and T are computable. Now, using a back and forth, it is not hard to see that one can construct a computable isomorphism between $(\mathcal{A}, \nu)$ and $(\mathcal{A}, \mu)$.

Now we assume that $\mathcal{A}$ is strongly autostable. We need to show that the type function $type_T$ is computable. Since $\mathcal{A}$ is strongly autostable, by the theorem of Nurtazin there exists a finite sequence $a_0, \dots, a_n$ of elements of $\mathcal{A}$ such that $(\mathcal{A}, a_0, \dots, a_n)$ is the prime model of the theory T' of $(\mathcal{A}, a_0, \dots, a_n)$ and the set of complete formulas of T' is computable. We claim that the set of complete formulas of T is also a computable set. Indeed, take a formula $\phi(\bar{x})$ of the language of T which is consistent with T. Find a complete formula $\psi(\bar{x}, a_0, \dots, a_n)$ of T' such that $T' \vdash \psi(\bar{x}, a_0, \dots, a_n) \to \phi(\bar{x})$. Then $\phi(\bar{x})$ is a complete formula of T if and only if

$$(\phi(\bar{x}) \leftrightarrow \exists y_1 \dots \exists y_n \psi(\bar{x}, y_1, \dots, y_n)) \in T.$$

This shows that the set of complete formulas of T is a computable set. Now $type_T(n) = m$ if and only if there exist exactly m formulas $\phi_1, \dots, \phi_m$ with exactly n variables such that all of these formulas are complete formulas of T and

$$\forall x_1 \dots \forall x_n (\phi_1 \bigvee \dots \bigvee \phi_m) \in T.$$

This shows that the type function $type_T$ of T is computable. The theorem is proved.

The following is a corollary of the theorem.

COROLLARY 4.9. *Let $\mathcal{A}$ be a model of a countably categorical theory T that admits effective elimination of quantifiers. Then the following are equivalent:*

1. *The dimension of $\mathcal{A}$ is* 1.
2. *There exists a finite sequence $a_0, \dots, a_n$ of elements of $\mathcal{A}$ such that $(\mathcal{A}, a_0, \dots, a_n)$ is the prime model of the theory T' of $(\mathcal{A}, a_0, \dots, a_n)$ and the set of atoms of T' is computable.*
3. *The type function t_T is computable.*

Proof. Since T admits effective elimination of quantifiers any constructivization of $\mathcal{A}$ is also a strong constructivization. The corollary is proved.

We note that there exists a strongly autostable countably categorical but not autostable model. Indeed, consider the structure (A, E), where E is an equivalence relation on A such that every E-equivalence class has either one or two elements. Clearly the system is countably categorical. Moreover, it is strongly autostable. However, it is not hard to prove that if E contains E has infinitely many equivalence classes of each size one and two then $\mathcal{A}$ is not autostable. One can guess that the noncomputability of the type function $type_T$ for a countably categorical theory may imply that the model of T has dimension greater than 1. However, the following result (which is in preparation) recently proved by Khoussainov, Lempp and Solomon gives a counterexample.

THEOREM 4.10. *There exists a countably categorical theory T with a noncomputable type function such that the model of T is autostable.*

Note that if T is countably categorical and has a constructive model then the type function of T is computable in $\mathbf{0}^{(\omega)}$. Hence the results above lead us to the following open question:

QUESTION 12. Does there exist a countably categorical theory T such that the type function of T computes $\mathbf{0}^{(\omega)}$ and T has a constructive autostable model?

The results that construct nonautostable algebraic systems of finite dimension do not control model-theoretic properties of the structures constructed. For example, all structures constructed by Goncharov (see [**9**] and [**10**]), Cholak, Goncharov, Khoussainov, Shore (see [**8**]), and Khoussainov and Shore (see [**24**]) have theories without prime models. Moreover all the known countably categorical models have dimensions equal to either 1 or ω. Hence the following question arises naturally.

QUESTION 13. If a countably categorical model is not autostable then is the model effectively infinite?

A positive answer to this question would show that countably categorical and nonautostable models do not have finite dimensions. Hence one may approach the question above by trying to give a counterexample. We pose this as the next question:

QUESTION 14. Does there exist, for a given $n > 1$, a countably categorical model of dimension n?

We finish this section with a comment followed by an open question. Using the theorem of Nurtazin and Theorem 4.8 we see that if a countably categorical decidable theory T has a noncomputable type function then the model of T has infinite dimension. One can ask whether this result can be generalized assuming that T is computable in $\mathbf{0}^{(n)}$ but the type function $type_T$ is not computable in $\mathbf{0}^{(n)}$. We pose this as the following question:

QUESTION 15. Assume that a countably categorical theory T has a constructive model. If T is computable in $\mathbf{0}^n$ and the type function $type_T$ is not computable in $\mathbf{0}^n$, then can we conclude that the model of T is not autostable?

4.3. Isomorphisms of Uncountably Categorical Models. Now we turn to the study of computable dimensions of uncountably categorical models. To provide some intuition, we present some examples of uncountably categorical models and their dimensions.

Let us consider the algebraic system (ω, S). The theory T of this system is uncountably categorical. The isomorphism type of a model $\mathcal{A}$ of T is determined by the number of its components. The saturated model of T has infinitely many components. All nonsaturated models of T are autostable. One can prove that the saturated model of T is not autostable, and is, in fact, effectively infinite.

Consider a second example. Let V be a vector space over a given infinite computable field F. Then the theory T of V (in the language that consists of $+$ for vector addition and unary operations f, $f \in F$, for multiplication by f) is uncountably categorical. It is a well known fact that the isomorphism type of a model $\mathcal{A}$ of T is characterized by the dimension of $\mathcal{A}$. The saturated model of T is the one of infinite dimension. Similarly to the example above, every finite dimensional vector space over F is autostable; the saturated model of T is not autostable, and is, in fact, effectively infinite.

Along the lines of the examples above, one can prove the following theorem about the models of the theory T of algebraically closed fields of a fixed characteristic which is uncountably categorical.

THEOREM 4.11. *Let T be the theory of algebraically closed fields of a fixed characteristic. Then a model $\mathcal{A}$ of T is autostable if and only if it has finite transcendence degree over its prime field.*

Note that all the theories in these examples are decidable and admit elimination of quantifiers. Hence one might expect that the nonsaturated models of such theories are autostable. Here we give a counterexample.

PROPOSITION 4.12. *There exists a decidable uncountably categorical theory T that admits elimination of quantifiers such that the prime mode of T is not autostable.*

Proof. To prove the proposition we provide an uncountably categorical theory T with prime model $\mathcal{A}$ such that the set of complete formulas of the theory of the model $(\mathcal{A}, a_1, \dots, a_n)$ is not computable for all $a_1, \dots, a_n \in A$. By the theorem of Nurtazin this will give the desired result. The theory T is basically the one constructed in [**25**] (page 204). The language of T consists of infinitely many unary predicates R_i. Each R_i contains exactly two elements. Distinct R_i and R_j are pairwise disjoint except for **designated triples** $< i, j, k >$ such that R_k consists of one element from each R_i and R_j. Moreover, for all designated distinct triples $< i, j, k >$ and $< i', j', k' >$ we have $(R_i \bigcup R_j \bigcup R_k) \bigcap (R_{i'} \bigcup R_{j'} \bigcup R_{k'}) = \emptyset$. So the theory is essentially determined by the list of designated triples and is uncountably categorical. If the list of designated triples is computable then the theory is decidable. Now consider the formula $R_i(x)$. This formula is a complete formula if and only if i is not a part of any designated triple. Assume that $< i, j, k >$ is a designated triple. Then each of the formulas $R_i(x) \& R_k(x)$ and $R_i(x) \& \neg R_k(x)$ is a complete formula. The list of designated triples is effectively enumerated in increasing order (and so is computable) by waiting to diagonalize each computable partial function ϕ_i at the formula R_{2i}. If $\phi_i(R_{2i}(x))$ converges at stage s we choose j, k so that $< 2i, 2j + 1, 2k + 1 >$ is bigger than all pairs already and declare the triple $< 2i, 2j + 1, 2k + 1 >$ to be designated. Hence no computable partial function ϕ can decide the set of complete formulas of one variable of the theory constructed. Now one notes that for the prime model $\mathcal{A}$ of the theory constructed and all $a_1, \dots, a_n \in A$, the set of complete formulas of the theory of the model $(\mathcal{A}, a_1, \dots, a_n)$ is not computable. The proposition is proved.

Thus, the following question and conjecture arise naturally:

QUESTION 16. Let T be a decidable uncountably categorical theory which has a strongly autostable prime model. Is every nonsaturated model of T strongly autostable?

CONJECTURE 4. The saturated model of any decidable uncountably categorical theory is effectively infinite.

If we omit the assumption of decidability for an uncountably categorical theory, then the situation becomes more complex. There has not been any research done on computable isomorphisms and dimensions of constructive models of uncountably categorical theories. For example, we do not know the spectra of dimensions of uncountably categorical models. Recall that all the models of an uncountably categorical theory T can be listed in a $\omega + 1$ chain of models:

$$chain(T): \quad \mathcal{A}_0 \preceq \mathcal{A}_1 \preceq \mathcal{A}_2 \preceq \ldots \preceq \mathcal{A}_\omega.$$

We formulate the following problem.

PROBLEM 17. Consider the model $\mathcal{A}_i$ of an uncountably categorical theory T in the $chain(T)$. Give necessary and sufficient conditions for $\mathcal{A}_i$ to be autostable.

In the study of this problem it would be interesting to see if one can control the dimension of uncountably categorical models. In particular, we ask the following question:

QUESTION 18. Does there exist an uncountably categorical nonautostable model of finite dimension?

Our comment about this question is as follows. As we noted in the previous section, Goncharov constructed nonautostable algebraic systems of finite computable dimension. In [**9**] [**10**] [**12**] [**24**] [**8**] [**20**] nonautostable algebraic systems of finite computable dimension have also been constructed to answer some open problems in the theory of constructive models. None of these algebraic systems are prime models of their own theories. In other words, the type structure of the theories of these algebraic systems is quite complicated. Therefore it is natural to ask whether it is possible to construct algebraic systems of finite computable dimension greater than 1 whose theories belong to a class of well-understood theories, e.g. uncountably categorical theories. We do not know the answer to this question. We state, however, that Hirschfeldt and Khoussainov in [**21**] noted that the construction in [**24**] can be modified to build a noncomputably categorical prime model of finite computable computable dimension:

THEOREM 4.13. *For every natural number $n > 1$ there exists an algebraic system $\mathcal{A}$ of computable dimension n such that $\mathcal{A}$ is the prime model of its own theory.*

We now make some comments about saturated models using terminology from dimension theory developed for uncountably categorical theories. Roughly speaking, the saturated model for a given uncountably categorical theory T is the most complicated one because the model has infinite dimension while all other models have finite dimension. Moreover, all models of T are elementarily embedded into the saturated model. In this sense one may suggest that the saturated model can also be complex from the computability-theoretic point of view. One way to show this would be to prove that the computable dimension of the saturated model is always infinite. All the known examples of uncountably categorical saturated models have, in fact, infinite computable dimension. Hence we naturally ask the following question:

QUESTION 19. Does there exist an uncountably categorical theory whose saturated model is autostable?

4.4. The Degree Spectra of Relations. Another central topic in the theory of constructive algebraic systems concerns the dependence of computability-theoretic properties of relations on constructivizations. The topic is closely related to autostability because of the following simple fact. Assume that R is an invariant relation on $\mathcal{A}$ (that is R is closed under automorphisms of $\mathcal{A}$) such that for two constructivizations ν and μ of $\mathcal{A}$ the sets $\nu^{-1}(R)$ and $\mu^{-1}(R)$ have different degrees. Then $\mathcal{A}$ is not autostable. Consider for example the linearly ordered set $(\omega, \leq)$ whose constructivization is the identity mapping. In this constructivization the successor function is computable. On the other hand, $(\omega, \leq)$ has a constructivization under which the successor function is not computable. In [**1**] Ash and Nerode singled out those relations whose computability is invariant with respect to all constructivization. Here is a definition.

DEFINITION 4.14. A relation R on a model $\mathcal{A}$ is **intrinsically computable (intrinsically c. e.)** if for all constructivizations ν of $\mathcal{A}$ the set $\nu^{-1}(R)$ is computable (c.e.).

Thus, for example in $(\omega, \leq)$ the successor function is not intrinsically computable. On the other hand, any computable, invariant relation in any autostable algebraic system is intrinsically computable. Thus, for autostable algebraic systems computability or computable enumerability of the invariant relations is independent on constructivizations.

One of the programs in the theory of constructive algebraic systems is to study the intrinsically computable (c.e.) relations. It turns out that for a large class of algebraic systems which have certain decidability properties one can characterize intrinsically c.e. relations. We define the following notion given by Ash and Nerode in [**1**].

DEFINITION 4.15. An n-ary relation R on a model $\mathcal{A}$ is **formally c.e.** if R is equivalent to a disjunction $\bigvee_i \phi_i(x_1, \dots, x_n, \bar{a})$ of a computable sequence of existential formulas ϕ_i with free variables $x_1, \dots, x_n$, where $\bar{a}$ is a finite sequence of elements from $\mathcal{A}$.

If R is formally c.e. then R is intrinsically c.e. To state the next theorem proved by Ash and Nerode we need to introduce the following notion. We say that $(\mathcal{A}, \nu)$ is **1-decidable** if the set of all existential formulas true in the expansion $(\mathcal{A}, \nu(0), \nu(1), \dots)$ is computable. Here is the theorem:

THEOREM 4.16. [**1**] *Let $(\mathcal{A}, \nu)$ be a 1-decidable algebraic system. Then for any R, the relation R is intrinsically c.e. if and only if R is formally c.e.*

Two natural problems are suggested by this theorem. One is to investigate c.e. intrinsic relations in constructive algebraic systems which are not 1-decidable. The other problem suggests to study those relations R which are not intrinsically c.e. In particular, one can be interested in computability-theoretic complexity of R under different constructivizations. An approach to these problems is suggested by the following definition.

DEFINITION 4.17. For a relation R on an algebraic system $\mathcal{A}$, the **degree spectrum of** R, $DgSp(R)$, is the set of all Turing degrees of $\nu^{-1}(R)$ under all constructivizations ν of $\mathcal{A}$.

There are a number of results that give conditions under which $DgSp(R)$ coincides with a given set of Turing degrees, e.g. the set of all c.e. degrees or of all degrees. Here we concentrate on the issue of finding conditions under which $DgSp(R)$ is finite. This search is basically motivated by our interest in whether we are able to control computable dimensions and the degree spectra of relations in building constructivizations of algebraic systems. Recasting the theorem of Goncharov in [**9**], Harizanov in [**16**] provided an example of a relation R in a system of computable dimension 2 such that $DgSp(R) = \{0, c\}$, where c is noncomputable and Δ_2^0. Later, the authors showed that there exists a relation R in a system of computable dimension 2 such that $DgSp(R) = \{0, c\}$, where c is the degree of a c.e. set [**12**]. This result was independently generalized by Khoussainov and Shore in [**24**]. In particular, Khoussainov and Shore proved that for any finite partially ordered set P there exists an intrinsically c.e. relation R in a system of computable dimension $|P|$ such that $DgSp(R)$ is isomorphic to P, where the ordering of $DgSp(R)$ is given by Turing reducibility. However, the following question has remained opened: Which finite sets $\{a_1, \ldots, a_n\}$ of computably enumerable degrees coincide with $DgSp(R)$, where R is a relation in an algebraic system of computable dimension n? Recently Khoussainov and Shore, and independently Hirschfeldt by different methods in [**20**], have been able to provide the following answer to this question.

THEOREM 4.18. *For any finite set $\{a_1, \ldots, a_n\}$ of computably enumerable degrees there exists an algebraic system $\mathcal{B}$ of computable dimension n and an intrinsically c.e. relation R in it such that $DgSp(R) = \{a_1, \ldots, a_n\}$.*

In light of these results the following question remains open:

QUESTION 20. Let $\{a_1, \ldots, a_n\}$ be a finite set of Turing degrees of Σ_n-sets. Does there exists a relation R in an algebraic system of computable dimension n such that $DgSp(R) = \{a_1, \ldots, a_n\}$?

A weaker version of this question not asking to control the dimension of the system is the following.

QUESTION 21. For a given finite collection $\{a_1, \ldots, a_n\}$ of Turing degrees of Σ_n-sets, does there exist a relation R in an algebraic system such that $DgSp(R) = \{a_1, \ldots, a_n\}$?

We end this section with the following question related to computable dimension and the two questions asked above. It is known that all constructed nonautostable models of finite algorithmic dimensions are Δ_3^0-autostable. Hence the following question, posed by Khoussainov and Shore, arises:

QUESTION 22. Is it true that for any $n \geq 3$ there exists a non Δ_n^0-autostable but Δ_{n+1}^0-autostable model of dimension 2?

5. Conclusion

In this paper we concentrated on open problems in two directions in the development of the theory of constructive algebraic systems. The first direction deals with positive algebras. In our discussion of positive algebras an emphasis was on showing the interplay between universal algebra and computability theory. We think that systematic development of the theory of positive algebras can bring fruitful results and deeper understanding of interactions between fundamental concepts of universal algebra and computability theory. Hence we proposed a systematic study of positive algebras as a new direction in the development of constructive algebraic systems. The second direction concerns the traditional topics in constructive model theory. First we proposed the study of constructive models of theories with few models such as countably categorical theories, uncountably categorical theories, and Ehrenfeucht theories. Next, we proposed the study of computable isomorphisms and computable dimensions of such models. We also discussed issues related to the computability-theoretic complexity of relations in constructive algebraic systems. We stress that we have not discussed many other equally important topics and open problems in other parts of this area. For a comprehensive survey of results and directions in this area we refer the reader to the papers in the Handbook of Recursive Mathematics [**6**].

At the end of this paper we would like to thank the referee for a careful reading of the paper and a numerous number of suggestions on the improvement of the text.

References

[1] C. J. Ash and A. Nerode, *Intrinsically Recursive Relations*, in J. N. Crossley (ed.), Aspects of Effective Algebra, Upside Down A Book Co., Yarra Glen, Australia, 1981, 26–41.

[2] C. J. Ash and T. Millar, *Persistently Finite, Persistently Arithmetic Theories*, Proc. Amer. Math. Soc. 89(1983), 487-492.

[3] J. Baldwin and A. Lachlan, *On Strongly Minimal Sets*, Journal of Symbolic Logic 36(1971), 79–96.

[4] C. C. Chang and H. J. Keisler, Model Theory, 3rd ed., Stud. Logic Found. Math., no. 73, North-Holland Pub. Co., Amsterdam, New York, 1990.

[5] Yu. L. Ershov, The Theory of Numerations, Monographs in Mathematical Logic and Foundations of Mathematics, Nauka, Moscow, 1977.

[6] Yu. L. Ershov, S. Goncharov, A. Nerode, and J. Remmel eds., Marek V. ssoc. ed., Handbook of Recursive Mathematics, Volume 1-2. North-Holland Pub Co., Amsterdam, New York, 1999.

[7] S. Goncharov, *Constructive Models of ω_1-categorical Theories*, Matematicheskie Zametki 23(6)(1978), 885–888.

[8] P. Cholak, S. S. Goncharov, B. Khoussainov, and R. A. Shore, *Computably Categorical Structures and Expansions by Constants*, Journal of Symbolic Logic 64(1999) 13–37.

[9] S. S. Goncharov, *Computable Univalent Numerations*, Algebra and Logic 19(1980), 507–551.

[10] S. S. Goncharov, *The Problem of Nonautoequivalent Constructivizations*, Algebra and Logic 19(1980), 621–639.

[11] S. S. Goncharov, A. V. Molokov, and N. C. Romanovsky, *Nilpotent Groups of Finite Algorithmic Dimensions*, Siberian Math. Journal 30(1989), 82–88.

[12] S. S. Goncharov and B. Khoussainov, *Degree Spectra of Decidable Relations*, Dokl. Akadem. Nauk 352(1997), 301–303.

[13] S. S. Goncharov, *The Problem of the Number of Non-Self-Equivalent Constructivizations*, Algebra and Logic 1(1980), 401–414.

[14] S. S.Goncharov, *Strong Constructivizibility of Homogeneous Models*, Algebra and Logic 17(1978), 363–388.

[15] G. Grätzer, Universal Algebra, Springer-Verlag Pub. Co., New York, Heidelberg, 2nd edition, 1979.

[16] V. Harizanov, *The Possible Turing Degree of the Nonzero Member in a two Element Degree Spectra*, Annals of Pure and Applied Logic 55(1991), 51–65.

[17] V. Harizanov, *Uncountable Degree Spectra*, Annals of Pure and Applied Logic 54(1991), 255–263.

[18] L. Harrington, *Recursively Presentable Prime Models*, Journal of Symbolic Logic 39(1974), 305–309.

[19] B. Herwig, S. Lempp, and M. Ziegler, *Constructive Models of Uncountable Categorical Theories*, Proc. American Math. Soc. 127(1999), 3711–3719.

[20] D. Hirschfeldt, Ph.D. Dissertation, Cornell University, 1999.

[21] D. Hirschfeldt and B. Khoussainov, *Dimensions of Prime Models*, Abstracts of the Malcev's Conference in Algebra, Novosibirsk, 1999.

[22] N. Kassimov, *Positive Algebras with Congruences of Finite Index*, Algebra and Logic 30(1989), 298–305.

[23] N. Kassimov, *Positive Algebras with Countable Congruence Lattices*, Algebra and Logic 31(1992), 21–37.

[24] B. Khoussainov and R. Shore, *Computable Isomorphisms, Degree Spectra of Relations and Scott Families*, Annals of Pure and Applied Logic 93(1998), 153–193.

[25] B.Khoussainov and R. Shore, *Effective Model Theory: The Number of Models and their Complexity*, In: Models and Computability, B. Cooper and J. Truss eds., Cambridge University Press, 1999, 193–24.

[26] B. Khoussainov, A. Nies, and R. Shore, *On Recursive Models of Theories*, Notre Dame Journal of Formal Logic 38(1997), 165–178.

[27] N. Khisamiev, *Strongly Constructive Models of a Decidable Theory*, Izv. Akad. Nauk Kazakh. SSR, Ser. Fiz.-Mat., 1, 1974, 83–84.

[28] K. Kudeiberganov, *On Constructive Models of Undecidable Theories*, Siberian Mathematical Journal 21(1980), 155–158.

[29] M. Lerman and J. Schmerl, *Theories with Recursive Models*, Journal of Symbolic Logic 44(1979), 59–76.

[30] A. I. Malcev, *Recursive Abelian Groups*, Dokl. Akademii Nauk SSSR 146(1961), 1009–1012.

[31] A. I. Malcev, Algebraic Systems, Nauka. Moskow., 1970.

[32] J. G. McKinsey, *The Decision Problem for Some Classes of Sentences Without Quantifiers*, Journal of Symbolic Logic 8(1945), p.61–76.

[33] T. Millar, *Omitting Types, Type Spectrums, and Decidability*, Journal of Symbolic Logic 48(1983), 171–181.

[34] M. Morley, *Decidable Models*, Israel Journal of Mathematics 25(1976), 233–240.

[35] A. T. Nurtazin, *Strong and Weak Constructivizations and Computable Families*, Algebra and Logic 13(1974), 311–324.

[36] M. Peretyat'kin, *On Complete Theories with Finite Number of Countable Models*, Algebra and Logic 12(1973), 550–576.

[37] J. B. Remmel, *Recursively Categorical Linear Orderings*, Proc. Amer. Math. Soc. 83(1981), 387–391.

[38] J. H. Schmerl. *A decidable $\aleph_0$-categorical theory with a nonrecursive Ryll-Nardzewsky function.* Fund. Math. 98(1978), 121–125.

[39] R. Soare, Recursively Enumerable Sets and Degrees, Springer-Verlag, Berlin, Heidelberg, New York, 1987.

Academy of Sciences, Siberian Branch, Mathematical Institute, 630090 Novosibirsk, Russia

E-mail address: `gonchar@math.nsc.ru`

University of Auckland, Department of Computer Science, Private Bag 92019, Auckland, New Zealand

E-mail address: `bmk@cs.auckland.ac.nz`

Contemporary Mathematics
Volume **257**, 2000

Independence results from ZFC in computability theory: some open problems

Marcia Groszek

ABSTRACT. There are at least two ways in which "independence results from ZFC in computability theory" arise. There are questions that are essentially set theoretic but morally fall under the umbrella of computability theory, for example questions about the degrees (of constructibility, or Kleene degrees, or ...) of reals. There are also questions in classical computability theory that turn out to be (possible) independence results; one way this can happen is when a question about a computability theoretic structure on all of $\mathbb{R}$ can be settled under CH by extending from one countable collection of reals to a larger one until all $\aleph_1$ reals are included, but the extension technique does not apply once uncountable collections are reached. We discuss one problem in each category.

1. Introduction

The title of this paper inevitably claims too much. We can generally recognize an open problem in computability theory or an open problem in ZFC. But how, short of solving the problem, can we recognize an independence result? This question is particularly sharp in the context of computability theory, where independence from ZFC seems strange and wonderful.

Of course classical computability theory is part of the study of the continuum and the nature of the continuum is highly dependent on set theoretic assumptions, so occasional independence from ZFC is not so strange after all. When we are searching for possible independence, the rare theorem that explicitly depends on set theoretic assumptions draws our attention. For example, in her paper in this volume[**10**], Julia Knight lists among open problems the question of which Scott sets are realized by models of Peano arithmetic. If CH holds, all of them are; the situation when CH fails could turn out to be independent of ZFC. We can also look for new independence results by starting in the neighborhood of known independence results.

But restricting computability theory to classical computability theory gives a very narrow perspective on the field. So, in a way, does considering computability theory as separated from set theory. Set theory and computability theory are not composed of two (almost) disjoint sets of people applying distinct methods to

1991 *Mathematics Subject Classification.* 03D28, 03D30, 03E35, 03E45.

distinct sets of questions. There is a great deal of overlap in the methods and concerns, as well as the practitioners, of the two fields, and in that overlap can be found essentially computability theoretic questions many of which are likely to lead to independence results from ZFC. For example, as of this writing, the first question on Arnold Miller's problem list[1] is partially answered by a theorem of Greg Hjorth [**7**] showing that the existence of a Σ^1_1 set universal for non-Borel Σ^1_1 sets is independent of ZFC. From this perspective, finding open problems that may lead to independence results is only too easy.

This paper asks (yet again) one question, an oldie-but-goodie: What partial orderings can be embedded into the degrees? By interpreting the word degrees as meaning "Turing degrees", we find a possible independence result in the neighborhood of known independence results and in a question whose answer is known under the continuum hypothesis. By interpreting the word degrees as meaning "degrees of constructibility", we find open problems whose answers are necessarily independence results; if we evade that necessity by making a reasonable set theoretic assumption, we find ourselves chasing the same hints we followed in our search for possible independence questions in the Turing degrees.

For a discussion of the history and context of these problems, broader and more extensive than space allows in this paper, read Richard Shore's beautiful expository papers [**19**] and [**18**].

2. Embeddings into the Turing degrees

The question of what structures can be embedded into the Turing degrees has been studied for nearly as long as the Turing degrees have been around. We can specify an embedding in various senses (for example as an initial segment, or as an upper semilattice or a lattice), of which embedding as a partial ordering is the weakest and easiest to obtain. It turns out that as long as the continuum hypothesis holds, we know exactly what structures we can and cannot embed (in all these senses) into the Turing degrees. If CH fails, we are in the realm of independence from ZFC.

2.1. Embeddings as initial segments. We have no open problems to pose regarding embeddings into the Turing degrees as initial segments; the known independence result of Marcia Groszek and Theodore Slaman in [**6**] is in some sense the final answer. However, we give a quick survey of some important results, partly as background for discussing the same question for the degrees of constructibility (where much less is known.)

The following theorems are high points on the path from Clifford Spector's minimal degree construction[**20**] to Uri Abraham and Richard Shore's complete solution of the problem of embeddability into the Turing degrees as an initial segment under CH[**1**]. Some important techniques, noteworthy for migrating into the area of degrees of constructibility, are Spector's use of perfect trees as conditions on the real being constructed and Manuel Lerman's use of lattice representations to handle non-distributive lattices.

THEOREM 2.1. *(Clifford Spector, 1956* [**20**]*.) There is a minimal (non-zero) Turing degree.*

[1]Miller's problem list [**15**] was first published in *Set Theory of the Reals*; the latest version is found on the web at http://www.math.wisc.edu/~miller.

THEOREM 2.2. *(Alistair Lachlan, 1968* [**11**].*) Every countable distributive lattice with least element can be embedded into the Turing degrees as an initial segment.*

THEOREM 2.3. *(Manuel Lerman, 1971* [**13**].*) Every finite lattice can be embedded into the Turing degrees as an initial segment.*

THEOREM 2.4. *(Alistair Lachlan and Robert Lebeuf, 1976* [**12**].*) Every countable upper semilattice with least element can be embedded into the Turing degrees as an initial segment.*

DEFINITION 2.5. A partial ordering is **locally countable** if each element has only countably many predecessors in the ordering, and **locally finite** if each element has only finitely many predecessors.

The Turing degrees are locally countable, and so any partial ordering embeddable into the Turing degrees must be locally countable.

THEOREM 2.6. *(Uri Abraham and Richard Shore, 1986* [**1**].*) Every locally countable upper semilattice with least element of size at most* $\aleph_1$ *can be embedded into the Turing degrees as an initial segment.*

Since every initial segment of the Turing degrees is a locally countable upper semilattice with least element of size at most continuum, this last result completely settles the question when the continuum hypothesis holds. The next result shows that if CH fails, the answer to the question is independent of ZFC. In some sense this completely answers the question.

THEOREM 2.7. *(Marcia Groszek and Theodore Slaman, 1983* [**6**].*) There is a model of ZFC in which there is a locally finite upper semilattice of size less than continuum that cannot be embedded into the Turing degrees as an upper semilattice (hence cannot be embedded as an initial segment.)*

Theorems 2.6 and 2.7 together answer the embedding question for initial segment embeddings, upper semilattice embeddings and lattice embeddings. Under CH all candidate structures (locally countable upper semilattices, or lattices, with least element having size at most continuum) can be embedded, and if CH fails there may be candidate structures of size greater than $\aleph_1$ that cannot be embedded.

2.2. Embeddings as partial orderings. Because embeddings as partial orderings are easier to produce than embeddings as initial segments, more about embeddings as partial orderings was known earlier (although, as we shall see, less is known now.) Stephen Kleene and Emil Post's foundational paper on the Turing degrees contains the proof of an embedding theorem.

THEOREM 2.8. *(Stephen Kleene and Emil Post, 1954* [**9**].*) Every finite partial ordering can be embedded into the Turing degrees.*

By 1963 Gerald Sacks had a complete solution to the problem when the continuum hypothesis holds.

THEOREM 2.9. *(Gerald Sacks, 1963* [**16**].*) Every locally countable partial ordering of size at most* $\aleph_1$ *can be embedded into the Turing degrees.*

Sacks also showed that the condition on the size of the partial ordering P can be relaxed to "size at most continuum" provided either every element of the partial ordering has at most $\aleph_1$ successors or the partial ordering is locally finite. He conjectured that these additional conditions can be eliminated.

CONJECTURE 2.10. (Gerald Sacks, 1963 [**16**].) Every locally countable partial ordering of size continuum can be embedded into the Turing degrees.

Sacks further conjectured that Conjecture 2.10 could be proved using the method of extension of embeddings that had worked to prove the embeddability of partial orderings of size at most $\aleph_1$. In order to do this in the most straightforward way, one would need to prove the following lemma: Given any locally countable partial ordering P, and any embedding φ of a downward closed suborder of P of size less than continuum into the Turing degrees, there is an extension of φ to a larger downward closed suborder of P. As a vote of confidence in this belief, Sacks conjectured a special case of this lemma.

DEFINITION 2.11. A set of degrees is **independent** if no finite join of elements of the set computes any other element of the set. An independent set is **maximal** if no proper superset is independent.

CONJECTURE 2.12. (Gerald Sacks, 1963 [**16**].) Any maximal independent set of Turing degrees must have size continuum.

This turns out to be independent of ZFC:

THEOREM 2.13. *(Marcia Groszek and Theodore Slaman, 1983* [**6**]*.) There is a model of ZFC with a maximal independent set of Turing degrees of size less than continuum.*[2]

Because it will be useful in the next section, here is a sketch of the proof of this theorem:

Begin with a model M of set theory in which the continuum hypothesis holds. (This assumption is convenient because it guarantees that the forcing we are about to do does not collapse cardinals.) First, force over M with the countable support product of $\aleph_1$ copies of Sacks forcing (forcing with perfect trees as conditions.) This adds a generic sequence of reals $G = \langle G_i \mid i \in \aleph_1 \rangle$.

It is not hard to show that the set X of the Turing degrees of the G_i is a maximal independent set in $M[G]$. It is independent because the G_i are mutually generic, so that any finite number can be adjoined to M and none of the others will be in the resulting model of set theory. It is also maximal. By genericity, any real r in the ground model is coded directly into some G_j (actually into uncountably many of them), so $r \leq_T G_j$ and adding the degree of r would destroy the independence of X. By a fusion argument (the common coin of perfect set forcing), any real s not in the ground model Turing computes one of the G_i relative to some ground model real r; since r is coded into some other G_j, $G_i \leq_T s \oplus G_j$, so adding the degree of s would destroy the independence of X.

Now force over $M[G]$ with the countable support product of κ copies of Sacks forcing, for some large κ of uncountable cofinality. This forcing does not collapse cardinals, and in the resulting model $M[G][H]$, the value of the continuum is κ. Also in the resulting model, the Turing degrees of the G_i remain a maximal independent set: Let s be any real in $M[G][H]$ that is not in $M[G]$. By a fusion argument, s Turing computes some G_i relative to a ground model real r. But we know that r is

[2]This theorem was proved independently in 1988 by Vladimir Kanovei in Russia[**8**]. Kanovei uses a non-well-founded iteration of Sacks forcing; his theorem further considers the spectrum of cardinalities of maximal independent sets.

coded into some other G_j, $G_i \leq_T s \oplus G_j$, so adding the degree of s would destroy the independence of X.

Theorem 2.13 of course does not settle Conjecture 2.10; we know that there is an independent set of Turing degrees of size continuum. It does show that the conjecture cannot be proved by a straightforward extension of embeddings argument.

2.2.1. *An open problem.* Sacks's Conjecture 2.10 remains open and is the first of our possible independence results from ZFC:

PROBLEM 1. Is every locally countable partial ordering of size at most continuum embeddable into the Turing degrees as a partial ordering?

The extension of Theorem 2.9 to locally finite partial orderings of size continuum is not proven by an extension of embeddings argument. Instead the proof is accomplished by embedding a *universal* locally finite partial ordering of size continuum, namely an upper semilattice freely generated under join by continuum many incomparable elements.

One possible approach to this problem is to produce an embedding of a universal locally countable partial ordering of size continuum. Producing such an object is not as straightforward as in the locally finite case, but there is a fairly understandable example ready to hand. By coding pairs of numbers as numbers, we can consider each real as the join of countably many columns. Under the transitive closure of the "is a column of" ordering, modulo equivalence, the reals form a universal locally countable partial ordering of size continuum.

A different possible solution to this problem would be to prove a modified extension of embeddings lemma, showing that if φ is an embedding of some size less than continuum suborder of a locally countable partial ordering P into the Turing degrees and φ has some special property, then φ can be extended to an embedding of a larger suborder of P having the same special property.

Another approach to the positive direction would be to first look at the special case of well-founded partial orderings.[3] The well-foundedness of the partial ordering might allow one to embed the ordering one level at a time, using a different version of an extension of embeddings lemma (extending embeddings of small ordinal height rather than small size.)

For a negative result, of course, one possibility is to produce a model containing a specific partial ordering that cannot be embedded. Another is to focus on a universal locally countable partial ordering of size continuum and produce a model in which that specific partial ordering cannot be embedded. One argument of this sort would have two pieces, first a proof that there were no simple embeddings of the chosen partial ordering into the Turing degrees ("simple" meaning those that would persist in any forcing extension due to Shoenfield absoluteness), and then an iterated forcing construction for destroying less simple embeddings one by one.

3. Embeddings into the degrees of constructibility

The degrees of constructibility are highly non-absolute, being trivial in some models of set theory and very rich in others, so in a sense almost all results about the degrees of constructibility are independence results from ZFC. This is perhaps not a very interesting way of getting independence results. We can adopt another

[3]Thanks to the referee for this suggestion.

perspective, and make a set theoretical assumption that guarantees that the degrees of constructibility are rich and settles many questions about their structure.

DEFINITION 3.1. If r is a real, the inner model $L[r]$ is the universe of sets constructible relative to r, and $\omega_1^r = \omega_1^{L[r]}$ is the least cardinal that is uncountable in that inner model. If for every real r, $\omega_1^r < \omega_1$, we say that "ω_1 is inaccessible from reals."

If ω_1 is inaccessible from reals, then (the genuine) ω_1 is an inaccessible cardinal in every $L[r]$. If P is a forcing notion in $L[r]$ whose conditions are reals, then the collection of dense subsets of P in $L[r]$ is actually countable, so there is a P-generic object over $L[r]$. This means that many forcing arguments that are phrased as relative consistency results can also be viewed as theorems under the assumption ω_1 is inaccessible from reals; for example, the constructions of [**17**], [**3**], [**4**] and [**5**] to produce models in which the degrees of constructibility are isomorphic to a given structure S show that if ω_1 is inaccessible from reals then S can be embedded into the degrees of constructibility as an initial segment.

In the rest of this section we will make the assumption that ω_1 is inaccessible from reals. By giving us generic objects for forcing partial orderings in inner models $L[r]$, this produces a very rich structure for the degrees of constructibility. Shore [**18**] seems to suggest this axiom as "morally" (although not technically) giving a positive answer to Robert Solovay's suggestion (quoted by Sacks[**17**]) that a sufficiently strong set theoretic assumption would answer all our questions about the structure of the degrees of constructibility. We will see later in Theorem 3.10 that some questions about the degrees of constructibility are independent of $ZFC+$ "ω_1 is inaccessible from reals".

3.1. Embeddings as initial segments. Even making the set theoretic assumption that ω_1 is inaccessible from reals, we still find the question of the possible initial segments of the degrees of constructibility to be less straightforward than the analogous question for the Turing degrees. The distinguishing factor is that while "recursive in" is not a recursive relation, "constructible in" is a constructible relation, so that information coded into the degrees of constructibility can be easily decoded. For instance, one can prove the following theorem in ZFC:

THEOREM 3.2. *(Uri Abraham and Richard Shore, 1986* [**2**].*) Not every countable upper semilattice is an initial segment of the degrees of constructibility.*

To prove this, consider a real r whose degree is the top of an initial segment of degrees forming a diamond lattice. (Under our set theoretic assumption, there is such a real; under ZFC alone, if there is no such real then the diamond cannot be embedded as an initial segment and so the theorem is true.) Any countable lattice with top point that codes r and does not have a diamond as an initial segment cannot be embedded into the degrees as an initial segment.

For this reason we restrict attention to constructible upper semilattices. Sacks adapted Spector's minimal Turing degree construction, and Zofia Adamowicz used the lattice representation methods developed by Lerman, to prove the following theorems.

THEOREM 3.3. *(Gerald Sacks, 1971* [**17**].*) There is a minimal non-zero degree of constructibility.*

THEOREM 3.4. *(Zofia Adamowicz, 1976* [**3**].*) Every finite lattice can be embedded into the degrees of constructibility as an initial segment.*

THEOREM 3.5. *(Zofia Adamowicz, 1977* [**4**].*) Every well-founded countable (in L) constructible upper semilattice with least element can be embedded into the degrees of constructibility as an initial segment.*

The limited class of structures covered by this theorem contrasts to the Turing degree case. Countable lattices are a natural starting point (Adamowicz actually extends the result to a somewhat larger class) and the criterion of constructibility is expected, but the requirement of well-foundedness is surprising. However, there are good reasons for this requirement, as shown by the following, perhaps equally surprising, theorem of Robert Lubarsky. This is (once we have restricted our attention to constructible structures) the first and still the only real negative result we have.

DEFINITION 3.6. A lattice (with top point) is **complete** if every subset of the lattice has a meet (an infimum) in the lattice.

THEOREM 3.7. *(Robert Lubarsky, 1987* [**14**].*) Any countable lattice that can be embedded into the degrees of constructibility as an initial segment must be complete.*

Of course, there is a sizable gap between complete and well-founded. One step toward closing that gap is given by the following result in the positive direction.

DEFINITION 3.8. A point x in a lattice is **compact** if every set of points whose infimum is below x has a finite subset whose infimum is below x. A lattice is **algebraic** if it is countable, complete and compactly generated (any point is the infimum of the compact points above it.)

THEOREM 3.9. *(Groszek and Shore, 1988* [**5**].*) Any countable (in L) constructible algebraic lattice with least element can be embedded into the degrees of constructibility as an initial segment.*

3.1.1. *An open problem.* This still leaves a significant gap between algebraic lattices and complete lattices. The astute reader will have noticed that our attention has precipitately narrowed from upper semilattices of size continuum to countable lattices. That is because we cannot even answer this question.

PROBLEM 2. (Assume ω_1 is inaccessible from reals.) What countable constructible lattices can be embedded into the degrees of constructibility as initial segments? In particular, can the lattice consisting of top point 1, bottom point 0, a decreasing ω-sequence of degrees $\langle a_n \rangle$ between 0 and 1, and a degree b between 0 and 1 incomparable to all the a_n be so embedded?

The particular test case here is described in Shore's paper [**18**]. Without having any suggestions as to how to solve this problem, we can say that new techniques will be required. In particular, the techniques that have been used for positive results are shown in Groszek and Shore[**5**] to work only for algebraic lattices. Either a new embedding technique, or a new idea like Lubarsky's analysis of the necessity of completeness, is needed.

3.2. Embeddings as partial orderings. In [**18**], Shore poses the question of embeddings into the degrees of constructibility as partial orderings. In this case we are very much in the same situation as with the Turing degrees; a good deal can be done under CH, but if CH fails, there is the same significant restriction on the possibility of extending embeddings of small partial orderings.

THEOREM 3.10. *There is a model of ZFC in which ω_1 is inaccessible from reals and there is a maximal independent set of degrees of constructibility of size less than continuum.*

This theorem is a variant of Groszek and Slaman's Theorem 2.13[**6**] and the proof is essentially the same. Begin with a model M of ZFC in which ω_1 is inaccessible from reals and the continuum hypothesis holds. (A classic model of "ω_1 is inaccessible from reals", produced by forcing over L to turn an inaccessible cardinal κ into the new ω_1, satisfies the hypotheses.) Then extend M to $M[G][H]$ in exactly the same way. The generic sequence of reals G that produced a maximal independent set of Turing degrees also produces a maximal independent set of degrees of constructibility. The argument that the G_i are Turing independent actually shows that they are constructibly independent, since they are mutually generic. And if adding any other real would destroy Turing independence (by introducing unwanted Turing dependencies) it would certainly destroy constructible independence.

It remains only to note that in the new model $M[G][H]$, ω_1 is still inaccessible from reals. It is enough to consider one-step extensions by countable support products of Sacks forcing. Consider any G_i and any product forcing condition p; $p(i)$ is a tree T that is a Sacks condition on G_i. Because ω_1 is inaccessible from reals, the collection of dense subsets of Sacks forcing in $L[T]$ is countable in M, and so in M there is a tree $S \subseteq T$ such that every branch of S is Sacks generic over $L[T]$. The extension of p obtained by substituting S for $p(i)$ forces G_i to be Sacks generic over $L[T]$, and so it forces $\omega_1^{G_i} \leq \omega_1^{L[T]} < \omega_1$. This density argument shows that $\omega_1^{G_i} < \omega_1$. Essentially the same argument can be used if we replace G_i by any countable sequence of the G_i together with any ground model real s. Now a fusion argument shows that *any* new real is constructible from a countable sequence of the G_i and a ground model real s. Therefore ω_1 is in fact inaccessible from reals.

3.2.1. *An open problem.* Shore's question, like Sacks's question for the Turing degrees, remains open. It is a candidate for another independence result, not merely from ZFC but from our stronger assumption that ω_1 is inaccessible from reals.

PROBLEM 3. (Assume ω_1 is inaccessible from reals.) Is every locally countable partial ordering of size at most continuum embeddable into the degrees of constructibility as a partial ordering?

The possible approaches to the analogous question for the Turing degrees could also be applied here. This question seems to be farther from solution, because we do not even understand the countable initial segments of the degrees of constructibility. However, embedding partial orderings evades the difficulty of tightly controlling the degrees that affects the initial segment question, so the embedding question for degrees of constructibility might not turn out to be harder than for Turing degrees. Considering that Sacks's conjecture has remained open for nearly forty years now, this is not saying much. In any case, a solution to either problem is likely to shed light on the other.

References

[1] Uri Abraham and Richard Shore, Initial segments of the Turing degrees of size $\aleph_1$. *Is. J. Math.* 55, 1986, 1–51.

[2] Uri Abraham and Richard Shore, The degrees of constructibility below a Cohen real. *J. Lon. Math. Soc.* 53, 1986, 193–208.

[3] Zofia Adamowicz, On finite lattices of degrees of constructibility of reals. *J. Symb. Logic* 41, 1976, 313–322.

[4] Zofia Adamowicz, Constructible semi-lattices of degrees of constructibility. *Set Theory and Hierarchy Theory V*, ed. Lachlan, Srebrny and Zarach, Lecture Notes in Mathematics 1969, Springer-Verlag, Berlin, 1977, 1–44.

[5] Marcia Groszek and Richard Shore, Initial segments of the degrees of constructibility. *Is. J. Math.* 63, 1988, 149-177.

[6] Marcia Groszek and Theodore Slaman, Independence results on the global structure of the Turing degrees. *Trans. Am. Math. Soc.* 277, 1983, 579-588.

[7] Greg Hjorth, Universal co-analytic sets. *Proc. of the Am. Math. Soc.* 124, 1996, 3867–3873.

[8] Vladimir Kanovei, On the cardinality of maximal independent sets of degrees (abstract of unpublished result). *Ninth USSR Conference in Mathematical Logic*, Leningrad *Nauka*, 1988, 69.

[9] Stephen Kleene and Emil Post, The upper semi-lattice of degrees of unsolvability. *Ann. Math.* 59, 1954, 379–407.

[10] Julia Knight, this volume.

[11] Alistair Lachlan, Distributive initial segments of the degrees of unsolvability. *Z. Math. Logik. Grund. Math.* 14, 1968, 457–572.

[12] Alistair Lachlan and Robert Lebeuf, Countable initial segments of the degrees of unsolvability. *J. Symb. Logic* 41, 1976, 289–300.

[13] Manuel Lerman, Initial segments of the degrees of unsolvability. *Ann. Math.* 93, 1971, 365–389.

[14] Robert Lubarsky, Lattices of c-degrees. *Ann. Pure and App. Logic* 36, 1987, 115-118.

[15] Arnold Miller, Some interesting problems. *Set Theory of the Realx*, ed. Judah, Israel Mathematical Conference Proceedings 6, American Math. Society, 1993, 645–654.

[16] Gerald Sacks, *Degrees of Unsolvability*. Annals of Math. Studies 55, Princeton University Press, Princeton NJ, 1963.

[17] Gerald Sacks, Forcing with perfect closed sets. *Axiomatic Set Theory*, ed. D. Scott, Proc. Symp. Pure Math. 13(1), Am. Math. Soc., 1971, 331–355.

[18] Richard Shore, Degrees of constructibility. *Set Theory of the Continuum*, ed. Judah, Just and Woodin, Mathematical Sciences Research Institute Publications 26, Springer-Verlag, New York NY, 1992, 123–135.

[19] Richard Shore, Conjectures and questions from Gerald Sacks's *Degrees of Unsolvability. Arch. Math. Logic* 36, 1997, 233–253.

[20] Clifford Spector, On degrees of recursive unsolvability. *Ann. Math.* 64, 1956, 581-592.

Department of Mathematics, 6188 Bradley Hall, Dartmouth College, Hanover NH 03755-3551

E-mail address: marcia.groszek@dartmouth.edu

Contemporary Mathematics
Volume **257**, 2000

Problems related to arithmetic

Julia F. Knight

ABSTRACT. This paper begins with a brief review of results on the complexity of models of various completions of PA. Then there are open problems suggested by the known results. The main problems concern the difficulty of recovering from a non-standard model $\mathcal{A}$ the various fragments of $Th(\mathcal{A})$. Since models of arithmetic are associated with Scott sets, there are also some problems on Scott sets.

1. Introduction

First order Peano Arithmetic, or PA, is a computably axiomatized theory with intended model $\mathcal{N} = (\omega, +, \cdot, S, 0)$. Here are the axioms.

Axioms for PA

1. $\forall x \forall y\, (S(x) = S(y) \rightarrow x = y)$,

2. $\forall x\ S(x) \neq 0$,

3. $\forall x\ x + 0 = x$,

4. $\forall x\, (x + S(y) = S(x + y))$,

5. $\forall x\ x \cdot 0 = 0$,

6. $\forall x\, (x \cdot S(y) = x \cdot y + x)$,

7_φ. $\forall \overline{u}\, (\, \varphi(\overline{u}, 0)\ \&\ \forall y\, [\, \varphi(\overline{u}, y) \rightarrow \varphi(\overline{u}, S(y))\,] \rightarrow \forall x\, \varphi(\overline{u}, x)\,)$.

The model $\mathcal{N}$ and its isomorphic copies are called *standard*, and other models are called *nonstandard*. In a nonstandard model, the values of the terms $S^{(n)}(0)$, for $n \in \omega$, are said to be *finite*, while other elements are *infinite*. The theory of $\mathcal{N}$, called *true arithmetic*, or TA, is one of many completions of PA. Every completion of PA codes a set $\mathcal{S} \subseteq P(\omega)$. Scott characterized the sets $\mathcal{S}$ that arise in this way—

1991 *Mathematics Subject Classification.* Primary: 03D45, 03C62, 03C57.
Key words and phrases. Models of arithmetic, Scott sets, enumerations.

they are the countable Scott sets. Every nonstandard model of PA also codes a Scott set.

In Section 2, we state some definitions and known results on models and completions of PA. In Section 3, we give some problems related to the difficulty in recovering from a nonstandard model $\mathcal{A}$ of PA, the fragments of the theory $Th(\mathcal{A})$. Finally, in Section 4, we mention a couple of questions on Scott sets. Other questions on Scott sets were stated at the workshop, but these were either answered immediately using known results, or solved during the week.

2. Background

To measure complexity, we identify sentences with their Gödel numbers. Then a theory T is a subset of ω. We consider models whose universe is a computable set of constants—a subset of ω, and we identify a model $\mathcal{A}$ with its atomic diagram $D(\mathcal{A})$. Then $\mathcal{A}$ is also a subset of ω. Solovay and Marker, independently, observed the following (see [**13**], or for a generalization, [**5**]).

THEOREM 2.1 (Solovay, Marker). *For any model $\mathcal{A}$ of PA, $\{deg(\mathcal{B}) : \mathcal{B} \cong \mathcal{A}\}$ is closed upwards.*

Since $\mathcal{N}$ is computable, Theorem 2.1 yields the following.

COROLLARY 2.2. *There are standard models of all Turing degrees.*

In contrast, Tennenbaum [**20**] proved the following.

THEOREM 2.3. *If $\mathcal{A}$ is a nonstandard model of PA, then $\mathcal{A}$ is not computable.*

Proof: For $a \in \mathcal{A}$, let

$$d_a = \{n \in \omega : \mathcal{A} \models p_n | a\} \ .$$

Then $d_a \leq_T \mathcal{A}$ (think of the Division Algorithm). Hence, it is enough to prove following.

Claim: For some $a \in \mathcal{A}$, d_a is not computable.

Proof of claim: Let X, Y be c.e. sets with no computable separator—S is a *separator* for X and Y if $X \subseteq S$ and $Y \cap S = \emptyset$. Take the natural formula $\varphi(u, v)$ saying

$$(\forall y < u) \ [(y \in X_u \rightarrow p_y | v) \ \& \ (y \in Y_u \rightarrow p_y \not| v)] \ ,$$

where X_u and Y_u are the usual stage u approximations for X and Y, respectively. Since $\exists v \, \varphi(u, v)$ holds for all finite elements of $\mathcal{A}$, it holds for some infinite element ν, by "overspill" (a simple consequence of the induction schema 7_φ). If $\mathcal{A} \models \varphi(\nu, a)$, then d_a separates X and Y, so d_a is not computable.

Definition: A *Scott set* is a set $\mathcal{S} \subseteq P(\omega)$ such that

1. if $X \in \mathcal{S}$ and $Y \leq_T X$, then $Y \in \mathcal{S}$,
2. if $X, Y \in \mathcal{S}$, then $X \oplus Y \in \mathcal{S}$ (where $X \oplus Y = \{2n : n \in X\} \cup \{2n + 1 : n \in Y\}$),
3. any infinite subtree of $2^{<\omega}$ in $\mathcal{S}$ has path in $\mathcal{S}$; equivalently, any consistent set of axioms in $\mathcal{S}$ has a completion in $\mathcal{S}$.

Definition: Let T be an extension of PA. A set $X \subseteq \omega$ is *representable* with respect to T if there is a formula $\varphi(x)$ such that

$$\begin{array}{ll} T \vdash \varphi(S^{(n)}(0)) & \text{for } n \in X \\ T \vdash \neg\varphi(S^{(n)}(0)) & \text{for } n \notin X. \end{array}$$

The family of all such X is denoted by *Rep*(T). For example, *Rep*(TA) is the family of arithmetical sets.

Here is Scott's result characterizing the families *Rep*(T) [**17**].

THEOREM 2.4. *For any $\mathcal{S} \subseteq P(\omega)$, the following are equivalent:*
(a) there is a completion T of PA such that Rep$(T) = \mathcal{S}$,
(b) $\mathcal{S}$ is a countable Scott set.

A B_n *formula* is a Boolean combination of Σ_n formulas. To prove Theorem 2.4, Scott showed that for each n, we can find a formula $\varphi(x)$ that is independent over PA and whatever B_n sentences are true. That is, for any completion T of PA and any set S,

$$\{\varphi(S^{(n)}(0)) : n \in S\} \cup \{\neg\varphi(S^{(n)}(0)) : n \notin S\} \cup (T \cap B_n) \cup PA$$

is consistent. This means that, having already chosen $T \cap B_n$, we may consistently let $\varphi(x)$ represent any desired set S.

Here is Scott's result on the families of sets coded in nonstandard models of PA [**17**].

THEOREM 2.5. *Let $\mathcal{A}$ be a nonstandard model of PA, and for each $a \in \mathcal{A}$, let $d_a = \{n : \mathcal{A} \models p_n | a\}$. Then*

$$\mathcal{SS}(\mathcal{A}) = \{d_a : a \in \mathcal{A}\}$$

is a Scott set.

Feferman [**3**] made the following observation (for TA).

PROPOSITION 2.6. *If T is a completion of PA and $\mathcal{A}$ is a (nonstandard) model of T, then Rep$(T) \subseteq \mathcal{SS}(\mathcal{A})$.*

Conversely, at least for any countable Scott set $\mathcal{S} \supseteq$ *Rep*(T), there is (a nonstandard) model $\mathcal{A}$ of T such that $\mathcal{SS}(\mathcal{A}) = \mathcal{S}$. If $\mathcal{S} \supseteq$ *Rep*(T), then $\mathcal{S}$ is said to be is *appropriate for* T.

Jockusch asked (in 1980) for a characterization of the Turing degrees of nonstandard models of TA. One idea was to look for a characterization in terms of jumps. However, in [**9**], it was shown that there are degrees **a** and **b**, both upper bounds for the arithmetical degrees, such that

$$\mathbf{a}'' = \mathbf{b}'' = \mathbf{0}^{(\omega)} ,$$

and **a** is the degree of a nonstandard model of TA, while **b** is not.

Solovay characterized the degrees of nonstandard models of TA. He also characterized the degrees of (nonstandard) models of an arbitrary completion T of PA. To state these results, we need some further definitions.

Definition: Let $\mathcal{S} \subseteq P(\omega)$. An *enumeration* is a binary relation R such that $\mathcal{S} = \{R_n : n \in \omega\}$, where $R_n = \{x : (n,x) \in R\}$. We refer to n as an *R-index* for the set R_n.

The following must have been known to Scott and Feferman.

PROPOSITION 2.7. *For a nonstandard model $\mathcal{A}$ of PA, let*

$$R = \{(a,n) : \mathcal{A} \models p_n | a\} .$$

Then

(a) R is an enumeration of $\mathcal{SS}(\mathcal{A})$,

(b) $R \leq_T D(\mathcal{A})$.

The enumeration R in Proposition 2.7 is called the *canonical enumeration* of $\mathcal{SS}(\mathcal{A})$.

Solovay's result on the degrees of nonstandard models of TA [**18**] was simplified by Marker [**14**] so that it reads as follows.

THEOREM 2.8 (Solovay, Marker). *The degrees of nonstandard models of TA are the degrees of enumerations of Scott sets containing the arithmetical sets.*

The model-construction half of the proof is an infinite-injury priority construction, which can be split into two finite-injury constructions (see [**7**]). The other half is clear from Propositions 2.6 and 2.7.

The extension of Theorem 2.8 to arbitrary arbitrary completions of PA is more complicated, involving a family of functions that give indices for fragments of the theory.

Definition: Let T be a complete theory, and let $T_n = T \cap \Sigma_n$. A *Solovay family* for T relative to X is a family of functions $(t_n)_{n\in\omega}$, where t_n is $\Delta^0_n(X)$, uniformly in n, such that $\lim_{s\to\infty} t_n(s)$ is an index for T_n as a set computable in X, and for all s, $t_n(s)$ is an index for a subset of T_n.

Remark: If $\mathcal{A}$ is a nonstandard model of TA, then there is a trivial Solovay family for TA relative to $\mathcal{A}$. There is a function $t(n)$, $\Delta^0_3(\mathcal{A})$, such that $t(n)$ is an index for $TA \cap \Sigma_n$ as a set computable in $\mathcal{A}$.

Here is Solovay's result on models of an arbitrary completion of PA [**19**].

THEOREM 2.9 (Solovay). *Let T be arbitrary completion of PA. The degrees of (nonstandard) models of T are the degrees of sets X such that*

(1) there is an enumeration R of a Scott set appropriate for T such that $R \leq_T X$,

(2) there is a Solovay family for T relative to X.

The model-construction half of the proof of Theorem 2.9 is an infinitely nested priority construction, with special features related to the functions t_n. For the other half, given a model $\mathcal{A}$, we have the canonical enumeration R of $\mathcal{SS}(\mathcal{A})$. The existence of the functions t_n may at first seem implausible. Using a complete $\Delta^0_n(\mathcal{A})$ oracle, we can enumerate true Σ_n sentences, with parameters. Let $\sigma_n(b)$ be a sentence saying

$$(\forall x < b)[\, p_x | b \rightarrow Sat_n(x)\,] .$$

We can write $\sigma_n(b)$ as a Σ_n sentence. We easily arrive at a sentence with n blocks of quantifiers, starting with $\exists$, followed by a formula with only bounded quantifiers. Using a result of Matijasevic [**15**], [**2**], we obtain a sentence which is genuinely Σ_1, even in the case where $n = 1$.

It is enough to define functions t_n giving R-indices of appropriate fragments of the theory—these are easily converted into indices relative to X. For each s, $t_n(s)$ is (in $\mathcal{A}$) a finite product of elements b_i such that $\mathcal{A} \models \sigma_n(b_i)$. We add a new element b to the product only if we find that $\mathcal{A} \models \sigma_n(b)$ and b is divisible by a new finite prime.

For a complete account of the proofs of Theorems 2.8 and 2.9, see [**7**].

There were a number of problems asking whether a "natural" upper bound for the set of degrees of elements of a Scott set could be a minimal upper bound for that set. These questions are all answered negatively in [**8**], where it is shown that for any completion T of PA, there is another completion T^*, of strictly lower degree, such that $Rep(T^*) = Rep(T)$, and for any nonstandard model $\mathcal{A}$ of PA, there is an isomorphic copy of strictly lower degree. Below, we state two further results from [**8**], related to the goals of [**9**]. In these results, we r estrict our attention to the degrees of enumerations of appropriate Scott sets, and then consider jumps.

The first result concerns the class of (nonstandard) models of a given completion of PA.

THEOREM 2.10. *Let T be a completion of PA. Let $\boldsymbol{a}$ and $\boldsymbol{b}$ be degrees of enumerations of Scott sets appropriate for T, and suppose T has a (nonstandard) model of degree $\boldsymbol{a}$. If there exists n such that $\boldsymbol{a}^{(n)} = \boldsymbol{b}^{(n)}$, then T also has a (nonstandard) model of degree $\boldsymbol{b}$.*

Theorem 2.10 follows easily from Theorem 2.9. Given a non-standard model $\mathcal{A}$ of degree $\mathbf{a}$, we have an enumeration R of a Scott set appropriate for T such that R is computable in $\mathcal{A}$. There is a Solovay family (t_k) for T relative to $\mathcal{A}$. There is an enumeration R^* of an appropriate Scott set, such that R^* has degree $\mathbf{b}$. We obtain a Solovay family (t_k^*) for T relative to R^* as follows. We assume that the the functions t_k give R-indices, and our functions t_k^* will give R^*-indices. For $1 \leq k \leq n$, and for all s, let $t_k^*(s)$ be the first R^*-index for T_k. For $k > n$, and for all s, let $t_k^*(s)$ be the first i such that $R_i^* = R_{t_k(s)}$. We are in a position to apply Theorem 2.9 to get a model $\mathcal{B}$ of T computable in R^*.

The second result concerns the class of copies of a given nonstandard model.

THEOREM 2.11. *Let $\mathcal{A}$ be a nonstandard model of PA, and let $\boldsymbol{a}$ and $\boldsymbol{b}$ be degrees of enumerations of $\mathcal{SS}(\mathcal{A})$. If $\boldsymbol{a}$ is the degree of a copy of $\mathcal{A}$, and $(\boldsymbol{b} \vee \boldsymbol{a})' = \boldsymbol{a}'$, then $\boldsymbol{b}$ is also the degree of a copy of $\mathcal{A}$.*

The proof of Theorem 2.11 uses ideas from the proof of Theorem 2.9.

Here we mention a further area of investigation.

Definition: The *n-diagram* of a model $\mathcal{A}$ is the set of B_n sentences of the complete diagram $D^c(\mathcal{A})$.

For the standard model $\mathcal{N}$, the n-diagram has degree $\emptyset^{(n)}$, for all $n \geq 1$. Arana is investigating the complexity of the n-diagrams for nonstandard models of PA. So far, he has extended Solovay's results [**1**].

Here is the result for TA.

THEOREM 2.12 (Arana). *The set of degrees of n-diagrams of nonstandard models of TA is the same for all n—the set of degrees of enumerations of Scott sets that include the arithmetical sets.*

For other completions of PA, the result is more complicated.

THEOREM 2.13 (Arana). *Let T be a completion of PA. The degrees of (nonstandard) models of T are the degrees of sets X such that*

(a) there is an enumeration R of an appropriate Scott set such that $R \leq_T X$,

(b) there are functions t_m, for $n < m$, $\Delta^0_{m-n}(X)$, uniformly in m, such that $lim_{s\to\infty} t_m(s)$ is an index for $T_m = T \cap \Sigma_m$ relative to X, and for all s, $t_m(s)$ is an index for a subset of T_m.

We purposely omit other natural questions on n-diagrams, and we encourage the reader to watch for results of Arana.

3. Questions on models and theories

Thanks to Solovay, we have characterizations of the degrees of nonstandard models TA and the degrees of models of other completions of PA. We may ask the following general question.

QUESTION 1. Are these good characterizations?

One feature of a good characterization is that it is not unnecessarily complicated. A second feature is that it can be used to settle specific questions.

Let us first consider the possibility of simplification.

For TA, Solovay's original result involved extra functions. Marker showed that these were unnecessary. It is difficult to imagine how Theorem 2.8 could be simplified further. As for Theorem 2.9, the statement is somewhat complicated, and there are no examples showing that the complications are necessary.

As we mentioned earlier, if $\mathcal{A}$ is a nonstandard model of TA, then there is a trivial Solovay family $(t_n)_{n\in\omega}$ for TA relative to $\mathcal{A}$. We may take the functions t_n to be constant, and there is a uniform effective procedure, using a $\Delta^0_3(\mathcal{A})$ oracle, for computing the value of t_n for all n.

Harrington [**4**] constructed an example of a nonstandard model $\mathcal{A}$ of PA such that $\mathcal{A} \equiv_T \emptyset'$ and $Th(\mathcal{A}) \equiv_T \emptyset^{(\omega)}$. For this example, and also for the variant in [**6**], there is a Solovay family $(t_n)_{n\in\omega}$ for $Th(\mathcal{A})$ relative to $\mathcal{A}$ in which the functions t_n are constant, but there is no m such that we can apply an effective procedure with a $\Delta^0_m(\mathcal{A})$ oracle to compute the value of t_n for all n. It seems likely that many completions T of PA have a (nonstandard) model $\mathcal{A}$ with no Solovay family for T relative to $\mathcal{A}$ consisting of constant functions. However, we have no examples.

PROBLEM 2. Construct a nonstandard model $\mathcal{A}$ of PA such that there is no Solovay family $(t_n)_{n\in\omega}$ for $Th(\mathcal{A})$ relative to $\mathcal{A}$ in which all t_n are constant. That is, find $\mathcal{A}$ such that for $T = Th(\mathcal{A})$, there is no uniform effective procedure that, for each n, uses $\Delta^0_n(\mathcal{A})$ and immediately locates an index for $T \cap \Sigma_n$ as a set computable relative to $\mathcal{A}$.

An example for Problem 2 would still not illustrate Theorem 2.9 fully. The next two problems ask for examples that go further.

PROBLEM 3. Construct a nonstandard model $\mathcal{A}$ of PA such that there is no Solovay family $(t_n)_{n\in\omega}$ for $Th(\mathcal{A})$ relative to $\mathcal{A}$ in which all t_n change value at most once.

PROBLEM 4. Construct a nonstandard model $\mathcal{A}$ of PA such that there is no Solovay family $(t_n)_{n\in\omega}$ for $Th(\mathcal{A})$ relative to $\mathcal{A}$ with a computable function b such that $t_n(s)$ changes value at most $b(n)$ times.

For Problems 2, 3, 4, we seem to need some new independence results. The following would be enough to give a solution to Problem 2.

Conjecture. Let φ_n be the natural Σ_n sentence saying that the set of true B_n sentences is "more inconsistent" with PA than the set of true B_{n-1} sentences (in the sense that there is a smaller proof of a contradiction). Then for any completion T of PA, there are further completions T^* and T^{**} such that

$$T \cap B_n \subseteq T^* \cap \Sigma_n \subseteq T^{**} \cap \Sigma_n$$

and $\varphi_n \in T^{**} - T^*$.

Problems 2, 3, and 4 were made public at the meeting, but now Arana is actively working on them. We hope that the interested reader will try the problems below, while waiting for Arana's results.

Now, we consider the usefulness of Theorems 2.8 and 2.9. Before Solovay's results, Lerman [**11**] had asked whether there is a nonstandard model $\mathcal{A}$ of TA with no enumeration R of just the family of arithmetical sets computable in $\mathcal{A}$. Lachlan and Soare [**10**] showed that there is such a model. They used Theorem 2.8. McAllister [**12**] extended the result of [**10**], showing that for a broad class of completions T of PA, there is a (nonstandard) model $\mathcal{A}$ with no enumeration of just $Rep(T)$ computable in $\mathcal{A}$—it is sufficient to to have some "jump ideal" $J \supseteq Rep(T)$ such that for all $X \in J$, there exists $C \in Rep(T)$ such that C is "1-generic" over X. McAllister did not use Theorem 2.8 or Theorem 2.9 (or their proofs).

PROBLEM 5 (Detlefsen). Determine whether for all completions T of PA, there is a (nonstandard) model $\mathcal{A}$ with no enumeration of $Rep(T)$ computable in $\mathcal{A}$?

There are not many applications of Theorems 2.8 and 2.9. The application of Theorem 2.8 that we mentioned above was inessential. Theorem 2.10 is one application of Theorem 2.9. The solution to Problem 5 could be another. It seems worthwhile to look for further applications. If we cannot find them, then perhaps we should also look for new, more useful characterizations.

The next problem asks whether Theorem 2.11 can be strengthened.

PROBLEM 6. Let $\mathcal{A}$ be a nonstandard model of PA, and suppose $\mathbf{a}$ and $\mathbf{b}$ are degrees of enumerations of $\mathcal{SS}(\mathcal{A})$. If $\mathbf{a}$ is the degree of a copy of $\mathcal{A}$, and $\mathbf{b}' = \mathbf{a}'$, then must $\mathbf{b}$ also be the degree of a copy of $\mathcal{A}$?

4. Scott sets

We have seen that theories and models of arithmetic are closely related to Scott sets. At the Boulder workshhop, Groszek and Slaman, and, independently, Kucera, answered a question of McAllister, showing that for any Scott set $\mathcal{S}$, if $X \in \mathcal{S}$ is non-computable, then there exists $Y \in \mathcal{S}$ such that $X \not\leq_T Y$ and $Y \not\leq_T X$. Kucera also answered some other questions related to McAllister's question. He showed that there is a Scott set $\mathcal{S}$ with a non-computable element X such that for all completions T of PA in $\mathcal{S}$, $X \leq_T T$.

In the real world—i.e., in the Scott set $P(\omega)$, there is no such A.

THEOREM 4.1. *If X is noncomputable, then there is a completion T of PA s.t. $X \not\leq_T T$ and $T' \leq_T X \oplus \emptyset'$.*

We give the proof because the way certain requirements are met may be of some interest. To control the jump of T, we try to keep elements *out*, instead of trying to put them in.

Proof: Let $(\alpha_k)_{k\in\omega}$ be a computable list of all sentences in the language of arithmetic. We have requirements of the following forms:
(a) $\varphi_e^T(e)\uparrow$, if possible
(b) $\varphi_e^T \neq \chi_X$,
(c) decide α_k.
Let T_0 be the set of axioms of PA. At step s, suppose we have a computable, consistent subset T_s of T, taking care of the first s requirements. Using X and $\emptyset'$, we extend T_s to T_{s+1} satisfying the next requirement, as follows:
(a) Suppose the requirement is $\varphi_e^T(e)\uparrow$, if possible.
There is a natural computable set C_e of sentences that, whose inclusion in T means that $\varphi_e^T(e)\uparrow$. We let

$$T_{s+1} = \begin{cases} T_s \cup C_e & \text{if this is consistent} \\ T_s & \text{otherwise .} \end{cases}$$

(b) Suppose the requirement is $\varphi_e^T \neq \chi_X$.
For each pair of numbers (k,v), there is a natural computable set $B(k,v)$ of sentences whose inclusion in T means that $\varphi_e^T(k) \neq v$. We let

$$T_{s+1} = T_s \cup B(k,v) ,$$

where k is first such that for $v = \chi_X(k)$, $T_s \cup B(k,v)$ is consistent.
(c) Suppose the requirement is to decide α_k.
We let

$$T_{s+1} = \begin{cases} T_s \cup \{\alpha_k\}, & \text{if this is consistent} \\ T_s \cup \{\neg\alpha_k\}, & \text{otherwise .} \end{cases}$$

Having defined T_s for all s, we let $T = \cup_s T_s$.

Claim: $T' \leq_T X \oplus \emptyset'$.

Using X and $\emptyset'$, we can determine the sequence of steps leading to T. In particular, we can see whether we kept e out of T'.

PROBLEM 7. Determine which abstract properties of Scott sets guarantee "real world" features such as the one in Theorem 4.1, or the existence of a minimal pair of completions of PA. Is one of the usual theories of "reverse mathematics" appropriate?

The final problem is not in computability, but it is an important, old problem on Scott sets.

PROBLEM 8. If $\mathcal{S}$ is a Scott set, must there be a nonstandard model $\mathcal{A}$ of PA such that $SS(\mathcal{A}) = \mathcal{S}$?

By a result of Nadel [**16**], the answer is positive for Scott sets of cardinality at most $\aleph_1$, so under CH, the answer is positive, but otherwise it is still open.

References

[1] Arana, A., "n-diagrams of models of PA", pre-print.

[2] Davis, M., "Hilbert's tenth problem is unsolvable," *Amer. Math. Monthly*, vol. 80(1973), pp. 233-269.

[3] Feferman, S., "Arithmetical definable models of formalizable arithmetic," *Notices of Amer. Math. Soc.*, vol. 5(1958), p. 679.

[4] Harrington, L., handwritten notes, 1979.

[5] Knight, J. F., "Degrees coded in jumps of orderings," *J. Symb. Logic*, vol. 51(1986), pp. 1034-1042.

[6] Knight, J. F., "Degrees of models with prescribed Scott set," *Classification: Proc. of Joint U.S.-Israel Workship*, ed. by Baldwin, 1987, pp. 182-191.

[7] Knight, J. F., "True approximations and models of arithmetic," *Models and Computability*, ed. by Cooper and Truss, Cambridge University Press, 1999, pp. 255-278.

[8] Knight, J. F., "Minimality and completions of PA," submitted to *J. Symb. Logic*.

[9] Knight, J. F., A. H. Lachlan, and R. I. Soare, "Two theorems on degrees of models of true arithmetic," *J. Symb. Logic*, vol. 49(1984) pp. 425-436.

[10] Lachlan, A. H., and R. I. Soare, "Models of arithmetic and upper bounds of arithmetic sets", *J. Symb. Logic*, vol 59(1994), pp. 977-983.

[11] Lerman, M., "Upper bounds for the arithmetical degrees," *Annals of Pure and Applied Logic*, vol. 29(1985), pp. 225-254.

[12] McAllister, A. M., "Completions of PA: models and enumerations of representable sets," *J. Symb. Logic*, vol. 63 (1998), pp. 1063-1082.

[13] Marker, D., "Degrees of models of true arithmetic," *Proc. of the Herbrand Symposium*, ed. by J. Stern, North-Holland, 1981, pp. 233-242.

[14] Macintyre, A., and D. Marker, "Degrees of recursively saturated models," *Trans. of the Amer. Math. Soc.*, vol 282(1984), pp. 539-554.

[15] Matijasevic, Yu., "On recursive unsolvability of Hilbert's tenth problem," *Proc. of Fourth International Congress on Logic, Methodology, and Philosophy of Science*, Bucharest, 1971, North Holland, 1973, 89-110.

[16] Nadel, M., "On a problem of MacDowell and Specker," *J. Symb. Logic*, vol. 45(1980), pp. 612-622.

[17] Scott, D., "Algebras of sets binumerable in complete extensions of arithmetic," *Recursive Function Theory*, ed. by Dekker, *Amer. Math. Soc.*, 1962, pp. 117-22.

[18] Solovay, R., preprint circulated in 1982.

[19] Solovay, R., personal correspondence, 1991.

[20] Tennenbaum, S., "Non-Archimedean models for arithmetic," *Notices of the Amer. Math. Soc.*, (1959), p. 270.

Department of Mathematics, University of Notre Dame, Notre Dame, IN 46556 USA

Contemporary Mathematics
Volume **257**, 2000

Embeddings into the Computably Enumerable Degrees

Manuel Lerman

ABSTRACT. We discuss the status of the problem of characterizing the finite (weak) lattices which can be embedded into the computably enumerable degrees. In particular, we summarize the current status of knowledge about the problem, provide an overview of how to prove these results, discuss directions which have been pursued to try to solve the problem, and present some related open questions.

1. Introduction

Degree structures were introduced by Post [**26**] as structures naturally generated from the notion of *relative computability*. Because the concept of computability is basic to mathematics, Post expected the study of computability, including the study of degree structures, to be important to mathematics. These structures have since been widely studied from both algebraic and logical viewpoints. Of interest to us is to discover what these structures tell us about the intrinsic nature of computability from oracles. In particular, we would like to know how close the information content of two degrees must be in order for them to look the same within a given degree structure. Cooper [**7**] has announced that there are nontrivial automorphisms of many degree structures, from which it follows that there are degrees with differing information content which look the same inside those degree structures. However, Slaman and Woodin (see [**25**] for a proof) have shown that if there is an automorphism taking the degree $\mathbf{a}$ to the degree $\mathbf{b}$, then $\mathbf{a}'' = \mathbf{b}''$, so the information content of $\mathbf{a}$ cannot differ too much from that of $\mathbf{b}$.

The degree structures introduced in Post [**26**] and in Kleene and Post [**10**] are defined as follows. Let A and B be sets of natural numbers. We say that A is *Turing reducible* to B (alternatively, A is *computable from* B) and write $A \leq_T B$ if there is a computer program with access to B as an oracle which can answer any membership question about A. $\leq_T$ is a prepartial ordering, allowing us to define an equivalence relation $A \equiv_T B$ if $A \leq_T B$ and $B \leq_T A$. The $\equiv_T$ equivalence classes are called *degrees*, and form a partially ordered set under the ordering induced by $\leq_T$. The corresponding poset $\mathcal{D} = \langle \mathbf{D}, \leq \rangle$ is called the *degrees of unsolvability*,

1991 *Mathematics Subject Classification.* Primary: 03D25.

Key words and phrases. Embedding, computably enumerable set, computably enumerable degree, lattice.

Research supported by NSF grant DMS-9625445.

or the *degrees* for short. A special subset of $\mathbf{D}$ is the set $\mathbf{R}$ of degrees containing sets which are the output of computer programs with oracle $\emptyset$; we let $\mathcal{R} = \langle \mathbf{R}, \leq \rangle$ denote this poset of *computably enumerable* (abbreviated as *c.e.*) degrees.

The questions asked in Post [**26**] and Kleene and Post [**10**] were algebraic in nature. There were questions about other natural functions and relations which could be defined on degree structures (any pair of elements has a join but not necessarily a meet, so the degrees form an *upper semilattice (usl)*; and there is a naturally defined *jump operator* which is order-preserving and takes each degree to a strictly larger degree), and about embeddings into the structures. Some of these questions have since been more fully answered, and other questions of an algebraic nature such as characterizing extensions of embeddings, homomorphisms, automorphisms and quotients have also been studied. For example, Slaman and Soare [**34**] have a complete characterization of extensions of embeddings in $\mathcal{R}$, while Cooper [**7**] has announced that both $\mathcal{R}$ and $\mathcal{D}$ have nontrivial automorphisms. Other questions of an algebraic nature will be posed in this paper.

Shoenfield [**29**] was the first to ask types of questions arising from logic about degree structures, conjecturing a characterization of $\mathcal{R}$ which implied that the elementary theory of $\mathcal{R}$ was $\aleph_0$-categorical. Shoenfield's conjecture was quickly refuted, but was followed by decidability and undecidability results about degree structures, primarily by Lachlan [**12, 13**]. However, it was Simpson's [**33**] characterization of the elementary theory of $\mathcal{D}$ followed quickly by results of Nerode and Shore [**23, 24**] and Shore [**30, 31, 32**] (among others) which turned the primary focus of the study of degree structures to questions arising from mathematical logic and to global algebraic questions. These questions involved the notions of decidability, definability, and elementary equivalence from mathematical logic, as well as homogeneity and isomorphisms; the latter questions had been raised earlier by Rogers [**27**]. Simpson called this focus 'global degree theory', and questions of this type motivate much of the current research on degree structures.

Both the algebraic and logical approaches to the study of degree structures have been productive, and the two approaches are closely related. The logical approach introduces constant symbols, relation symbols, and function symbols which determine a language and are to be interpreted over a universe; the algebraic approach begins with a universe, and designates constants, relations and functions which determine the structure. Of course, the symbols in the language of logic interpret the corresponding objects over the universe of study.

We now concentrate on algebraic structures with universe $\mathbf{R}$. There are several choices of language which make sense. We have already introduced the language consisting of the binary relation symbol $\leq$ which gives rise to the structure $\mathcal{R}$. Our main concern will be decidability questions about $Th(\mathcal{R})$, the elementary theory of $\mathcal{R}$, and about fragments of this elementary theory. Harrington and Shelah [**9**] (see [**4**] for a proof) showed that $Th(\mathcal{R})$ is undecidable. The fragments of the theory which will be considered will be defined in terms of quantifier complexity. Thus the $\exists$-theory consists of the sentences in prenex normal form, all of whose quantifiers are existential, the $\forall\exists$-theory consists of the sentences in prenex normal form which begin with a block of universal quantifiers, followed by a block of existential quantifiers, followed by a quantifier free formula, etc. Sacks [**28**] showed that every finite poset can be embedded into $\mathcal{R}$, from which it easily follows that the $\exists$-theory of $\mathcal{R}$, $\exists \cap Th(\mathcal{R})$, is decidable. Lempp, Nies, and Slaman [**16**] showed that $\exists\forall\exists \cap Th(\mathcal{R})$ is undecidable. However, it has not yet been determined whether or not $\forall\exists \cap Th(\mathcal{R})$ is decidable.

One approach towards studying the decidability of $\forall\exists \cap Th(\mathcal{R})$ is to look at fragments of this theory which properly extend $\exists \cap Th(\mathcal{R})$. These are obtained through an algebraic approach; one expands the language by adding naturally and simply defined constant symbols and/or relation symbols and/or function symbols, and studies the existential theory in the expanded language. The most natural choices to add to the language are the constant symbols 0 and 1 denoting the least and greatest elements of **R**, and the two-place function symbol $\vee$ denoting the least upper bound operator (the least upper bound of two degrees is the degree of the disjoint sum of two sets; the degree is independent of the choice of sets). This corresponds to studying the c.e. degrees as an upper semilattice with least and greatest elements; we denote this structure as $\mathcal{R}_{U,0,1}$. (The subscripts denote the extra relations, functions, and constants added to the language; we will consider other structures, and the subscripts should make clear the nature of the structure.) Standard techniques easily show, however, that $\exists \cap Th(\mathcal{R}_{U,0,1})$ is decidable. This proof reduces the problem to showing that every finite upper semilattice can be embedded into $\mathcal{R}_{U,0,1}$ preserving least and greatest element if they exist. While lattice structure cannot be placed on $\mathcal{R}$, there are pairs of degrees which have a meet. Thus the next natural step, if one is taking an algebraic approach, is to try to determine which finite lattices can be embedded into $\mathcal{R}$ (as lattices). The natural corresponding logical approach is to introduce predicates $M_n(a_1, \ldots, a_n, b)$ for all $n \geq 2$ such that M_n implies the universal formula stating that every element which is $\leq$ every a_i is also $\leq b$; one then asks about embedding diagrams in the language $\langle \leq, \vee, \{M_n : n \in \mathcal{N}\}\rangle$. We call structures in this language with the natural axioms *weak lattices*, and the weak lattice with universe **R** is denoted as $\mathcal{R}_W$. The techniques for studying the embedding questions produced by algebraic and logical approaches are similar, so we will concentrate on the lattice embedding question; in fact, it seems that there are no additional complications, technique-wise, if we ask that the embedding preserve least element, so will study the question asking which finite lattices can be embedded into $\mathcal{R}_{W,0}$.

There are other predicates which will need to be considered to fully reach $\forall\exists \cap Th(\mathcal{R})$, but we will only allude to these predicates in this paper. We note that instead of working up from $\exists \cap Th(\mathcal{R})$, one can try to work down from $\forall\exists \cap Th(\mathcal{R})$. The first natural question which arises from the latter approach is the algebraic *extension of embeddings question* which was answered by Slaman and Soare [**34**].

There is an extensive literature dealing with special cases of the lattice embedding problem. We refer the reader to [**19**] for a more thorough summary than we present here, and will just summarize the major advances from our point of view.

The first lattice embedding result was obtained independently by Lachlan [**11**] and Yates [**38**] who embedded M_2 (see Figure 1), the boolean algebra generated by two atoms, into $\mathcal{R}_{W,0}$. Those ideas, together with some simple lattice theory, were sufficient to show that all finite (and many infinite) distributive lattices are embeddable into $\mathcal{R}_{W,0}$. New ideas were needed for the nondistributive case, and were introduced by Lachlan [**14**] who embedded the two 5-element nondistributive lattices M_3 and N_5 (see Figure 1) into $\mathcal{R}_{W,0}$. These two lattices differ greatly in nature; N_5 is *principally decomposable*, i.e., given any two successive elements $b < a$ of the lattice, $\{c : c \leq a \;\&\; c \not\leq b\}$ has a minimum, while M_3 is not principally decomposable. (Principally decomposable lattices are also known as lattices without critical triples.) The general belief at that time was that all finite lattices could be embedded into $\mathcal{R}_{W,0}$. In pursuing embedding strategies, Lerman proposed a lattice, S_8 (see Figure 2), which could not be embedded using Lachlan's techniques. Subsequently, Lachlan and Soare [**15**] showed that S_8 could not be embedded into $\mathcal{R}_W$.

Following the result yielding the nonembeddability of S_8, Ambos-Spies and Lerman [**2, 3**] tried to understand the embeddability and nonembeddability results. In a series of two papers, they first generalized the Lachlan-Soare nonembeddability result to obtain a nonembedding condition NEC, and then proved a general embeddability theorem. The latter paper introduced a condition EC which was sufficient for embeddability, and left open the question as to whether EC and NEC were complementary conditions. In terms of actual computability-theoretic constructions, there was a technical obstacle which seemed not to be covered by EC or NEC, and the question of complementarity was more a lattice-theoretic question. In 1995, Lempp and Lerman [**17**] found a (principally decomposable) nonembeddable lattice L_{20} which failed to satisfy EC. Subsequently, in a series of two papers, Lerman [**20, 21**] found a necessary and sufficient condition for embeddability for the class of principally decomposable finite lattices. This condition is not effective–rather it is computable from a $\mathbf{0}''$ oracle. We will try to describe this condition in a simple way, and discuss unsuccessful attempts at obtaining an equivalent computable condition.

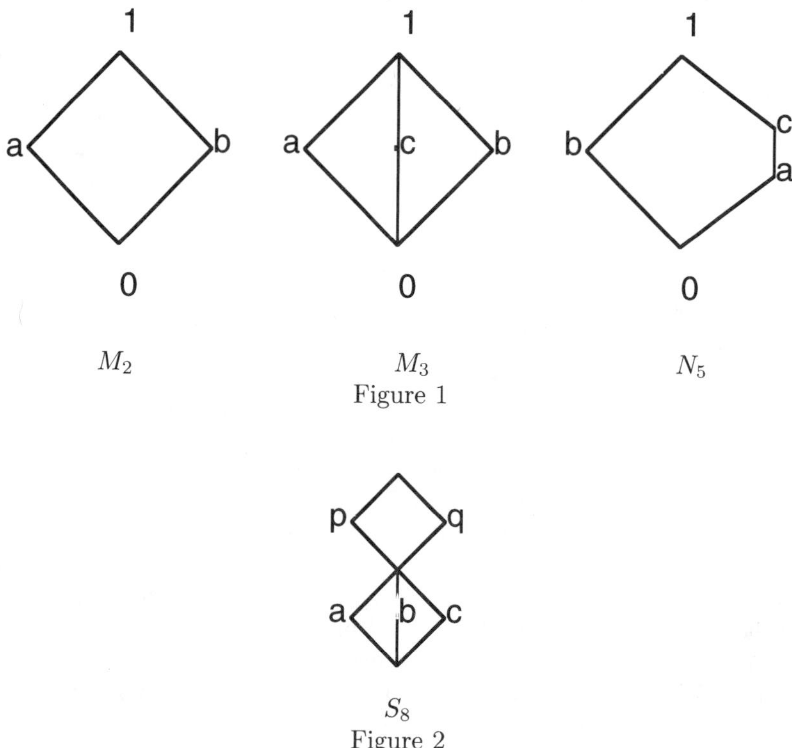

M_2 M_3 N_5

Figure 1

S_8

Figure 2

2. Pinball Machines

Pinball machines were introduced in [**18**] to pictorially describe embedding constructions. The numbers which may enter sets, and the sets which these numbers enter are the labels of balls emanating from holes. These balls roll down a track and, at various stages, are impeded by gates at which they temporarily reside. This prevents them from entering a basket (and so having the numbers enter their target sets). New balls may also be introduced at the gates. There are rules governing the release of balls from a hole, and the release of balls by gates, and these rules completely determine the entry of numbers into sets. While the pinball machine for the whole construction is infinite, it suffices to focus on machines with one hole and finitely many gates for embeddings of principally decomposable lattices. Such a machine is pictured below (Figure 3). There are natural rules which govern the action of any pinball machine, and rules which are needed for the satisfaction of requirements. We discuss the former rules in this section, and the latter rules in the next section. The first four rules introduced in this section deal with configurations of the machine, and the remaining three rules deal with transition, i.e., the movement of balls. Fix a pinball machine with hole H and gates $G_0, \ldots, G_n$.

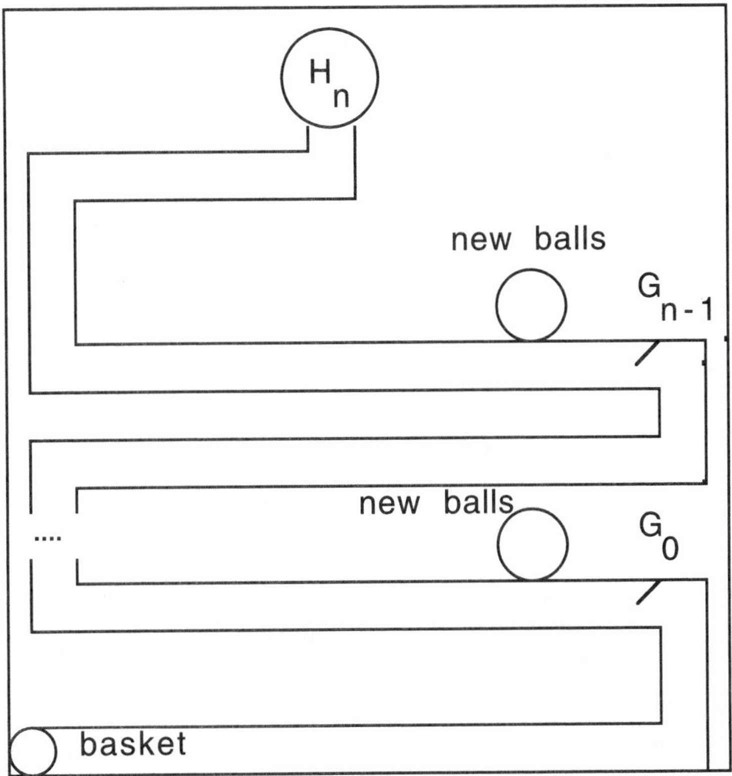

pinball machine
Figure 3

The first two rules specify possible ball locations at a given stage.

RULE 1. *The game begins with all balls in holes, (so none reside at gates or are in the basket). At all subsequent stages, all balls which have been placed in play either reside at gates or are in the basket. New balls may be placed into play at each stage, and may be placed into old blocks or new blocks.*

RULE 2. *At the end of the game, all balls placed in play are in the basket.*

We define equivalence relations on the balls in play, one for each gate for which the balls placed into play have not yet passed. The relation for G_i will collect all balls which are to pass G_i simultaneously into the same equivalence class, called a *G_i-block*. While this is predictive of the future action of the game, it simplifies some of the later rules. There are two rules governing blocks.

RULE 3. *Balls in the same G-block must reside at the same gate (not necessarily gate G).*

RULE 4. *(a) A ball is in some G_i-block iff it currently resides at gate G_j for some $j \geq i$.*
(b) All balls which are in the same G_i block (where $i < n$) and do not currently reside at G_i must be in the same G_{i+1}-block.

The next set of rules are *transition rules* which govern the movement of balls between successive configurations. The first of these rules tells us which balls move, and their next location. The second rule tells us that the order of balls is preserved. The third rule prohibits blocks from changing when we pass to a new configuration.

RULE 5. *Fix the smallest i such that G_i is currently inhabited. The balls in play for the current configuration which change location for the next configuration are precisely the balls which are in the first G_i-block of G_i, and these move to G_{i-1} if $i > 0$ and to the basket if $i = 0$.*

RULE 6. *Balls not entering the basket between configurations maintain their order.*

RULE 7. *(a) Balls which are in the same G_i-block for the current configuration and reside at gate G_j for some $j \geq i$ for the next configuration remain in the same G_i-block for the next configuration.*
(b) Balls currently in play which are not in the same G_i-block for the current configuration cannot be in the same G_i-block for the next configuration.

3. Rules for Requirements

We now consider requirements for the embedding construction, and impose rules on the pinball machine game which, if obeyed, will ensure the satisfaction of the requirements. Suppose that $\langle L, \leq, \vee, \wedge \rangle$ is a finite lattice which we wish to embed into $\mathcal{R}_{W,0}$ preserving join and meet. We will use lower-case letters for elements of L, and the corresponding upper-case letters will denote the sets of integers whose degrees determine the embedding into $\mathcal{R}_{W,0}$.

The first type of embedding requirement will ensure that order is preserved. Thus if $a \leq b \in L$, we wish to ensure that $A \leq_T B$. We do so by requiring that if a ball labeled with target A and number x enters the basket, then x is simultaneously placed into all sets C such that $A \leq_T C$. This is enough to ensure the satisfaction of the order-preserving requirements, and does not require the imposition of any rules on the game.

The second type of embedding requirement will ensure that incomparability is preserved. The full embedding game is an amalgamation of the finite embedding games being presented here. Each of these finite games deals with a single incomparability requirement corresponding to a condition $a \not\leq b \in L$ and a computable partial functional Φ. Thus if $a \not\leq b \in L$, we wish to ensure that $A \neq \Phi(B)$. The following rule will ensure the satisfaction of this requirement.

RULE 8. *If $a \not\leq b$, then the initial configuration has a ball with target $\leq_T A$, but no ball with target $\leq_T B$.*

The third type of embedding requirement will ensure that joins are preserved. Suppose that $a \vee b = c \in L$. As $a, b \leq c$, the satisfaction of the order-preserving requirements will ensure that $A \oplus B \leq_T C$. Thus it suffices to ensure that $C \leq_T A \oplus B$. The standard way to satisfy such a requirement is to use the *method of traces*. The strongest form of this method ensures that whenever a number x is targeted for C, another number y is targeted for A or B, and y must enter its target set at a stage s which is no later than the stage at which x enters C; furthermore, if x does not enter C at stage s, then a new trace z for x is appointed at stage s, and is burdened with the same restrictions as was y. The success of this method relies on showing that any number x has only finitely many traces during the course of the construction. The following rule corresponds to the method of traces.

RULE 9. *If $a \vee b = c$, then for every configuration, if a ball has target $\leq_T C$, then there is a predecessor ball whose target is either $\leq_T A$ or $\leq_T B$.*

The final type of embedding requirement will ensure that meets are preserved. Suppose that $a \wedge b = c \in L$. It will follow from the order-preserving requirements that $C \leq_T A, B$. Thus we must ensure that for every c.e. set U, if $U \leq A, B$ then $U \leq_T C$. In fact, we need not restrict our attention to c.e. sets. Thus for each pair of computable partial functionals Φ and Ψ, we construct a computable partial functional Δ such that whenever $\Phi(A) = \Psi(B)$ and $\Phi(A)$ is total, then $\Delta(C) = \Phi(A)$. The strategy is to define $\Delta(C; x)[s+1] = \Phi(A; x)[s]$ whenever we encounter a suitable stage s at which $\Phi(A; y)[s] = \Psi(B; y)[s]$ for all $y \leq x$, and then to allow numbers to enter only A or only B to injure the equality until the equality recovers, unless a number which injures the axiom $\Delta(C; x)[s+1]$ enters C in the interim. Thus at all stages $t+1$ at which $\Delta(C; x)[t+1] \downarrow$, either $\Delta(C; x)[t+1] = \Phi(A; x)[s]$ or $\Delta(C; x)[t+1] = \Psi(B; x)[s]$. The implementation of this condition as a pinball machine rule requires the introduction of the new notion of a G-block B being *G-prohibited.* We will define this notion after presenting the rule, but indicate the intuition behind the notion now; if G corresponds to $p \wedge q = r$, then B is G-prohibited if the introduction of a new ball with target $\leq r$ into the block and its later entry into the basket will cause injury to the computations from both P and Q on some argument x, without correcting the computation from R. The following rule formalizes G_i-prohibitions for $i \leq n$.

RULE 10. *If a G_i-block B is released by gate G_i, then B is not G_i-prohibited at the time of release.*

Before defining prohibitions, we present two examples to provide some intuition.

EXAMPLE 3.1. Consider the lattice N_5 pictured in Figure 1. The incomparability requirement $C \not\leq_T A$ requires the placement of a ball with target C on the pinball machine, and as $a \vee b \geq c$, Rule 9 requires this ball to be preceded by a ball with target B or A; as Rule 8 precludes target A, our first sequence of targets will be $\langle B, C\rangle$. The meet condition $c \wedge b = 0$ at gate G requires that the two balls with targets B and C respectively be in different blocks, else the simultaneous entry of both balls into their respective target sets would injure the requirement (thus the single block would have a G-prohibition). The block consisting of a single ball with target B comes first, and the block consisting of a single ball with target C has a G-prohibition. Once the ball with target B enters the basket, a new ball with target A is appointed, placed in the same G-block as the ball with target C, and the G-prohibition is removed. (Note that the number on the new ball targeted for A can be chosen sufficiently large to avoid injuring the initial computation from oracle A.) At this point, the simultaneous entry of these two balls into the basket will give rise to a target sequence $\langle A, C\rangle$, and entry of balls into these sets cannot injure a computation from oracle B, so the requirement will not be injured.

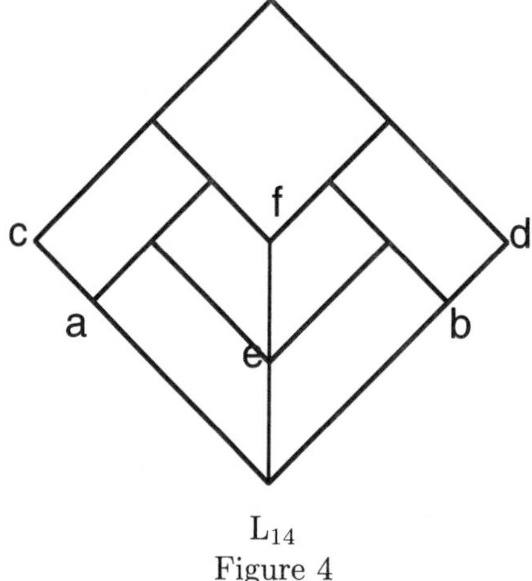

L_{14}
Figure 4

EXAMPLE 3.2. Consider the lattice L_{14} of Figure 4. We first note that $a \vee b \geq f, d$ and $d \vee e \geq f$. Thus in order to satisfy the incomparability requirement corresponding to $f \not\leq e \vee b$, we may assume that our starting array on the pinball machine has balls targeted, in order, for A, D, and F. Let the meet requirement $(f \vee c) \wedge d = 0$ be assigned to gate G, the first gate that the above sequence of balls will encounter when released from the hole. As $a, f \leq f \vee c$, these balls must all be in different G-blocks else the meet requirement will be injured, and a G-prohibition is placed in all but the first block. When the first ball enters its target set, a suitable next configuration of the pinball machine would have balls at gate G targeted, in order, for B, D, A, and F, where the first and third balls are new, and the second and fourth balls are old. Furthermore, we can place the first two balls in one G-block, and the last two balls in another G-block; the G-prohibition is removed from the first block but is inherited by the second block. Now suppose that the first two balls enter the basket simultaneously. A possible next sequence would have three balls targeted, in order, for A, E, and F, where the middle ball is a new ball and the other two balls are old balls. All three balls may be placed in the same G-block (from which the G-prohibition is removed), and so will be released simultaneously by gate G.

The above strategy will work if G is the only gate. Suppose now that there is a second gate G' which corresponds to the meet requirement $(b \vee e) \wedge (a \vee e) = e$. Consider the G'-blocks for the configuration with targets B, D, A, F. The G'-blocks containing A and F must have G'-prohibitions, as a new ball targeted for E will be appointed in a location where computations from $A \oplus E$ and $B \oplus E$ will be injured but the new E ball has a number too large to correct the computation from the meet. In this situation, we can pass to a configuration with four balls targeted, in order, for A, E, A, and F, where the first two balls are new balls and the last two balls are old balls, and place all the balls in the same G-block but the second and third balls in different G'-blocks (the latter having a G'-prohibition), and will be able to successfully continue the game when the balls are simultaneously released by G and roll to G'.

We now present the three gate prohibition rules. Some of these rules deal with two consecutive configurations of the pinball machine. The *current configuration* refers to the configuration being specified at the current stage, and the *preceding configuration* refers to the configuration which was specified at the immediately preceding stage. Blocks of the current configuration are *old* if they contain elements from the preceding configuration, and are *new* otherwise. An old G-block B has an *old* G-prohibition if the G-block of the preceding configuration containing elements of B has a G-prohibition. A mathematically precise specification of the rules can be found in [**20**]. The first rule governs the removal of old G-prohibitions, and the last two rules govern the imposition of new G-prohibitions, one rule for old G-blocks and one rule for new G-blocks. Let B be a G-block of the current configuration. Suppose that $p \wedge q = r$ is the meet requirement assigned to gate G, and let K be the upward-closure of the lattice elements representing targets for the balls in the G-blocks preceding B, together with those for the balls in B itself.

RULE 11. *If B is an old G-block which has an old G-prohibition, then the G-prohibition is removed if and only if the first inhabited gate of the current configuration is either gate G or a gate above gate G, and either $p \notin K$ or $q \notin K$.*

RULE 12. *If the first inhabited gate is above gate G, then a new G-prohibition is imposed on block B if and only if $p, q \in K$ but $r \notin K$.*

RULE 13. *If the first inhabited gate is either gate G or a gate below gate G, then a new G-prohibition is imposed on block B if and only if B is a new G-block and the last old G-block which precedes B in the current configuration has an old G-prohibition which is not removed by Rule 11.*

A necessary and sufficient condition, $GBTA$, for embedding principally decomposable finite lattices into the c.e. degrees is presented in Lerman [**21**], and is rephrased here in terms of the rules for games.

DEFINITION 3.3. $GBTA$: For every pinball machine game as above (one hole and a finite sequence of gates) there is a finite array of configurations for that pinball machine game which satisfies Rules 1–13. (Such an array is called a *good blocked target array (gbta).*)

Note that $GBTA$ is an $\forall\exists$ condition, so can be decided by an oracle of degree $\mathbf{0}''$. The sufficiency of $GBTA$ is proved by a standard embedding construction (pinball machine model); note that if the condition is true, then a finite search will effectively produce the sequence of configurations, given the game (i.e., if an $\forall\exists$ sentence has a Skolem function, then it has a computable Skolem function). The necessity of $GBTA$ is obtained by reversing the portion of the embedding construction corresponding to any choice of a hole and a finite sequence of gates, and showing that the failure of the opponent to win such a game can be translated into a sequence of configurations for that game which satisfies Rules 1–R13. The restriction to principally decomposable lattices is needed for the nonembedding games, as we must ensure that we can always get by with complete trace functions for join requirements that are finite at every stage. For ranked lattices (defined in Lerman [**20**]), all tracing functions are finite. In [**21**], it is shown that, for principally decomposable lattices, while not all complete trace functions are finite at every stage if the lattice is not ranked, one can replace the trace function at any given stage with a complete finite trace function. If the lattice is not principally decomposable, then there is no way to avoid using infinite complete trace functions, and the nonembedding game cannot be implemented.

4. Attempts to determine whether GBTA is equivalent to an effective condition

Several approaches have been tried to show that $GBTA$ is equivalent to an effective condition. These include:

• Trying to approximate to a given lattice through smaller subsets of the lattice, and to find a relationship between the configurations for the smaller subsets and the configurations for the lattice itself. As subsets, we have looked at sublattices, subsets which preserve upper semilattice operations, and, most atomically, we have looked at configurations obtained by adding one requirement at a time. In all cases, we were unable to find a uniform way of extending configurations for the subset to configurations for the larger set of requirements which either produces an obstruction implying nonembeddability, or produces a sequence of configurations satisfying all rules for the larger set of requirements.

• Defining a concept of *normal form* for configurations, and showing that we can effectively compute, for each game, a bound on the length of the normal form sequences which are relevant to that game. The first attempt to define a normal form was to try to use sequences which satisfied the tracing requirements and were minimal in this regard. Such sequences place a restriction on the ability of meet requirements to correct computations, and we later found an example of a sequence of requirements which has a gbta, but does not have a gbta in which all sequences satisfy this proposed definition of normal form. We then turned unsuccessfully to sequences without repetitions, and, having failed to define a suitable notion of normal form, to seeing if there was a uniform way of piecing together uniformly bounded configurations to obtain a full configuration. These attempts were also unsuccessful. Both Nerode, and later Khoussainov, suggested that the problem might be coded into the decision procedure for $S2S$; but the only way we saw to approach this required us to first show that there is a bound on the size of configurations for a given game, so we encounter the same obstacle. The basic problem, even for a single game, is that the rules for prohibition functions are not local (restricted to a small interval of the configuration) rules.

Shore has noted that the methods used to code in an undecidable fragment of a theory cannot be used to obtain undecidability results; rather, one must directly try to code the halting problem, the word problem for groups, or some other basic undecidability result. These undecidability results involve analogs of *sequences of configurations*, but changes in passing from configuration to configuration are local, while the pinball machine game configurations are not, making it difficult to see how such a coding might go. As nonlocal conditions seem to be more complicated than local conditions, this might be taken as evidence for an intuitive guess that the lattice embedding problem is undecidable.

5. Open Questions Related to Embeddings

As was mentioned in the introduction, one can approach the study of $\mathcal{R}$ by asking the questions a logician would ask, or by asking the questions an algebraist would ask. From a logician's point of view, the main open question is:

QUESTION 1. Is $\forall\exists \cap Th(\mathcal{R})$ decidable?

A suggested approach was to enrich the language, and to try to decide the $\exists$-theory in a richer language. This gave rise to the notion of *weak lattice*, and the question:

QUESTION 2. Is there an algorithm to determine which finite weak lattices are embeddable into $\mathcal{R}_W$?

As the techniques to embed lattices and weak lattices seem to be identical, and as we are more familiar with lattices, we are led to the question which this paper has focused on, namely:

QUESTION 3. Is there an algorithm to determine which finite lattices are embeddable into $\mathcal{R}_W$?

The pursuit of a negative answer to Question 1 might proceed by trying to code an undecidable problem into the existential theory of a structure obtained by enriching the language. For as we come closer to the full $\forall\exists \cap Th(\mathcal{R})$ and find necessary and sufficient conditions for the corresponding decision problems, we should expect more flexibility in our ability to code undecidable problems into these necessary and sufficient conditions. Thus it is reasonable to further extend the language, and to try to find a condition which characterizes the decision problem for the existential theory of the enriched structure. Lerman [**19**] has shown that $GBTA$ can be converted to a necessary and sufficient condition for an existential sentence of $\mathcal{R}_W$ to be true, and this condition remains unchanged for $\mathcal{R}_{W,0}$. Englert (work in progress) has a similar result for $\mathcal{R}_{W,1}$.

QUESTION 4. Is there an algorithm to determine which finite weak lattices are embeddable into $\mathcal{R}_{W,0,1}$?

Similar questions can be asked for other enriched languages whose $\exists$-theory is a subset of the $\forall\exists \cap Th(\mathcal{R})$. Predicates of this type were crucial in the extension of embeddings result of Slaman and Soare [**34**].

The methods described in this paper can be used to study embeddings into structures other than $\mathcal{R}$. We mention two problems of this type which have been studied. The first is the problem of effectively characterizing the finite (weak) lattices which can be embedded into every nontrivial initial segment of $\mathcal{R}$, i.e., every interval of the form $[\mathbf{0}, \mathbf{a}]$ where $\mathbf{a} \neq \mathbf{0}$ is a c.e. degree. Weinstein [**37**] and Downey [**8**] showed that every such (weak) lattice must be principally decomposable, and Lerman [**21**] noted that a principally decomposable (weak) lattice L is embeddable into every nontrivial initial segment of $\mathcal{R}$ iff L satisfies $GBTA$. Thus the problem again reduces to the question of whether $GBTA$ is equivalent to an effective condition.

The other structure is that of the upper semilattice $\mathcal{I}$ of ideals of $\mathcal{R}$, which has been studied by Calhoun [**6**]. Calhoun showed that the lattice S_8 (which Lachlan and Soare showed could not be embedded into $\mathcal{R}$) is embeddable into $\mathcal{I}$. Thus the obstruction to embedding witnessed by S_8 is not an obstruction for $\mathcal{I}$. However, the obstruction to embedding witnessed by the lattice L_{20} (which Lempp and Lerman showed could not be embedded into $\mathcal{R}$) still seems to remain, and Calhoun and Lerman have a preliminary sketch which indicates that L_{20} cannot be embedded into $\mathcal{I}$. In fact, they feel that it is highly likely that a principally decomposable finite (weak) lattice is embeddable into $\mathcal{I}$ iff it satisfies $GBTA$.

6. Open Questions: Homomorphisms

We discussed the algebraic approach to studying degree structures in the Introduction. One topic which falls under this heading and has not received much attention is that of homomorphic images of degree structures (with the exception of automorphism questions) and homomorphisms and endomorhisms of degree structures. We will focus primarily on $\mathcal{R}$, but note that similar questions can be asked about all degree structures.

We begin by trying to determine whether a degree structure has any nontrivial homomorphic images, i.e., whether the structure is *simple*. The definition of simplicity is meant to rule out all but the trivial homomorphic images, i.e., the maps onto the structure itself, and the maps onto a single element.

DEFINITION 6.1. A degree structure $\mathcal{S}$ with universe S is *simple* if for every homomorphism f with domain S, either $f(S) = S$ or $|f(S)| = 1$.

QUESTION 5. Which degree structures are simple?

The map j^n taking a degree to its *nth* jump is a poset homomorphism. Shore [S] has shown that if $n \geq 3$, then the original degree structure and its homomorphic image are not isomorphic. Furthermore, any such map on the standard degree structures has infinite range. Hence the standard degree structures, viewed as posets, are not simple. A substantial literature, due to Martin [**22**], Steel [**36**], and Slaman–Steel [**35**] exists studying poset homomorphisms on degree structures which take every degree $\mathbf{a}$ to a degree $\mathbf{b} \geq \mathbf{a}$. Such homomorphisms, however, are not upper semilattice homomorphisms.

The only investigation of upper semilattice homomorphisms of which we are aware is for the upper semilattice $\mathcal{R}_U$ of c.e. degrees.

DEFINITION 6.2. Let $\mathcal{P} = \langle P, \leq \rangle$ be a poset. A *downward closed* (*upward closed*, resp.) subset of $\mathcal{P}$ is a set S such that whenever $a \in S$, $b \in P$ and $b \leq a$ ($b \geq a$, resp.) then $b \in S$. S is an *ideal* (*filter*, resp.) of $\mathcal{P}$ if S is downward closed (upward closed, resp.) and if $a, b \in S$, then there is a $c \in S$ such that $c \geq a, b$ ($c \leq a, b$, resp.). The ideal S (filter, resp.) of $\mathcal{P}$ is *prime* if it is proper and its complement is a filter (ideal, resp.) and is *maximal* if it is proper and no ideal (filter, resp.) properly extends S.

Ambos-Spies, Jockusch, Shore and Soare [**1**] showed that the cappable degrees ($\mathbf{c}$ is *cappable* if there is a $\mathbf{d} > \mathbf{0}$ such that $\mathbf{c} \wedge \mathbf{d} = \mathbf{0}$) form a prime ideal $\mathbf{M}$ whose complement, the promptly simple degrees $\mathbf{PS}$ is a prime filter. Thus we can define a preorder relation $\cong_M$ on $\mathbf{R}$, setting $\mathbf{a} \leq_{\mathbf{M}} \mathbf{b}$ if there is a cappable degree $\mathbf{c}$ such that $\mathbf{a} \leq \mathbf{b} \cup \mathbf{c}$. We then define $\mathbf{a} \cong_{\mathbf{M}} \mathbf{b}$ iff $\mathbf{a} \leq_{\mathbf{M}} \mathbf{b}$ and $\mathbf{b} \leq_{\mathbf{M}} \mathbf{a}$, and note that the map taking a degree in $\mathbf{R}$ to its congruence class in $\mathbf{R}/\mathbf{M}$ is a usl homomorphism. Thus $\mathcal{R}_U$ is not simple, as $\mathbf{0}$ and $\mathbf{0}'$ are in different congruence classes. $\mathbf{R}/\mathbf{PS}$ is similarly defined. Leonhardi, Lerman and Yi (unpublished) have shown that the above homomorphism from $\mathbf{R}$ to $\mathbf{R}/\mathbf{M}$ is not a weak lattice homomorphism. However, the map taking $\mathbf{M}$ to 0 and $\mathbf{PS}$ to 1 is a weak lattice homomorphism of $\mathcal{R}_W$, so $\mathcal{R}_W$ is also not simple. Calhoun [**5**] has shown that $\mathcal{R}$ has an infinite antichain of prime ideals.

An old question, mentioned in both the talks of Nies and Shore at this conference, is whether there is a definable prime ideal of $\mathcal{R}$ other than $\mathcal{M}$. We repeat this question here, phrased somewhat differently.

Problem 6. Is there a nice characterization of all definable weak lattice homomorphisms of $\mathcal{R}$ onto $\{0, 1\}$?

Other related problems, posed for $\mathcal{R}$ but appropriate for other degree structures are:

Problem 7. Characterize all homomorphic images (quotients) of $\mathcal{R}$.

Problem 8. Characterize all definable homomorphisms of $\mathcal{R}$.

Problem 9. Characterize all (definable) maximal ideals of $\mathcal{R}$.

We conclude with an observation of Lempp that there is no usl homomorphism mapping the d-c.e. degrees onto $\mathcal{R}$. This follows easily from the fact that every d-c.e. degree can be cupped to $\mathbf{0}'$, while not every c.e. degree can be cupped to $\mathbf{0}'$.

References

[1] K. Ambos-Spies, C. G. Jockusch, R. A. Shore, and R. I. Soare, *An algebraic decomposition of the recursively enumerable degrees and the coincidence of several degree classes with the promptly simple degrees*, Trans. Amer. Math. Soc. **281** (1984), 109–128.

[2] K. Ambos-Spies and M. Lerman, *Lattice embeddings into the recursively enumerable degrees*, J. Symbolic Logic **51** (1986), 257–272.

[3] K. Ambos-Spies and M. Lerman, *Lattice embeddings into the recursively enumerable degrees: II*, J. Symbolic Logic **54** (1989), 735–760.

[4] K. Ambos-Spies and R. A. Shore, *Undecidability and 1-types in the recursively enumerable degrees*, Ann. Pure Appl. Logic **63** (1993), 3–37.

[5] W. C. Calhoun, *Incomparable prime ideals of recursively enumerable degrees*, Ann. Pure Appl. Logic **63** (1993), 39–56.

[6] W. C. Calhoun, *The lattice of ideals of recursively enumerable degrees*, Ph.D. Dissertation, University of California at Berkeley, 1990.

[7] S. B. Cooper, *The Turing universe is not rigid*, to appear.

[8] R. G. Downey, *Lattice non-embeddings and initial segments of the recursively enumerable degrees*, Ann. Pure Appl. Logic **49** (1990), 97–119.

[9] L. Harrington and S. Shelah, *The undecidability of the recursively enumerable degrees* (research announcement), Bull. Amer. Math. Soc. (N. S.) **6** (1982), 79–80.

[10] S. C. Kleene and E. L. Post, *The upper semi-lattice of degrees of recursive unsolvability*, Ann. Math. (2) **59** (1954), 379–407.

[11] A. H. Lachlan, *Lower bounds for pairs of recursively enumerable degrees*, Proc. London Math. Soc. **16** (1966), 537–569.

[12] A. H. Lachlan, *Distributive initial segments of the degrees of unsolvability*, Z. Math. Logik Grund. Math. **14** (1968), 457–472.

[13] A. H. Lachlan, *The elementary theory of recursively enumerable sets*, Duke Math Jour. **35** (1968), 123–146.

[14] A. H. Lachlan, *Embedding nondistributive lattices in the recursively enumerable degrees*, In: Conference in Mathematical Logic, 1970, W. Hodges ed., Lecture Notes in Mathematics 255, Springer-Verlag, Berlin, Heidelberg, New York, 1972, 149–177.

[15] A. H. Lachlan and R. I. Soare, *Not every finite lattice is embeddable in the recursively enumerable degrees*, Advances in Math. **37** (1980), 74–82.

[16] S. Lempp, A. Nies and T. A. Slaman, *The Π_3-theory of the enumerable Turing degrees is undecidable*, Trans. Amer. Math. Soc. **350** (1998), 2719–2736.

[17] S. Lempp and M. Lerman, *A finite lattice without critical triple that cannot be embedded into the enumerable Turing degrees*, Ann. Pure Appl. Logic **87** (1997), 167–185.

[18] M. Lerman, *Admissible ordinals and priority arguments*, In: Cambridge Summer School in Mathematical Logic – Proceedings 1971, A. R. D. Mathias and H. Rogers eds., Lecture Notes in Mathematics, 337, Springer–Verlag Pub. Co., Berlin, Heidelberg, New York, 1973, 311–344.

[19] M. Lerman, *Embeddings into the recursively enumerable degrees*, In: Computability, Enumerability, Unsolvability: Directions in Recursion Theory, S. B. Cooper, T. A. Slaman, and S. S. Wainer eds., London Math. Soc. Lecture Note Series, 224, Cambridge University Press, Cambridge, 1996, 185–204.

[20] M. Lerman, *A necessary and sufficient condition for embedding ranked finite partial lattices into the computably enumerable degrees*, Ann. Pure Appl. Logic **94** (1998), 143–180.

[21] M. Lerman, *A necessary and sufficient condition for embedding principally decomposable finite partial lattices into the computably enumerable degrees*, Ann. Pure Appl. Logic, to appear.

[22] D. A. Martin, *The axiom of determinateness and reduction principle in the analytical hierarchy*, Bull. Amer. Math. Soc. **74** (1968), 687–689.

[23] A. Nerode and R. A. Shore, *Second order logic and first order theories of reducibility orderings*, In: The Kleene Symposium, K. J. Barwise, H. J. Keisler, and K. Kunen eds., North-Holland Pub. Co., Amsterdam, New York, Oxford, 1980, 181–200.

[24] A. Nerode and R. A. Shore, *Reducibility Orderings: Theories, definability and automorphisms*, Ann. Math. Logic **18** (1980), 61–89.

[25] A. Nies, R. A. Shore, and T. A. Slaman, *Interpretability and definability in the recursively enumerable degrees*, Proc. London Math. Soc. (3) **77** (1998), 241–291.

[26] E. L. Post, *Recursively enumerable sets of positive integers and their decision problems*, Bull. Amer. Math. Soc. **50** (1944), 284–316.

[27] H. Rogers, Jr., Theory of Recursive Functions and Effective Computability, McGraw-Hill, New York, 1967.

[28] G. E. Sacks, Degrees of Unsolvability, Annals of Mathematics Studies, 55, Princeton Univ. Press, Princeton, 1963.

[29] J. R. Shoenfield, *Application of Model Theory to degrees of unsolvability*, In: Symposium on the Theory of Models, J. W. Addison, L. Henkin, and A. Tarski eds., North-Holland Pub. Co., Amsterdam, 1965, 359–363.

[30] R. A. Shore, *The homogeneity conjecture.* Proc. Nat. Acad. Sci. U. S. A. **76** (1979), 4218–4219.

[31] R. A. Shore, *The theory of the degrees below* $\mathbf{0}'$, J. London Math. Soc. (2) **24** (1981), 1–14.

[32] R. A. Shore, *On homogeneity and definability in the first-order theory of the Turing degrees.* Jour. Symb. Logic **47** (1982), 8–16.

[33] S. G. Simpson, *First-order theory of the degrees of recursive unsolvability*, Ann. Math. (2) **105** (1977), 121–139.

[34] T. A. Slaman and R. I. Soare, *Algebraic aspects of the computably enumerable degrees*, Proc. Nat. Acad. Sci. USA **92** (1995), 617–621.

[35] T. A. Slaman and J. R. Steel, *Definable functions on degrees*,

[36] J. R. Steel, *A classification of jump operators*, Jour. Symb. Logic **47** (1982), 347–358.

[37] B. Weinstein, *On embeddings of the 1-3-1 lattice into the recursively enumerable degrees*, Doctoral Dissertation, University of California at Berkeley, 1988.

[38] C. E. M. Yates, *A minimal pair of recursively enumerable degrees*, Jour. Symb. Logic **31** (1966), 159–168.

Department of Mathematics, University of Connecticut, Storrs, CT 06269-3009 USA

E-mail address: mlerman@@math.uconn.edu

Contemporary Mathematics
Volume **257**, 2000

Definability in the c.e. degrees: Questions and results

André Nies

ABSTRACT. We ask questions and state results about definability in the partial order $\mathcal{R}_T$ of computably enumerable Turing degrees. The main open question is whether $\mathcal{R}_T$ is biinterpretable with $\mathbb{N}$ in parameters. Some of the results can be viewed as approximations to an affirmative answer. For instance, we have proved that all uniformly computably enumerable sets of nonzero c.e. Turing degrees can be defined from parameters by a fixed formula. As a consequence we obtain a new $\emptyset$-definable ideal.

1. Introduction

The biinterpretability conjecture in parameters for an arithmetical structure $\boldsymbol{A}$ (in brief, BI-conjecture) states that there is a parameter defined copy M of $(\mathbb{N}, +, \times)$ and a parameter definable 1-1 map $f : \boldsymbol{A} \to M$. This has far reaching consequences for $\boldsymbol{A}$, for instance that all automorphisms are arithmetical (and therefore there are only countable many), and that each orbit in A^n is $\emptyset$-definable (in other words, $\boldsymbol{A}$ is a prime model of its theory). If the BI-conjecture holds, we can, in fact, require that $f(\boldsymbol{a})$ is an index for $\boldsymbol{a}$. In a different direction, we can actually make f the identity, so that each degree in $\boldsymbol{A}$ can be viewed as a number in a parameter defined copy M of $(\mathbb{N}, +, \times)$. The conjecture for arithmetical degree structures goes back to Harrington, as well as Slaman and Woodin (see [**8**]).

Throughout this article, "definable" means parameter definable. Let $\mathcal{R}_T$ denote the upper semilattice of computably enumerable (c.e.) Turing degrees. As approximations to the BI-conjecture for $\boldsymbol{A}$, we try to find a definable configuration $g : B \to M$, g 1-1, where the arithmetical set $B \subseteq A$ is large in some sense. We will argue in Section 3 that if B is an automorphism base (and this is proved in a certain way) then the approximation implies the full conjecture. So the hope for $\mathcal{R}_T$ is that eventually a B which is large enough in that sense will be found. For example, by a result of Ambos-Spies, each nontrivial interval $[\mathbf{o}, \boldsymbol{c}]$ would do. (In [**7**] a different type of approximation to BI, without parameters, is given, namely BI up to second jump.)

Even the approximations to BI have interesting consequences. A main result is that for u.c.e. sets $B \subseteq \mathcal{R}_T - \{\mathbf{o}\}$, $g : B \to M$ exists. In particular, B is (parameter)

1991 *Mathematics Subject Classification.* Primary: 03D25.
Key words and phrases. C.e. Degrees, biinterpretability, u.c.e. sequences.
Partially supported by NSF grant DMS–9803482.

definable via a fixed formula. From this we derive that the finite sets are uniformly definable (see Definition 4.4 below), and prove that there is a new $\emptyset$-definable ideal, namely the ideal generated by the nonbounding degrees. This answers a question posed by the author at the 1999 AMS workshop in Boulder.

Acknowledgments. The author would like to thank Richard Shore for many helpful discussions. He also incorporated an idea from work of Shore and Slaman (unpublished) who proved a special case of Corollary 5.4 below.

2. Biinterpretability with $\mathbb{N}$ in parameters

In the following all structures and languages are relational. In particular, we view $(\mathbb{N}, +, \times)$ as a structure with two ternary relations. A structure $\boldsymbol{A}$ over a finite symbol set is called *arithmetical* if there is an onto map $\beta : \mathbb{N} \to A$ such that the preimages of the relations of $\boldsymbol{A}$ under β are arithmetical. For instance, $\mathcal{R}_T$ is arithmetical via $\beta(e) = \deg_T(W_e)$, since the preimage of the ordering is Σ^0_4. A relation on $\boldsymbol{A}$ is called arithmetical if its preimage is.

DEFINITION 2.1. *A copy of* $\mathbb{N}$ *coded in* $\boldsymbol{A}$ *with parameters is a structure*

$$N = (D, R_+, R_\times) \cong (\mathbb{N}, +, \times)$$

such that D *and the relations* $R_+, R_\times$ *are parameter definable in* $\boldsymbol{A}$.

DEFINITION 2.2. *An arithmetical structure* $\boldsymbol{A}$ *is biinterpretable with* $\mathbb{N}$ *in parameters if there is a parameter coded copy* N *of* $(\mathbb{N}, +, \times)$ *and a definable 1-1 map* $f : A \to N$.

Since $\boldsymbol{A}$ is arithmetical, this definition coincides with the usual definition of biinterpretability of structures $\boldsymbol{A}, \boldsymbol{B}$ (see [**1**], p. 222). BI in parameters has many interesting consequences.

PROPOSITION 2.3. *Suppose that the arithmetical structure* $\boldsymbol{A}$ *is biinterpretable with* $\mathbb{N}$ *in parameters. A relation* R *on* $\boldsymbol{A}$ *is arithmetical iff* R *is parameter-definable.* $\diamondsuit$

Recall that a countable structure $\boldsymbol{A}$ is a prime model (of its theory) iff each orbit under the action of $\mathrm{Aut}(\boldsymbol{A})$ on A^n is $\emptyset$-definable, and that, by definition, $\boldsymbol{A}$ is a minimal model if $\boldsymbol{A}$ has no proper elementary submodel.

PROPOSITION 2.4. *Suppose that the arithmetical structure* $\boldsymbol{A}$ *is biinterpretable with* $\mathbb{N}$ *in parameters. Then:*

(i) $\boldsymbol{A}$ *is a prime model and a minimal model.*
(ii) *The automorphisms of* $\boldsymbol{A}$ *are uniformly definable, and thus the group* $Aut(\boldsymbol{A})$ *can be interpreted in* $\boldsymbol{A}$ *without parameters in a natural way. There is a finite automorphism base for* $\boldsymbol{A}$.
(iii) *A copy of* $(\mathbb{N}, +, \times)$ *can be coded in* $\boldsymbol{A}$ *without parameters.*

Our main open problem is the following.

QUESTION 2.5. *Is* $\mathcal{R}_T$ *biinterpretable with* $\mathbb{N}$ *in parameters?*

Even an affirmative answer to one of the following easier problems would be of interest.

QUESTION 2.6. *Is* $\mathcal{R}_T$ *prime, or at least* ω*-homogeneous? Is* $\mathcal{R}_T$ *a minimal model?*

We briefly look at related results for other degree structures. BI with $\mathbb{N}$ in parameters is known to hold for $\mathcal{D}(\leq \emptyset')$ by a result of Slaman and Woodin (unpublished). On the other hand, BI with $\mathbb{N}$ fails for $\mathcal{R}_m$. In fact, $|\mathrm{Aut}(\mathcal{R}_m)| = 2^\omega$, and also $\mathcal{R}_m$ is not a minimal model since $\exists e[\mathbf{o}, e) \prec \mathcal{R}_m - \{\mathbf{1}\}$ (see [**4**]). The following question remains open:

QUESTION 2.7. *Is $\mathcal{R}_m$ prime?*

In [**2**] we prove

THEOREM 2.8. *The many-one degrees of arithmetical sets form a prime model.*

Thus in some cases a degree structure can be shown to be prime without proving biinterpretability with $\mathbb{N}$ in parameters.

For other arithmetical degree structures the question remains open. For instance:

QUESTION 2.9. *Is the structure $\mathcal{R}_{wtt}$ of c.e. weak truth-table degrees biinterpretable with $\mathbb{N}$ in parameters?*

The necessary condition that a copy of $(\mathbb{N}, +, \times)$ can be coded in $\mathcal{R}_{wtt}$ without parameters (cf. (iii) of Proposition 2.4) has been obtained in [**3**]. However, all the coding methods outlined below, which may lead to BI for $\mathcal{R}_T$, fail in $\mathcal{R}_{wtt}$ because of distributivity.

3. Automorphism bases

DEFINITION 3.1. *A subset B of a structure $\boldsymbol{A}$ is an automorphism base if the only automorphism of $\boldsymbol{A}$ fixing B pointwise is the identity.*

For example, $B \subseteq (\mathbb{Q}, <)$ is automorphism base iff B is dense in $\mathbb{Q}$.

THEOREM 3.2 (Ambos-Spies). *For each $\mathbf{c} \in \mathcal{R}_T - \{\mathbf{o}\}$, $[\mathbf{o}, \mathbf{c}]$ is an automorphism base.*

A modified proof of this result [**2**] uses the following general method to show that B is an automorphism base of $\boldsymbol{A}$: one provides a definable (relative to B) and 1-1 map

$$H : \boldsymbol{A} \to \tau B,$$

where τB is the collection of objects of "type" τ constructed from B. Before making this more precise, we give an example of such a map of the required form, which, however, may fail to be 1-1. Let $\widehat{\boldsymbol{x}}$ denote the interval $[\mathbf{o}, \boldsymbol{x}]$. Let $B = \widehat{\mathbf{c}}$, and

$$H(\boldsymbol{w}) = \{(\widehat{\boldsymbol{y}} \cap \widehat{\mathbf{c}}, \widehat{\boldsymbol{z}} \cap \widehat{\mathbf{c}}) : \boldsymbol{y} \vee \boldsymbol{z} = \boldsymbol{w}\}. \tag{3.1}$$

In general, τB-maps are defined by the following:

1. For each formula $\varphi(x, y)$, $H_\varphi(\boldsymbol{x}) = \{\boldsymbol{y} \in B : \varphi(\boldsymbol{x}, \boldsymbol{y})\}$ is a τB-map.
2. If $H_1, \dots, H_n$ are τB-maps, then for each formula $\varphi(x, y_1, \dots, y_n)$, $H(\boldsymbol{x}) = \{(H_1(\boldsymbol{y}_1), \dots, H_n(\boldsymbol{y}_n)) : \varphi(\boldsymbol{x}, \boldsymbol{y}_1, \dots, \boldsymbol{y}_n)\}$ is also a τB-map.

As explained in the introduction, definable automorphism bases promise to be useful in proving the BI–conjecture for $\mathcal{R}_T$:

PROPOSITION 3.3. *Suppose $B \subseteq \mathcal{R}_T$ is definable, and $H : \mathcal{R}_T \to \tau B$ is a 1-1 τB-map. Then to show BI in parameters for $\mathcal{R}_T$, it suffices to produce a standard N and a 1-1 definable $g : B \to N$.*

Sketch of proof. The map $f = (\tau g) \circ H$ is a 1-1 definable map $\mathcal{R}_T \to N$, where the 1-1 map τg is obtained roughly as follows. Notice that, since B is arithmetical, an object $t \in \tau B$ can be described by an arithmetical construction. For H as in (3.1), for instance, t is described by a Σ_4^0 set of pairs of indices for Σ_4^0 sets of indices for degrees. Then $(\tau g)(t)$ is obtained by modeling this construction inside N, replacing the atomic constituents of t by their images under g. $\diamondsuit$

If B is an automorphism base, but we know no proof of this by the method of τB-maps, instead of BI we obtain, as a weaker statement, the existence of a finite automorphism base from the existence of a 1-1 definable $g : B \to N$: The automorphism base consists of the parameters used to code N and g.

4. Schemes and uniform definability

We recall some definitions. See [**7**] for more details. A scheme for coding in an L-structure $\boldsymbol{A}$ is given by a list of L-formulas

$$\varphi_1, \dots, \varphi_n$$

with a shared parameter list $\overline{p}$, together with a correctness condition $\alpha(\overline{p})$. If a scheme S_X is given, $X, X_0, X_1, \dots$ denote objects coded via S_X by a list of parameters satisfying the correctness condition.

EXAMPLE 4.1. *A scheme S_N for coding models of some finitely axiomatized fragment PA^- of Peano arithmetic (in the language $L(+, \times)$) is given by the formulas*

$$\varphi_{num}(x, \overline{p}), \varphi_+(x, y, z; \overline{p}), \varphi_\times(x, y, z; \overline{p}) \tag{4.1}$$

and a correctness condition $\alpha_0(\overline{p})$ which expresses, among others things, that φ_+ and $\varphi_\times$ define binary operations on the nonempty set $\{x : \varphi_{num}(x; \overline{p})\}$, and that $\{x : \varphi_{num}(x; \overline{p})\}$ with the corresponding operations satisfies the finitely many axioms of PA^-.

For our applications it is sufficient to work with Robinson arithmetic Q. Then all coded models N have a standard part isomorphic to $\mathbb{N}$. If this standard part equals N we say that N is *standard.*

EXAMPLE 4.2. *A scheme S_g for defining a function g is given by a formula $\varphi_1(x, y; \overline{p})$ defining the relation between arguments and values, and a correctness condition $\alpha(x, y; \overline{p})$ which says that a function is defined: $\forall x \exists^{\leq 1} y \varphi_1(x, y; \overline{p})$.*

EXAMPLE 4.3. *A scheme for defining n-ary relations on $\boldsymbol{A}$ is given by a formula $\varphi(x_1, \dots, x_n; \overline{p})$ and a correctness condition $\alpha(\overline{p})$.*

The object, coding, it becomes possible to

DEFINITION 4.4. (i) *A class $\mathcal{C}$ of n-ary relations on $\boldsymbol{A}$ is* uniformly definable *in $\boldsymbol{A}$ if, for some scheme S for coding relations, $\mathcal{C}$ is the class of relations coded via S as the parameters range over tuples in $\boldsymbol{A}$ which satisfy the correctness condition.*

(ii) *$\mathcal{C}$ is* weakly uniformly definable *if $\mathcal{C}$ is contained in a uniformly definable class.*

For instance, if $\boldsymbol{A}$ is a linear order and $\mathcal{C}$ is the set of closed intervals, then $\mathcal{C}$ is uniformly definable via the scheme consisting of $\varphi_1(x; a, b) \Leftrightarrow a \leq x \leq b$ and the correctness condition $\alpha(a, b) \Leftrightarrow a \leq b$.

5. Maps onto sets of degrees

Suppose S_N is a scheme as in Example 4.1. For standard N and $B \neq \emptyset$, the existence of a definable 1-1 $g : B \to N$ is clearly equivalent to the existence of a definable onto map $f : N \to B$. In the following, it is more natural to work with onto maps.

THEOREM 5.1 (Nies [**5**]). *There are schemes S_M, S_f as in Example 4.1 and Example 4.2 with the following property. Suppose $\mathcal{A} = \{\boldsymbol{a}_i : i \in \mathbb{N}\}$, $(\boldsymbol{a}_i)$ is u.c.e. and $\forall i\ \boldsymbol{a}_i \neq \mathbf{o}$. Then there is a standard M and a map f such that $f : M \to \mathcal{A}$ is onto and in fact $\boldsymbol{a}_i = f(i^M)$.* ◇

In particular, for every u.c.e. sequence $(\boldsymbol{a}_i)$ of pairwise distinct nonzero degrees, there is a uniformly coded copy L of $(\mathbb{N}, +, \times)$ such that $\forall i\ i^L = \boldsymbol{a}_i$.

U.c.e. sequences can behave in a complicated way. For instance, the author has shown that there is a strictly decreasing u.c.e. sequence with infimum $\mathbf{o}$.

The schemes S_M, S_f from Theorem 5.1 will be used in the following. We derive some consequences.

COROLLARY 5.2. *The u.c.e. set of nonzero degrees are weakly uniformly definable.* ◇

COROLLARY 5.3. *If there is a u.c.e. set B of nonzero degrees which forms an automorphism base, then there is a finite automorphism base.* ◇

As a special case of Theorem 5.1, consider $\mathcal{A} = [\boldsymbol{d}, \mathbf{1}]$ for some $\boldsymbol{d} \neq \mathbf{o}$.

COROLLARY 5.4. *For each noncomputable c.e. D there is a standard M and a map f such that $\forall i\ f(i^M) = deg(W_i \oplus D)$.* ◇

This was proved by Shore and Slaman for sets D of promptly simple degree.

In [**7**] *comparison maps* were introduced, maps between two coded models of PA$^-$ which extend the isomorphism between the standard parts. They were used to single out parameters coding a standard models in a first order way. We show that whenever the first model is standard, then the comparison map is definable, even if the two models are coded via different schemes.

PROPOSITION 5.5. *Suppose S_K and S_L are schemes to define models of PA$^-$ as in Example 4.1 such that $\mathbf{o}$ is not in the domain of any coded model. Then there is a scheme S_h to code functions, depending on S_K, S_L, such that $\forall$ L standard $\forall K\ \exists h[h$ is the isomorphism between K and the standard part of $L]$.*

Proof. We use an auxiliary scheme $S_{\widetilde{h}}$ to code isomorphisms between finite initial segments. Given K, L and $n \in \mathbb{N}$, apply Theorem 5.1 to the sequence which begins $0^K, 0^L, 1^K, 1^L, \ldots, n^K, n^L$, and is $0^K, 0^L, 0^K, 0^L, \ldots$ from then on. We obtain M and f. Let $\widetilde{h}$ be the relation $\{(f(\boldsymbol{x}), f(\boldsymbol{x} +_M 1^M)) : \boldsymbol{x} \text{ even in } M\}$. As a correctness condition on parameters $\overline{\boldsymbol{q}}$ which code K, L, M, f we require that the relation so defined is indeed an isomorphism between initial segments of K, L. Clearly all finite isomorphisms can be coded with $S_{\widetilde{h}}$.

We obtain the scheme S_h using the formula $\varphi(x, y; \overline{q})$ which expresses that for some $\widetilde{h}$, $\widetilde{h}(\boldsymbol{x}) = \boldsymbol{y}$. ◇

REMARK 5.6. As a consequence, for *any* scheme S_K such that some $\widetilde{K}$ is standard and $\mathbf{o}$ is not in the domain of any coded model, we can add a correctness condition which holds of parameters coding K iff K is standard: the condition expresses that

$$\forall \widetilde{K}\ \exists h\ [h \text{ is isomorphism between } K \text{ and an initial segment of } \widetilde{K}].$$

Furthermore, the isomorphism between any standard K, L coded via S_K, S_L is uniformly definable.

The following is an approximation to the statement (ii) of Proposition 2.4.

PROPOSITION 5.7. *The class of partial 1-1 functions on* $\mathcal{R}_T$ $\{\Phi \upharpoonright \mathcal{A} : \Phi \in Aut(\mathcal{R}_T)\ \&\ \mathcal{A} \subseteq \mathcal{R}_T - \{\mathbf{o}\}$ *u.c.e.* $\}$ *is weakly uniformly definable.*

Proof. Given Φ and $\mathcal{A}$, choose M, f for $\mathcal{A}$ by Theorem 5.1. Let $\widetilde{M}, \widetilde{f}$ be the standard model and map coded by the images under Φ of the coding parameters. Then $\widetilde{f} : \widetilde{M} \to \Phi(\mathcal{A})$ is onto. By Remark 5.6, choose a uniformly definable isomorphism $h : M \to \widetilde{M}$. Then for any $\boldsymbol{x} \in \mathcal{A}$, $\Phi(\boldsymbol{x}) = \boldsymbol{y} \Leftrightarrow \exists \boldsymbol{w} \in M[\boldsymbol{x} = f(\boldsymbol{w})\ \&\ \boldsymbol{y} = \widetilde{f}(h(\boldsymbol{w}))]$. ◇

THEOREM 5.8. *The finite sets of degrees are uniformly definable.*

Proof. It suffices to show that the finite nonempty sets of nonzero degrees are uniformly definable, since we can add an extra parameter to include the remaining finite sets: suppose a scheme $\varphi(x;\overline{p}), \alpha(\overline{p})$ uniformly defines the nonempty finite sets. Let $\widetilde{\varphi}(x;\overline{p},q)$ be a formula such that, if F is the set defined by $\overline{\boldsymbol{p}}$ via $\varphi(x;\overline{p})$, then $\widetilde{\varphi}(x;\overline{p},q)$ defines $\emptyset$ if $\boldsymbol{q} = \mathbf{o}$, $F \cup \{\mathbf{o}\}$ if $\boldsymbol{q} = \mathbf{1}$, and F itself otherwise.

Clearly a finite nonempty set of nonzero degrees $\boldsymbol{A}$ meets the hypotheses of Theorem 5.1. To ensure that we only define finite sets, we add as a correctness condition on (parameters coding) M, f that M is standard (see the remark after Proposition 5.5) and that f is almost constant. ◇

For the c.e. many–one degrees, results corresponding to Proposition 5.7 and Theorem 5.8 were obtained in [**6**].

COROLLARY 5.9. *If* $D \subseteq \mathcal{R}_T$ *is* $\emptyset$*-definable, then so is* $[D]_{id}$*, the ideal of the upper semilattice* $\mathcal{R}_T$ *generated by* D*.*

Proof. By Theorem 5.8 and the fact that $[D]_{id} = \{\boldsymbol{x} : \exists\, F \subseteq D \text{ finite } [\boldsymbol{x} \leq \sup(F)]\}$. (Here $\sup(\emptyset) = \mathbf{o}$.) ◇

At the 1999 AMS workshop in Boulder, the author asked how to obtain more examples of $\emptyset$-definable ideals in $\mathcal{R}_T$. We use the previous corollary to show there is a $\emptyset$-definable ideal which does not coincide with the ideal **M** of cappable or the ideal **NCup** of noncuppable degrees (the author is grateful to T. Slaman for telling him about the ideal of noncuppable degrees). Recall that **NCup** is a proper subideal of **M**. Let **NB** be the class of degrees which do not bound a minimal pair. Note that $\mathbf{NB} \subseteq \mathbf{M}$.

THEOREM 5.10. *There is a noncuppable degree which is not below a finite supremum of degrees in* **NB**. ◇

Since the ideal $\mathbf{I} = [\mathbf{NB}]_{id}$ is contained in **M**, all the three ideals are pairwise distinct. We conjecture (but haven't had the patience to prove) that in fact also the $\emptyset$-definable ideals $\mathbf{I} \cap \mathbf{NCup}$ and $[\mathbf{I} \cup \mathbf{NCup}]_{id}$ are new.

QUESTION 5.11. *Are there infinitely many $\emptyset$-definable ideals in $\mathcal{R}_T$?*

Recall that the set $\mathcal{R}_T - \mathcal{M}$ of prompt degrees is a strong filter.

QUESTION 5.12. *Find a further $\emptyset$-definable strong filter in $\mathcal{R}_T$.*

Finally we discuss a further type of set B such that a definable map from a coded copy of $\mathbb{N}$ onto B exists. This set B is large in the sense that it is downward dense within an initial interval.

THEOREM 5.13. *There are $\mathbf{c} \neq \mathbf{o}$, $B \subseteq [\mathbf{o}, \mathbf{c}]$, a coded copy L of $(\mathbb{N}, +, \times)$ and an onto parameter definable map $d : L \to B$ such that $\forall \boldsymbol{y} \leq \mathbf{c}\ [\boldsymbol{y} \neq \mathbf{o} \Rightarrow \exists \boldsymbol{x} \in B[\boldsymbol{x} \neq \mathbf{o}\ \&\ \boldsymbol{x} \leq \boldsymbol{y}]]$.*

Sketch of proof. We enumerate C and let B be the degrees of c.e. set splits of C. We construct L distinguishing elements of B in $[\mathbf{o}, \mathbf{c}]$ in the following sense: If $\boldsymbol{v} \in B, \boldsymbol{u} \leq \mathbf{c}$ and $\boldsymbol{u} \not\leq \boldsymbol{v}$, then for some $\boldsymbol{g} \in L$, $\boldsymbol{v} \leq \boldsymbol{g}\ \&\ \boldsymbol{u} \not\leq \boldsymbol{g}$. Then $\boldsymbol{v} = \inf(\{\mathbf{c}\} \cup \{\boldsymbol{g} \in L : \boldsymbol{v} \leq \boldsymbol{g}\})$, so for a definable map d, $\boldsymbol{v} = d(\boldsymbol{e})$, where $\boldsymbol{e}$ is an arithmetical index in L for $\{\boldsymbol{g} \in L : \boldsymbol{v} \leq \boldsymbol{g}\}$. The domain of L is a Slaman-Woodin set $\mathrm{SW}(\boldsymbol{q}, \boldsymbol{p}; \mathbf{o}, \boldsymbol{r})$. The advantage of the restriction to degrees $\boldsymbol{v}$ of set splits V is that one can control the size of V changes in reaction to C enumeration of a requirement P_e trying to make C noncomputable. If a candidate z for P_e enters C, it may enter a split V and therefore a set G where $\boldsymbol{g} \geq \boldsymbol{v}$ and $\boldsymbol{g}$ is in L. So the requirement P_e needs to maintain a P restraint. If V were merely given by a reduction to C, then the potential enumeration of z into C could cause larger and larger numbers to enter V, so the P restraint for a single candidate would become unbounded. Further ideas are used to make L an antichain and for the downward density of B. $\diamondsuit$

We conjecture that in addition $\mathbf{c}$ can be made nonbounding. In this case one obtains a strictly descending definable (but necessarily not u.c.e.) sequence $(\boldsymbol{b}_i)$ such that $\forall \boldsymbol{x} \leq \mathbf{c}[\boldsymbol{x} \neq \mathbf{o} \Rightarrow \exists i\ \boldsymbol{b}_i \leq \boldsymbol{x}]$.

References

[1] W. Hodges. *Model Theory.* Encyclopedia of Mathematics. Cambridge University Press, Cambridge, 1993.

[2] A. Nies. Global properties of degree structures. To appear in the Proceedings of the 11th Conference on Logic, Methodology and Philosophy of Science.

[3] A. Nies. Interpreting **N** in the c.e. weak truth table degrees. Submitted, also in Habilitationsschrift, Univ. Heidelberg, 1998.

[4] A. Nies. Model theory of the computably enumerable many-one degrees. To appear in the Logic Journal of the IGPL, http://www.dcs.kcl.ac.uk/journals/IGPL.

[5] A. Nies. Parameter definable subsets of the computably enumerable degrees. To appear.

[6] A Nies. The last question on recursively enumerable many-one degrees. *Algebra i Logika*, 33(5):550–563, 1995. English Translation, Consultants Bureau, NY, July 1995.

[7] A. Nies, R. Shore, and T. Slaman. Interpretability and definability in the recursively enumerable turing-degrees. *Proc. Lond. Math. Soc.*, 3(77):241–291, 1998.

[8] Theodore A. Slaman. Degree structures. In *Proceedings of the International Congress of Mathematicians, Kyoto, 1990*, volume I, pages 303–316, Heidelberg, 1991. Springer-Verlag.

DEPARTMENT OF MATHEMATICS, THE UNIVERSITY OF CHICAGO, 5734 S. UNIVERSITY AVE., CHICAGO, IL 60637, USA, WEBSITE HTTP://WWW.MATH.UCHICAGO.EDU/~NIES

E-mail address: nies@math.uchicago.edu

Contemporary Mathematics
Volume **257**, 2000

Strong Reducibilities, Again

Piergiorgio Odifreddi

The subject of strong reducibilities has flourished in recent years, and it is now perceived less and less as a marginal curiosity, and more and more as a central enterprise of Recursion Theory. In particular, it is now recognized that strong reducibilities not only provide test cases for techniques and methods that find their best applications in the study of Turing degrees, but also that they are a subject worthy of independent study.

We have devoted to the structures of degrees induced by strong reducibilities a substantial amount of the two volumes of *Classical Recursion Theory* (Odifreddi [1989], [1999]), to which we refer for notation, background information, proofs and references. In a chapter for the *Handbook of Computability Theory* (Odifreddi [1999a]) we have also given a survey of results and a new list of 20 open problems in the area, together with a report on the progress made on the old list of 26 open problems proposed in Odifreddi [1981]. We can thus restrict ourselves here to an indication of the most important directions of research, and of the associated fundamental questions.

Global Degree Structures

If Recursion Theory is the study of the universe of functions and sets of natural numbers, reducibilities are the main instruments used for the investigation. Turing reducibility is usually considered as the central notion, and plays a role similar to the naked eye in the investigation of the real world. Both provide the intuitive picture on which the naive conception of the world is based.

Galileo was the first scientist to take seriously the suggestion that the world consists of more than meets the eye, and following him modern science has developed pictures of the microworld and the macroworld based on images obtained by extending the power of sight by various kinds of microscopes and telescopes. Strong and weak reducibilities play a similar role in Recursion Theory, and at the extremes we find the pictures provided by the many-one degrees $\mathcal{D}_m$ and the hyperdegrees $\mathcal{D}_h$.

The universe of functions and sets of natural numbers as seen through the eye and these instruments does not appear to be invariant under changes of scale, since Ershov [1975] and Slaman and Woodin [199?] have proved that the pictures they provide are as different as they can be, as the following summary of results shows:

- **Automorphisms**. $\mathcal{D}_m$ admits $2^{2^{\aleph_0}}$ automorphisms, while $\mathcal{D}_h$ admits no nontrivial automorphism.

2000 *Mathematics Subject Classification.* Primary 03D30, 03D28.

- **Invariance**. $\mathbf{0}_m$ is the only m-degree invariant under every automorphism, while every hyperdegree is invariant.
- **Homogeneity**. Any two principal filters of $\mathcal{D}_m$ are isomorphic, while there are no distinct isomorphic principal filters of $\mathcal{D}_h$.

In particular, since the structure of many-one degrees has the greatest possible number of automorphisms allowed by its cardinality, and the structure of hyperdegrees the smallest one, we can expect that intermediate reducibilities provide intermediate pictures. This seems to have been confirmed for the Turing degrees by announcements of Cooper [1997] on the positive side, and Slaman and Woodin [199?] on the negative side, to the effect that the Turing degrees have exactly countably many automorphisms.

On the basis of the guess that the various reducibilities provide a complete spectrum of possibilities, we conjecture that *the number of automorphisms of truth-table degrees is exactly* $2^{\aleph_0}$*, or at least* $\aleph_0$. A similar conjecture can be made for the number of automorphisms of the weak truth-table degrees.

Speaking of which, no elementary difference is known between the structures of truth-table and weak truth-table degrees. This is an exceptional event, since all other pairs of common degree structures are known to be elementary inequivalent (see Odifreddi [1989], p. 590). This exceptionality is underlined by the result of Shore [1982] that the structures of truth-table and weak truth-table degrees have isomorphic principal filters. More precisely, the cones above $\mathbf{0}'_{tt}$ and $\mathbf{0}'_{wtt}$ are isomorphic, the isomorphism actually being the identity. This prompts the conjecture that *the structures of truth-table and weak truth-table degrees are isomorphic, or at least elementarily equivalent*.

The cone just considered, of the truth-table degrees above $\mathbf{0}'_{tt}$, is not elementarily equivalent to the full structure of truth-table degrees. For example, while there are minimal truth-table degrees (see Odifreddi [1989], p. 588), Mohrherr [1984] has proved that $\mathbf{0}'_{tt}$ has no minimal cover. This can now be seen as a relativization of a result proved by Homer [1987], according to which there is no minimal polynomial time Turing degree. Since the proof only uses the fact that there is a recursive function majorizing every polynomial and having a polynomial time computable graph, the result generalizes to any version of Turing reducibility restricted by a class of time bounds which can be majorized by a recursive function whose graph is computable in a time bound belonging to the given class (see Odifreddi [1999], p. 196).

Homer's result does not extend to truth-table reducibility because, by Nerode [1957], the latter is equivalent to Turing reducibility with a recursive time bound. In particular, the usual reducibilities considered in Complexity Theory, from polynomial time to primitive recursive, are special cases of truth-table reducibility. The existence of minimal truth-table degrees is made possible by the fact that the recursive functions are not majorized by a recursive function, while the nonexistence of minimal covers for $\mathbf{0}'_{tt}$ is a consequence of the fact that there is instead such a function recursive in $\mathbf{0}'_{tt}$.

These observations prompt the conjecture, proposed among others by Shore and Slaman [1992], that *the structures of truth-table degrees above* $\mathbf{0}'_{tt}$ *and of all polynomial time Turing degrees are isomorphic, or at least elementarily equivalent*. As a partial confirmation of the guess, Mohrherr [1984] and Shinoda (see Odifreddi [1999], p. 195) have proved that the two structures are both dense.

We turn now to the interplay of the degrees with the jump and ω-jump operators. First, it is known that the jump operator is definable in the structures of hyperdegrees, arithmetical degrees and Turing degrees,[1] but not in the structure of many-one degrees, because $\mathbf{0}_m$ is the only definable m-degree. It is thus natural to ask whether *the jump operator is definable in the structure of the truth-table degrees*, although it is difficult to make a guess. The answer is obviously related to the number of automorphisms, since the more automorphisms there are, the fewer definable degrees there can be.

Second, the various reducibilities provide a full spectrum of relationships between the ω-jump and the usual jump in terms of the n-least upper bound, defined as the smallest n-th jump of the upper bounds:

- For the hyperdegrees, $\mathbf{0}_h^{(\omega)}$ is the least upper bound of the $\mathbf{0}_h^{(n)}$'s;
- for the arithmetical degrees, $\mathbf{0}_a^{(\omega)}$ is the 1-least upper bound of the $\mathbf{0}_a^{(n)}$'s (Odifreddi [1983]);
- for the Turing degrees, $\mathbf{0}^{(\omega)}$ is the 2-least upper bound of the $\mathbf{0}^{(n)}$'s (Enderton and Putnam [1970], Sacks [1971]).

It is natural to conjecture that the trend extends to strong reducibilities as well. In particular, that *the ω-jump for the truth-table degrees is definable in terms of iterations the jump.*

Local Degree Structures

The usual local structures considered in degree theory consist of the r.e. degrees, the degrees below $\mathbf{0}'$ and the Δ_2^0 degrees. Many questions arise when these structures are compared horizontally, i.e. different structures for the same reducibility, or vertically, i.e. the same structure for different reducibilities.

In the extreme cases of many-one and Turing reducibilities, there are only two structures to consider. Indeed, the r.e. m-degrees and the m-degrees below $\mathbf{0}'_m$ coincide by Post [1944], while the Turing degrees below $\mathbf{0}'$ and the Δ_2^0 Turing degrees coincide by Shoenfield [1959]. For intermediate reducibilities the three structures are different, by Cooper and Lachlan (see Odifreddi [1999], p. 723).

The structure of the r.e. m-degrees has been completely characterized from an algebraic point of view by Denisov [1978], and from a complexity point of view by Nies [1994]. In particular, local versions of the global results are known to hold. For example, there are $2^{\aleph_0}$ automorphisms, $\mathbf{0}_m$ and $\mathbf{0}'_m$ are the only degrees invariant under every automorphism, and any two nontrivial principal filters are isomorphic. The main open problem here is whether *the r.e. m-degrees form a prime model of their theory*, in the sense of being elementarily embedded in every model of the theory. Related to this, Nies [199?] has announced that the arithmetical m-degrees are indeed a prime model of their theory.

The Δ_2^0 m-degrees are not elementarily equivalent to the r.e. m-degrees, since Ershov [1970] has proved that they do not have a greatest element. More precisely, it is possible to define a notion of mini-jump which always goes up in m-degree, and under which the Δ_2^0 sets are closed. Apparently, however, this is the only difference

[1] For the first two structures, unnatural definitions have been provided by Slaman and Woodin [199?], and no natural definition is known. For the last structure, the original natural definition of Cooper [1990] has recently been shown to be incorrect by Slaman and Shore, who have also provided an unnatural definition. However, Cooper has announced that a simple modification of his original definition should work, essentially by the same proof.

between the structures of r.e. and Δ_2^0 m-degrees, since Denisov [1978] has announced that, by removing the greatest element from the former, one obtains a structure isomorphic to the latter. However, no proof of this result has ever been published.

Coles, Downey and LaForte [199?] have found an appropriate version of the mini-jump for the truth-table degrees as well, so that the structure of Δ_2^0 tt-degrees is not elementarily equivalent to any of the structures of r.e. tt-degrees and of tt-degrees below $\mathbf{0}'_{tt}$. It is not known, however, whether *the structures of the r.e. tt-degrees and of the tt-degrees below $\mathbf{0}'_{tt}$ are elementarily equivalent.* A peculiar property proved by Marchenkov [1976] is that the minimal r.e. tt-degrees are bounded below $\mathbf{0}'_{tt}$. However, Shore has noticed that the proof also works for the minimal tt-degrees below $\mathbf{0}'_{tt}$, so that a possible elementary inequivalence of the two structures has to be looked for elsewhere.

Turning to weak truth-table degrees, the structure of Δ_2^0 wtt-degrees is not elementarily equivalent to any of the structures of r.e. wtt-degrees and of wtt-degrees below $\mathbf{0}'_{wtt}$, again by the existence of an appropriate version of the mini-jump found by Coles, Downey and LaForte [199?]. And the structures of r.e. wtt-degrees and of wtt-degrees below $\mathbf{0}'_{wtt}$ are not elementarily equivalent, because the former is dense by Ladner and Sasso [1975], while the latter has minimal degrees by Haught and Shore [1990a].

Having thus completed our look at the horizontal relationships among local structures of degrees, we briefly turn to their vertical relationships. The structures of r.e. m-degrees, tt-degrees and wtt-degrees are pairwise not elementarily equivalent, because the minimal r.e. degrees are not bounded for the m-degrees by Ershov and Lavrov [1973], are bounded for the tt-degrees by Marchenkov [1976], and do not exist for the wtt-degrees by Ladner and Sasso [1975].

Lachlan [1970] has proved that the structure of m-degrees below $\mathbf{0}'_m$ is distributive, and hence not elementarily equivalent to any of the structures of tt-degrees below $\mathbf{0}'_{tt}$ and wtt-degrees below $\mathbf{0}'_{wtt}$, which are not distributive by Haught and Shore [1990], [1990a]. However, it is not known whether *the structure of tt-degrees below $\mathbf{0}'_{tt}$ is elementarily equivalent to the structure of wtt-degrees below $\mathbf{0}'_{wtt}$*. In particular, it is not known whether the minimal wtt-degrees below $\mathbf{0}'_{wtt}$ are bounded below $\mathbf{0}'_{wtt}$.

For the same reason as above, the structure of Δ_2^0 m-degrees is not elementarily equivalent to any of the structures of Δ_2^0 tt-degrees and wtt-degrees. However, it is not known whether *the structures of Δ_2^0 tt-degrees and wtt-degrees are elementarily equivalent.*

We conclude our survey of local degree structures by asking whether *the tt-degree of a hypersimple set is noncupping in the r.e. tt-degrees*, i.e. not part of a pair joining to $\mathbf{0}'_{tt}$. The existence of noncupping r.e. tt-degrees follows from Downey [1988]. Moreover, the btt-degree of a simple set is noncupping in the r.e. btt-degrees by Fenner and Shaefer [1999] and Downey [199?], and the wtt-degree of a hypersimple set is noncupping in the r.e. wtt-degrees by Downey and Jockusch [1987].

Bibliography

Coles, R., Downey, R.G., and LaForte, G.

[199?] A hierarchy for Δ_2^0 sets based on the wtt- and tt-jump operators, to appear.

Cooper, S.B.

[1990] The jump is definable in the structure of the degrees of unsolvability, *Bull.*

Am. Math. Soc. 23 (1990) 151–158.

[1997] Beyond Gödel's Theorem: the failure to capture information content, in *Complexity, Logic and Recursion Theory*, Sorbi ed., Dekker, 1997, pp. 93–122.

Degtev, A.N.

[1979] Some results on uppersemilattices and m-degrees, *Alg. Log.* 18 (1979) 664–679, transl. 18 (1979) 420–430.

Denisov, S.D.

[1978] The structure of the uppersemilattice of recursively enumerable m-degrees and related questions, I, *Alg. Log.* 17 (1978) 643–683, transl. 17 (1978) 418–443.

Downey, R.G.

[199?] A note on *btt*-degrees, to appear.

Downey, R.G., and Jockusch, C.

[1987] T-degrees, jump classes, and strong reducibilities, *Trans. Am. Math. Soc.* 301 (1987) 103–136.

Enderton, H.B., and Putnam, H.

[1970] A note on the hyperarithmetical hierarchy, *J. Symb. Log.* 35 (1970) 429–430.

Ershov, Y.L.

[1970] A hierarchy of sets III, *Alg. Log.* 9 (1970) 34–51, transl. 9 (1970) 20–31.

[1975] The uppersemilattice of enumerations of a finite set, *Alg. Log.* 14 (1975) 258–284, transl. 14 (1975) 159–175.

Ershov, Y.L., and Lavrov, I.A.

[1973] The uppersemilattice $L(\gamma)$, *Alg. Log.* 12 (1973) 167–189, transl. 12 (1973) 93–106.

Fenner, S., and Schaefer, M.

[1999] Bounded immunity and *btt*-reductions, *Math. Log. Quart.* 45 (1999) 3–21.

Haught, C.A., and Shore, R.A.

[1990] Undecidability and initial segments of the (r.e.) *tt*-degrees, *J. Symb. Log.* 55 (1990) 987–1006.

[1990a] Undecidability and initial segments of the *wtt*-degrees below $\mathbf{0}'$, *Springer Lect. Not. Math.* 1432 (1990) 223–244.

Homer, S.

[1987] Minimal degrees for polynomial reducibilities, *J. Ass. Comp. Mach.* 34 (1987) 480–491.

Lachlan, A.H.

[1970] Initial segments of many-one degrees, *Can. J. Math.* 22 (1970) 75–85.

Ladner, R.E., and Sasso, L.P.

[1975] The weak truth-table degrees of recursively enumerable sets, *Ann. Math. Log.* 8 (1975) 429–448.

Marchenkov, S.S.

[1976] On the comparison of the uppersemilattice of r.e. m-degrees and tt-degrees, *Mat. Zam.* 20 (1976) 19–26, transl. 20 (1976) 567–570.

Mohrherr, J.

[1984] Density of a final segment of the truth-table degrees, *Pac. J. Math.* 115 (1984) 409–419.

Nerode, A.

[1957] General topology and partial recursive functionals, *Talks Cornell Summ. Inst. Symb. Log.*, Cornell, 1957, pp. 247–251.

Nies, A.

[1994] The last question on recursively enumerable many-one degrees, *Alg. Log.* 33 (1994) 550–563.

[199?] Model theoretic properties of degree structures, *Proc. 11th Conf. Log. Meth. Phil. Sci.*, to appear.

Odifreddi. P.G.

[1981] Strong reducibilities, *Bull. Am. Math. Soc.* 4 (1981) 37–86.

[1983] On the first-order theory of the arithmetical degrees, *Proc. Am. Math. Soc.* 87 (1983) 505–507.

[1989] *Classical Recursion Theory*, North Holland, 1989. Second edition, 1999.

[1999] *Classical Recursion Theory*, volume II, North Holland, 1999.

[1999a]Reducibilities, in *Handbook of Recursion Theory*, Griffor editor, North Holland, 1999, pp. 89–120.

Sacks, G.E.

[1971] Forcing with perfect closed sets, *Proc. Symp. Pure Math.* 17 (1971) 331–355.

Shoenfield, J.R.

[1959] On degrees of unsolvability, *Ann. Math.* 69 (1959) 644–653.

Shore, R.A.

[1982] The theories of the truth-table and Turing degrees are not elementarily equivalent, in *Logic Colloquium '80*, Van Dalen et al. eds., North Holland, 1982, pp. 231–237.

Shore, R.A., and Slaman, T.A.

[1992] The p-T-degrees of the recursive sets: lattice embeddings, extensions of embeddings and the two-quantifier theory, *Theor. Comp. Sci.* 97 (1992) 263–284.

Slaman, T.A., and Woodin, H.W.

[199?] *Definability in degree structures*, to appear.

Department of Mathematics, University of Torino, Italy
E-mail address: `piergior@di.unito.it`

Contemporary Mathematics
Volume **257**, 2000

Finitely axiomatizable theories and Lindenbaum algebras of semantic classes

Mikhail Peretyat'kin

ABSTRACT. A numerated Boolean algebra $(\mathcal{B}, \nu)$ is called a Σ_n^0-*algebra* if its operations $\cup$, $\cap$, and $-$ are presentable by recursive functions on ν-numbers, while the equality predicate is a Σ_n^0-relation. $(\mathcal{B}, \nu)$ is called a Σ_n^0-*universal* algebra if it is a Σ_n^0-algebra and, moreover, for any Σ_n^0-algebra $(\mathcal{B}', \nu')$, there is an element $a \in \mathcal{B}$ such that the algebras $(\mathcal{B}', \nu')$ and $(\mathcal{B}, \nu)[a]$ are isomorphic, where $(\mathcal{B}, \nu)[a]$ denotes the initial segment of $(\mathcal{B}, \nu)$ below a. Similar definitions apply to other classes of hierarchies.

Universal Boolean algebras exist for the classes Σ_n^0, Π_n^0, Δ_n^0, Σ_n^1, and Π_n^1 for $n > 0$, and they are pairwise non-isomorphic, except that the universal Boolean algebras for the classes Π_n^0 and Δ_n^0 for $n > 0$ are different as numerated algebras but isomorphic as abstract algebras. Using finitely axiomatizable theories, some applications to Lindenbaum algebras are given, and a number of conjectures concerning the isomorphism types of Lindenbaum algebras of semantic classes of models are formulated.

In this paper, we develop some applications of universal Boolean algebras to investigate Lindenbaum algebras of theories of semantic classes of models. William Hanf was the earlier explorer in this direction. For a long time, he intensively developed methods to investigate Lindenbaum algebras of first-order theories. As a result of his deep research, based also on results in Boolean algebras by Tarski and Vaught, Hanf obtained in 1975 a complete solution to the basic problem of logic concerning the isomorphism type of the Lindenbaum algebra of the predicate calculus of a finite rich signature [8], [9]. This problem was considered by Alfred Tarski in the 1930's, who himself spent some time to investigate approaches to its solution.

Hanf gave the key definition of a positively numerated positively universal Boolean algebra, and he well understood the important role of this concept just for the solution of the problem of characterizing the Lindenbaum algebra of predicate logic. Later, a substantial theory of Boolean algebras with the positive universality condition was developed by V. L. Selivanov [18], [19], who proved the existence and uniqueness of a positively numerated positively universal Boolean algebra, and on this basis, there appeared a simpler and more transparent proof of the characterization of the Lindenbaum algebra of predicate logic. The principal result of V. L. Selivanov is that the positive universality condition for positively numerated Boolean algebras is in fact the only condition among those considered by Hanf in [8] and [9] which play a role in characterizing the Lindenbaum algebra of predicate logic. Moreover, he found a representation of a positively numerated positively universal Boolean algebra by a concrete formula (7.1), which, by standard methods of algorithm theory, can be generalized to other classes in various hierarchies. For the research into universal Boolean algebras, distinctions between isomorphism types of Boolean algebras via their Feiner characteristics also play an essential role.

It is known that there exist continuum many isomorphism types of countable Boolean algebras. However, in the majority of cases, we encounter Boolean algebras which possess a simple characterization in terms of quotients modulo the Fréchet ideal and a universality condition over a class of a hierarchy. We describe in this paper these isomorphism types of Boolean algebras, which are important for our purposes and which appear to be suitable for the characterization of the Lindenbaum algebras of various semantic classes of models. Some applications of such algebras are presented, and a rather large number of conjectures about the isomorphism types of the Lindenbaum algebras of interesting semantic classes of models is formulated.

Preliminaries. We consider theories in first-order predicate logic with equality and use general concepts of model theory, algorithm theory, constructive models and Boolean algebras found in Chang and Keisler [2], Soare [20], Ershov and Goncharov [7], and Goncharov [6]. A finite signature σ is called *rich* if it contains at least one n-ary predicate or function symbol for $n > 1$, or at least two unary functional symbols. By $SL(\sigma)$, we denote the set of all sentences (i. e., closed formulas) in the signature σ. We consider Boolean algebras in the signature $\sigma_{BA} = \{\cup, \cap, -, 0, 1\}$.

1. Numerated Boolean algebras

A *numeration* of a finite or countable Boolean algebra $\mathcal{B}$ is a surjective mapping $\nu : \mathbb{N} \xrightarrow{onto} |\mathcal{B}|$. A Boolean algebra $\mathcal{B}$ together with a numeration ν is called a *numerated Boolean algebra* and is denoted by $(\mathcal{B}, \nu)$.

Two numerated Boolean algebras $(\mathcal{B}_1, \nu_1)$ and $(\mathcal{B}_2, \nu_2)$ are called *constructively isomorphic* or simply *isomorphic* if there exist an isomorphism μ between $\mathcal{B}_1$ and $\mathcal{B}_2$ and two general recursive functions $f(x)$ and $g(x)$ such that the following diagram is commutative:

$$\begin{array}{ccc} \mathbb{N} & \underset{g}{\overset{f}{\rightleftarrows}} & \mathbb{N} \\ \nu_1 \downarrow & & \downarrow \nu_2 \\ |\mathcal{B}_1| & \xrightarrow{\mu} & |\mathcal{B}_2| \end{array}$$

In constructive model theory, two numerated Boolean algebras which are constructively isomorphic can be regarded as two numerated representations of the same structure up to isomorphism since the above recursive functions $f(x)$ and $g(x)$ guarantee effective transitions between numbers in these two numerations for any algorithmic and model-theoretic construction over those algebras.

Let Ξ be a class of some hierarchy. A numerated Boolean algebra $(\mathcal{B}, \nu)$ is called a Ξ-*algebra* if all the operations from its signature are uniformly presentable by general recursive functions over ν-numbers while equality in $\mathcal{B}$ is a Ξ-relation in the numeration ν, i.e., there exist general recursive functions $f(x,y)$, $g(x,y)$, $h(x)$ and a relation $P(x,y)$ in the class Ξ which represent the Boolean algebra as follows (for any $m, n \in \mathbb{N}$):

$$\begin{aligned} \nu(m) \cup \nu(n) &= \nu(f(m,n)), \\ \nu(m) \cap \nu(n) &= \nu(g(m,n)), \\ -\nu(m) \qquad &= \nu(h(m)), \\ \nu(m) = \nu(n) &\Leftrightarrow P(m,n). \end{aligned} \tag{1.1}$$

A numerated Boolean algebra $(\mathcal{B}, \nu)$ is called a *positively numerated* Boolean algebra if it is a Σ_1^0-algebra; a *negatively numerated* Boolean algebra if it is a Π_1^0-algebra; and a *constructive* Boolean algebra if it is a Δ_1^0-algebra.

Note the following important fact.

Lemma 1.1. [13] *Let* $(\mathcal{B}, \nu)$ *be an arbitrary Boolean* Π_1^0*-algebra. Then,* $\mathcal{B}$ *has a constructive numeration.*

PROOF. See [13] or [6]. □

2. The Lindenbaum algebra of a theory

We recall the concept of a Lindenbaum algebra of a theory. Let T be a theory in a signature σ. On the set of sentences $SL(\sigma)$, an equivalence relation $\sim$ is defined by

$$\Phi \sim \Psi \Leftrightarrow T \vdash \Phi \leftrightarrow \Psi.$$

The logical operations $\vee$, &, and $\neg$, generate Boolean operations $\cup$, $\cap$, and $-$ on the quotient set $SL(\sigma)/_\sim$ by

$$[\Phi]_\sim \cup [\Psi]_\sim =_{df} [\Phi \vee \Psi]_\sim, \quad [\Phi]_\sim \cap [\Psi]_\sim =_{df} [\Phi \,\&\, \Psi]_\sim, \quad -[\Phi]_\sim =_{df} [\neg\Phi]_\sim,$$
$$0 =_{df} [(\forall x)(x \neq x)]_\sim, \quad 1 =_{df} [(\forall x)(x = x)]_\sim.$$

(One can easily check, that these operations are well-defined on the $\sim$-classes.)

By virtue of this construction, we obtain an algebra of the form

$$\mathcal{L}(T) = \left(SL(\sigma)/_{\sim};\ \cup, \cap, -, 0, 1 \right),$$

which is indeed a Boolean algebra. It is called the *Lindenbaum algebra* of the theory T and is denoted by $\mathcal{L}(T)$. Model-theoretically speaking, the algebra $\mathcal{L}(T)$ represents the structure of all completions of the theory T. In particular, this algebra $\mathcal{L}(T)$ consists of two elements if and only if T is a complete theory.

Let T be a recursively enumerable theory in a signature σ. For the Lindenbaum algebra $\mathcal{L}(T)$, there is a natural numeration γ induced by the Gödel enumeration G of the set of sentences $SL(\sigma)$. We call γ the *Gödel enumeration* of the algebra $\mathcal{L}(T)$.

If T is a theory in the signature σ such that the Lindenbaum algebra $\mathcal{L}(T)$ has a Gödel enumeration γ, one can show that the algebra $(\mathcal{L}(T), \gamma)$ is positively numerated if and only if the theory T is recursively axiomatizable. Moreover, the Boolean algebras realizable as Lindenbaum algebras of recursively axiomatizable theories are described by the following statement:

Theorem 2.1. *Let $(\mathcal{B}, \nu)$ be an arbitrary numerated Boolean algebra. Then there is a recursively axiomatizable theory T such that its numerated Lindenbaum algebra $(\mathcal{L}(T), \gamma)$ is constructively isomorphic to the algebra $(\mathcal{B}, \nu)$ if and only if $(\mathcal{B}, \nu)$ is a positively numerated Boolean algebra.*

Proof. By standard methods. □

Let T and T' be two recursively axiomatizable theories. An isomorphism $\mu : \mathcal{L}(T) \to \mathcal{L}(T')$ between the Lindenbaum algebras of these theories is called a *recursive isomorphism* if it is a constructive isomorphism with respect to the Gödel enumerations γ and γ' of these algebras.

We now formulate a statement about the reduction of signatures for finitely axiomatizable theories which is necessary for our purposes.

Lemma 2.2. *Let F be a finitely axiomatizable theory in a finite signature σ_0, and let σ be any finite rich signature. Then one can construct effectively in F and σ a finitely axiomatizable theory $H = \mathrm{Reduct}(F, \sigma)$ in the signature σ together with a recursive isomorphism $\mu : \mathcal{L}(F) \to \mathcal{L}(H)$ between the Lindenbaum algebras of these theories, which preserves all properties of a list MDL $\supseteq$ MQL $\cup \{p_0\}$, where MQL is defined later in Section 3, while p_0 is the property "theory has a finite model".*

Proof. See [17]. □

Note. In fact, there is a closed proof just for the statement of Lemma 2.2, independently of complicated proof of Theorem 3.3 formulated below. The corresponding paper is in preparation now.

By the procedure $F \mapsto \mathrm{Reduct}(F, \sigma)$, all results obtained below can be transferred to any finite rich signature. In the following, we will therefore always assume this reduction of signatures and restrict our attention to a particular finite rich signature.

3. Theorems on local predicate expressiveness

The following important theorem about the expressiveness of predicate logic was obtained by William Hanf.

Theorem 3.1. [9] *For any recursively axiomatizable theory T, there is a finitely axiomatizable theory $F = \mathbb{L}_\sigma(T)$ in a finite rich signature σ such that the Lindenbaum algebras of the theories T and F are recursively isomorphic. Moreover, the theory F and the isomorphism μ can be constructed effectively by the recursively enumerable index of the theory T.*

The above theorem of Hanf can also be formulated as follows, which will be important for applications:

Theorem 3.2. [9] *Let $(\mathcal{B}, \nu)$ be an arbitrary numerated Boolean algebra. Then there is a finitely axiomatizable theory T such that its numerated Lindenbaum algebra $(\mathcal{L}(T), \gamma)$ is constructively isomorphic to the algebra $(\mathcal{B}, \nu)$ if and only if $(\mathcal{B}, \nu)$ is a positively numerated Boolean algebra.*

PROOF. Immediate by Theorems 2.1 and 3.1. □

The expressive power of formulas of predicate logic is described by the following stronger theorem, the proof of which is presented in [17]:

Theorem 3.3. [Predicate Expressiveness Theorem] *Let T be an arbitrary recursively axiomatizable theory without finite models, and σ a finite rich signature. Then, effectively in a recursively enumerable index of T, one can construct a finitely axiomatizable, model complete theory $F = \mathbb{F}_\sigma(T)$ in the signature σ and a recursive isomorphism $\mu : \mathcal{L}(T) \to \mathcal{L}(F)$ between their Lindenbaum algebras such that any completion T' of the theory T and any corresponding completion F' of the theory F (where $F' = \mu(T')$) coincide in terms of the following list of model-theoretic properties:*

(a) *stability, superstability, ω-stability, stability in cardinality α,*

(b) *existence of a prime model and the value of its algorithmic dimension (with respect to strong constructivizations); and the number of atomic models of cardinality $\alpha > \omega$,*

(c) *number of countable minimal models (Jónsson models), and the values of their algorithmic dimensions; number of minimal models of cardinality $\alpha > \omega$,*

(d) *existence of a countable strongly constructivizable homogeneous model; existence of a countable strongly constructivizable ω^+-homogeneous model; existence of an α^+-homogeneous model of cardinality $\alpha \geqslant \omega$,*

(e) *existence and strong constructivizability of a countable saturated model; existence of a saturated model of cardinality* $\alpha > \omega$,

(f) *existence of a model with first-order definable elements and its strong constructivizability; existence of a model with almost first-order definable (i. e., algebraic) elements and the value of its algorithmic dimension,*

(g) *number of countable rigid models (i. .e., with only the trivial automorphism) and the values of their algorithmic dimensions; number of rigid models of cardinality* $\alpha > \omega$,

(h) *non-maximality of the spectrum function.*

This theorem represents a construction in [17] which is called *universal.* The list of model-theoretic properties contained in Theorem 3.3 is called *universal* and is denoted by *MQL*.

4. Semantic similarity of theories

Let L be a list of model-theoretic properties. The statement of Theorem 3.3 suggests the definition of a relation $\equiv_L$ among theories as follows. For two recursively axiomatizable theories T_1 and T_2, $T_1 \equiv_L T_2$ will denote that there is a recursive isomorphism $\mu : \mathcal{L}(T_1) \to \mathcal{L}(T_2)$ between the Lindenbaum algebras of these theories such that for any complete extension T_1' of the theory T_1 and any corresponding complete extension T_2' of the theory T_2 (where $T_2' = \mu(T_1')$), the theories T_1' and T_2' have identical properties in terms of this list L. If $T_1 \equiv_L T_2$ holds according to this definition, we will call the theories T_1 and T_2 *semantically similar* over the list L.

NOTE. Thus the universal list *MQL* determines a semantic similarity relation $\equiv_{MQL}$. We note a fine point here, namely, that the theories T_1 and T_2 (or one of them) may have finite models, while in the statement of Theorem 3.3 (representing the universal list), finite models were excluded. However, the actual definition of the list *MQL* in [17] is such that the relation $\equiv_{MQL}$ is well defined in all cases, even when the theories T_1 and T_2 have finite models. The above complete extensions T_1' and T_2' of the theories will have identical properties in terms of the list *MQL* if and only if the theories $T_1' \oplus SI$ and $T_2' \oplus SI$ have identical properties in terms of the list *MQL*, where *SI* is the complete ω_1-categorical (infinitely axiomatizable) theory of a successor relation in the signature $\sigma = \{\lhd^2, c\}$ without cycles and with one initial element c. In any case, it is thus possible to add the summand *SI* to only one of the theories, and to compare $T_1' \oplus SI$ with T_2', or T_1' with $T_2' \oplus SI$. Moreover, $T \equiv_{MQL} T \oplus SI$ holds true for any theory T. Such a reduction to the case with the summand *SI* allows us a better understanding of the relation $\equiv_{MQL}$, even when the theories have finite models.

This note also applies to any smaller list $L \subseteq MQL$ and to the corresponding semantic similarity relation $\equiv_L$.

5. Semantic classes of sentences

In this section, we define the notion of a semantic class of sentences based on the semantic similarity relation over a list. For this section, we fix a finite rich signature σ.

First, we introduce some auxiliary definitions.

By $SL(\sigma)$, we denote the set of all sentences of the signature σ. By $[\Phi]_\sigma$, we denote the theory in the signature σ generated by the sentence Φ as an axiom. A set $E \subseteq SL(\sigma)$ is called *closed under deduction* if for $\Phi, \Psi \in SL(\sigma)$, $\vdash (\Phi \leftrightarrow \Psi)$ implies $\Phi \in E$ iff $\Psi \in E$. For the remainder of this section, we deal only with sets of sentences in $SL(\sigma)$ which are closed under deduction.

On the family of all sets $E \subseteq SL(\sigma)$ closed under deduction, the following two natural operations are defined:

$$C(E) = \mathrm{Compl}(E) = SL(\sigma) \setminus E,$$

$$N(E) = \mathrm{Neg}(E) = \{\Psi \in SL(\sigma) \mid \vdash \Psi \leftrightarrow (\neg\Phi) \text{ for some } \Phi \in E\}.$$

The operations Compl and Neg have order 2 and commute with each other. For a set $E \subseteq SL(\sigma)$ we set $I(E) = E$ and $CN(E) = C(N(E))$.

Consider a list of properties L. We say that formulas Φ and Ψ of the signature σ are *semantically similar over the list* L if the theories $[\Phi]_\sigma$ and $[\Psi]_\sigma$ are semantically similar over the list L. We denote this fact symbolically by $\Phi \equiv_L \Psi$. A set $E \subseteq SL(\sigma)$ is called *semantically closed over a list* L if for any sentences $\Phi, \Psi \in SL(\sigma)$,

$$\Phi \equiv_L \Psi \Rightarrow (\Phi \in E \Leftrightarrow \Psi \in E).$$

We now proceed to the main definitions. Let $\Lambda \in \{I, C, N, CN\}$. A set $E \subseteq SL(\sigma)$ is called *semantically Λ-closed over a list* L if there exists a set $E' \subseteq SL(\sigma)$ which is semantically closed over the list L such that $E = \Lambda(E')$. A set $E \subseteq SL(\sigma)$ is called *semantic over a list* L if it is semantically Λ-closed over the list L for some $\Lambda \in \{I, C, N, CN\}$. A class of models M of a finite rich signature is called *semantic* over a list L if its theory $\mathrm{Th}(M)$ is a semantic set over the list L.

Note two simple facts.

Lemma 5.1. *If a set $E \subseteq SL(\sigma)$ is semantic over a list L then E is closed under deduction.*

PROOF. By direct inspection of the definitions. □

Lemma 5.2. *Let L and L' be two lists of properties such that $L \subseteq L'$. If a set $E \subseteq SL(\sigma)$ is semantic over the list L then E is semantic over the larger list L' as well.*

PROOF. Obvious. □

A semantic set $E \subseteq SL(\sigma)$ over the empty list $L = \varnothing$ is called an *absolutely semantic set.* By Lemma 5.1, such a set will be semantic over any list of model-theoretic properties.

Now, we introduce some notation for algorithmic complexity estimates.

For $A, B \subseteq \mathbb{N}$, we denote by $A \approx B$ that A and B are recursively isomorphic (i. e., that there is a bijection π: $\mathbb{N} \to \mathbb{N}$ which is a general recursive function such that $\pi(A) = B$). For $A \subseteq \mathbb{N}$, we define

$$\Sigma_n^0 \leqslant_m A \Leftrightarrow (\forall X \in \Sigma_n^0)(X \leqslant_m A),$$
$$A \approx \Sigma_n^0 \Leftrightarrow A \in \Sigma_n^0 \,\&\, (\forall X \in \Sigma_n^0)(X \leqslant_m A).$$

Similar definitions apply to other classes of a hierarchy.

Fix a Gödel enumeration Φ_i, $i \in \mathbb{N}$, for the set of all sentences of a signature σ which is assumed to be fixed in this section. For $E \subseteq SL(\sigma)$, we denote by $\mathrm{Nom}\,(E)$ the set of Gödel numbers $\{i \mid \Phi_i \in E\}$.

One can prove the following assertion.

Lemma 5.3. *Let $E_0 \subseteq SL(\sigma)$ be a set closed under deduction, and let $E_1 = N(E_0)$, $E_2 = C(E_0)$, $E_3 = CN(E_0)$. Then*

(a) $\mathrm{Nom}\,(E_i) \approx \mathrm{Nom}\,(E_{i+1})$, $i \in \{0, 2\}$,

(b) $\mathrm{Nom}\,(E_i) \approx \mathbb{N} \setminus \mathrm{Nom}\,(E_j)$, $i \in \{0, 1\}$, $j \in \{2, 3\}$.

PROOF. By standard methods. □

6. The complexity of particular semantic classes

In this section, we fix a finite rich signature σ. The following notation for classes of models of this signature is used:

$M_{all}(\sigma)$ — the class of all models in the signature σ,

$M_{fin}(\sigma)$ — the class of all finite models,

$M_{dec}(\sigma) = D$ — the class of all models with decidable theory,

$M_{f.a.}(\sigma) = F$ — the class of all models with finitely axiomatizable theory,

$M_{s.c}(\sigma) = \mathscr{C}$ — the class of all strongly constructivizable models,

$M_{r.e.}(\sigma)$ — the class of all recursively enumerable models,

$M_c(\sigma)$ — the class of all constructivizable models,

$M_{p.r.}(\sigma)$ — the class of all primitive recursive models,

P — the class of all prime models,

S — the class of all countable saturated models,

W — the class of models with ω-stable theory.

The following list of statements summarizes the results on the complexity estimates of semantic classes of sentences of a fixed finite rich signature σ. They are proved in [1], [2], [12], [15], [16], [21], and [22].

$$\mathrm{Th}(M_{all}) = \{n \mid \Phi_n \text{ is valid }\} \approx \Sigma^0_1$$

$$\mathrm{Th}(M_{fin}) = \{n \mid \Phi_n \text{ holds in all finite models }\} \approx \Pi^0_1$$

$$M_{fin}(\sigma) \subseteq K \subseteq M_{r.e}(\sigma) \Rightarrow \mathrm{Th}(K) \geqslant_m \Pi^0_1$$

$$M_{p.r}(\sigma) \subseteq K \subseteq M_c(\sigma) \Rightarrow \mathrm{Th}(K) \geqslant_m \varnothing^{(\omega)}$$

$$\{n \mid \Phi_n \text{ determines a complete theory }\} \approx \Pi^0_2$$

$$\{n \mid \Phi_n \text{ determines a complete stable theory }\} \approx \Pi^0_2$$

$$\{n \mid \Phi_n \text{ has a finite number of completions }\} \approx \Sigma^0_3$$

$$\{n \mid \Phi_n \text{ determines a decidable theory }\} \approx \Sigma^0_3$$

$$\mathrm{Th}(M_{dec}(\sigma)) \approx \Pi^0_3$$

$$\mathrm{Th}(M_{s.c}(\sigma)) \approx \Pi^0_3$$

$$\mathrm{Th}(M_{f.a}(\sigma)) \approx \Pi^0_3$$

$$\{n \mid \Phi_n \text{ is complete and has a prime model }\} \approx \Pi^0_3$$

$$\{n \mid \Phi_n \text{ is complete and has no prime model }\} \approx \Sigma^0_3$$

$$\{n \mid \Phi_n \text{ is complete and has a strongly constructivizable prime model}\} \approx \Sigma^0_4$$

$$\{n \mid \Phi_n \text{ is complete and has a non-s. c. prime model}\} \approx \Pi^0_4$$

$$\{n \mid \Phi_n \text{ is complete and has a countable saturated model }\} \approx \Pi^1_1$$

$$\{n \mid \Phi_n \text{ is complete and } \omega\text{-stable }\} \approx \Pi^1_1$$

$$\mathrm{Th}(P) \approx \Pi^1_1$$

$$\mathrm{Th}(S), \mathrm{Th}(W) \approx \Pi^1_2$$

$$\mathrm{Th}(P \cap D), \mathrm{Th}(P \cap F), \mathrm{Th}(P \cap \mathscr{C}), \mathrm{Th}(P \cap \mathscr{C} \cap F) \approx \Pi^0_4$$

$$\mathrm{Th}(S \cap D), \mathrm{Th}(S \cap F), \mathrm{Th}(S \cap \mathscr{C}), \mathrm{Th}(S \cap \mathscr{C} \cap F) \approx \Sigma^1_1$$

$$\mathrm{Th}(W \cap D), \mathrm{Th}(W \cap F), \mathrm{Th}(W \cap \mathscr{C}), \mathrm{Th}(W \cap \mathscr{C} \cap F) \approx \Sigma^1_1$$

Here, the statement "Φ_n is complete" as well as "Φ_n determines a complete theory" means, of course, that the theory $[\Phi_n]_\sigma$ is complete. One can easily check that the above classes of sentences are indeed semantic classes over appropriate lists of model-theoretic properties. Moreover, some of those semantic classes are theories of semantic classes of models.

7. Hanf's Theorem in global form

In this section, based on Hanf's Theorem 3.1 (expressing some "local" properties of predicate logic), we will obtain Hanf's theorem in another form, which represents the "global" structure of the Lindenbaum algebra of predicate logic.

The following definition is due to Hanf:

DEFINITION. Let $(\mathcal{B}, \nu)$ be a positively numerated Boolean algebra. We say that this algebra is *universal over the class of positively numerated Boolean algebras* if for any positively numerated Boolean algebra $(\mathcal{B}', \nu')$, there is an element $a \in \mathcal{B}$ such that the numerated algebras $(\mathcal{B}', \nu')$ and $(\mathcal{B}, \nu)[a]$ are constructively isomorphic where $(\mathcal{B}, \nu)[a]$ denotes the initial segment of $(\mathcal{B}, \nu)$ below a. If a numerated Boolean algebra $(\mathcal{B}, \nu)$ is positively numerated and is universal over the class of positively numerated Boolean algebras, we say that $(\mathcal{B}, \nu)$ is a *positively universal* Boolean algebra.

The following important result holds:

Theorem 7.1. [Hanf, 8, 9] (a) *There exists a positively numerated Boolean algebra which is positively universal.*

(b) [Selivanov, 19] *Any two positively numerated, positively universal Boolean algebras are constructively isomorphic.*

PROOF. (a) Proof by V. L. Selivanov: We denote by $\mathfrak{B}$ some fixed countable atomless Boolean algebra. The algebra $\mathfrak{B}$ possesses a constructivization μ which is in fact a strong constructivization.

Consider an arbitrary filter $\mathcal{F}$ of the algebra $\mathfrak{B}$. We denote by $(\mathfrak{B}/_{\mathcal{F}}, \mu/_{\mathcal{F}})$ the numerated quotient algebra of the algebra $\mathfrak{B}$ modulo the filter $\mathcal{F}$, where $\mu/_{\mathcal{F}}$ is the numeration of this quotient algebra induced by the numeration μ. We denote by $\mathcal{F}_m$ the filter of the Boolean algebra $\mathfrak{B}$ generated in $\mathfrak{B}$ by the set of elements $\{\mu(i) \mid i \in W_m\}$, where W_m is the recursively enumerable set with Post index m. Clearly, the sequence $\mathcal{F}_m$, $m \in \mathbb{N}$, contains all r. e. filters of the Boolean algebra $\mathfrak{B}$. Therefore, the sequence $(\mathfrak{B}/_{\mathcal{F}_m}, \mu/_{\mathcal{F}_m})$, $m \in \mathbb{N}$, includes (up to isomorphism) all possible positively numerated Boolean algebras.

Consider the following numerated Boolean algebra:

$$(\mathfrak{B}^*, \mu^*) = \bigoplus_{m \in \mathbb{N}} (\mathfrak{B}/_{\mathcal{F}_m}, \mu/_{\mathcal{F}_m}). \tag{7.1}$$

Since this is the sum of an effective sequence of positively numerated Boolean algebras, $(\mathfrak{B}^*, \mu^*)$ is a positively numerated Boolean algebra. It follows directly from (7.1) that it is a positively universal Boolean algebra.

(b) Proof by V. L. Selivanov: Let $(\mathcal{B}, \nu)$ be an arbitrary positively numerated Boolean algebra which is positively universal. It suffices to prove that $(\mathcal{B}, \nu)$ is constructively isomorphic to the numerated algebra (7.1). By the definition of positive universality for $(\mathcal{B}, \nu)$, there is an element $a \in \mathcal{B}$ such that the numerated Boolean algebra $(\mathcal{B}, \nu)[a]$ is constructively isomorphic to the algebra (7.1). Using the decomposition $(\mathcal{B}, \nu) = (\mathcal{B}, \nu)[-a] \oplus (\mathcal{B}, \nu)[a]$ and denoting the algebra

$(\mathcal{B},\nu)[-a]$ by $(\mathcal{B}',\nu')$, we have the following isomorphism

$$(\mathcal{B},\nu) \cong (\mathcal{B}',\nu') \oplus \bigoplus_{m\in\mathbb{N}} \left(\mathfrak{B}/_{\mathcal{F}_m}, \mu/_{\mathcal{F}_m}\right). \tag{7.2}$$

One can see, $(\mathcal{B}',\nu')$ is a positively numerated Boolean algebra. So there is a number e such that the algebra $(\mathcal{B}',\nu')$ is isomorphic to the algebra $\left(\mathfrak{B}/_{\mathcal{F}_e}, \mu/_{\mathcal{F}_e}\right)$. Using standard methods of algorithm theory, it is possible to construct a general recursive permutation $f(x)$ of the set of natural numbers $\mathbb{N}$ such that $f(0) = e$, and $W_{f(n+1)} = W_n$ for all $n \in \mathbb{N}$. As a result, we obtain

$$\begin{aligned}
&(\mathfrak{B}^*, \mu^*) \cong \\
&\textstyle\bigoplus_{m\in\mathbb{N}}\left(\mathfrak{B}/_{\mathcal{F}_m}, \mu/_{\mathcal{F}_m}\right) \cong \bigoplus_{m\in\mathbb{N}}\left(\mathfrak{B}/_{\mathcal{F}_{f(m)}}, \mu/_{\mathcal{F}_{f(m)}}\right) \cong \\
&\left(\mathfrak{B}/_{\mathcal{F}_e}, \mu/_{\mathcal{F}_e}\right) \oplus \textstyle\bigoplus_{m\in\mathbb{N}}\left(\mathfrak{B}/_{\mathcal{F}_{f(m+1)}}, \mu/_{\mathcal{F}_{f(m+1)}}\right) \cong \\
&(\mathcal{B}',\nu') \oplus \textstyle\bigoplus_{m\in\mathbb{N}}\left(\mathfrak{B}/_{\mathcal{F}_m}, \mu/_{\mathcal{F}_m}\right) \cong \\
&(\mathcal{B},\nu).
\end{aligned}$$

Therefore, Part (b) of Theorem 7.1 is proved as well. □

We now characterize the Lindenbaum algebras of the predicate calculi of various finite rich signatures. Here we denote by $PC(\sigma)$ the predicate calculus of the signature σ, i. e., the theory of this signature which is determined by the empty set of axioms.

Theorem 7.2. [Hanf, 9] *Let σ be a finite rich signature. Then the Lindenbaum algebra of the predicate calculus of the signature $\mathcal{L}(PC(\sigma))$ together with its Gödel numeration γ is a positively numerated Boolean algebra, which is positively universal.*

PROOF. Obviously, $(\mathcal{L}(PC(\sigma)),\gamma)$ is a positively numerated Boolean algebra. Now let $(\mathcal{B},\nu)$ be an arbitrary positively numerated Boolean algebra. Consider a finitely axiomatizable theory F in the signature σ whose Lindenbaum algebra $(\mathcal{L}(F),\gamma')$ is constructively isomorphic to the algebra $(\mathcal{B},\nu)$. Such a theory exists by Theorem 3.2. Since F is a finitely axiomatizable theory, it is a finite extension of the predicate calculus. Therefore, there is an element $a \in \mathcal{L}(PC(\sigma))$ such that

$$(\mathcal{L}(PC(\sigma)),\gamma)[a] \cong (\mathcal{L}(F),\gamma') \cong (\mathcal{B},\nu).$$

This shows that the algebra $(\mathcal{L}(PC(\sigma)),\gamma)$ is positively universal. □

One more application of Theorem 7.2:

Theorem 7.3. *Let σ and σ' be any two finite rich signatures. Then there exists a recursive isomorphism μ: $\mathcal{L}(PC(\sigma)) \to \mathcal{L}(PC(\sigma'))$ between the Lindenbaum algebras of the predicate calculi of these signatures.*

PROOF. By Theorems 7.1 and 7.2. □

8. Boolean algebras with a universality condition

Now we turn to the study of universal Boolean algebras. First, we introduce the basic definition of this section. Suppose that Ξ is a class in the arithmetical or analytical (or even some other) hierarchy.

DEFINITION. Let $(\mathcal{B}, \nu)$ be a numerated Boolean algebra. We say that this algebra *universal over the class of Boolean* Ξ*-algebras* if, for any numerated Boolean Ξ-algebra $(\mathcal{B}', \nu')$, there is an element $a \in \mathcal{B}$ such that the numerated algebras $(\mathcal{B}', \nu')$ and $(\mathcal{B}, \nu)[a]$ are isomorphic to each other. Let $(\mathcal{B}, \nu)$ be a numerated Boolean algebra. We say that this algebra is Ξ*-universal* if it is a Ξ-algebra and is universal over the class of Boolean Ξ-algebras.

We note the following important result about the existence of universal Boolean algebras:

Theorem 8.1. *There exist numerated Boolean algebras which are* Ξ*-universal over the following hierarchy classes* Ξ*:*

$$\Sigma^0_n, n>0, \ \Pi^0_n, n>0, \ \Delta^0_n, n>0, \ \Sigma^1_n, n>0, \ \Pi^1_n, n>0.$$

PROOF. Construct a computable sequence of Ξ-algebras, containing all isomorphism types of numerated Ξ-algebras, and then use the formula of the form (7.1). □

Theorem 8.2. *There is no Boolean algebra which is universal over the class* Δ^1_n *for any* $n > 0$.

PROOF. This follows from the fact that the class Δ^1_n is not exhausted by sets hyperarithmetical over a single set $A \in \Delta^1_n$. □

Theorem 8.3. *Let* Ξ *be one of the classes* Σ^0_n, Π^0_n, Δ^0_n, Σ^1_n, Π^1_n, *for some* $n > 0$. *Then, if* $(\mathcal{B}, \nu)$ *and* $(\mathcal{B}^\circ, \nu^\circ)$ *are* Ξ*-universal Boolean algebras, then* $(\mathcal{B}, \nu) \cong (\mathcal{B}^\circ, \nu^\circ)$.

PROOF. Similar to the proof of Theorem 7.1 (b). □

Theorem 8.4. *The* Ξ*-universal Boolean algebras (for* $\Xi = \Sigma^0_n, \Delta^0_n, \Sigma^1_n, \Pi^1_n$ *for* $n > 0$*) are pairwise nonisomorphic as abstract algebras.*

PROOF. By their Feiner characteristics [4], and the results in [4] and [5], and also by virtue of elementary properties of the classes of the analytical hierarchy. □

Theorem 8.5. *For any* $n > 0$*, the following statements hold:*

(a) *If* $(\mathcal{B}, \nu)$ *is a* Π^0_n*-universal Boolean algebra and* $(\mathcal{B}', \nu')$ *is* Δ^0_n*-universal then* $(\mathcal{B}, \nu) \ncong (\mathcal{B}', \nu')$.

(b) *If* $(\mathcal{B}, \nu)$ *is a* Π^0_n*-universal Boolean algebra while* $(\mathcal{B}', \nu')$ *is a* Δ^0_n*-universal Boolean algebra then* $\mathcal{B} \cong \mathcal{B}'$ *(as abstract algebras).*

PROOF. (a) By a (simple) example of a numerated Boolean Π^0_n-algebra $(\mathcal{B}, \nu)$ which is not a Δ^0_n-algebra. □

(b) By relativizing the proof of Lemma 1.1 with respect to the oracle $\varnothing^{(n-1)}$. □

Theorem 8.6. *The Ξ-universal Boolean algebras (for $\Xi = \Sigma^0_n, \Phi^0_n, \Delta^0_n, \Sigma^1_n, \Pi^1_n$ for $n > 0$) are pairwise nonisomorphic as numerated Boolean algebras.*

PROOF. Immediate by Theorems 8.4 and 8.5. □

9. Perfect Boolean algebras and their characteristics

The *Fréchet ideal* of a Boolean algebra $\mathcal{B}$ is the set $\mathcal{F}(\mathcal{B})$ consisting of all finite unions of atoms of the algebra $\mathcal{B}$.

In the following two lemmas, we establish that for an atomic Boolean algebra, the quotient algebra modulo the Fréchet ideal uniquely determines the isomorphism type of the Boolean algebra; moreover, any countable Boolean algebra can be the quotient of a countable atomic Boolean algebra modulo its Fréchet ideal.

Lemma 9.1. [14] *Let $\mathcal{B}_1$ and $\mathcal{B}_2$ be two arbitrary countable atomic Boolean algebras. Then*

$$\mathcal{B}_1/_{\mathcal{F}(\mathcal{B}_1)} \cong \mathcal{B}_2/_{\mathcal{F}(\mathcal{B}_2)} \Rightarrow \mathcal{B}_1 \cong \mathcal{B}_2.$$

PROOF. By standard methods in Boolean algebras, using the Vaught isomorphism criteria for countable Boolean algebras. □

Lemma 9.2. [6] *Let $\mathcal{B}'$ be any Boolean algebra of cardinality no more than ω. Then there is a countable atomic Boolean algebra $\mathcal{B}$ such that $\mathcal{B}/_{\mathcal{F}(\mathcal{B})} \cong \mathcal{B}'$.*

PROOF. By standard methods in Boolean algebras. □

The results of the two lemmas above can be expanded to the more general case of repeated quotients modulo the Fréchet ideal. For this purpose, we have to introduce a number of technical definitions.

For a Boolean algebra $\mathcal{B}$, we define the iterated sequence of Fréchet ideals of this algebra by induction on ordinals as follows:

$$\begin{aligned}
&\mathcal{F}_0(\mathcal{B}) = \{0\},\\
&\mathcal{F}_{\alpha+1}(\mathcal{B}) = \mathcal{F}(\mathcal{B}/_{\mathcal{F}_\alpha(\mathcal{B})}\ ),\\
&\mathcal{F}_\beta(\mathcal{B}) = \cup\{\mathcal{F}_\delta(\mathcal{B}) \mid \delta < \beta\} \text{ if } \beta \text{ is a limit ordinal.}
\end{aligned}$$

Let α be an ordinal. A Boolean algebra $\mathcal{B}$ is called *α-atomic* if for any $\beta < \alpha$, the quotient algebra $\mathcal{B}/_{\mathcal{F}_\beta(\mathcal{B})}$ is an atomic Boolean algebra.

The following restriction on the iterated Fréchet ideals holds:

Lemma 9.3. *Let $\mathcal{B}$ be a Boolean algebra, and let α be an ordinal such that $\mathcal{B}/_{\mathcal{F}_\alpha(\mathcal{B})}$ is the one-element Boolean algebra (i.e., $\mathcal{F}_\alpha(\mathcal{B}) = |\mathcal{B}|$), and moreover, $\mathcal{F}_\beta(\mathcal{B}) \neq |\mathcal{B}|$ for all $\beta < \alpha$. Then, the ordinal α cannot be a limit ordinal.*

We note the following three statements about the existence and uniqueness of a Boolean algebra with a given quotient algebra modulo the iterated Fréchet ideal (keeping in mind the above lemma):

Lemma 9.4. *Let α be any countable ordinal, and let $\mathcal{B}_1$ and $\mathcal{B}_2$ be two arbitrary countable α-atomic Boolean algebras. Then,*

$$\mathcal{B}_1/_{\mathcal{F}_a(\mathcal{B}_1)} \cong \mathcal{B}_2/_{\mathcal{F}_a(\mathcal{B}_2)} \Rightarrow \mathcal{B}_1 \cong \mathcal{B}_2.$$

PROOF. By standard methods in Boolean algebras. □

Lemma 9.5. *Let α be any countable ordinal, and let $\mathcal{B}'$ be a Boolean algebra of cardinality ω. Then there is a countable α-atomic Boolean algebra $\mathcal{B}$ such that $\mathcal{B}/_{\mathcal{F}_\alpha(\mathcal{B})} \cong \mathcal{B}'$.*

PROOF. By standard methods in Boolean algebras. □

Lemma 9.6. *Let α be any countable ordinal, and let $\mathcal{B}'$ be a finite Boolean algebra. Then there is a countable α-atomic Boolean algebra $\mathcal{B}$ such that $\mathcal{B}/_{\mathcal{F}_\alpha(\mathcal{B})} \cong \mathcal{B}'$.*

PROOF. By standard methods in Boolean algebras. □

By the above lemmas, we can give a characterization for a wide class of Boolean algebras.

We consider a countable Boolean algebra $\mathcal{B}$ and the sequence of its quotient algebras modulo iterated Fréchet ideals. We consider the least p such that the quotient algebra $\mathcal{B}^{(p)} = \mathcal{B}/_{\mathcal{F}_P(\mathcal{B})}$ is either finite, or contains an atomless nonzero element. Suppose now that the algebra $\mathcal{B}^{(p)}$ is finite (2^k), or atomless (AL), or Ξ-universal for a class of a hierarchy Ξ as schematically shown in Fig. 9.1. If we denote by τ the isomorphism type of Boolean algebra $\mathcal{B}^{(p)}$, then the pair (p, τ) is called the *characteristic* of the Boolean algebra $\mathcal{B}$, which is denoted as $\mathcal{B} \equiv (p, \tau)$. (We abbreviate this by $\mathcal{B} \equiv \tau$ in the case when $p = 0$.)

In this scheme, the number of steps p can be a finite or infinite ordinal. From the above properties of Fréchet ideals, it follows that the characteristic (p, τ) determines the countable Boolean algebra $\mathcal{B}$ uniquely up to isomorphism.

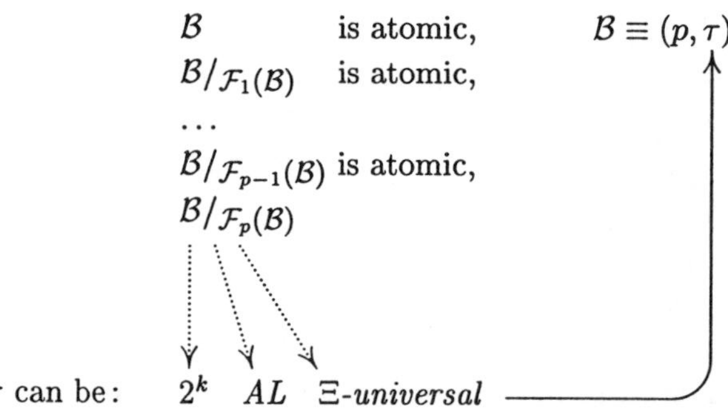

Fig. 9.1. Determinining the type of a Boolean algebra

A Boolean algebra $\mathcal{B}$ is called *recursively perfect* if it is a countable algebra and has one of the following characteristics

(9.1.a) $(p, 2^k),\ \ p < \omega_1^{sc},\ \ k \in \mathbb{N} \setminus \{0\},$

(9.1.b) $(p, AL),\ \ p < \omega_1^{sc},$

(9.1.c) $(p, \Sigma_n^r),\ (p, \Pi_n^r),\ (p, \Delta_n^0),\ p < \omega_1^{sc},\ n \in \mathbb{N} \setminus \{0\},\ r \in \{0, 1\}.$

(Here ω_1^{sc} is the least nonconstructive ordinal.)

From the above lemmas, it follows that all possible characteristics for $p < \omega_1^{sc}$ are realized; moreover, they determine the appropriate Boolean algebras uniquely up to isomorphism. (The case $p \geqslant \omega_1^{sc}$ is also possible but not of interest to us here.)

Theorem 9.7. (a) *For any characteristic ξ of the form* (9.1.a-c), *there is a countable Boolean algebra $\mathcal{B}$ having characteristic ξ.*

(b) *Let $\mathcal{B}_1$ and $\mathcal{B}_2$ be two countable Boolean algebras having the same characteristic ξ of the form* (9.1.a-c). *Then $\mathcal{B}_1$ and $\mathcal{B}_2$ are isomorphic.*

Of course, the above classification does not encompass all countable numerated Boolean algebras. Nevertheless, the algebras of these characteristics are interesting because they represent natural Boolean algebras as well as Lindenbaum algebras of theories of semantic classes of models. (One might call this fact the *generalized Rogers phenomenon.*) In this sense, the Boolean algebras of these characteristics form a special class of Boolean algebras (similarly to the way the sequence of sets which are m-universal in classes of the arithmetical and analytical hierarchies form a special class of sets in classical algorithm theory).

We now give examples of recursively perfect Boolean algebras.

Theorem 9.8. [Selivanov] *Let $\mathcal{B}(\Xi)$ be the Boolean algebra generated by the sets of the class Ξ. Then*

(a) $\mathcal{B}(\Sigma_1^0) \equiv (1, \Sigma_3^0)$,

(b) $\mathcal{B}(\Sigma_n^0) \equiv (1, \Sigma_{n+2}^0)$.

Proof. (a) Obviously $\mathcal{B}(\Sigma_1^0)$ is a countable atomic Boolean algebra. There is a numeration ν of the algebra $\mathcal{B}(\Sigma_1^0)$ based on the representation of Boolean expressions with r. e. sets. One can check that its quotient algebra modulo the Fréchet ideal is a Boolean Σ_3^0-algebra. It remains to prove that for any Boolean Σ_3^0-algebra $(\mathcal{B}, \mu)$ there is an element a such that $\mathcal{B} \cong \mathcal{B}(\Sigma_1^0)/_{\mathcal{F}(\Sigma_1^0)}[a]$. This fact is a direct consequence of Lachlan's result [11] (see also [20]) that every Boolean Σ_3^0-algebra is isomorphic to the algebra of supersets of some hh-simple set modulo the ideal of finite sets.

(b) By relativizing the proof of part (a) with respect to the oracle $\varnothing^{(n-1)}$, keeping in mind that Lachlan's result also relativizes (as noted in [11]). □

10. Lindenbaum algebras of semantic classes and open questions

Here we use the notation for semantic classes of models introduced in Section 6. In addition, we shall consider the following two semantic classes:

N — the class of models with non-finitely axiomatizable theory.

A — the class of autostable models.

Recall here that a model $\mathfrak{M}$ is called *autostable* if it is strongly constructivizable and has algorithmic dimension 1 with respect to strong constructivizations (see [7]). In this paper, we consider the class A only in combination with the class P of prime models. By [10], a prime model is autostable if it is strongly constructivizable; moreover, the set of atomic formulas of different number of variables is uniformly recursive. (Actually, it suffices for this set to be recursively enumerable). The equivalent result was also obtained by S. S. Goncharov (see [7]).

Now the following results characterizing the Lindenbaum algebras of semantic classes of models are obtained:

Theorem 10.1. [Hanf, 9]
$\mathcal{L}(M_{all}) \equiv \Sigma_1^0$.

Theorem 10.2. [Lempp, Peretyat'kin, Solomon]
(a) $\mathcal{L}(N) \equiv \Sigma_3^0$,
(b) $\mathcal{L}(F) \equiv (1, \Sigma_4^0)$.

Theorem 10.3. [Lempp, Peretyat'kin, Solomon]
$\mathcal{L}(M_{fin}) \equiv (1, \Sigma_2^0)$.

Note that Theorem 10.1 just represents another formulation of Theorem 7.2. Therefore, a proof of Theorem 10.1 was presented in Section 7 (minus the proof of Theorem 3.1). The statements of Theorem 10.2 are easy consequences of results included in the abstract [18]. The proofs of Theorems 10.2 and 10.3 are technically sophisticated and can thus not be included in this text due to lack of space.

We now turn to the discussion of problems in this direction.

First, we note the following rather simple fact:

Lemma 10.4. *Let Ξ be a class of a hierarchy, M a class of models in a finite signature. Then,* $\mathrm{Th}(M) \in \Xi$ *if and only if the Lindenbaum algebra with the Gödel enumeration* $(\mathcal{L}(\mathrm{Th}(M)), \gamma)$ *is a Boolean Ξ-algebra.*

PROOF. By direct inspection. □

Thus the results of Section 6 show that a number of semantic classes of sentences (including theories of semantic classes of models) form m-universal sets in a class of a hierarchy. Lemma 10.4 shows that the Lindenbaum algebras of these semantic classes of models are Ξ-algebras over appropriate classes of hierarchies. The concept of a Ξ-universal Boolean algebra is stronger than the concept of a Ξ-algebra, so it is natural to establish the Ξ-universality of these algebras as well. The results of Theorems 10.1-10.3 show that, in general, it is possible to prove the universality condition (using a construction of finitely axiomatizable theories).

Rough estimates suggest that the Lindenbaum algebras of various combinations of classes $N, F, D, \mathscr{G}, A$ as well as these classes intersected with the classes P, S, W are all recursively perfect. This problem is of great interest since it reveals a deep algorithmic nature of classical first-order predicate logic.

We now formulate a list of conjectures about the characteristics of the Lindenbaum algebras of the most interesting Boolean combinations of the semantic classes $N, F, D, \mathscr{G}, A$ and P, S, W. For the sake of brevity, we write $\mathcal{L}(M)$ instead of $\mathcal{L}(\mathrm{Th}(M))$ for a class of models M.

Open questions (conjectures)

1. $\mathcal{L}(D) \equiv \mathcal{L}(\mathscr{G}) \equiv \Pi^0_3$
2. $\mathcal{L}(N \cap D) \equiv \Pi^0_4$
3. $\mathcal{L}(N \cap \mathscr{G}) \equiv \Pi^0_4$
4. $\mathcal{L}(F \cap \mathscr{G}) \equiv (1, \Sigma^0_4)$
5. $\mathcal{L}(P) \equiv \Pi^1_1$
6. $\mathcal{L}(P \cap N) \equiv \Pi^1_1$
7. $\mathcal{L}(P \cap F) \equiv (1, \Sigma^0_5)$
8. $\mathcal{L}(P \cap D) \equiv \Pi^0_4$
9. $\mathcal{L}(P \cap \mathscr{G}) \equiv \Pi^0_4$
10. $\mathcal{L}(P \cap A) \equiv \Pi^0_3$
11. $\mathcal{L}(P \cap D \cap \overline{\mathscr{G}}) \equiv \Pi^0_5$
12. $\mathcal{L}(P \cap \mathscr{G} \cap \overline{A}) \equiv \Pi^0_4$
13. $\mathcal{L}(P \cap N \cap D) \equiv \Pi^0_4$
14. $\mathcal{L}(P \cap N \cap \mathscr{G}) \equiv \Pi^0_4$
15. $\mathcal{L}(P \cap N \cap A) \equiv \Pi^0_4$
16. $\mathcal{L}(P \cap N \cap D \cap \overline{\mathscr{G}}) \equiv \Pi^0_5$
17. $\mathcal{L}(P \cap N \cap \mathscr{G} \cap \overline{A}) \equiv \Pi^0_4$
18. $\mathcal{L}(P \cap F \cap \mathscr{G}) \equiv (1, \Sigma^0_5)$
19. $\mathcal{L}(P \cap F \cap A) \equiv (1, \Sigma^0_4)$
20. $\mathcal{L}(P \cap F \cap \mathscr{G} \cap \overline{A}) \equiv (1, \Sigma^0_5)$
21. $\mathcal{L}(S) \equiv \Pi^1_2$
22. $\mathcal{L}(S \cap N) \equiv \Pi^1_2$
23. $\mathcal{L}(S \cap F) \equiv (1, \Sigma^1_1)$
24. $\mathcal{L}(S \cap D) \equiv \Sigma^1_1$
25. $\mathcal{L}(S \cap \mathscr{G}) \equiv \Sigma^1_1$
26. $\mathcal{L}(S \cap D \cap \overline{\mathscr{G}}) \equiv \Sigma^1_1$
27. $\mathcal{L}(S \cap N \cap D) \equiv \Sigma^1_1$
28. $\mathcal{L}(S \cap N \cap \mathscr{G}) \equiv \Sigma^1_1$
29. $\mathcal{L}(S \cap N \cap D \cap \overline{\mathscr{G}}) \equiv \Sigma^1_1$
30. $\mathcal{L}(S \cap F \cap \mathscr{G}) \equiv (1, \Sigma^1_1)$
31. $\mathcal{L}(W) \equiv \Pi^1_2$
32. $\mathcal{L}(W \cap N) \equiv \Pi^1_2$
33. $\mathcal{L}(W \cap F) \equiv (1, \Sigma^1_1)$
34. $\mathcal{L}(W \cap D) \equiv \Sigma^1_1$
35. $\mathcal{L}(W \cap \mathscr{G}) \equiv \Sigma^1_1$
36. $\mathcal{L}(W \cap D \cap \overline{\mathscr{G}}) \equiv \Sigma^1_1$
37. $\mathcal{L}(W \cap N \cap D) \equiv \Sigma^1_1$
38. $\mathcal{L}(W \cap N \cap \mathscr{G}) \equiv \Sigma^1_1$
39. $\mathcal{L}(W \cap N \cap D \cap \overline{\mathscr{G}}) \equiv \Sigma^1_1$
40. $\mathcal{L}(W \cap F \cap \mathscr{G}) \equiv (1, \Sigma^1_1)$

In addition, there are similar questions for other natural semantic classes connected to stability, superstability, rigid models and models with first-order definable elements, minimal models, non-maximality of the spectrum function, as well as other properties of the list MQL.

Finally, we formulate some particular questions in this direction.

Question 41. Let M be any semantic class of models over a list $L \subseteq MQL$ such that $\mathrm{Th}(M) \approx \Sigma_2^0$ or $\mathrm{Th}(M) \approx \Pi_2^0$. Is it true in this case that the Lindenbaum algebra $\mathcal{L}(\mathrm{Th}(M))$ is Σ_2^0-universal or Π_2^0-universal?

Question 42. Find a characterization of all semantic classes of models M over the list MQL such that $\mathcal{L}(\mathrm{Th}(M))$ is Σ_2^0-universal or Π_2^0-universal.

Question 43. Let σ be a finite rich signature. Consider the following more complex structure of predicate calculus

$$\Big((\mathcal{L}_1(PC(\sigma)), S^1), \gamma\Big),$$

where $\mathcal{L}_1(PC(\sigma))$ is the Lindenbaum algebra of predicate calculus in the signature σ over formulas with one free variable x_0, γ is its Gödel enumeration, while the unary predicate $S(x)$ distinguishes the formulas which are equivalent to those without free variables. Is it true that

(a) the structure $\big((\mathcal{L}_1(PC(\sigma)), S^1), \gamma\big)$ is Σ_1^0-universal over the class of all Σ_1^0-algebras of the form $((\mathcal{B}, S^1), \nu)$ where $(\mathcal{B}, \nu)$ is a positive numerated Boolean algebra, while $S(x)$ distinguishes a recursively enumerable subalgebra of $\mathcal{B}$ in this numeration ν?

(b) first-order arithmetic is interpretable in the structure $((\mathcal{L}_1(PC(\sigma)), S^1), \gamma)$?

Question 44. Let GR be the first-order theory of groups. Is it true that the Lindenbaum algebra $\mathcal{L}(GR)$ is Σ_1^0-universal?

Question 45. Let GR_{fin} be the first-order theory of the class of all finite groups. Is it true that the Lindenbaum algebra $\mathcal{L}(GR_{fin})$ has characteristic $(1, \Sigma_2^0)$?

As to Question 43, the generalized Rogers phenomenon suggests that the structure considered there must either be of some well characterized form, or it must be very complicated. Parts (a) and (b) of this question represent possible solutions for these two possibilities.

11. References

1. W. W. Boone and H. Rogers, Jr., *On a problem of J. H. C. Whitehead and a problem of Alonzo Church*, Math. Scand., Vol. 19, 1966, pp. 185–192.
2. C. C. Chang, and H. J. Keisler, *Model Theory*, Elsevier Science Publishers, New York, 1992.
3. A. Church, *A note on the "Entscheidungsproblem"*, J. Symbolic Logic, Vol. 1, No. 1, 1937, pp. 40–41. Correction: ibid., pp. 101–102.
4. L. Feiner, *Hierarchies of Boolean algebras*, J. Symbolic Logic, Vol. 35, No. 3, 1970, pp. 365–373.

5. S. S. GONCHAROV, *Some properties of constructivizations of Boolean algebras*, Siberian Math. J., Vol. 16, No. 2, 1975, pp. 264–278.
6. S. S. GONCHAROV, *Countable Boolean algebras and decidability*, Plenum, New York, 1996.
7. S. S. GONCHAROV and YU. L. ERSHOV, *Constructive models*, Plenum, New York, 1999.
8. W. HANF, *Primitive Boolean algebras*, Proc. Sympos. Pure Math., Amer. Math. Soc., Providence, R. I., 1974, pp. 75–90.
9. W. HANF, *The Boolean algebra of Logic*, Bull. American Math. Soc., Vol. 31, 1975, pp. 587–589.
10. L. HARRINGTON, *Recursively presented prime models*, J. Symbolic Logic, Vol. 39, No. 2, 1974, pp. 305–309.
11. A. H. LACHLAN, *On the lattice of recursively enumerable sets*, Trans. Amer. Math. Soc., 130(1968), pp. 1-37.
12. A. MOSTOWSKI, *On recursive models of formalized arithmetic*, Bull. Acad. Pol. Sci., 7, 1957, pp. 705–710.
13. C. P. ODINTSOV and V. L. SELIVANOV, *Arithmetical hierarchy and ideals of numerated Boolean algebras*, Siberian Math. J., Vol. 30, No. 6, 1989, pp. 140–149.
14. M. G. PERETYAT'KIN, *Constructive models and numerations of Boolean algebra of recursive sets*, Algebra i Logika, Vol. 10, No. 5, 1971, pp. 535–557.
15. M. G. PERETYAT'KIN, *Turing Machine computations in finitely axiomatizable theories*, Algebra i Logika, Vol. 21, No. 4, 1982, pp. 272–295.
16. M. G. PERETYAT'KIN, *Finitely axiomatizable totally transcendental theories*, Trudy Inst. Math., Siberian Branch of Russian Academy of Sciences, Novosibirsk, Vol. 2, 1982, pp. 88–135.
17. M. G. PERETYAT'KIN, *Finitely axiomatizable theories*, Plenum, New York, 1997, 297 pages.
18. V. L. SELIVANOV, *Universal Boolean algebras and their applications*, International Conference on Algebra (abstracts), Russia, Barnaul, 20–25 August 1991, p. 127.
19. V. L. SELIVANOV, *On Recursively Enumerable Structures*, Forschungsberichte Mathematische Logik, Univ. Heidelberg, Math. Institut, Bericht Nr. 10, Juli 1994, pp. 1–20 (Section: Universal structures, pp. 8–10).
20. R. I. SOARE, *Recursively Enumerable Sets and Degrees*, Springer-Verlag, Berlin–Heidelberg–New York, 1986.
21. B. A. TRAKHTENBROT, *The impossibility of an algorithm for the decision problem for finite domains*, Doklady Akademii Nauk SSSR, Vol. 70, No. 4, 1950, pp. 569–572.
22. R. L. VAUGHT, *Sentences true in all constructive models*, J. Symbolic Logic, Vol. 25, No. 1, 1961, pp. 39–58.

E-mail: `peretya@math.kz`

Contemporary Mathematics
Volume **257**, 2000

Towards an Analog of Hilbert's Tenth Problem for a Number Field.

Alexandra Shlapentokh

Abstract. We consider the analogs of Hilbert's Tenth Problem (HTP) over subrings of number fields and related issues of Diophantine definability. One of the most interesting open problems in the field is the analog of HTP for the field of rational numbers. We will discuss how one can approach this problem and some recent results concerning undecidability of HTP over subrings of number fields where all the primes contained in the prime sets of high density are allowed in the denominator.

1. Historical background and main results.

The interest in the questions of Diophantine definability and decidability goes back to a question which was posed by Hilbert: given an arbitrary polynomial equation in several variables over $\mathbb{Z}$, is there a uniform algorithm to determine whether such an equation has solutions in $\mathbb{Z}$? This question, otherwise known as Hilbert's 10th problem, has been answered negatively in the work of M. Davis, H. Putnam, J. Robinson and Yu. Matijasevich. (See **[2]** and **[3]**.) Since the time when this result was obtained, similar questions have been raised for other fields and rings. In other words, let R be a recursive ring. Then, given an arbitrary polynomial equation in several variables over R, is there a uniform algorithm to determine whether such an equation has solutions in R? In this paper we will give a brief survey of the existing results in the case of R being a subring of a number field, i.e., a finite extension of $\mathbb{Q}$. Our discussion will make use of some number theoretic facts. We have assembled these facts and all the necessary notations in the appendix at the end of this paper where we also provide the references.

1.1. One Equals Finitely Many.

Perhaps, we should start with the following easy observation. If the ring R under consideration is an integral domain and its field of fractions K is not algebraically closed then for the purposes of solving Hilbert's Tenth Problem over R (abbreviated as "HTP" in the rest of the paper), a finite system of polynomial equations is equivalent to a single polynomial equation.

1991 *Mathematics Subject Classification.* Primary 03D35, 11U05. Secondary 11D57, 11D72, 11R04.

The author was supported in part by NSA Grant MDA904-98-1-0510.

In other words, for every finite system of polynomial equations there exists a single polynomial equation which has solutions in the ring if and only if the original system has solutions in the ring. Moreover, the procedure constructing this single polynomial equation for a given system is recursive and does not depend on the system. (However, the procedure does depend on R.) It is enough to demonstrate this construction for a system of two equations:

$$\begin{cases} f(x_1, \dots, x_m) = 0, \\ g(x_1, \dots, x_m) = 0. \end{cases}$$

Let $h(x) = \sum_{i=0}^{n} A_i x^i \in R[x]$ be a polynomial without roots in K. (This is the part of the proof requiring the assumption that K is not algebraically closed.) Note also that the polynomial

$$\bar{h}(x) = x^n h(x^{-1}) = \sum_{i=0}^{n} A_i x^{n-i}$$

also does not have roots in K. Finally, consider the equation

$$H(x_1, \dots, x_m) = \sum_{i=0}^{n} A_i f(x_1, \dots, x_m)^i g(x_1, \dots, x_m)^{n-i} = 0.$$

Suppose now that for some $a_1, \dots, a_m \in R$, $H(a_1, \dots, a_m) = 0$ but

$$f(a_1, \dots, a_m) \neq 0.$$

Then $H/f^n = 0$. In other words

$$\sum_{i=1}^{n} A_i \left(\frac{g}{f}\right)^{n-i} = \bar{h}(\frac{f}{g}) = 0.$$

If, on the other hand, $g(a_1, \dots, a_m) \neq 0$, while $H(a_1, \dots, a_m) = 0$, then, of course, we obtain $h(f/g) = 0$. Hence, $H = 0 \Leftrightarrow f = 0 \wedge g = 0$.

Arguably the two most interesting and difficult problems in the area concern $R = \mathbb{Q}$ and R equal to the ring of algebraic integers of an arbitrary number field.

One way to resolve the question of Diophantine decidability negatively over a ring of characteristic 0 is to construct a Diophantine definition of $\mathbb{Z}$ over such a ring. This notion is defined below.

1.2. Definition. Let R be a ring and let $A \subset R$. Then we say that A has a Diophantine definition over R if there exists a polynomial $f(t, x_1, \dots, x_n) \in R[t, x_1, \dots, x_n]$ such that for any $t \in R$,

$$\exists x_1, \dots, x_n \in R, f(t, x_1, ..., x_n) = 0 \Longleftrightarrow t \in A.$$

If the quotient field of R is not algebraically closed, from the argument above it follows that we can allow a Diophantine definition to consist of several polynomials without changing the nature of the relationship.

The usefulness of Diophantine definitions stems from the following lemma.

1.3. Lemma. Let $R_1 \subset R_2$ be two recursive rings such that the quotient field of R_2 is not algebraically closed. Assume that HTP is undecidable over R_1, and R_1 has a Diophantine definition over R_2. Then HTP is undecidable over R_2.

PROOF. Indeed, let $f(T, X_1, \dots, X_n)$ be a Diophantine definition of R_1 over R_2 and let $P(T_1, \dots, T_m) \in R_1[T_1, \dots, T_m]$. Then

$$\exists t_1, \dots, t_m \in R_1, P(t_1, \dots, t_m) = 0$$

if and only if

$$\exists t_1, \dots, t_m, x_{1,1}, \dots, x_{m,n} \in R_2, \begin{cases} P(t_1, \dots, t_m) = 0, \\ f(t_i, x_{i,1}, \dots, x_{i,n}) = 0, i = 1, \dots, m. \end{cases}$$

First of all we note here that since the quotient field of R_2 is not algebraically closed, the system above can be replaced by an equivalent single polynomial equation. Secondly, if we had an algorithm to determine whether that polynomial equation had solutions in R_2, we would have an algorithm to determine whether the original polynomial equation had solution in R_1. Given our assumptions on R_1, we can now conclude that HTP is undecidable over R_2.

1.4. Combining Diophantine Definitions. Suppose $R_3 \subset R_2 \subset R_1$ are integral domains whose fraction fields are not algebraically closed. Assume further that R_2 has a Diophantine definition $f_1(t, x_1, \dots, x_{m_1})$ over R_1 and R_3 has a Diophantine definition $f_2(t, y_1, \dots, y_{m_2})$ over R_2. Then the following system of equations would correspond to a Diophantine definition of R_3 over R_1:

$$\begin{cases} f_2(t, y_1, \dots, y_{m_2}) = 0, \\ f_1(t, x_1, \dots, x_{m_1}) = 0, \\ f_1(y_i, x_{i,1}, \dots, x_{i,m_1}) = 0, i = 1, \dots, m_2 \end{cases}$$

Diophantine definitions have been obtained for $\mathbb{Z}$ over the rings of algebraic integers of some number fields. Jan Denef has constructed a Diophantine definition of $\mathbb{Z}$ for totally real extensions of $\mathbb{Q}$ (i.e. fields all of whose embeddings into $\mathbb{C}$ are real). Jan Denef and Leonard Lipshitz extended Denef's results to the totally complex extensions of degree 2 of the totally real fields. Thanases Pheidas and the author of this paper have independently constructed Diophantine definitions of $\mathbb{Z}$ for number fields with exactly one pair of complex conjugate embeddings. Finally Harold N. Shapiro and the author of this paper showed that the subfields of all the fields mentioned above "inherit" the Diophantine definitions of $\mathbb{Z}$. (These subfields include all the abelian extensions.) The problem is still open for a general number field. The proofs of the results listed above can be found in [**4**], [**6**], [**5**], [**12**], [**13**], and [**16**].

A similar approach can in theory be applied to $\mathbb{Q}$. In other words, one could show that HTP is undecidable over $\mathbb{Q}$ by showing that $\mathbb{Z}$ has a Diophantine definition over $\mathbb{Q}$. Unfortunately, one of the consequences of a series of conjectures by Barry Mazur and Colliot-Thélène, Swinnerton-Dyer and Skorobogatov is that $\mathbb{Z}$ does not have a Diophantine definition over $\mathbb{Q}$, and thus one would have to look to some other method for resolving HTP over $\mathbb{Q}$. (Mazur's conjectures can be found in [**9**] and [**10**]. However, Colliot-Thélène, Swinnerton-Dyer and Skorobogatov have found a counterexample to the strongest of the conjectures in the papers cited above. Their modification of Mazur's conjecture in view of the counterexample can be found in [**1**].) Given the difficulty of the Diophantine problem for $\mathbb{Q}$ (and number fields in general), one might adopt a gradual approach, i.e consider the following problem.

1.5. An intermediate Problem for $\mathbb{Q}$ and Number Fields. Let W be a recursive set of rational primes. Let

$$O_{\mathbb{Q},W} = \{x \in \mathbb{Q} \mid x = \frac{a}{b}, a, b \in \mathbb{N}, \forall p \not\in W, p \nmid b\}.$$

Then we can ask whether HTP is decidable for $O_{\mathbb{Q},W}$ or whether $\mathbb{Z}$ has a Diophantine definition over $O_{\mathbb{Q},W}$. We can answer these questions for *finite* W. More precisely, we know that for finite W, $\mathbb{Z}$ does have a Diophantine definition over $O_{\mathbb{Q},W}$ and therefore HTP is undecidable over $O_{\mathbb{Q},W}$. (See [**17**].) Unfortunately, we have been unsuccessful in obtaining the analogous results for infinite W. On the other hand we have been more successful in solving the analogous problem in some extensions of $\mathbb{Q}$. Before we state the results which have been obtained for some extensions, we should remark on the following. First of all, if we do succeed in constructing a Diophantine definition of $\mathbb{Z}$ over $O_{\mathbb{Q},W}$ with infinite W, we would like to have a measure of our success. In other words, we need a way to measure infinite sets of primes. Fortunately, such a measure exists. It is called Dirichlet density and it is defined in the appendix. Secondly, we have to clarify what we mean by the analogous problem in the extensions. Rings $O_{\mathbb{Q},W}$ have natural analogs in the extensions which are called the rings of W-integers. We have constructed a Diophantine definition of $\mathbb{Z}$ in the rings of W-integers, where W is infinite and rather large from the point of view of Dirichlet density. Below we give the definition of the rings of W-integers of number fields and the statements of the best undecidability results to date. (By "best" we mean the closest we have been able to come to resolving HTP over some number field.) This definition uses the notions of "primes in number fields" and "order at a prime" which are discussed in the appendix.

1.6. Definition. Let M be a number field and let W be a set of its primes. Then a ring

$$O_{M,W} = \{x \in M \mid \text{ord}_{\mathfrak{p}} x \geq 0 \, \forall \mathfrak{p} \not\in W\}$$

is called a ring of *W-integers* . (The term W-integers usually presupposes that W is finite, but we will use this term for infinite W also.) If $W = \emptyset$, then $O_{M,W} = O_M$ – the ring of algebraic integers of M. If W contains all the primes of M, then $O_{M,W} = M$.

1.7. Theorem. *Let K be a totally real field or a totally complex extension of degree 2 of a totally real field. Then for any $\varepsilon > 0$, there exists a set W of primes of K whose Dirichlet density is bigger than $1 - [K : \mathbb{Q}]^{-1} - \varepsilon$ and such that $\mathbb{Z}$ has a Diophantine definition over $O_{K,W}$. (Thus, Hilbert's Tenth Problem is undecidable over $O_{K,W}$.)*

1.8. Theorem. *Let K be as above and let $\varepsilon > 0$ be given. Let $S_{\mathbb{Q}}$ be the set of all the rational primes splitting in K. (If the extension is Galois but not cyclic, $S_{\mathbb{Q}}$ contains all the rational primes.) Then there exists a set of K-primes W such that the set of rational primes $W_{\mathbb{Q}}$ below W differs from $S_{\mathbb{Q}}$ by a set contained in a set of Dirichlet density less than ε and such that $\mathbb{Z}$ has a Diophantine definition over $O_{K,W}$. (Again this will imply that Hilbert's Tenth Problem is undecidable in $O_{K,W}$.)*

The proofs of these theorems can be found in **[18]**, **[15]** and **[14]**.

1.9. HTP for a number field vs HTP for $\mathbb{Q}$. If we were successful in showing that HTP was undecidable for some number field, what implication would it have for HTP over $\mathbb{Q}$? It is not hard to show that *the undecidablity of HTP over any number field K implies the undecidability of HTP over $\mathbb{Q}$.* Indeed, suppose HTP is decidable over $\mathbb{Q}$ and we want to determine whether a polynomial equation

$$P(X_1 \dots, X_m) = \sum A_{i_1,\dots,i_m} X_1^{i_1} \dots X_m^{i_m} = 0$$

over K has solutions over K. We can easily convert the problem concerning a polynomial equation over K with solutions in K to a problem concerning a system of polynomial equations over $\mathbb{Q}$ with solutions in $\mathbb{Q}$. Indeed, let $\omega_1, \dots, \omega_n$ be a basis of K over $\mathbb{Q}$. Then the polynomial equation above has solutions in K if and only if the following polynomial system has solutions over $\mathbb{Q}$:

$$\sum \left(\sum_{j=1}^{n} a_{j,i_1,\dots,i_m} \omega_j \right) \left(\sum_{j=1}^{n} x_{1,j} \omega_j \right)^{i_1} \dots \left(\sum_{j=1}^{n} x_{j,m} \omega_j \right)^{i_m} = 0,$$

where $A_{i_1,\dots,i_m} = \sum_{j=1}^{n} a_{j,i_1,\dots,i_m} \omega_j$. If we were to multiply all the sums out and use the fact that any product of ω's can be rewritten as a linear combination of ω's with coefficients in $\mathbb{Q}$, we would obtain an equation of the form

$$\sum_{j=1}^{n} P_j(x_{1,1}, \dots, x_{n,m}) \omega_j = 0,$$

where all the coefficients of $P_1, \dots, P_n$ are in $\mathbb{Q}$ and all the variables range over $\mathbb{Q}$ also. The last equation is, of course equivalent to the system $P_j = 0, j = 1, \dots, n$. This system however can be replaced by a single equivalent equation.

In the rest of the paper we will describe some ideas involved in the proofs of Theorems 1.7 and 1.8, and the difficulties of moving beyond them. We will first divide the problems at hand into two categories, vertical and horizontal ones, and consider their relative difficulties.

2. Vertical and Horizontal Problems.

We will start with the description of the "horizontal" problem.

2.1. Horizontal Problem. Let $R_1 \subset R_2$ be two integral domains with the same quotient field. Construct a Diophantine definition of R_1 over R_2.

Now the "vertical" problem.

2.2. Vertical Problem. Let R_1 be an integral domain with a quotient field F_1. Let R_2 be an integral domain with a quotient field F_2. Assume further that F_2/F_1 is a field extension and $R_2 \cap F_1 = R_1$. Construct a Diophantine definition of R_1 over R_2.

2.3. Vertical and Horizontal Problems for Number Fields. To apply these concepts to our situation let K be a number field, let W be a set of primes of K, let S be a set of rational primes and assume $O_{K,W} \cap \mathbb{Q} = O_{\mathbb{Q},S}$. Next consider the following diagram.

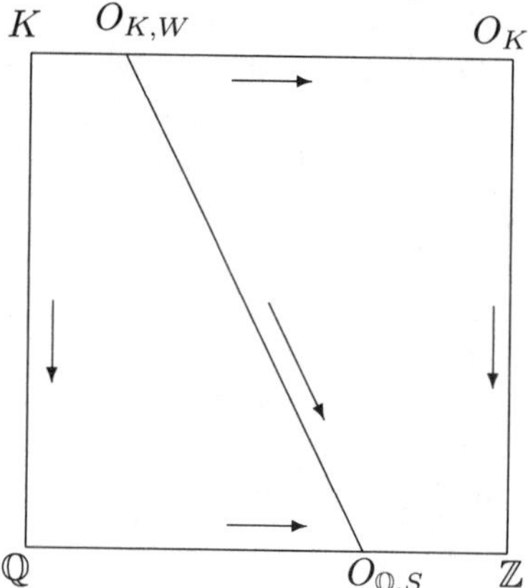

The following problems will be examples of **horizontal** problems: *construct a Diophantine definition of* $\mathbb{Z}$ *over* $\mathbb{Q}$*, or over* $O_{\mathbb{Q},S}$*; of* $O_{\mathbb{Q},S}$ *over* $\mathbb{Q}$*; of* O_K *over* K *or* $O_{K,W}$*; of* $O_{K,W}$ *over* K.

The following problems will be examples of **vertical** problems: *construct a Diophantine definition of* $\mathbb{Z}$ *over* O_K*,* $O_{\mathbb{Q},S}$ *over* $O_{K,W}$*,* $\mathbb{Q}$ *over* K.

When one first considers HTP over $O_{K,W}$, for K a totally real number field, one's first impulse might be to attempt to solve the horizontal problem, i.e. to construct a Diophantine definition of O_K over $O_{K,W}$. This would show the undecidability of HTP over $O_{K,W}$, since HTP is known to be undecidable over O_K. However, except for the case of a finite W, this direct approach has not been fruitful. In the past we had more success with vertical problems, i.e. problems of constructing Diophantine definitions of $\mathbb{Z}$ over the rings of algebraic integers of number fields described at the beginning of this paper. It turned out that it was by strengthening the vertical results that we were be able to solve some horizontal problems. Our new vertical results are contained in the two theorems below. There are two theorems because separate statements are needed to describe the density of primes sets in the extensions and in $\mathbb{Q}$. (See appendix for more details on this aspect of Dirichlet density.)

2.4. Theorem. Let K be a totally real number field or a totally complex extension of degree 2 of a totally real number field, let W_K be a set of primes of K and let $\varepsilon > 0$. Then there exists a set of K-primes $\bar{W}_K$ such that the difference between W_K and $\bar{W}_K$ is contained in a set of Dirichlet density less than ε and such that $O_{K,\bar{W}_K} \cap \mathbb{Q}$ has a Diophantine definition over $O_{K,\bar{W}_K}$.

2.5. Theorem. Let $W_{\mathbb{Q}}$ be any set of rational primes. Then for any $\varepsilon > 0$ and any K, as described in Theorem 2.4, there exists a set of rational primes $\bar{W}_{\mathbb{Q}}$ such that the difference between $\bar{W}_{\mathbb{Q}}$ and $W_{\mathbb{Q}}$ is contained in a set of primes of density less than ε, and $O_{\mathbb{Q},\bar{W}_{\mathbb{Q}}}$ has a Diophantine definition in its integral closure in K (i.e the ring of all the elements of K satisfying monic irreducible polynomials over $O_{\mathbb{Q},\bar{W}_{\mathbb{Q}}}$).

(See **[18]**, **[14]**, **[15]** for proofs of the theorems.)

As we have mentioned before, we were able to solve the horizontal problem in the cases when the sets of primes allowed in the denominator of divisors were finite. We will now exploit these results and Remark 1.4 concerning a combination of Diophantine definitions together with the vertical results for a carefully selected W. We select W so that $\mathbb{Q} \cap O_{K,W} = O_{\mathbb{Q},S}$, where S is **finite**. This is accomplished by not allowing W to contain all factors of any rational prime except possibly for finitely many exceptions. S will contain the exceptions. This restriction will allow W to be of density no more than $1 - [K : \mathbb{Q}]^{-1}$. On the other hand, for such a W, Theorem 1.7 holds.

Since O_K is a free $\mathbb{Z}$-module (see the appendix for more details), an immediate corollary of Theorem 1.7 is the fact that O_K has a Diophantine definition over $O_{K,\bar{W}}$, providing us with an answer to the original horizontal problem. Indeed, let $f(t, x_1, \dots, x_m)$ be a Diophantine definition of $\mathbb{Z}$ over $O_{K,\bar{W}}$. Let $\omega_1, \dots, \omega_n$ be an integral basis of O_K over $\mathbb{Z}$. Next consider the following system of equations:

$$\begin{cases} y = \sum_{i=1}^{n} t_i\omega_i \\ f(t_i, x_{i,1}, \dots, x_{i,m}) = 0, i = 1, \dots, n. \end{cases}$$

It is clear that this system has solutions over $O_{K,\bar{W}}$ if and only if $y \in O_K$.

3. Solving the Extended Vertical Problem.

In this section we sketch the construction for an extended vertical problem. The problem we will consider is the problem of giving a Diophantine definition of $O_{K,W} \cap \mathbb{Q}$ over $O_{K,W}$ for a totally real field K and an infinite set W of its primes. We will first reduce the problem to a case of a cyclic extension of number fields.

3.1. Reducing the problem to the cyclic case. Let M be the Galois closure of K over $\mathbb{Q}$. (In other words, let M be the smallest Galois extension of $\mathbb{Q}$ containing K.) Let W_M be the collection of all M-primes lying above the primes in W. (See the appendix for some remarks on "primes above" and "below".) Then it is enough to construct a Diophantine definition of $O_{M,W_M} \cap \mathbb{Q}$ over O_{M,W_M}. Indeed, observe that $O_{K,W} \subset O_{M,W_M}$ since, as has been noted in the appendix, O_{M,W_M} is the integral closure of $O_{K,W}$ in M. Further let $f(t, x_1, \dots, x_m)$ be a Diophantine definition of $O_{M,W_M} \cap \mathbb{Q}$ over O_{M,W_M}. Then for any $t \in O_{K,W}$, $f(t, x_1, \dots, x_m) = 0$ has solutions in O_{M,W_M} if and only if $t \in O_{K,W} \cap \mathbb{Q}$. Thus the only remaining problem would be to replace $f(t, x_1, \dots, x_m) = 0$ by an equivalent equation over K with variables ranging over $O_{K,W}$. This can be done using a method similar to the one we used in showing that the undecidability of HTP over a number field would imply the undecidability of HTP over $\mathbb{Q}$. The only difference is that given a basis $\Omega = \{\omega_1, \dots, \omega_{[M:K]}\}$ of M over K, O_{M,W_M} is not necessarily generated by elements of Ω as a module over $O_{K,W}$. But there exists an element $D \in O_K$ such that for every $x \in O_{M,W_M}$, $Dx = \sum_{i=1}^{[M:K]} a_i\omega_i$, where $a_i \in O_{K,W}$. Finally note that if K is totally real, then so is its Galois closure over $\mathbb{Q}$.

Thus, without loss of generality we can assume that K is Galois over $\mathbb{Q}$. Consider the Galois group of K over $\mathbb{Q}$ and all of its cyclic subgroups. Each cyclic subgroup corresponds to a cyclic subextension of K. Let $F_1, \dots, F_r$ be all of these cyclic subextensions. Note that $\bigcap_{i=1}^{r} F_i = \mathbb{Q}$, since elements of the intersection are fixed by every element of the Galois group. Suppose that we could write down a Diophantine definition $f_i(t, x_{i,1}, \dots, x_{i,m_i})$ of $O_{K,W} \cap F_i$ over $O_{K,W}$. Assume that $t \in O_{K,W}$ and the system $f_i(t, x_{i,1}, \dots, x_{i,m_i}) = 0, i = 1, \dots, r$ has solutions in $O_{K,W}$. Then $t \in \bigcap_{i=1}^{r} F_i = \mathbb{Q}$. Thus, the problem is reduced to the construction of a Diophantine definition of $O_{K,W} \cap F$ over $O_{K,W}$ for a cyclic subextension F of K.

3.2. A unit norm equation. We will now make use of the fact that K is totally real and that the extension K/F is cyclic. Let L be a totally complex extension of degree 2 of $\mathbb{Q}$ and let E be a totally real cyclic extension of $\mathbb{Q}$ of prime degree $p > 2$. Assume that p is relatively prime to $[K : \mathbb{Q}]$. Then $[LKE : EK] = 2, [LEK : LK] = p, [LEF : EF] = 2, [LEF : LF] = p$. Further, the fields KE, FE are totally real and the fields LK, LKE, FL, FLE are totally complex. Next consider the system of norm equations

$$\begin{cases} \mathbf{N}_{LEK/EK}(x) = 1, \\ \mathbf{N}_{LEK/LK}(x) = 1, \end{cases}$$

where $x \in LEK$. Let $\bar{x}$ be the conjugate of x over EK. (If $x \in EK$, then $\bar{x} = x$.) Then $x\bar{x} = 1$. If the divisor of x is of the form $\frac{\prod \mathfrak{p}_i^{a_i}}{\prod \mathfrak{q}_j^{b_j}}$, where $a_i, b_j \in \mathbb{N}$, then the divisor of $\bar{x}$ is of the form $\frac{\prod \bar{\mathfrak{p}}_i^{a_i}}{\prod \bar{\mathfrak{q}}_j^{b_j}}$, where $\mathfrak{p}$ and $\bar{\mathfrak{p}}$, $\mathfrak{q}$ and $\bar{\mathfrak{q}}$ must be factors of the same prime in EK. This is so because the elements of the Galois group of the extension LEK/EK send a factor of a prime in EK to a factor of the same EK-prime. Since the divisor of 1 is trivial, each $\mathfrak{p}_i$ must be equal to $\bar{\mathfrak{q}}_j$ with $a_i = b_j$. In other words, the divisor of x must either be trivial (so that x is a unit of LE) or of the form $\frac{\prod \mathfrak{p}^{a_i}}{\prod \bar{\mathfrak{p}}_i^{a_i}}$. Now, Dirichlet's Unit Theorem tells us that the rank of the unit group in EK and LEK is the same. This implies that the only LEK units with EK-norm equal to 1 are the roots of unity in LEK. Thus, if we assume that x is not a root of unity, the divisor of x will not be trivial. To eliminate the roots of unity from the picture we can do the following. Since any power of a solution to the norm equations above will also be a solution to the norm equations above, we can assume without loss of generality that $x = y^r$, where r is large enough so that for any roots of unity $\xi \in LEK$, $\xi^r = 1$, and y is also a solution to the norm equations.

The second norm equation, by a similar argument, tells us that the divisor of x is of the form $\frac{\prod \mathfrak{p}^{a_i}}{\prod \tilde{\mathfrak{p}}_i^{a_i}}$, where $\tilde{\mathfrak{p}}_i$ is a factor of the same LK-prime as $\mathfrak{p}_i$. Hence, we conclude that the only primes which can appear in the divisor of x are primes of LEK lying above primes of LK and EK splitting in the extensions LEK/LK and LEK/EK. Given the fact that both extensions are cyclic of distinct prime degrees, we can conclude that the only primes appearing in the divisor of x lie above primes of K splitting completely in the extension LEK/K. Further, given our assumption on x and roots of unity, each solution x is completely determined by its divisor. (The ratio of two solutions will be a solution. Thus a ratio of two solutions with the same divisor will be a unit which is a solution.) This situation gives us a clue on how to select primes for W. W can contain all the primes of K which do not split completely in the extension LEK/K. By The Chebotarev Density Theorem, the density of all such primes is $1 - 1/2p$, where p can be made arbitrary large. However, if we restrict W just to these primes we will not get any solutions to the norm equations that are not roots of unity. So we need to add one more prime $\mathfrak{P}$. This prime $\mathfrak{P}$ must satisfy the following conditions:

1. $\mathfrak{P}$ must lie above an F-prime $\mathfrak{P}_F$ *not splitting* in the extension K/F. (This is where we need K/F to be cyclic. Otherwise we might have no primes which do not split.)
2. $\mathfrak{P}_F$ should split completely in the extension FEL/F.
3. $\mathfrak{P}$ should split completely in the extension EKL/K.

Given our assumptions on $\mathfrak{P}_F$ and $\mathfrak{P}$ and the degrees of the extensions, no factor of $\mathfrak{P}_F$ in FEL will split in the extension KEL/FEL. Such a prime $\mathfrak{P}$ of K exists by an argument using The Chebotarev Density Theorem and some Galois theory. If we choose W as described above, and x is a solution to our system while being an element of the integral closure of $O_{K,W}$ in LEK, then the divisor of x consists of the factors of $\mathfrak{P}$ in LKE only. But by the remark above, factors of $\mathfrak{P}$ in LEK are in essence the same as the factors of $\mathfrak{P}_F$ in FEL. Since x is determined completely by its divisor, this means that $x \in FEL$.

Let $\delta_1 = 1, \ldots, \delta_{2p}$ be a basis of LEK over K consisting of algebraic integers and consider the following system of equations.

$$\begin{cases} \prod_{j=1}^{p} \sum_{i=1}^{2p} c_i \sigma_j(\delta_i) = 1, \\ \left(\sum_{i=1}^{2p} c_i \delta_j\right)\left(\sum_{i=1}^{2p} c_i \bar{\delta}_j\right) = 1, \end{cases}$$

where $\sigma_1, \ldots, \sigma_p$ are all the elements of the $\mathrm{Gal}(KEL/LK)$ and δ_j and $\bar{\delta}_j$ are conjugates over EK. If this system of equations is satisfied for some $c_1, \ldots, c_{2p} \in O_{K,W}$, then $x = \sum_{i=1}^{2p} c_i \delta_j$ satisfies the system of norm equations above and $x \in FEK$. Hence, $c_1, \ldots, c_{2p} \in F \cap O_{K,W}$. The equation above will have its coefficients in KEL, but as we have indicated before, we can rewrite these equations so that the coefficients are in K. Thus, we have constructed a system of equations whose solutions in $O_{K,W}$ are actually in $O_{K,W} \cap F$. Of course we still need to show that the system above does have solutions in $O_{K,W}$. That argument (which we omit) relies on Dirichlet's Unit Theorem again. Now we use these equations to construct a Diophantine definition of the intersection.

3.3. Divisibility in $O_{K,W}$ and Height Bounds. Let $a, b \in O_K$ and assume that the divisor of b (which has no denominator since b is an algebraic integer) has no primes from W. Then if $b \mid a$ in $O_{K,W}$, $b \mid a$ in O_K. In other words b is "composed" of the primes which are not allowed in the denominator in $O_{K,W}$ and thus in order for $a/b \in O_{K,W}$, all the primes of b must be canceled by the primes of a. That in turn means that $a/b \in O_K$. How could one obtain such a b? Given the discussion above, we can assume that except for one prime, all the primes of W do not split in the extension KE/K. Let γ be an integral generator of KE of K and let $F(T)$ be γ's monic irreducible polynomial over K. Then, assuming that W does not contain any primes dividing the divisors of the discriminant or the coefficients of this polynomial (there are only finitely many such primes), for any $x \in K$ and for any prime $\mathfrak{q} \in W \setminus \{\mathfrak{P}\}$, $\mathrm{ord}_{\mathfrak{q}} x \leq 0$. Let $t, w \in O_{K,W\setminus\{\mathfrak{P}\}}$ and assume that $\frac{w}{F(t)} \in O_{K,W\setminus\{\mathfrak{P}\}}$. If $w/F(t) \in O_{K,W\setminus\{\mathfrak{P}\}}$, then all the primes dividing the numerator of the divisor of $F(t)$ must have been canceled by the primes in the numerator of the divisor of w, because none of the primes in the numerator of the divisor of $F(t)$ is in $W \setminus \{\mathfrak{P}\}$ and thus allowed in the denominator of the divisors of the elements of $O_{K,W\setminus\{\mathfrak{P}\}}$. Hence, (assuming that $w = \tilde{w}^h$, where h is the class number of K), $w = w_1 w_2$, where $w_1 \in O_K$, the divisor of w_1 is not divisible by any prime of $W \setminus \{\mathfrak{P}\}$ and cancels all the primes in the numerator of the divisor of $F(t)$. This way we can also obtain a useful bound on the height of t in terms of w_1. (Actually we will require a bound on $\mathbf{N}_{K/\mathbb{Q}}(\beta)t$, where $t = \alpha/\beta, \alpha, \beta \in O_K$. To obtain this bound we will need $w/F(t-i) \in O_{K,W\setminus\{\mathfrak{P}\}}$, for $i = 0, \ldots, [K : \mathbb{Q}]$.)

3.4. Some Useful Equations. Let $y \in O_K$ and assume that

$$y - a = vw,$$

where $a, v \in O_{FEL}$, $w \in O_{KEL}$ and the height of v is "sufficiently large" relative to the height of x. Then we can conclude that $y \in FEL \cap K = F$. To get an idea as to why this is true, consider the following. Let $\Gamma = \{\gamma_1 = 1, \ldots, \gamma_r\} \subset O_{KEL}$ be a basis of KEL over FEL. To simplify the argument, assume that Γ is an integral basis of O_{KEL} over O_{FEL}, or in other words O_{KEL} is generated by Γ as a module over O_{EFL}. Then $y = a_1 + \sum_{i=2}^{n} a_i\gamma_i$, where $a_i \in O_{FEL}$, and $\frac{y-a}{v} = (a_1 - a)/v + \sum_{i=2}^{n}(a_i/v)\gamma_i$, where $a_i/v \in O_{EFL}$. A bound on the height of y corresponds to a bound on the height of $\mathbb{Q}$-norms of $a_1, \ldots, a_n$. On the other hand, $a_i/v \in O_{FEL}$ implies that the height of $\mathbf{N}_{K/\mathbb{Q}}(v)$ is less than the height of the corresponding $\mathbb{Q}$-norm of a_i, *unless* $a_i = 0$. Thus, we can force a_i's to be 0 for $i = 2, \ldots, n$. This will of course push y into FEL.

Suppose now that $z \in O_{K,W}$. Then $z = \alpha/\beta$, where $\alpha, \beta \in O_K$. Further, $z \in O_{K,W} \cap F$ if and only if $\mathbf{N}_{K/\mathbb{Q}}(\beta)z \in O_{K,W} \cap F$. On the other hand, $\mathbf{N}_{K/\mathbb{Q}}(\beta)z \in O_K$.

So consider the following equation where x_1, x_2 satisfy the norm equations above, and $\frac{x_1-1}{F(z-i)} \in O_{K,W\setminus\{\mathfrak{P}\}}$ for sufficiently many values of i;

$$z - (x_2 - 1)/(x_1 - 1) = (x_1 - 1)b.$$

We can multiply this equation by $\mathbf{N}_{K/\mathbb{Q}}(\beta)$ to obtain

$$\mathbf{N}_{K/\mathbb{Q}}(\beta)z - \mathbf{N}_{K/\mathbb{Q}}(\beta)(x_2 - 1)/(x_1 - 1) = (x_1 - 1)\mathbf{N}_{K/\mathbb{Q}}(\beta)zb.$$

The last equation is not quite in the form which we described above, but a few applications of the Strong Approximation Theorem will transform it into the desired form. Thus we will be able to conclude that $z \in F$.

Conversely, if $z \in O_{K,W} \cap F$, can we make sure that the equations above have solutions? Actually we can easily do this if $z \in \mathbb{N}$. Since every $z \in O_{K,W} \cap F$ can be written as a ratio of two linear combinations of some basis elements of F over $\mathbb{Q}$ with integer coefficients, it is enough. To see how to satisfy the equation above for $z \in \mathbb{N}$ observe that solutions of our system of unit norm equations form a multiplicative group. In other words if ε is a solution, so is ε^k. But

$$\frac{\varepsilon^{kn} - 1}{\varepsilon^n - 1} \cong k \text{ modulo}(\varepsilon^n - 1)$$

in $\mathbb{Z}[\varepsilon]$. Or in other words,

$$\frac{\varepsilon^{kn}-1}{\varepsilon^n-1}-k=(\varepsilon^n-1)u,$$

where $u \in O_K$.

3.5. Going Further Up and Left. All the known vertical results rely on the following fact concerning the rank of the unit group of totally real fields, totally complex extensions of degree two of the totally real fields and fields with one pair of complex conjugate embeddings. Each of these fields has either a subextension or an extension where the unit group has the same rank or rank which is exactly one more than the rank of the unit group of the field under consideration. If the difference in ranks is exactly one, then all the solutions to the corresponding unit norm equation are essentially powers of one fixed unit. These powers can the be used to generate integers as in the discussion above. If the ranks of the unit group of a field and its subextension is the same then some fixed power of any unit of the field under consideration is in the subextension. This fact can be used to construct a Diophantine definition of the ring of algebraic integers of the subextension, again using methods similar to the ones described above. Unfortunately, the ranks of the unit groups do not have the desired properties for the fields other than the ones mentioned above. Thus, a new, as yet unknown approach is required to move further up.

Getting closer to a field, or in other words, putting more primes into W is difficult, because the whole technique is predicated on W not having all the conjugates of any rational prime. The Chebotarev Density Theorem imposes a limit on the density of such a prime set. The nature of this technique also makes it unusable over $\mathbb{Q}$. Again, it is likely that a radical departure from the current method would be required to increase the density of W and to solve the intermediate problem for $\mathbb{Q}$.

4. Open Questions and Conjectures.

We will end our discussion with a list of some open questions and conjectures.

4.1. Rings of W-integers: vertical problems. Is HTP undecidable over the ring of algebraic integers of an arbitrary number field? Does $\mathbb{Z}$ have a Diophantine definition over the ring of algebraic integers of an arbitrary number field? More generally, let $O_{K,W}$ be a ring of W-integers of a number field K. Let E be a subfield of K. Is HTP undecidable over $O_{K,W}$? Does $O_{K,W} \cap E$ have a Diophantine definition over K?

4.2. Rings of W-integers: horizontal problems. Is HTP undecidable over $\mathbb{Q}$ and other number fields? Does $\mathbb{Z}$ have a Diophantine definition over $\mathbb{Q}$ and other number fields ? More generally, let K be a number field. Let $V \subset W$ be sets of primes of K. Does $O_{K,V}$ have a Diophantine definition over $O_{K,W}$?

4.3. Conjecture. Let K be a totally real extension of $\mathbb{Q}$ (including $\mathbb{Q}$ itself) or a totally complex extension of degree 2 of a totally really field. Let W be a set of primes of K. Then for any $\varepsilon > 0$ there exists a set $\bar{W}$ of primes of K such that the difference between $\bar{W}$ and W is contained in a set of Dirichlet density less than ε and $\mathbb{Z}$ has a Diophantine definition over $O_{K,\bar{W}}$.

5. Appendix.

We suggest [7] and [8] as two possible references for the material in this section.

5.1. Totally Real and Totally Complex Number Fields. Let K be a number field. Assume that $[K : \mathbb{Q}] = n$. Then K has n embeddings into $\mathbb{C}$. Let $\sigma : K \longrightarrow \mathbb{C}$ be an embedding of K into $\mathbb{C}$. If $\sigma(K) \subset \mathbb{R}$ then σ is called a real embedding. Otherwise σ is called a complex embedding. Complex embeddings can be paired off via conjugation. If all embeddings of K into $\mathbb{C}$ are real, then K is called totally real. If none of the embeddings are real, then K is called totally complex.

5.2. Rings of Algebraic Integers. Let K be a number field. Then O_K, the ring of algebraic integers of K, is the ring of all the elements of K that are roots of monic irreducible polynomials over $\mathbb{Z}$. O_K has an integral basis over $\mathbb{Z}$. In other words, there exists a set $\Omega = \{\omega_1, \dots, \omega_n\} \subset O_K$ such that O_K is generated by Ω as a free abelian group. (This is a consequence of the fact that $\mathbb{Z}$ is a principal ideal domain.) Ω is called an *integral basis* of O_K over $\mathbb{Z}$.

5.3. Units of Number Fields. If ε and $\varepsilon^{-1} \in O_K$ then ε is called a unit of K. Units of K form a finitely generated multiplicative group. By Dirichlet's Unit Theorem, its rank is equal to $r + s - 1$, where r is the number of real embeddings of K and s is the number of pairs of complex embeddings.

5.4. Primes and Divisors in a Number Field. Let $x \in O_K$ and consider the principal ideal generated by x in O_K. This ideal xO_K can be factored uniquely as a product of finitely many prime ideals of O_K. Let $(x) = \prod \mathfrak{P}^{n(\mathfrak{P})}$, where $\mathfrak{P}$ is a prime ideal of O_K and $n(\mathfrak{P}) \in \mathbb{N}\backslash\{0\}$, be this factorization. Then $\prod \mathfrak{P}^{n(\mathfrak{P})}$ is called the *divisor* of x. Now let $w \in K$. We can write $w = x/y, x, y \in O_K$. Let $\mathfrak{D}_x, \mathfrak{D}_y$ be the divisors of x and y respectively. Define the divisor of w to be the formal quotient obtained by performing the obvious cancelations in $\frac{\mathfrak{D}_x}{\mathfrak{D}_y} = \prod \mathfrak{P}^{n(\mathfrak{P})}, n(\mathfrak{P}) \in \mathbb{Z}\backslash\{0\}$. (It can be shown that the divisor of w, thus defined, does not depend on the choice of x, y.) Next let $\mathfrak{T}$ be a prime of K (i.e. a prime ideal of O_K). Then we define

$$\mathrm{ord}_{\mathfrak{T}} w = \begin{cases} n(\mathfrak{P}), \text{ if } \mathfrak{T} = \mathfrak{P}, \text{ for some } \mathfrak{P} \text{ occurring in the divisor }, \\ 0, \text{ if } \mathfrak{T} \text{ does not occur in the divisor.} \end{cases}$$

Note that the divisors of algebraic integers have no denominators and the divisors of units are trivial. Let F be a subextension of K and let $\mathfrak{p}$ be a prime of F. Consider an O_K-ideal $\mathfrak{p}O_K$. In O_K this ideal might factor $\mathfrak{p}O_K = \prod \mathfrak{P}_i^{a_i}$. Then we will say that $\mathfrak{P}_i$ is a K-factor of $\mathfrak{p}$, $\mathfrak{P}_i$ lies above $\mathfrak{p}$, and $\mathfrak{p}$ lies below $\mathfrak{P}_i$.

Let $x \in F$ and let $\prod_{\mathfrak{p}} \mathfrak{p}^{n(\mathfrak{p})}, n(\mathfrak{p}) \in \mathbb{Z} \setminus \{0\}$ be the divisor of x in F. In K, let

$$\mathfrak{p}O_K = \prod_{\text{all factors of } \mathfrak{p} \text{ in } O_K} \mathfrak{P}^{e(\mathfrak{P})}.$$

(It can be shown that the sets of K-factors of distinct primes of F have empty intersections and every prime of K is a factor of a prime of F.) The divisor of x in K will be of the form

$$\prod_{\mathfrak{p}} \prod_{\mathfrak{P}|\mathfrak{p}} \mathfrak{P}^{e(\mathfrak{P})n(\mathfrak{p})}.$$

From this factorization formula we can derive a useful corollary. *Let* $x \in K$ *and suppose that for some prime* $\mathfrak{p}$ *and its two distinct factors in* K, $\mathfrak{P}_1$ *and* $\mathfrak{P}_2$, $ord_{\mathfrak{P}_1} x \geq 0$ *while* $ord_{\mathfrak{P}_2} x < 0$. *Then* $x \notin F$. If $\mathfrak{p}$ is a prime of F which has only one factor K, then we say that $\mathfrak{p}$ does not split in the extension K/F. Otherwise $\mathfrak{p}$ is said to "split" in the extension. If the number of factors of $\mathfrak{p}$ in K is equal to $[K : F]$–the maximum possible number of factors, $\mathfrak{p}$ is said "to split completely".

Now, let W_F be a set of primes of F and let W_K be the set of all the primes of K lying above the primes of W_F. Then O_{K,W_K} is the integral closure of O_{F,W_F} in K, i.e. O_{K,W_K} is the ring of all the elements of K satisfying monic irreducible polynomials with coefficients in O_{F,W_F}. In particular, $O_{F,W_F} \subset O_{K,W_K}$. Note that the denominators of the divisors of elements of O_{K,W_K} can have primes of W_K only. The units of this ring will be the elements whose divisors have primes of W_K only in the numerators.

One of the most important theorems describing the interaction between different primes of a number field is The Strong Approximation Theorem which can be found in [**11**].

5.5. Dirichlet Density. We will now discuss a few important facts concerning Dirichlet density. First we have to define a norm of a prime in K. Let $\mathfrak{P}$ be a prime of K, then $O_K/\mathfrak{P}$ is a finite field, called *the residue field* of $\mathfrak{P}$. Let $p = \mathfrak{P} \cap \mathbb{Z}$ be the rational prime below $\mathfrak{P}$. Then $O_K/\mathfrak{P}$ is of characteristic p and contains $\mathbb{Z}/p$. Let $f(\mathfrak{P}/p) = [O_K/\mathfrak{P} : \mathbb{Z}/p]$. Then $f = f(\mathfrak{P}/p)$ is called the relative degree of $\mathfrak{P}$ over p. Define the norm of $\mathfrak{P}$, denoted by $\mathbf{N}\mathfrak{P}$, to be p^f. Then given a set W of primes of K, define $\delta(W)$, the *Dirichlet density* of W, to be

$$\lim_{s \to 1^+} \frac{\sum_{\mathfrak{P} \in W} (\mathbf{N}\mathfrak{P})^{-s}}{\sum_{\text{all } \mathfrak{P}} (\mathbf{N}\mathfrak{P})^{-s}}.$$

From this definition it is not hard to see that

$$\delta(W) = \delta(W \cap \{ \text{ primes of relative degree 1 over } \mathbb{Q}\}).$$

Note also that if V is the set of rational primes below the primes of W, then in general $\delta(V) \neq \delta(W)$. For example, if $K/\mathbb{Q}$ is a cyclic extension of prime degree p, then by The Chebotarev Density Theorem (see [**8**] or [**7**]), the density of rational primes not splitting in this extension is $(p-1)/p$, but the density of K-primes lying above these rational primes is 0 since these primes are not of relative degree 1 over the primes in $\mathbb{Q}$.

5.6. Height of an algebraic number. Let $m, n \in \mathbb{N}$ be relatively prime. Then the height of m/n is the $\max(|m|, |n|)$. Let x be an algebraic number and let $P(X) = a_0 + a_1 X + \ldots + a_{n_1} X^{n-1} + X^n$ be its monic irreducible polynomial over $\mathbb{Q}$. Then the height of x is the $\max(h(a_0), \ldots, h(a_{n-1}))$, where $h(a_i)$ is the height of a_i.

References

1. Jean-Louis Colliot-Thélène, Alexei Skorobogatov, and Peter Swinnerton-Dyer. Double fibres and double covers: Paucity of rational points. *Acta Arithmetica*, 79:113–135, 1997.
2. Martin Davis. Hilbert's tenth problem is unsolvable. *American Mathematical Monthly*, 80:233–269, 1973.
3. Martin Davis, Yurii Matijasevich, and Julia Robinson. Positive aspects of a negative solution. In *Proc. Sympos. Pure Math.*, volume 28, pages 323– 378. Amer. Math. Soc., 1976.

4. Jan Denef. Hilbert's tenth problem for quadratic rings. *Proc. Amer. Math. Soc.*, 48:214–220, 1975.
5. Jan Denef. Diophantine sets of algebraic integers, II. *Transactions of American Mathematical Society*, 257(1):227– 236, 1980.
6. Jan Denef and Leonard Lipschitz. Diophantine sets over some rings of algebraic integers. *Journal of London Mathematical Society*, 18(2):385–391, 1978.
7. M. Fried and M. Jarden. *Field Arithmetic.* Springer Verlag, New York, 1986.
8. G. Januz. *Algebraic Number Fields.* Academic Press, New York, 1973.
9. Barry Mazur. The topology of rational points. *Experimental Mathematics*, 1(1):35–45, 1992.
10. Barry Mazur. Questions of decidability and undecidability in number theory. *Journal of Symbolic Logic*, 59(2):353–371, June 1994.
11. O. T. O'Meara. *Introduction to Quadratic Forms.* Springer Verlag, Berlin, 1973.
12. Thanases Pheidas. Hilbert's tenth problem for a class of rings of algebraic integers. *Proceedings of American Mathematical Society*, 104(2), 1988.
13. Harold Shapiro and Alexandra Shlapentokh. Diophantine relations between algebraic number fields. *Communications on Pure and Applied Mathematics*, XLII:1113–1122, 1989.
14. Alexandra Shlapentokh. Defining integrality at prime sets of high density in number fields. *Duke Mathematical Journal.* To appear.
15. Alexandra Shlapentokh. On diophantine definability and decidability in large subrings of totally real number fields and their totally complex extensions of degree 2. Preprint.
16. Alexandra Shlapentokh. Extension of Hilbert's tenth problem to some algebraic number fields. *Communications on Pure and Applied Mathematics*, XLII:939–962, 1989.
17. Alexandra Shlapentokh. Diophantine classes of holomorphy rings of global fields. *Journal of Algebra*, 169(1):139–175, October 1994.
18. Alexandra Shlapentokh. Diophantine definability over some rings of algebraic numbers with infinite number of primes allowed in the denominator. *Inventiones Mathematicae*, 129:489–507, 1997.

Department of Mathematics, East Carolina University, Greenville, NC 27858
E-mail address: shlapentokh@math.ecu.edu

Contemporary Mathematics
Volume **257**, 2000

Natural Definability in Degree Structures

Richard A. Shore

ABSTRACT. A major focus of research in computability theory in recent years has involved definability issues in degree structures. There has been much success in getting general results by coding methods that translate first or second order arithmetic into the structures. In this paper we concentrate on the issues of getting definitions of interesting, apparently external, relations on degrees that are order-theoretically natural in the structures $\mathcal{D}$ and $\mathcal{R}$ of all the Turing degrees and of the r.e. Turing degrees, respectively. Of course, we have no formal definition of natural but we offer some guidelines, examples and suggestions for further research.

1. Introduction

A major focus of research in computability theory in recent years has involved definability issues in degree structures. The basic question is, which interesting apparently external relations on degrees can actually be defined in the structures themselves, that is, in the first order language with the single fundamental relation of relative computability, the basic partial ordering $\leq$ on degrees? Most of the work has focused on the Turing degrees and on the structures $\mathcal{R}$ and $\mathcal{D}$ consisting of the recursively enumerable degrees and all the degrees, respectively. We will do the same in this paper.

At the level of establishing abstract definability of relations there has been great success. In both structures, any relation invariant under the double jump whose definability is not ruled out simply by the absolute limitations imposed by the structures being themselves subsystems of first or second order arithmetic, respectively, is actually definable in the structures.

DEFINITION 1.1. *An n-ary relation* $P(\mathbf{x}_1, \dots, \mathbf{x}_n)$ *on* $\mathcal{R}$ *is* invariant under the double jump *if, whenever* $\mathcal{R} \models P(\mathbf{x}_1, \dots, \mathbf{x}_n)$ *and* $\mathbf{x}_1'' \equiv_T \mathbf{y}_1'', \dots, \mathbf{x}_n'' \equiv_T \mathbf{y}_n''$, *it is also true that* $\mathcal{R} \models P(\mathbf{y}_1, \dots, \mathbf{y}_n)$. *We say that* P *is* invariant in $\mathcal{R}$ *if whenever* $\mathcal{R} \models P(\mathbf{x}_1, \dots, \mathbf{x}_n)$ *and* φ *is an automorphism of* $\mathcal{R}$, $\mathcal{R} \models P(\varphi(\mathbf{x}_1), \dots, \varphi(\mathbf{x}_n))$.

1991 *Mathematics Subject Classification.* Primary 03D25, 03D39; Secondary 03D55, 03E45.

Key words and phrases. degrees, Turing degrees, recursively enumerable degrees, natural definabiltiy.

Partially supported by NSF Grant DMS-9802843. I would also like to thank T. A. Slaman and S. G. Simpson for their helpful comments.

THEOREM 1.2. (Nies, Shore and Slaman [1998]) *Any relation on $\mathcal{R}$ which is invariant under the double jump is definable in $\mathcal{R}$ if and only if it is definable (on indices) in first order arithmetic.*

The route to this result is rather complicated. It begins by coding $\mathbb{N}$, the standard model of arithmetic, in $\mathcal{R}$ in the sense of interpretations of one structure in another as in Hodges [1993]. The coding uses parameters in such a way that, when they are chosen to satisfy some definable condition, the formulas define such structures. One then divides by a definable equivalence relation to get a single model.

THEOREM 1.3. (Nies, Shore and Slaman [1998]) *(i) There is a uniformly definable class $\mathcal{C}_{st}$ of coded standard models of arithmetic.*
(ii) Let $\tilde{\mathbf{N}} = \{(\mathbf{x}, \overline{\mathbf{p}}) : M(\overline{\mathbf{p}}) \in \mathcal{C}_{st} \wedge \mathbf{x} \in M(\overline{\mathbf{p}})\}$. The equivalence relation Q on $\tilde{\mathbf{N}}$ given by

$$(\mathbf{x}, \overline{\mathbf{p}})Q(\mathbf{y}, \overline{\mathbf{q}}) \Leftrightarrow (\exists n \in \omega)[\mathbf{x} = n^{M(\overline{\mathbf{p}})} \wedge \mathbf{y} = n^{M(\overline{\mathbf{q}})}]$$

is definable in $\mathcal{R}$.
(iii) A standard model of arithmetic $\mathbf{N}$ can be defined on the set of equivalence classes $\tilde{\mathbf{N}}/Q$ without parameters.

THEOREM 1.4. (Nies, Shore and Slaman [1998]) *There is a definable map f : $\mathcal{R} \to \mathbf{N}$ such that $(\forall \mathbf{a})[\deg(W^{(2)}_{f(\mathbf{a})}) = \mathbf{a}^{(2)}]$.*

PROOF. (Sketch) To give a first-order definition of f, we have to provide an appropriate definable relation R_f which holds between degrees $\mathbf{a}$ and tuples $(i, \overline{\mathbf{p}})$ representing an equivalence class in $\mathbf{N}$. Note that $\mathbf{a}^{(2)}$ is the least degree $\mathbf{v}$ such that each set in $\Sigma^0_3(A)$ is r.e. in $\mathbf{v}$. (If the last statement holds for $\mathbf{v}$, then $A^{(2)}$ and $\overline{A^{(2)}}$ are r.e. in $\mathbf{v}$.) We argue that $\Sigma^0_3(A) = S(\mathbf{a})$, the class of sets coded in a particular way that depends on $\mathbf{a}$. Thus using a first-order way to obtain, from the degree $\mathbf{a}$, representations of $S(\mathbf{a})$ "inside" $\mathbf{N}$ we can define R_f since finding an index for such a least $\mathbf{v}$ is an arithmetical process. R_f then provides the required map from degrees $\mathbf{a}$ to codes of numbers i in $\mathbf{N}$ such that $\mathbf{a}'' = W''_i$. □

A similar result also holds for $\mathcal{D}$ and is the result of a long line of research.

THEOREM 1.5. (Simpson [1977]; Nerode and Shore [1980,1980a]; Jockusch and Shore [1984]; Slaman and Woodin [2000]; Nies, Shore and Slaman [1998]; Shore and Slaman [2000a]) *Any relation on $\mathcal{D}$ which is invariant under the double jump is definable in $\mathcal{D}$ if and only if it is definable in second order arithmetic.*

PROOF. (Sketch) One can begin by coding arbitrary countable relations by Slaman and Woodin [1986] or directly coding models of arithmetic in lattice initial segments with quantification over sets provided by quantification over ideals via their representation as exact pairs as in Simpson [1977] or Nerode and Shore [1980,1980a]. An analysis of the complexity of sets that can be coded in these models by degrees below a given $\mathbf{x} > \mathbf{0}''$ in terms of the degree of $\mathbf{x}$ itself (or more precisely in terms of $\mathbf{x}''$ or $\mathbf{x}'''$) allows one to define a map between degrees $\mathbf{x}$ and sets of degree $\mathbf{x}''$ coded in such models of arithmetic. Given such a coding, one translates definitions in second order arithmetic into $\mathcal{D}$ by using it together with an interpretation of second order arithmetic in $\mathcal{D}$. □

Whatever further success might be won along these lines they have not provided and will not provide "natural" definitions of degrees or relations on the degrees. This investigation is the provenance of another area of long term interest in the study of $\mathcal{R}$ and $\mathcal{D}$: the relationships between order-theoretic properties of degrees and external properties of other sorts. Of particular interest in $\mathcal{R}$ have been set-theoretic properties described in terms of the lattice of r.e. sets and dynamic properties of the enumerations of the r.e. sets. In both $\mathcal{R}$ and $\mathcal{D}$ we have relations with rates of growth of functions recursive in various degrees and relations with definability considerations in arithmetic as expressed by the jump operator as well as others. In $\mathcal{D}$, a central role has also been played by the jump operator itself, its analogs and their iterations into the transfinite. We will also discuss two notions that play fundamental roles in computability theory but often seem difficult to capture in terms of degree theoretic properties alone: uniformity and recursion.

Of course, we have no formal definition determining which definitions are natural. We can say, with Justice Potter Stewart, that we know the unnatural ones when we see them. The examples given above are prime candidates. One clear offense is that they simply copy the standard definitions in (second order) arithmetic into the degrees by translating all of arithmetic into the language of ($\mathcal{D}$) $\mathcal{R}$. Natural definitions, on the other hand should be more directly expressed in the language of partial orderings and preferably be related to structural or algebraic properties already of interest. The artistic merit of such definitions can also be displayed in the notions and constructions used to establish them. They may also have redeeming social value in that the proofs reveal or exploit specific properties of the degrees in the class being defined, for example, in terms of rates of growth, dynamic properties of enumerations or of the external relation being characterized. We discuss some examples and suggest questions for further research. As might be expected, borderline cases will arise. In the end, the naturalness of proposed definitions along with the beauty of the constructions and ideas needed will be judged by the community of readers and researchers in the field.

2. The recursively enumerable degrees

The jump operator itself is, of course, not defined in $\mathcal{R}$ but there are many important results connecting the jump classes of r.e. degrees with the lattice theoretic structure of the r.e. sets; with approximation procedures for functions recursive in different jumps of the given set; and with the growth rates of functions recursive in the sets themselves for several of these jump classes.

DEFINITION 2.1. *An r.e. degree* $\mathbf{a}$ *is* high$_n$ *iff* $\mathbf{a}^{(n)} = \mathbf{0}^{(n+1)}$ *(its* n^{th} *jump is as high as possible). The degree* $\mathbf{a}$ *is* low$_n$ *if* $\mathbf{a}^{(n)} = \mathbf{0}^{(n)}$ *(its* n^{th} *jump is as low as possible). If* $n = 1$, *we usually omit the subscript.*

EXAMPLE 2.2. An r.e. degree $\mathbf{a}$ is high iff it contains a maximal set in $\mathcal{E}^*$ (the lattice of r.e. sets modulo the ideal of finite sets) iff there is a function f of degree a which dominates every recursive function iff every function h recursive in $0''$ is approximable by one $g \leq \mathbf{a}$ in the sense that $h(x) = \lim_s g(x, s)$.

EXAMPLE 2.3. An r.e. degree $\mathbf{a}$ is low$_2$ iff every r.e. set of degree $\mathbf{a}$ has a maximal superset iff there is a function f recursive in $0'$ which dominates every $g \leq \mathbf{a}$ iff every function $h \leq \mathbf{a}''$ is approximable by a function recursive in g in the sense that $h(x) = \lim_s \lim_t g(x, s, t)$.

(Proofs of these facts about high and low_2 r.e. sets and other similar ones can be found in Soare [1987, XI].)

These interrelations have played an important role in the study of both the lattice of r.e. sets (see Soare [1999]) and the degrees below $0'$ (see Cooper [1999]). The results provide definitions in $\mathcal{R}'$ ($\mathcal{R}$ with an added predicate for two degrees having the same jump) but no natural degree theoretic definitions within $\mathcal{R}$ itself.

QUESTION 2.4. *Are any of the jump classes $\mathbf{H}_n, \mathbf{L}_{n+1}$ naturally definable in $\mathcal{R}$? In particular, what about $\mathbf{L}_2$ and $\mathbf{H}_1$ where we have the strongest connections with rates of growth and significant techniques already developed to exploit the known characterizations of these classes?*

If we omit the requirement of naturalness then this question was raised in Soare [1987] and answered affirmatively for $n \geq 1$ by Nies, Shore and Slaman [1998] with Theorem 1.2 and Proposition 4.3. The first natural, truly internal definitions in $\mathcal{R}$ of apparently external properties of r.e. degrees arose from the study of Maass' [1982] notion of prompt simplicity.

DEFINITION 2.5. *A coinfinite r.e. set A is* promptly simple *if there is there is a nondecreasing recursive function p and a recursive one-to-one function f enumerating A (i.e. $A = \operatorname{rg} f$) such that for every infinite r.e. W_e there is an s and an x such that x is enumerated in W at stage s (in some standard uniform enumeration of all the r.e. sets) and is also enumerated in A by stage $p(s)$, i.e. $x = f(n)$ for some $n \leq p(s)$. An r.e. degree* **a** *is* promptly simple *if it contains a promptly simple r.e. set. We let* **PS** *denote the set of promptly simple r.e. degrees.*

DEFINITION 2.6. *i)* $\mathbf{M} = \{\mathbf{a} | \exists \mathbf{b}(\mathbf{a} \wedge \mathbf{b} = \mathbf{0})\}$ *is the set of* cappable *r.e. degrees, i.e. those which are halves of minimal pairs.* **NC**, *the set of* noncappable *r.e. degrees, is its complement in $\mathcal{R}$.*
ii) $\mathbf{LC} = \{\mathbf{a} | \exists \mathbf{b}(\mathbf{a} \vee \mathbf{b} = \mathbf{0}' \,\&\, \mathbf{b}' = \mathbf{0}')\}$ *is the set of* low cuppable degrees, *i.e. those which can be cupped (joined) to $\mathbf{0}'$ by a low degree* **b**.
iii) $\mathbf{SP\bar{H}}$ *is the set of r.e. sets definable in $\mathcal{E}$, the lattice of r.e. sets, as the non-hyperhypersimple r.e. sets A with the* splitting property, *i.e. for every r.e. set B there are r.e. sets B_0, B_1 such that $B_0 \cup B_1 = B$; $B_0 \cap B_1 = \emptyset$; $B_0 \subseteq A$; and if W is r.e. but $W - B$ is not, then $W - B_0$ and $W - B_1$ are also not r.e.*

THEOREM 2.7. (Ambos-Spies et al. [1984]) *The four classes* $\mathbf{PS}, \mathbf{NC}, \mathbf{LC}$ *and* $\mathbf{SP\bar{H}}$ *all coincide and together with their complement* **M** *partition $\mathcal{R}$ as follows:*

i) **M** *is a proper ideal in $\mathcal{R}$, i.e. it is closed downward in $\mathcal{R}$ and if* $\mathbf{a}, \mathbf{b} \in \mathbf{M}$ *then* $\mathbf{a} \vee \mathbf{b} \in \mathbf{M}$ *and* $\mathbf{a} \vee \mathbf{b} < \mathbf{0}'$.

ii) **NC** *is a strong filter in $\mathcal{R}$, i.e. it is closed upward in $\mathcal{R}$ and if* $\mathbf{a}, \mathbf{b} \in \mathbf{NC}$ *then there is a* $\mathbf{c} \in \mathbf{NC}$ *with* $\mathbf{c} \leq \mathbf{a}, \mathbf{b}$.

We would also like to point out two hidden uniformities in these results. Their proofs provide a recursive function f such that if $\deg(W_e) \in \mathbf{M}$ then W_e and $W_{f(e)}$ form a minimal pair. They also shows that $\mathbf{NC} = \mathbf{ENC}$, the effectively noncappable degrees, i.e. those **a** such that there is an r.e. $A \in \mathbf{a}$ and a recursive function f such that for all e, $W_e \leq_T A$ uniformly in e; $W_{f(e)} \leq_T W_e$ uniformly in e; and if W_e is not recursive then $W_{f(e)}$ is not recursive either.

QUESTION 2.8. *As Slaman has pointed out, the noncuppable degrees (those degrees* **a** *for which there is no* $\mathbf{b} < \mathbf{0}'$ *such that* $\mathbf{a} \vee \mathbf{b} = \mathbf{0}'$*) trivially also form an ideal in $\mathcal{R}$. Are there any other (naturally) definable ideals or filters in $\mathcal{R}$?*

QUESTION 2.9. *If $\mathcal{B}$ is a (particular) definable subset of $\mathcal{R}$ is there a way to define the ideal generated by $\mathcal{B}$?*

The only other known example of a natural definition in $\mathcal{R}$ is that of contiguity, a notion relating Turing reducibility and weak truth-table reducibility, $\leq_{wtt}$. As wtt-reducibility is the restriction of Turing reducibility to operators with a recursive bound on their use functions, this characterization provides an example of defining a notion apparently dependent on external computational procedures solely in terms of Truing reducibility. Indeed, in this case we have two equivalent definitions in $\mathcal{R}$ with the second being a later refinement of the first.

DEFINITION 2.10. *An r.e. degree* $\mathbf{a}$ *is* contiguous *if and only if for every r.e. $A, B \in \mathbf{a}$, $A \equiv_{wtt} B$. It is* strongly contiguous *if and only if for every $A, B \in \mathbf{a}$, $A \equiv_{wtt} B$ (i.e.* $\mathbf{a}$ *consists of a single wtt-degree).*

THEOREM 2.11. (Downey and Lempp [1997]) *An r.e. degree* $\mathbf{a}$ *is contiguous if and only if it is strongly contiguous if and only if it is* locally distributive, *i.e. if* $\mathbf{b} \vee \mathbf{c} = \mathbf{a} > \mathbf{d}$ *then* $\exists \mathbf{b}_0 \leq \mathbf{b} \exists \mathbf{c}_0 \leq \mathbf{c}(\mathbf{b}_0 \vee \mathbf{c}_0 = \mathbf{d})$.

THEOREM 2.12. (Ambos-Spies and Fejer [2000]) *An r.e. degree* $\mathbf{a}$ *is contiguous if and only if it is not the top of an embedding of the pentagon into $\mathcal{R}$ (if and only if there is an r.e. $A \in \mathbf{a}$ with the strong universal splitting property).*

An interesting property that seems to capture the ability to do multiple permitting in a way intermediate between the permitting arguments that just use nonrecursiveness and those that use nonlow$_2$-ness is that of being array nonrecursive.

DEFINITION 2.13. (Downey, Jockusch and Stob [1990]) *An r.e. degree* $\mathbf{a}$ *is* array nonrecursive *iff there is an r.e. $A \in \mathbf{a}$ and a very strong array F_n such that $\forall e \exists n (W_e \cap F_n = A \cap F_n)$ or equivalently there are disjoint r.e. $B, C \leq_T A$ such that $B \cup C$ is coinfinite and no $D \in \mathbf{0}'$ separates B and C. (The sequence F_n is a* very strong array *if there is a recursive function f such that $f(n)$ is the canonical index for F_n, $\bigcup F_n = \mathbb{N}$, the F_n are pairwise disjoint and of strictly increasing cardinality.)*

QUESTION 2.14. (Walk): *Are the array recursive degrees (naturally) definable in $\mathcal{R}$?*

In fact, Walk had suggested that the array nonrecursive degrees might be those bounded by a contiguous degree. This turns out to be false but the work needed to prove this provides a definable automorphism base for $\mathcal{R}$, i.e. a set $\mathbf{MC}$ of degrees such that any automorphism of $\mathcal{R}$ which is fixed on $\mathbf{MC}$ is the identity on $\mathcal{R}$.

THEOREM 2.15. (Cholak, Downey and Walk [2000]) *There is a maximal contiguous degree and indeed the set,* $\mathbf{MC}$, *of maximal contiguous degrees forms an automorphism base for $\mathcal{R}$.*

We close this section with a question about relativizations. It was a long standing question (originally asked in Sacks [1963a]) as to whether $\mathcal{R}^{\mathbf{a}}$, the degrees r.e. in and above $\mathbf{a}$, are isomorphic (or elementary equivalent to) $\mathcal{R}^{\mathbf{b}}$, those r.e. in and above $\mathbf{b}$, for arbitrary $\mathbf{a}$ and $\mathbf{b}$. The question about isomorphisms was answered negatively in Shore [1982] and for elementary equivalence in Nies, Shore and Slaman [1998]. The first result distinguishes them on the basis of which infinite lattices (of degrees related to **a** and **b**) can be embedded and the second on the basis of which sets (again of degrees related to **a** and **b**) can be definably coded in models of arithmetic represented in the structures. Of course, neither of these approaches provides a natural difference between the structures.

QUESTION 2.16. (Simpson) *Is there a natural sentence ϕ of degree theory that is true in some $\mathcal{R}^{\mathbf{a}}$ but not in all $\mathcal{R}^{\mathbf{b}}$? If so, for which* **a** *and* **b** *can one find such sentences?*

3. The Turing degrees

We begin our discussion of natural definability in $\mathcal{D}$ with the first example of a natural definition of an interesting, seemingly external, class of degrees, those of the arithmetic sets.

DEFINITION 3.1. *A degree* **a** *is a* minimal cover *if there is a* **b** *such that* $\mathbf{a} > \mathbf{b}$ *and there is no* **c** *strictly between* **a** *and* **b**. *If* **b** *is such a degree for* **a** *we say that* **a** *is a* minimal cover of **b**.

DEFINITION 3.2. $\mathcal{A} = \{\mathbf{d}|\exists n(\mathbf{d} \leq \mathbf{0}^{(n)})\}$. $\mathcal{C}_\omega = \{\mathbf{c}|\forall \mathbf{z}(\mathbf{z} \vee \mathbf{c}$ *is not a minimal cover of* $\mathbf{z}\}$. $\overline{\mathcal{C}}_\omega = \{\mathbf{d}|\exists \mathbf{c} \in \mathcal{C}_\omega(\mathbf{d} \leq \mathbf{c})\}$.

THEOREM 3.3. (Jockusch and Shore [1984]) $\mathcal{A} = \overline{\mathcal{C}}_\omega$ *and the relation* **a** *is arithmetic in* **b** *is naturally definable in* $\mathcal{D}$ *(by relativization).*

This result was proved by combining a completeness and join theorem for certain operators with known structural results for $\mathcal{D}$ and the r.e. degrees. We begin with the definitions of the operators studied. (We restrict ourselves here to the cases that $\alpha \leq \omega$ but both of the following definitions have been usefully generalized into the transfinite and the corresponding theorems proven.)

DEFINITION 3.4. *The* $1 - REA$ operators J *(from* $2^{\mathbb{N}}$ *to* $2^{\mathbb{N}}$*) are those of the form* $J(A) = J_e(A) = A \oplus W_e^A$. *The* $n - REA$ operators J *are those of the form* $J_{\langle e_1,\ldots,e_n\rangle} = J_{e_n} \circ J_{e_{n-1}} \circ \ldots \circ J_{e_1}$. *The* $\omega - REA$ operators J *are those of the form* $J(A) = \bigoplus\{J_{f\restriction n}(A)|n \in \omega\}$ *for some recursive* f.

DEFINITION 3.5. *The* $n -$ r.e. operators J *are those of the form* $J(A)(x) = \lim \phi_e^A(x,s)$ *for a (total recursive in* A*) function* ϕ_e^A *which has* $\phi_e^A(x,0) = 0$ *for all* x *and for which there are at most* n *many* s *such that* $\phi_e^A(x,s) \neq \phi_e^A(x,s+1)$. *The* $\omega -$ r.e. operators J *are those of the form* $J(A)(x) = \lim \phi_e^A(x,s)$ *for a (total recursive in* A*) function* ϕ_e^A *which has* $\phi_e^A(x,0) = 0$ *for all* x *and for which there are at most* $f(x)$ *many* s *such that* $\phi_e^A(x,s) \neq \phi_e^A(x,s+1)$ *for some recursive function* f.

Now for the completeness theorem generalizing those of Friedberg [1957] and MacIntyre [1977].

THEOREM 3.6. (Jockusch and Shore [1984]) *For any* $\alpha - REA$ *operator* J *and any* $C \geq_T 0^{(\alpha)}$ *there is an* A *such that* $J(A) \equiv_T C$.

Applying this theorem to specific operators provides suggestions for producing interesting definable subsets of $\mathcal{D}$. We consider the minimal degree construction but clearly others are possible.

THEOREM 3.7. (Sacks [1963]) *There is an* $\omega - r.e.$ *operator* J *such that* $J(A)$ *is a minimal cover of* A *for every* A.

COROLLARY 3.8. $\mathbf{0}^{(\omega)}$ *is the base of a cone of minimal covers, i.e. every* $\mathbf{a} \geq \mathbf{0}^{(\omega)}$ *is a minimal cover.*

QUESTION 3.9. *Is* $\mathbf{0}^{(\omega)}$ *the least degree which is the base of a cone of minimal covers?*

We conjecture that the answer to this question is no and suggest that one build a tree T such that every path through T is a minimal cover by virtue of being $J(A)$ for some A while simultaneously making the degree of T incomparable with $\mathbf{0}^{(\omega)}$. If this is possible, we ask instead what can one say about the degrees which are bases of cones of minimal covers?

An improvement to the completeness theorem that includes a join operation provides an approach to our first natural definability result for $\mathcal{D}$.

THEOREM 3.10. (essentially as in Jockusch and Shore [1984]) *For any* $\alpha-r.e.$ *operator* J $(1 \leq \alpha \leq \omega)$ *and any* $D \not\leq_T 0^{(\beta)}$ *for every* $\beta < \alpha$ *there is an* A *such that* $J(A) \equiv_T D \vee 0^{(\alpha)} \equiv_T A^{(\alpha)}$.

Thus for every nonarithmetic degree $\mathbf{x}$ there is a $\mathbf{z}$ such that $\mathbf{x} \vee \mathbf{z}$ is a minimal cover of $\mathbf{z}$. On the other hand, $\mathbf{0}^{(n)} \vee \mathbf{z}$ is not a minimal cover of $\mathbf{z}$ for any $\mathbf{z}$ by Jockusch and Soare [1970]. This establishes Theorem 3.3.

Cooper [1990, 1993, 1994 and elsewhere] suggested a similar approach to the problem of defining the jump operator. His plan was to use Theorem 3.10 to define $\mathbf{0}'$ by finding a suitable $2-r.e.$ operator (rather than an $\omega-r.e.$ one as above) that would produce a degree with an order-theoretic property that no r.e. degree could have (again even relative to any degree below it). He defined the following notions and classes.

DEFINITION 3.11. $\mathbf{d}$ *is* splittable over $\mathbf{a}$ avoiding $\mathbf{b}$ *if either* $\mathbf{a}, \mathbf{b} \not\leq \mathbf{d}$ *or* $\mathbf{b} \leq \mathbf{a}$ *or there are* $\mathbf{d}_0, \mathbf{d}_1$ *such that* $\mathbf{a} < \mathbf{d}_0, \mathbf{d}_1 < \mathbf{d}$, $\mathbf{d}_0 \vee \mathbf{d}_1 = \mathbf{d}$ *and* $\mathbf{b} \not\leq \mathbf{d}_0, \mathbf{d}_1$. $\mathcal{C}_1 = \{\mathbf{c} | \forall \mathbf{a}, \mathbf{b}(\mathbf{a} \vee \mathbf{c}$ *is splittable over* $\mathbf{a}$ *avoiding* $\mathbf{b}\}$. $\overline{\mathcal{C}}_1 = \{\mathbf{d} | \exists \mathbf{c} \in \mathcal{C}_1(\mathbf{d} \leq \mathbf{c})\}$.

Now, one of the needed results was already well known.

THEOREM 3.12. (Sacks [1963]) *Every r.e. degree* $\mathbf{d}$ *is in* $\mathcal{C}_1$.

For the other direction Cooper [1990, 1993] claimed as his main theorem that there is a suitable $2-r.e.$ set and so $2-r.e.$ operator J such that for every C there are $\mathbf{a}$ and $\mathbf{b}$ such that $\mathbf{d} \equiv_T \deg(J(C))$ is not splittable over $\mathbf{a}$ avoiding $\mathbf{b}$.

Such a result would provide a natural definition of $\mathbf{0}'$ as the maximum degree in $\overline{\mathcal{C}}_1$ and, by relativization, a natural definition of the jump operator. Unfortunately, there is no such $2-r.e.$ operator. Nor, indeed, any $n-REA$ one.

THEOREM 3.13. (Shore and Slaman [2000a]) *If* $\mathbf{a}, \mathbf{b} \leq_{\mathbf{T}} \mathbf{d}$, $\mathbf{b} \not\leq_{\mathbf{T}} \mathbf{a}$ *and* $\mathbf{d}$ *is* $n-REA$ *in* $\mathbf{a}$, *then* $\mathbf{d}$ *can be split over* $\mathbf{a}$ *avoiding* $\mathbf{b}$.

A number of other specific suggestions of natural properties that might define $\mathbf{0}'$ have been made and most have been refuted. Posner [1980] conjectured that $\mathbf{0}'$ is the least degree $\mathbf{x}$ such that every $\mathbf{d} \geq \mathbf{x}$ is the join of two minimal degrees or [1981] such that $\mathcal{D}(\leq \mathbf{d})$ is complemented for every $\mathbf{d} \geq \mathbf{x}$. One can refute the first conjecture by constructing a function $f : \omega \to \{0, 1, 2\}$ such that $f|_T 0'$ and $(\forall g : \omega \to \{0, 1\})(\forall x[f(x) \in \{0, 1\} \to f(x) = g(x)] \to \deg(g)$ is minimal). Given any $h \geq_T f$ one defines g_0, g_1 as in the construction of f such that $g(a_i) = h(i)$ and $g_1(a_i) = 1 - h(i)$ where $\langle a_i : i \in \omega \rangle$ lists the numbers x such that $f(x) = 2$. Thus

$h \equiv_T g_1 \oplus g_2$. The construction of f follows Lachlan [1971] with added steps to make f incomparable with $0'$. Slaman and Steel [1989] ask if $\mathbf{0}'$ is the least $\mathbf{x}$ such that for every $\mathbf{d} \geq \mathbf{x}$ there is an $\mathbf{a}$ such that every nonzero $\mathbf{b} \leq \mathbf{a}$ is a complement for every $\mathbf{c} \geq \mathbf{x}$ in $\mathcal{D}(\leq \mathbf{d})$ but now expect that this proposal will also fail.

At the Boulder meeting we proposed that direct natural definitions of the $\mathbf{0}^{(n)}$ might be proved by strengthening Theorem 3.10 to handle all $n - REA$ operators and then finding an appropriate property of some $n - REA$ degrees along the lines suggested by Cooper for $2 - r.e.$ Thus one should first answer the following question:

QUESTION 3.14. (Jockusch and Shore [1984]) *For each $\alpha > 1$, is there, for each $\alpha - REA$ operator J ($\alpha > 1$) and each D such that $D \not\leq_T \mathbf{0}^{(\beta)}$ for every $\beta < \alpha$, an A such that $J(A) \equiv_T D \vee A \equiv_T D \vee 0^{(\alpha)}$.*

For $\alpha = 1$, the result follows from Posner and Robinson [1981] for which a new proof is supplied in Jockusch and Shore [1984] as the paradigm for that of Theorem 3.10. The case of $\alpha = \omega$ is singled out in Jockusch and Shore [1985] because of the role it played in the proof of Theorem 3.3. This case was solved by Kumabe and Slaman (personal communication). During the Boulder conference, Shore and Slaman discussed refining the construction to apply to $\alpha < \omega$ (a more delicate question as it turns out). They have now answered that question.

THEOREM 3.15. (Shore and Slaman [2000]) *For each $(n+1) - REA$ operator J $(n \geq 0)$ and $D \not\leq_T 0^{(n)}$ there is an A such that $J(A) \equiv_T D \vee A = D \vee 0^{(n+1)}$.*

Thus, if one can find a (natural) property of some $(n+1) - REA$ degree that is not enjoyed by any $n - REA$ degree and both of these facts relativize, then one would have a (natural) direct definition of $\mathbf{0}^{(n)}$. Indeed, it was in a attempt to produce such a definition of $\mathbf{0}'$ by trying to construct a $2 - REA$ degree $\mathbf{d}$ and $\mathbf{a}, \mathbf{b} \leq \mathbf{d}$ such that $\mathbf{d}$ is not splittable over $\mathbf{a}$ avoiding $\mathbf{b}$ that Shore and Slaman were led to Theorem 3.13. They then realized that there was a property that Slaman and Woodin had shown to be definable that would play the required role, that of being a double jump.

THEOREM 3.16. (Slaman and Woodin [2000]) $\mathbf{0}''$ *and, indeed, the operator taking* $\mathbf{a}$ *to* $\mathbf{a}''$ *is definable in* $\mathcal{D}$.

Slaman and Woodin had discovered this result in 1990 but never published or even publicized it as Cooper had already announced that $\mathbf{0}'$ and the jump operator itself were definable in $\mathcal{D}$. It does, however, supply the missing ingredient for a definition of the jump.

THEOREM 3.17. (Shore and Slaman [2000]) $\mathbf{0}'$ *and the jump operator are definable in* $\mathcal{D}$.

PROOF. The claim is that $\mathbf{0}'$ is the greatest degree $\mathbf{x}$ such that there is no $\mathbf{b}$ for which $\mathbf{b}'' = \mathbf{x} \vee \mathbf{b}$. In one direction, it is clear that if $\mathbf{x} \leq \mathbf{0}'$ then $\mathbf{x} \vee \mathbf{b} \leq \mathbf{b}' < \mathbf{b}''$ for every $\mathbf{b}$ and so $\mathbf{x}$ has the desired property. For the other direction, if $\mathbf{x} \not\leq \mathbf{0}'$ then by Theorem 3.10 there is a $\mathbf{b}$ such that $\mathbf{b}'' = \mathbf{x} \vee \mathbf{b}$. Finally, $\mathbf{b}''$ is definable (from $\mathbf{b}$) by Theorem 3.16 while, of course, the join is definable from $\leq_T$. This then defines the degree $\mathbf{0}'$. The jump operator (applied to any $\mathbf{z}$) is defined by relativizing the theorems and definitions to $\mathbf{z}$. □

Interestingly, the proof of Theorem 3.16 proceeds by applying metamathematical arguments involving forcing and absoluteness. Concomitantly, the definition of $\mathbf{0}''$ that these arguments provide involves an explicit translation of isomorphism facts to definability facts via a coding of (second order) arithmetic. Thus, unfortunately, we must say that the definition provided is not natural. We do, however, expect that a more directly expressible appropriate property of $2-REA$ degrees will be found that will supply the desired natural definition. Indeed, Cooper has announced that a modification of his original $2-r.e.$ operator does provide such a definition of the jump and of the relation "**a** is r.e. in **b**". In any case, we would like to see such direct definitions for all the $\mathbf{0}^{(n)}$.

QUESTION 3.18. *For each n what properties are enjoyed by some $(n+1)-REA$ or even $(n+1)-r.e.$ degree which are not enjoyed by any $n-REA$ degree? (Note that by Jockusch and Shore [1984] every $n-r.e.$ degree is $n-REA$.)*

This problem suggests a general area of investigation.

QUESTION 3.19. *What special properties do the $n-REA$ or $n-r.e.$ degrees have within the whole structure $\mathcal{D}$?*

QUESTION 3.20. *Is the class of $n-REA$ degrees and the relation "**a** is $n-REA$ in **b**" (naturally) definable for each n?*

Now if one has a natural definition of the relation "**a** is r.e. in **b**" then, of course, one can define these relations by iterating the assumed definition for $n=1$. It then becomes debatable at what point (if any) such iterations cease to provide natural definitions. In any case, it would be of interest to provide direct definitions for each n that directly exploit some aspect of being $n-REA$.

The next question might well be about the union of these classes: $\mathbf{REA}^{<\omega} = \{\mathbf{a} | \exists n \in \omega (\mathbf{a} \text{ is } n-REA)\}$. It is not, perhaps, immediately clear, even assuming the definability of "**a** is r.e. in **b**", that $\mathbf{REA}^{<\omega}$ is definable in $\mathcal{D}$ let alone naturally. How do we quantify over $n \in \omega$? Well, it would clearly suffice to quantify over finite sequences (assuming the definability of "**a** is r.e. in **b**"). This we can certainly do by Slaman-Woodin [1986] coding. Should this count as a natural definition? It is phrased in order-theoretic terms but is perilously close to defining arithmetic. Perhaps some more traditional methods can be employed. A similar question arises for defining the set $\{\mathbf{0}^{(n)} | n \in \omega\}$ from a definition of the jump operator. In this case Simpson has supplied an answer.

THEOREM 3.21. (Simpson [1977]) *If $\langle \mathbf{b}_n | n \in \omega \rangle$ is a sequence of degrees such that $\mathbf{0}'' \leq \mathbf{b}_1 \leq \ldots \leq \mathbf{b}_n \leq \ldots$, then there an initial segment $0 < \mathbf{a}_1 < \ldots < \mathbf{a}_n < \ldots$ of $\mathcal{D}$ such that $\mathbf{a}_i'' = \mathbf{a}_i \vee \mathbf{0}'' = \mathbf{b}_i$ for $i \geq 1$.*

COROLLARY 3.22. *$\{\mathbf{0}^{(n)} | n \in \omega\} = \{\mathbf{d} | \mathbf{d} = \mathbf{0} \vee \mathbf{d} = \mathbf{0}' \vee$ there is a finite initial segment of the degrees with top degree **a** which is linearly ordered and such that the double jump of each non zero element is the jump its immediate predecessor and **d** is the double jump of some $\mathbf{c} < \mathbf{a}\}$.*

The only part of this definition that is not obviously in the language of $\mathcal{D}$ with jump is the assertion that the linearly ordered initial segment with top **a** is *finite*. The crucial ingredient here is Spector's exact pair theorem that we shall see mercilessly exploited below (as in Jockusch and Simpson [1976]).

DEFINITION 3.23. *An ideal of $\mathcal{D}$ is a subset I of $\mathcal{D}$ which is closed downward and under join. It is a* jump ideal *if it is also closed under jump. We say that a pair $\mathbf{x}, \mathbf{y}$ of degrees is an* exact pair *for an ideal I if $I = \{\mathbf{z} | \mathbf{z} \leq \mathbf{x}$ and $\mathbf{z} \leq \mathbf{y}\}$.*

THEOREM 3.24. (Spector [1956]) *Every countable ideal I of $\mathcal{D}$ has an exact pair.*

Using this theorem one can refer to countable ideals or quantify over them in a first order way by using the corresponding exact pairs. Given this it is easy to say that a linear ordering is finite by saying that every countable initial segment has a greatest element.

A similar line of argument provides a natural definition of the intermediate r.e. degrees (from the relation of being r.e. in).

COROLLARY 3.25. *The set of intermediate r.e. degrees, i.e. $\{\mathbf{x}|\mathbf{x}$ is r.e. and $\forall n(\mathbf{0}^{(n)} < \mathbf{x}^{(n)} < \mathbf{0}^{(n+1)})\}$, is definable as $\{\mathbf{x}|\mathbf{x}$ is r.e. and for every finite initial segment of the degrees with top degree $\mathbf{a}$ which is linearly ordered and such that the double jump of each nonzero element is the jump of its immediate predecessor there is another finite initial segment of the degrees such that the double jump of the first nonzero element is $\mathbf{x}''$ and the double jump of each later element is the jump of its immediate predecessor and the double jumps of the second sequence are interleaved with those of the first$\}$.*

Can we get a similar natural definition for $\mathbf{REA}^{<\omega}$? One route is suggested by the following question.

QUESTION 3.26. *If $\mathbf{a}$ is $n-REA$ as witnessed by the sequence $\mathbf{a}_1 < \mathbf{a}_2 < \ldots < \mathbf{a}_n = \mathbf{a}$, is there an initial segment $\mathbf{0} < \mathbf{b}_1 < \ldots < \mathbf{b}_n$ and a $\mathbf{c}$ such that $\mathbf{a}_i = \mathbf{b}_i \vee \mathbf{c}$?*

This analysis suggests a general area for investigation that asks to what extent one can control the jump or double jump of all degrees in some initial segment of $\mathcal{D}$.

QUESTION 3.27. *What is the range of the jump or double jump operator on initial segments, i.e. for which sets C of degrees above $\mathbf{0}'(\mathbf{0}'')$ can we find an initial segment I of $\mathcal{D}$ such that the range of the jump operator on I and/or $I \vee \mathbf{0}'$ $(I \vee \mathbf{0}'')$ is C? Or prove that some interesting classes of degrees are so representable.*

QUESTION 3.28. *Which sets of degrees are of the form $I \vee \mathbf{c}$ for an initial segment I of $\mathcal{D}$ and a degree $\mathbf{c}$? Or prove that some interesting classes of degrees are so representable.*

Actually, in the case of $n-REA$ we might be able to do even better and supply a natural definition that does not appear to mention recursion even indirectly.

PROPOSITION 3.29. (Jockusch and Shore [1984]) *$\mathcal{C}_\omega$ is closed under the relation $n-REA$ in, i.e. if $\mathbf{a} \in \mathcal{C}_\omega$ and $\mathbf{b}$ is $n-REA$ in $\mathbf{a}$ then $\mathbf{b} \in \mathcal{C}_\omega$.*

QUESTION 3.30. *Does $\mathcal{C}_\omega = \mathbf{REA}^{<\omega}$?*

The questions presented here about $\{\mathbf{0}^{(n)}|n \in \omega\}$ and $\mathbf{REA}^{<\omega}$ suggest another array of related issue involving uniformity and upper bounds as represented by the problems of defining $\mathbf{0}^{(\omega)}$ and the ω-REA degrees.

DEFINITION 3.31. *If S is a set of functions then a degree $\mathbf{a}$ is a* uniform upper bound (uub) *for S if there is a function $f \leq_T \mathbf{a}$ such that $\{f^{[i]}|i \in \omega\} = S$ where $f^{[i]}(x) = f(\langle i, x\rangle)$. It is a* subuniform upper bound (suub) *for S if there is a function $f \leq_T \mathbf{a}$ such that $\{f^{[i]}|i \in \omega\} \supseteq S$.*

The classic example here is $0^{(\omega)}$ which is a uub for the $0^{(n)}$. There is a long history of attempts to characterize or define in $\mathcal{D}$ (or $\mathcal{D}$ with the jump operator) the degrees of uub's, suub's and other types of upper bounds for the recursive and arithmetic sets and functions as well as other (jump) ideals of $\mathcal{D}$. We mention a couple of results and questions.

THEOREM 3.32. (Jockusch [1972]) *A degree* $\mathbf{a}$ *is the degree of a (sub)uniform upper bound of the recursive functions iff it is a uub for the recursive sets iff* $\mathbf{a}' \geq \mathbf{0}''$.

THEOREM 3.33. (Jockusch [1972]; Jockusch and Simpson [1976]; Lachlan and Soare [1994]; Lerman[1985]) *A degree* $\mathbf{a}$ *is the degree of a uniform upper bound of the arithmetic functions iff there is an* $f \in \mathbf{a}$ *which dominates every partial (total) arithmetic function iff there is a* $\mathbf{d}$ *which is an upper bound for the arithmetic degrees and* $\mathbf{a} = \mathbf{d}'$ *iff there is a* $\mathbf{d}$ *which is an upper bound for the arithmetic degrees and* $\mathbf{a}' = \mathbf{d}''$.

QUESTION 3.34. *Are there natural definitions of the property of being the degree of a uniform upper bound for other specific (jump) ideals, of being the degree of a subuniform upperbound for* $\mathcal{A}$ *or other (jump) ideals?*

The hierarchy generated by iterating the jump into the transfinite and taking some kind of uniform upper bound at limit levels has been an object of intense study since Kleene and Spector. Major contributions were also made by Putnam and his students culminating in Hodes [1982]. Jockusch and Simpson [1976] is a remarkable collection of natural definitions (using the jump operator) of many specific natural upper bounds for jump ideals as well as the ideals themselves. They also tie quantification over the functions in such ideals to standard subsystems of second order arithmetic in a definable way. We mention a few of the results.

THEOREM 3.35. (Sacks [1971]; Enderton and Putnam [1970]) $\mathbf{0}^{(\omega)}$ *is the largest degree below* $(\mathbf{a} \vee \mathbf{b})^{(2)}$ *for every exact pair* $\mathbf{a}, \mathbf{b}$ *for* $\mathcal{A}$. *It is also the least degree of the form* $(\mathbf{a} \vee \mathbf{b})^{(2)}$ *for any exact pair* $\mathbf{a}, \mathbf{b}$ *for* $\mathcal{A}$.

Of course, given the jump operator we may also say that an ideal (as given by an exact pair) is a jump ideal. The following theorem is the key to defining many interesting properties of jump ideals in $\mathcal{D}$.

THEOREM 3.36. (Jockusch and Simpson [1976]) X *is* Δ^1_n *over the set* M_I *of functions with degrees in a countable jump ideal* I *if and only if* $\mathbf{h} \leq (\mathbf{a} \vee \mathbf{b})^{(n+1)}$ *for every exact pair* $\mathbf{a}, \mathbf{b}$ *for* I. X *is analytical (i.e.* Σ^1_k *for some* k*) over this set of functions if and only if* $\mathbf{h}$ *is arithmetic in* $\mathbf{a} \vee \mathbf{b}$ *for every exact pair* $\mathbf{a}, \mathbf{b}$ *for* I.

COROLLARY 3.37. (Jockusch and Simpson [1976]) *The property that* M_I *is (a collection of functions corresponding to) an* ω*-model of* Δ^1_n*-Comprehension (*Δ^1_n*-CA) is naturally definable in* $\mathcal{D}$ *by "every* $\mathbf{d}$ *below* $(\mathbf{a} \vee \mathbf{b})^{(n+1)}$ *for every exact pair for* I *is itself in* I*".*

COROLLARY 3.38. (Jockusch and Simpson [1976]) *The jump ideals* $\mathcal{A}$ *and* $\mathcal{H}$ *(the hyperarithmetic degrees) and the relations "*$\mathbf{a}$ *is arithmetic in* $\mathbf{b}$*" and "*$\mathbf{a}$ *is hyperarithmetic (i.e.* Δ^1_1*) in* $\mathbf{b}$*" are naturally definable in* $\mathcal{D}$ *with the jump operator. (For example,* $\mathbf{a}$ *is arithmetic in* $\mathbf{b}$ *iff it is in every countable jump ideal containing* $\mathbf{b}$ *while it is hyperarithmetic in* $\mathbf{b}$ *iff it is every countable jump ideal containing* $\mathbf{b}$ *which is a model of* Δ^1_1*-CA.)*

We have already seen (Theorem 3.3) a definition of $\mathcal{A}$ in $\mathcal{D}$ that does not explicitly refer to the jump operator. It can also be used to produce an alternative definition of $\mathcal{H}$ as well.

PROPOSITION 3.39. (Jockusch and Shore [1984]) $\mathbf{a}$ *is hyperarithmetic in* $\mathbf{b}$ *iff it is in every countable ideal* I *containing* $\mathbf{b}$ *such that every degree* $\mathbf{d}$ *which is arithmetic in every upper bound for* I *is actually in* I.

The interest in alternate definitions of concepts is, we would say, a positive indicator of the naturalness of the definitions. Providing new coding methods for producing definitions can be of technical interest but rarely sheds any additional light on the classes or degrees being defined. In contrast, new natural definitions typically exploit different properties of $\mathcal{D}$ or of the objects being defined and so provide additional insights. Jockusch and Simpson provide an extensive array of definitions for one especially interesting degree.

THEOREM 3.40. (Jockusch and Simpson [1976]) *The degree of Kleene's* $\mathcal{O}$ *(the complete* Π^1_1 *set) is the largest degree which is below* $(\mathbf{a} \vee \mathbf{b})^{(3)}$ *whenever* $\mathbf{a}, \mathbf{b}$ *is an exact pair for* $\mathcal{H}$. *It is also the largest degree which is below* $\mathbf{a}^{(3)}$ *for every minimal upper bound* $\mathbf{a}$ *for* $\mathcal{H}$; *the smallest degree of the form* $(\mathbf{a} \vee \mathbf{b})^{(3)}$ *for some exact pair* $\mathbf{a}, \mathbf{b}$ *for* $\mathcal{H}$ *or of the form* $\mathbf{a}^{(3)}$ *for a minimal upper bound* $\mathbf{a}$ *for* $\mathcal{H}$. *By relativization, the hyperjump is naturally definable in* $\mathcal{D}$ *with the jump operator.*

The proofs here use, for example, the nontrivial fact that $X \leq_T \mathcal{O}$ if and only if X is Δ^1_2 over $\mathcal{H}$. Continuing the connection with subsystems of second order arithmetic Jockusch and Simpson also define the ideals I such that M_I is a β-model of Δ^1_n-CA for $n \geq 2$ or of full comprehension by exploiting the fact that these are the ones that are closed under hyperjump. We note that an approach along the lines of controlling initial segments allows us to drop the comprehension assumption.

DEFINITION 3.41. *A nonempty set* M *of functions from* $\mathbb{N}$ *to* $\mathbb{N}$ *is a* β-model (of arithmetic) *if it is absolute for* Σ^1_1 *formulas, i.e. if* $S \subseteq (\mathbb{N}^{\mathbb{N}})^3$ *is* Σ^1_1, $f_1, f_2 \in M$ *and* $(\exists f)S(f_1, f_2, f)$ *then there is an* $f \in M$ *such that* $S(f_1, f_2, f)$.

THEOREM 3.42. *The property that* I *is a jump ideal and* M_I *is a* β*-model is definable in* $\mathcal{D}$ *with the jump operator. Indeed, for* I *a jump ideal* M_I *is a* β*-model if and only if for every linearly ordered initial segment* J *of* $\mathcal{D}$ *with top in* I, *if there is an exact pair defining an initial segment* K *of* J *with no least element of* J *above* K *then there is such an exact pair in* I.

PROOF. First suppose that M_I is a β-model and J is an initial segment of $\mathcal{D}$ with top $\mathbf{a} \in I$. The assertion that there is an exact pair $\mathbf{x}, \mathbf{y}$ for some initial segment K of J with no least element of J above it is Σ^1_1. Thus, if true, there are witnesses for $\mathbf{x}$ and $\mathbf{y}$ in I as I is a β-model. For the other direction suppose S is Σ^1_1, $f_1, f_2 \in M_I$ and $S(f_1, f_2, f)$ holds for some arbitrary f. Now consider the Kleene-Brouwer ordering $KB(S)$ of sequence numbers associated with the Σ^1_1 in $f_1 \oplus f_2$ predicate $S(f_1, f_2, f)$ of f. The ordering is recursive in $f_1 \oplus f_2$ and so (as I is a jump ideal) there is an initial segment J of I with top $\mathbf{a} \in I$ which is isomorphic to $KB(S)$ by an isomorphism in M_I. As there is an f satisfying this predicate, the ordering is not well ordered. Thus there is an initial segment K of J with no least element above it. By Theorem 3.24, there is an exact pair for K. By our assumption there is an exact pair for K in I. Thus there is a descending chain in J arithmetic in this exact pair and $\mathbf{a}$ and so one in M_I. Finally, arithmetically in any such descending chain we can find a g such that $S(f_1, f_2, g)$ and so there is one in M_I as required. (The information needed here about Σ^1_1 predicates and the Kleene-Brouwer ordering is classical but can be found for example in Simpson [1999, Ch. V]. Information about β-models in general can be found in Simpson [1999, Ch. VII].) □

Jockusch and Simpson carry their analysis into the ramified analytic hierarchy which is defined like the constructible hierarchy L in set theory but only involves partial functions on ω as elements and has its definitions restricted to ones analytic over the previously defined level (with parameters allowed although they are not actually necessary). The Δ^1_n master codes H (for an M_I) are the (Turing) complete Δ^1_n over M_I sets, i.e. $X \leq_T H$ iff X is Δ^1_n over M_I. Thus, for example, if $a \in \mathcal{O}$ is a notation for a limit ordinal then H_a is a Δ^1_1 master code for $\{f | (\exists b <_{\mathcal{O}} a)(f \leq_T H_b)\}$ and $\mathcal{O}$ is a Δ^1_2 master code for $\mathcal{H}$.

THEOREM 3.43. (Jockusch and Simpson [1976]) *One can prove analogs of the first definition of $\mathcal{O}$ for the Δ^1_n master codes ($n \geq 2$) as the largest degrees below $(\mathbf{a} \vee \mathbf{b})^{(n+1)}$ for $\mathbf{a}, \mathbf{b}$ exact pairs for the appropriate ideals carried out through the ramified analytic hierarchy. For example, $\mathcal{O}_\omega$, the recursive join of the nth hyperjumps, $\mathcal{O}_n$, of $\mathbf{0}$, is the largest degree $\mathbf{d}$ such that $\mathbf{d} \leq (\mathbf{a} \vee \mathbf{b})^{(3)}$ for every exact pair $\mathbf{a}, \mathbf{b}$ for the ideal generated by the $\mathcal{O}_n$. If λ is the smallest admissible limit of admissibles then $\mathcal{O}_\lambda$, the degree of the complete $\Sigma_1(L_\lambda)$ set, is the largest degree $\mathbf{d}$ such that $\mathbf{d} \leq (\mathbf{a} \vee \mathbf{b})^{(4)}$ for every exact pair $\mathbf{a}, \mathbf{b}$ for the ideal generated by the $\mathcal{O}_\alpha$ for $\alpha < \lambda$.*

The extent to which these results provide natural definitions in $\mathcal{D}$ deserves some further investigation. Jockusch and Simpson proceed through the ramified analytic hierarchy by assuming at limit levels, analogously to what is done in the definition of L, that the ideal I_λ consisting of the union of all the previous levels is given as a predicate of the language and then working with this ideal as an added predicate. It is an interesting question as to when these recursions can be naturally defined in $\mathcal{D}$ or when the ideals themselves have natural independent definitions in $\mathcal{D}$.

Continuing in this vein, Simpson [1977] mentions the following definability result which is remarkable for the depth of the information about constructibility that is needed to establish the definition.

THEOREM 3.44. (Simpson) *There is a natural definition of the relation "$\mathbf{a}$ is constructible from $\mathbf{b}$" in $\mathcal{D}$ with the jump.*

PROOF. (Sketch) Let $\mathcal{C}^{\mathbf{a}}_1$ be the maximal Π^1_1 in $\mathbf{a}$ set with no perfect subset. $\mathbf{b}$ is constructible from $\mathbf{a}$ iff $\mathbf{b}$ is hyperarithmetic in some element $\mathbf{c}$ of $\mathcal{C}^{\mathbf{a}}_1$. Moreover, $\mathbf{c} \in \mathcal{C}^{\mathbf{a}}_1$ iff $\forall \mathbf{b}(\omega_1^{\mathbf{b} \vee \mathbf{a}} = \omega_1^{\mathbf{c} \vee \mathbf{a}} \rightarrow \mathbf{a} \leq_h \mathbf{b})$. We already know how to say that $\mathbf{a} \leq_h \mathbf{b}$ and so only need to define the relation $\omega_1^{\mathbf{x}} = \omega_1^{\mathbf{y}}$. Now for arbitrary $\mathbf{x}$ and $\mathbf{y}$, $\omega_1^{\mathbf{x}} = \omega_1^{\mathbf{y}} \Leftrightarrow \exists \mathbf{z}(\omega_1^{\mathbf{x}} = \omega_1^{\mathbf{x} \vee \mathbf{z}} = \omega_1^{\mathbf{z}} = \omega_1^{\mathbf{y} \vee \mathbf{z}} = \omega_1^{\mathbf{z}})$. Thus it suffices to define this relation if $\mathbf{x} \leq_h \mathbf{y}$. In this case, it is clear that $\omega_1^{\mathbf{x}} = \omega_1^{\mathbf{y}}$ iff $\mathcal{O}^{\mathbf{x}} \not\leq_h \mathbf{y}$ and we also already know that $\mathcal{O}^{\mathbf{x}}$, the hyperjump of $\mathbf{x}$, is also definable in $\mathcal{D}$ with the jump operator. □

This result provides a natural statement about $\mathcal{D}$ (with jump) that is independent of ZFC: There is a nonconstructible subset of $\mathbb{N}$. By Corollary 3.38 or

Proposition 3.39, another such statement provided by Simpson is "there is a cone of minimal covers in the hyperdegrees" which is true assuming $0^{\#}$ exists and false in set-forcing extensions of L by Simpson [1975]. These examples prompt a more general question and a specific one.

QUESTION 3.45. (Simpson) *What other natural statements about $\mathcal{D}$ are independent of ZFC?*

QUESTION 3.46. (Simpson) *Is there is a natural definition of the sharp operator in $\mathcal{D}$?*

Along these lines we also mention the analog of Question 2.16. It was also a long standing question first raised as Rogers' [1967] homogeneity conjecture if $\mathcal{D}(\geq \mathbf{a})$, the degrees greater than or equal to $\mathbf{a}$, are isomorphic (or even elementary equivalent, to $\mathcal{D}(\geq \mathbf{b})$ for every $\mathbf{b}$. A negative answer for isomorphisms based on how complicated the base of a cone of minimal covers must be was provided by Shore [1979] and for elementary equivalence by a translation of this issue via codings of models of second order arithmetic in Shore [1982a]. However, we still have no natural difference between any two cones.

QUESTION 3.47. (Simpson) *Is there a natural sentence ϕ of degree theory that is true in some $\mathcal{D}(\geq \mathbf{a})$ but not in every $\mathcal{D}(\geq \mathbf{b})$? If so, for which $\mathbf{a}$ and $\mathbf{b}$ can one find such sentences?*

We close this section by asking for natural definitions of other degrees and ideals that correspond to various types of models of arithmetic.

QUESTION 3.48. *Are the degrees of complete extensions or models of Peano arithmetic or the models of true arithmetic naturally definable in $\mathcal{D}$? Is a countable ideal I of $\mathcal{D}$ being a Scott set, i.e. a model of WKL_0, naturally definable in $\mathcal{D}$?*

4. Parameters and additional predicates

We conclude with a few remarks and questions on (natural) definability in extensions of $\mathcal{R}$ and $\mathcal{D}$. One might first consider definability from parameters. This certainly provides an interesting and important line of investigation. Indeed, in the case of $\mathcal{D}$ the situation (omitting considerations of naturalness) has been fully analyzed by Slaman and Woodin (see Slaman [1991]). Every relation definable in second order arithmetic with parameters is definable in $\mathcal{D}$ with parameters. However, such results are not likely to produce natural definitions in $\mathcal{R}$ or $\mathcal{D}$ itself unless the parameters can later be eliminated (by quantification perhaps) or be defined themselves (as was the case for $\mathbf{0}'$). On the other hand, there are interesting results defining finite sets of degrees in a specified interval of $\mathcal{R}$ such as in Stob [1983] where r.e. degrees $\mathbf{a}_0$ and $\mathbf{a}_1$ are constructed such that $\mathbf{a}_0$ is the unique complement of $\mathbf{a}_1$ in the interval $[\mathbf{0}, \mathbf{a}_0 \vee \mathbf{a}_1]$. This suggests the following question:

QUESTION 4.1. (Li) *Are there definable properties in $\mathcal{R}$ which determine a single degree in some nontrivial interval.*

Another interesting line of investigation is that of adding (a priori) external relations on $\mathcal{R}$ or $\mathcal{D}$ and determining what can then be defined. This is perhaps a more promising line in that the additional relations (like the jump) may eventually be defined or we might be willing to view them as natural themselves in some context. One such class of relations frequently brought into $\mathcal{R}$ is the following.

DEFINITION 4.2. $\mathbf{a}$ *is* n-jump equivalent *to* $\mathbf{b}$, $\mathbf{a}\sim_n\mathbf{b}$, *iff* $\mathbf{a}^{(n)} = \mathbf{b}^{(n)}$.

PROPOSITION 4.3. (Nies, Shore and Slaman [1998]) $\mathbf{H}_1$ *is naturally definable from* $\sim_2$ *in* $\mathcal{R}$: $\mathbf{a} \in \mathbf{H}_1 \Leftrightarrow (\forall \mathbf{y})(\exists \mathbf{z} \leq \mathbf{a})(\mathbf{z} \sim_2 \mathbf{a})$.

QUESTION 4.4. *Are there any other interesting natural definitions in* $\mathcal{R}$ *that use the relations* $\sim_n$*?*

QUESTION 4.5. *Are any of the relations* $\sim_n$ *themselves naturally definable in* $\mathcal{R}$*? In particular what about* $\sim_2$*?*

An example of an interesting but apparently external relation on $\mathcal{D}$ that has been used for relative definability results is that of a degree of a complete extension of Peano arithmetic. More generally notions and results can be phrased in terms of the relation $\mathbf{a} \ll \mathbf{b}$, i.e. every nonempty $\Pi_1^{0,A}$ class has a member recursive in $\mathbf{b}$. (Note that, by Simpson [1977a, p. 649]), $\mathbf{0} \ll \mathbf{b}$ if and only if $\mathbf{b}$ is the degree of a complete extension of PA if and only if $\mathbf{b}$ is the degree of a set which separates an effectively inseparable pair of r.e. sets.) Here we have an result defining $\mathbf{0}'$ from this relation.

THEOREM 4.6. (Kucera [1988]) $\mathbf{0}' = \inf\{\mathbf{a} \vee \mathbf{b} | \mathbf{0} \ll \mathbf{a}, \mathbf{b} \ \& \ (\mathbf{a} \wedge \mathbf{b}) = \mathbf{0}\}) = \inf\{\mathbf{a} | \mathbf{0} \ll \mathbf{a} \ \& \ \forall \mathbf{c}(\mathbf{0} < \mathbf{c} \leq \mathbf{a} \rightarrow \exists \mathbf{b}(\mathbf{0} \ll \mathbf{b} \leq \mathbf{a} \ \& \ \mathbf{c} \nleq \mathbf{b}))$.

QUESTION 4.7. *Is the relation* $\mathbf{a} \ll \mathbf{b}$ *(naturally) definable in* $\mathcal{D}$*?*

QUESTION 4.8. *What other interesting natural predicates on* $\mathcal{D}$ *or* $\mathcal{R}$ *might be profitably added to the language?*

References

Ambos-Spies, K. and Fejer, P. [2000], Embeddings of N_5 and the contiguous degrees, to appear.

Ambos-Spies, K. Jockusch, C. G. Jr., Shore, R. A. and Soare, R. I. [1984], An algebraic decomposition of the recursively enumerable degrees and the coincidence of several degree classes with the promptly simple degrees, *Trans. Amer. Math. Soc.* **281**, 109-128.

Cholak, P. Downey, R. G. and Walk, S. M. [2000] Maximal Contiguous Degrees, to appear.

Cooper, S. B. [1990], The jump is definable in the structure of the degrees of unsolvability (research announcement), *Bull. Am. Math. Soc.* **23**, 151-158.

Cooper, S. B. [1993], On a conjecture of Kleene and Post, *University of Leeds, Department of Pure Mathematics Preprint Series* No. 7.

Cooper, S. B. [1994], Rigidity and definability in the noncomputable universe, in *Logic, Meth. and Phil. of Science IX*, Proc. Ninth Int. Cong. Logic, Meth. and Phil. of Science, Uppsala, Sweden August 7-14, 1991, D. Prawitz, B. Skyrms and D. Westerstahl eds., North-Holland, Amsterdam, 209-236.

Cooper, S. B. [1999], The degrees below $\mathbf{0}'$, in *Handbook of Computability Theory*, E. Griffor ed., North-Holland, Amsterdam, 1999.

Downey, R. G., Jockusch, C. G. Jr. and Stob, M. [1990], Array nonrecursive sets and multiple permitting arguments, in *Recursion Theory Week*, K. Ambos-Spies, G. H. Müller and G. E. Sacks eds., Springer-Verlag, Berlin, 141-174.

Downey, R. G. and Lempp, S. [1997], Contiguity and distributivity in the enumerable Turing degrees, *J. Symb. Logic* **62**, 1215-1240.

Enderton, H. B. and Putnam, H. [1970], A note on the hyperartihmetical hierarchy, *J. Symb. Logic* **35**, 429-430.

Friedberg, R. M. [1957], A criterion for completeness of degrees of unsolvability, *J. Symb. Logic* **22**, 159-60.

Hodes, H. [1982], Uniform upper bounds on ideals of Turing degrees, *J. Symb. Logic* **43**, 601-612.

Hodges, W. [1993], *Model Theory*, Cambridge University Press, Cambridge, England.

Jockusch, C. G. Jr. [1972], Degrees in which the recursive sets are uniformly recursive, *Can. J. Math.* **24**, 1092-1099.

Jockusch, C. G. Jr. and Shore, R. A. [1984], Pseudo-jump operators II: transfinite iterations, hierarchies and minimal covers, *J. Symb. Logic* **49**, 1205-1236.

Jockusch, C. G. Jr. and Shore, R. A. [1985], REA operators, r.e. degrees and minimal covers, in *Recursion Theory, Proc. Symp. Pure Math.* **42**, A. Nerode and R. A. Shore, eds., American Mathematical Society, Providence, RI, 1985, 3-11.

Jockusch, C. G. Jr. and Simpson, S. G. [1976], A degree theoretic definition of the ramified analytic hierarchy, *Ann. Math. Logic* **10**, 1-32.

Jockusch, C. G. Jr. and Soare, R. I., Minimal covers and arithmetical sets, *Proc. Amer. Math. Soc.* **25**, 856-859.

Kucera, A. [1988], On the role of $\mathbf{0}'$ in recursion theory, in *Logic Colloquium '86*, F. R. Drake and J. K. Truss eds., North-Holland, Amsterdam, 133-141.

Lachlan, A. H. [1971], Solution to a problem of Spector, *Can. J. Math.* **23**, 247-256.

Lachlan, A. H. and Soare, R. I. [1994], Models of arithmetic and upper bounds of arithmetic sets, *J. Symb. Logic* **59**, 977-983.

Lerman, M. [1983], *Degrees of Unsolvability*, Springer-Verlag, Berlin.

Lerman, M. [1985], Upper bounds for the arithmetical degrees, *Ann. Pure and Applied Logic* **29**, 225-254.

MacIntyre, J. M. [1977], Transfinite extensions of Friedberg's completeness criterion, *J. Symb. Logic* **42**, 1-10.

Maass, W. [1982], Recursively enumerable generic sets, *J. Symb. Logic*, **47**, 809-823.

Nerode, A. and Shore, R. A. [1980], Second order logic and first order theories of reducibility orderings in *The Kleene Symposium*, J. Barwise, H. J. Keisler and K. Kunen, eds., North-Holland, Amsterdam, 181-200.

Nerode, A. and Shore, R. A. [1980a], Reducibility orderings: theories, definability and automorphisms, *Ann. Math. Logic* **18**, 61-89.

Nies, A., Shore, R. A. and Slaman, T. A. [1998], Interpretability and definability in the recursively enumerable degrees, *Proc. London Math. Soc.* (3) **77**, 241-291.

Posner, D. B. [1980], A survey of non-RE degrees $\leq 0'$, in *Recursion Theory: its Generalizations and Applications, Proc. Logic Col. '79, Leeds, August 1979*, F. R. Drake and S. S. Wainer eds., Cambridge University Press, Cambridge, England, 110-139.

Posner, D. B. [1981], The upper semilattice of degrees $\leq \mathbf{0}'$ is complemented, *J. Symb. Logic*, **46**, 714-722.

Rogers, H. Jr. [1967], Some problems of definability in recursive function theory, in *Sets, Models, and Recursion Theory, Proceedings of the Summer School in Mathematical Logic and 10th Logic Colloquium, Leicester, August-Sept. 1965*, J. N. Crossley ed., North-Holland, Amsterdam, 183-201.

Sacks, G. E. [1963], On the degrees less than $\mathbf{0}'$, *Ann. of Math. (2)* **77**, 211-231.

Sacks, G. E. [1963a], *Degrees of unsolvability*, Annals of Math. Studies **55**, Princeton Univ. Press, Princeton NJ.

Sacks, G. E. [1971], Forcing with perfect closed sets, in *Axiomatic Set Theory, Proc. Symp. Pure Math.* **13**, part I, D. S. Scott ed., AMS, Providence RI, 331-355.

Shore, R. A. [1979], The homogeneity conjecture, *Proc. Nat. Ac. Sci.* **76**, 4218-4219.

Shore, R. A. [1982], Finitely generated codings and the degrees r.e. in a degree **d**, *Proc. Am. Math. Soc.* **84**, 256-263.

Shore, R. A. [1982a], On homogeneity and definability in the first order theory of the Turing degrees, *J. Symb. Logic* **47**, 8-16.

Shore, R. A. and Slaman, T. A. [2000], Defining the Turing jump, to appear.

Shore, R. A. and Slaman, T. A. [2000a], A splitting theorem for n-REA degrees, to appear.

Simpson, S. G. [1975], Minimal Covers and Hyperdegrees, *Trans. AMS* **209**, 45-64.

Simpson, S. G. [1977], First order theory of the degrees of recursive unsolvability, *Ann. Math. (2)*, **105**, 121-139.

Simpson, S. G. [1977a], Degrees of unsolvability: A survey of results, in *Handbook of Mathematical Logic*, J. Barwise ed., North-Holland, Amsterdam, 631-652.

Simpson, S. G. [1999], *Subsystems of Second Order Arithmetic*, Springer-Verlag, Berlin.

Slaman, T. A. [1991], Degree structures, in *Proc. Int. Cong. Math., Kyoto 1990*, Springer-Verlag, Tokyo, 303-316.

Slaman, T. A. and Steel, J. R. [1989], Complementation in the Turing degrees, *J. Symb. Logic* **54**, 160-176.

Slaman, T. A. and Woodin, H. W. [1986], Definability in the Turing degrees, *Illinois J. Math.* **30**, 320-334.

Slaman, T. A. and Woodin, H. W. [2000], Defining the double jump, to appear.

Soare, R. I. [1987], *Recursively enumerable sets and degrees*, Springer-Verlag, Berlin.

Soare, R. I. [1999], The lattice of recursively enumerable sets, in *Handbook of Computability Theory*, E. Griffor ed., North-Holland, Amsterdam, 1999.

Spector, C. On degrees of recursive unsolvability, *Ann. Math. (2)* **64**, 581-592.

Stob, M. [1983], WTT-degrees and T-degrees of r.e. sets, *J. Symb. Logic* **48**, 921-930.

DEPARTMENT OF MATHEMATICS, CORNELL UNIVERSITY, ITHACA NY 14853
E-mail address: shore@math.cornell.edu
URL: http://math.cornell.edu/~shore/

Contemporary Mathematics
Volume **257**, 2000

Recursion Theory in Set Theory

Theodore A. Slaman

1. Introduction

Our goal is to convince the reader that recursion theoretic knowledge and experience can be successfully applied to questions which are typically viewed as set theoretic. Of course, we are not the first to make this point. The detailed analysis of language, the absoluteness or nonabsoluteness of the evaluation of statements, and the interaction between lightface and relativized definability are thoroughly embedded in modern descriptive set theory. But it is not too late to contribute, and recursion theoretic additions are still welcome. We will cite some recent work by Slaman, Hjorth, and Harrington in which recursion theoretic thinking was applied to problems in classical descriptive set theory.

It is the parameter-free or lightface theory that seems closest to our recursion theoretic heart. Where another might see a continuous function, we see a function which is recursive relative to a real parameter. In the same way, we can see the Borel sets through the hyperarithmetic hierarchy and the co-analytic sets by means of well-founded recursive trees. We will make our way through most of the relevant mathematical terrain without invoking concepts which are not natively recursion theoretic. At the end, we will mention some problems which are similarly accessible.

We owe a debt to Sacks's (1990) text on higher recursion theory and to Kechris's (1995) text on descriptive set theory. These are valuable resources, and we recommend them to anyone who wishes to learn more about what we will discuss here. In the following, we will cite theorems from the nineteenth and early twentieth centuries without giving the original references; the motivated reader can find the history of descriptive set theory well documented in these texts.

2. The Classical Theory

Here is the framework. We speak exclusively about subsets of Baire space, ${}^{\omega}\omega$, and refer to an ω sequence of natural numbers as a real number. A basic open set $B(\sigma)$ is determined by a finite sequence σ from $\omega^{<\omega}$: $x \in B(\sigma)$ if and only if σ is

1991 *Mathematics Subject Classification.* primary 03E15, secondary 03D65.

Key words and phrases. recursion theory, descriptive set theory, analytic.

Slaman wishes to thank L. Harrington, G. Hjorth, J. Steel, and W. H. Woodin for the information and advice that they provided during the preparation of this paper. Slaman was also partially supported by NSF Grant DMS-97-96121.

an initial segment of x. A function $f : {}^{\omega}\omega \to {}^{\omega}\omega$ is continuous if any finite initial segment of $f(x)$ is determined by a finite initial segment of x. If we think of this correspondence between the argument and the domain as having been coded by a real number, then f is recursive relative to that real. Conversely, if f is is recursive relative to some real parameter, then f is continuous.

DEFINITION 2.1. 1. The *Borel* subsets of ${}^{\omega}\omega$ are those sets which can be obtained from open sets by a countable iteration of countable union and complementation.
2. The *analytic* sets are the continuous images of the Borel sets.

The classical notions correspond to levels in the descriptive hierarchy of second order arithmetic.

DEFINITION 2.2. 1. A is a $\mathbf{\Sigma}^1_1$ set if and only if membership in A is definable as follows.

$$x \in A \iff (\exists w)(\forall n)R(n, x \restriction n, w \restriction n, a \restriction n) \tag{1}$$

where a is a fixed element of ${}^{\omega}\omega$, w ranges over ${}^{\omega}\omega$, n ranges over ω, and R is recursive.
2. A is a $\mathbf{\Delta}^1_1$ set if and only if both A and its complement are $\mathbf{\Sigma}^1_1$ sets.

Here is the connection. A set C is closed if and only if there is an $a \in {}^{\omega}\omega$ and there is a recursive predicate R, such that for all x,

$$x \in A \iff (\forall n)R(n, x \restriction n, a \restriction n).$$

By a classical fact, for every analytic set A, there is a closed set C such that for all x, $x \in A$ if and only if there is a witness w such that $(x, w) \in C$. Thus, a set is analytic if and only if it is $\mathbf{\Sigma}^1_1$.

Similarly, by a classical theorem of Suslin, the Borel sets are exactly those analytic sets whose complements are analytic. Consequently, B is Borel if and only if it is $\mathbf{\Delta}^1_1$.

DEFINITION 2.3. The *projective* sets are obtained from the Borel sets by closing under continuous images and complements.

Similarly to the above, the projective sets are those sets which can be defined in second order arithmetic using real parameters.

Initially, the projective sets were studied topologically. Much of the progress was limited to the Borel sets and the $\mathbf{\Sigma}^1_1$ sets, for which a variety of regularity properties were established.

2.1. Perfect set theorems. Recall, a set P is *perfect* if it is nonempty, closed, and has no isolated points. Equivalently, P is perfect if and only if there is a (perfect) tree $T \subseteq \omega^{<\omega}$ such that every element of T has incompatible extensions in T and P is $[T]$, the collection of infinite paths through T.

THEOREM 2.4. 1. (Cantor–Bendixson) *Every uncountable closed subset of ${}^{\omega}\omega$ has a perfect subset.*
2. (Alexandrov, Hausdorff) *Every uncountable Borel subset of ${}^{\omega}\omega$ has a perfect subset.*
3. (Suslin) *Every uncountable analytic subset of ${}^{\omega}\omega$ has a perfect subset.*

Suslin's Theorem follows directly from the representation of analytic sets given in 1. Suppose that A is uncountable and is defined by

$$x \in A \iff (\exists w)(\forall n)R(n, x \restriction n, w \restriction n, a \restriction n).$$

We build a tree T of pairs $(\tau, \sigma) \in \omega^{<\omega} \times \omega^{<\omega}$ such that if $(\tau, \sigma) \in T$ then there are uncountably many x extending τ for which there is a w extending σ such that $(\forall n)R(n, x \restriction n, w \restriction n, a \restriction n)$. We use the fact that A is not countable to ensure that the projection of T onto the first coordinates of its elements is a perfect tree T_1. Each path x through T_1 is an element of A, as it is associated with a witness w to that fact in T.

2.2. Representations of Borel sets. A diagonal argument shows that there is no universal Borel set. However, in a different sense, the Borel sets are closer to having a universal element than one might have thought.

THEOREM 2.5 (Luzin–Suslin). *For every Borel set B, there is a closed set C and a continuous function f which maps C bijectively to B.*

PROOF. Whether x belongs to B is determined at a countable ordinal in the jump hierarchy relative to x and the Borel code b for B. Let C be the set of triples (x, b, s) such that s is the Skolem function verifying the relevant hyperarithmetic statement about x and b. □

COROLLARY 2.6. *Every uncountable Borel subset of ${}^{\omega}\omega$ is a continuous injective image of the sum of ${}^{\omega}\omega$ with a countably infinite discrete set.*

PROOF. By Theorem 2.5, it is enough to show that every uncountable closed set is a continuous injective image of the sum of ${}^{\omega}\omega$ with a countably infinite discrete set. This follows from the Cantor–Bendixson analysis of closed sets. □

Now, we prove the converse.

THEOREM 2.7 (Luzin–Suslin). *Suppose that B is a Borel subset of ${}^{\omega}\omega$, and that f is a continuous function that is injective on B. Then the range of f applied to B is a Borel set.*

PROOF. Clearly, $f''B$ is a $\mathbf{\Sigma}^1_1$ set. Let b be a real parameter used in the Borel definition of B.

Note, if $x \in f''B$ then

$$f^{-1}(x)(n) = m \iff (\exists z)[z \in B \text{ and } z(n) = m \text{ and } f(z) = x]$$

So, $f^{-1}(x)$ is uniformly $\Sigma^1_1(x, f, b)$ definable and similarly $\Delta^1_1(x, f, b)$ definable. Consequently, $x \in f''B$ if and only if there is an ordinal β less than $\omega_1^{x,f,b}$ and a z in $L_\beta[x, f, b]$ such that $L_\beta[x, f, b]$ satisfies the conditions that $z \in B$ and $f(z) = x$. This gives a $\mathbf{\Pi}^1_1$ definition of $f''B$.

Consequently, $f''B$ is a $\mathbf{\Pi}^1_1$ set, and therefore is a Borel set. □

COROLLARY 2.8. *The Borel sets are exactly the injective continuous images of the sum of ${}^{\omega}\omega$ with a countably infinite discrete set.*

2.3. Sierpiński's Problem. Sierpiński (1936) raised the question whether there is an analogous version of Theorem 2.7 for the analytic sets.

QUESTION 2.9 (Sierpiński (1936)). Does there exist a subset U of ${}^\omega\omega$ such that for every uncountable $\mathbf{\Sigma}^1_1$ set A, there is a continuous function f that maps U bijectively to A?

Of course, the complete $\mathbf{\Sigma}^1_1$ set has all of the desired properties except for the key property that f should be injective.

2.3.1. *Solution to Sierpiński's problem.*

THEOREM 2.10 (Slaman (1999)).

- *There is no $\mathbf{\Sigma}^1_1$ set which satisfies all of the Sierpiński properties.*
- *There is a set U which satisfies all of the Sierpiński properties.*

PROOF. The proof of the first claim is short enough to include here, but we will only point to (Slaman, 1999) or (Hjorth, n.d.) for the proof of the second claim.

To prove the first claim, we proceed by contradiction. Let U be an $\mathbf{\Sigma}^1_1$ set such that for every uncountable $\mathbf{\Sigma}^1_1$ set A, there is a continuous function f that maps U bijectively to A.

Let f be a continuous bijection from U to ${}^\omega\omega$. As above, for $x \in {}^\omega\omega$, $f^{-1}(x)$ is $\Delta^1_1(x, f)$. Then for all $y \in {}^\omega\omega$, $y \in U$ if and only if $y = f^{-1}(f(y))$, and so U is $\mathbf{\Delta}^1_1$.

Now apply the Luzin–Suslin theorem. Since the complete Σ^1_1 set of reals is a continuous bijective image of U, it is $\mathbf{\Delta}^1_1$, an impossibility. □

2.3.2. *Hjorth's theorem.* Slaman (1999) raised the question, is there a projective U as in Theorem 2.10? Hjorth (n.d.) gave the best possible example.

THEOREM 2.11 (Hjorth (n.d.)). *There is a $\mathbf{\Pi}^1_1$ set U which provides a positive solution to Sierpiński's problem. In fact, the set of reals which are hyperjumps*

$$\mathcal{H} = \{\mathcal{O}^x : x \in {}^\omega\omega\}$$

is such a set U.

It takes us beyond the usual recursion theoretic horizon, but Sierpiński's question is sensible at any level of the projective hierarchy.

THEOREM 2.12 (Hjorth). *The following statements are equivalent.*

1. *Every uncountable $\mathbf{\Pi}^1_1$ set contains a perfect set. Equivalently, for all $z \in {}^\omega\omega$, $\aleph_1$ is inaccessible in $L[z]$.*
2. *Every uncountable $\mathbf{\Sigma}^1_2$ set is the continuous injective image of $\mathcal{H}$.*

Though not contained in its entirety, much of the proof of Theorem 2.12 contained in (Hjorth, n.d.).

2.3.3. *Harrington's theorem.* In Theorem 2.10, we observed that no analytic set could be a universal injective preimage for all of the uncountable analytic sets. We argued that if U is analytic and ${}^\omega\omega$ is a continuous injective image of U, then U is Borel. Then, we concluded that U could not be universal.

Steel raised the question, could an analytic set be a universal injective preimage for all of the properly analytic sets? Harrington provided the answer.

THEOREM 2.13 (Harrington). *The following statements are equivalent.*

1. *There is a $\mathbf{\Sigma}^1_1$ set A such that every non-Borel analytic set is a continuous injective image of A.*

2. *For every real x, the real $x^{\#}$ exists. Equivalently, $\mathbf{\Sigma}^1_1$-determinacy holds.*

In fact, under $\mathbf{\Sigma}^1_1$-determinacy, every non-Borel $\mathbf{\Sigma}^1_1$ set satisfies the property of A in (1).

It may seem that we have wandered far from our recursion theoretic home base, but that is not the case. The principal ingredients in Harrington's proof are the Kleene Fixed Point Theorem and Steel forcing over Σ_1-admissible sets. What could be more recursion theoretic?

3. Directions for further investigation

In the previous section, we saw that a recursion theorist can travel into the set theoretic domain without becoming completely lost. Now, we turn to some open questions which can be found there.

3.1. Harrington's question.

DEFINITION 3.1. For subsets A and B of ${}^\omega\omega$, we say $A \geq_W B$ if there is a continuous function f such that for all x in ${}^\omega\omega$, $x \in B$ if and only if $f(x) \in A$.

When $A \geq_W B$, we say that B is *Wadge reducible to* B. Wadge reducibility between sets of reals is analogous to many-one reducibility between sets of natural numbers.

THEOREM 3.2 (Wadge). *Under the Axiom of Determinacy, for all subsets A and B of ${}^\omega\omega$, either $A \geq_W B$ or $\overline{B} \geq_W A$, where $\overline{B}$ is the complement of B in ${}^\omega\omega$.*

COROLLARY 3.3. *For any two properly $\mathbf{\Sigma}^1_1$ sets A and B, $A \equiv_W B$.*

Steel (1980), and later Harrington in more generality, sharpened Corollary 3.3 to add injectivity to the reducing function.

THEOREM 3.4 (Steel (1980)). *Under the Axiom of Determinacy, for any two properly $\mathbf{\Sigma}^1_1$ sets A and B there is an injective continuous function f such that for all x, $x \in B$ if and only if $f(x) \in A$.*

COROLLARY 3.5. *Under the Axiom of Determinacy, for any two properly $\mathbf{\Sigma}^1_1$ subsets A and B of ${}^\omega\omega$, there is a Borel permutation π of ${}^\omega\omega$ such that $\pi[A] = B$.*

THEOREM 3.6 (Harrington (1978)). *Suppose that every $\mathbf{\Sigma}^1_1$ subset A of ${}^\omega\omega$ is either Borel or $\geq_W$-complete. Then the Axiom of Determinacy holds for $\mathbf{\Sigma}^1_1$ sets.*

Harrington's proof of Theorem 3.6 involves a fair amount of set theory. Though off topic for us, Harrington has raised the interesting question of whether Theorem 3.6 is provable from the usual axioms of second order arithmetic. More on topic is his question of whether determinacy follows from the weaker hypothesis that the Wadge degrees of the $\mathbf{\Sigma}^1_1$ sets are linearly ordered.

QUESTION 3.7 (Harrington). Suppose that any two $\mathbf{\Sigma}^1_1$ non-Borel subsets of ${}^\omega\omega$ are $\geq_W$ comparable. Does the Axiom of Determinacy hold for $\mathbf{\Sigma}^1_1$ sets?

3.2. Hjorth's question. Let D denote a countable discrete set.

QUESTION 3.8 (Hjorth (n.d.)). Suppose that for any uncountable $\mathbf{\Sigma}^1_2$ set B and $\mathbf{\Pi}^1_1$ non-Borel set C, B is the continuous image of $C \oplus D$. Then, must $\mathbf{\Sigma}^1_1$-determinacy hold?

3.3. Comparing the Wadge and Sierpiński orderings.

DEFINITION 3.9. For subsets A and B of ${}^\omega\omega$, say that $A \geq_S B$ if there is a function f which is partial recursive in some real parameter whose restriction to A is a bijection from A to B.

QUESTION 3.10. 1. What is the structure of $\geq_S$?
2. Is there a relationship between $\geq_W$ and $\geq_S$?

We first heard the following question from Hjorth, who also told us that it is not original to him.

QUESTION 3.11. What is the structure of $\geq_S$ on countable metric spaces?

3.4. Steel's question.

DEFINITION 3.12. Say that two subsets A and B are *homeomorphically equivalent* if there is a homeomorphism $f : {}^\omega\omega \to {}^\omega\omega$ such that $f[A] = B$.

QUESTION 3.13 (Steel). Is there a natural classification of the analytic sets up to homeomorphic equivalence?

References

Harrington, L. A. (1978). Analytic determinacy and $0^\#$, *J. Symbolic Logic* **20**: 685–693.
Hjorth, G. (n.d.). Continuous images of coanalytic sets, *Logic Colloquium'98 Prague*. to appear.
Kechris, A. S. (1995). *Classical Descriptive Set Theory*, Vol. 156 of *Graduate Texts in Mathematics*, Springer–Verlag, Heidelberg.
Sacks, G. E. (1990). *Higher Recursion Theory*, Perspectives in Mathematical Logic, Springer–Verlag, Heidelberg.
Sierpiński, W. (1936). Problem 70, *Fund. Math.* **26**: 334.
Slaman, T. A. (1999). On a question of Sierpiński, *Fund. Math.* **159**(2): 153–159.
Steel, J. R. (1980). Analytic sets and Borel isomorphisms, *Fund. Math.* **108**(2): 83–88.

UNIVERSITY OF CALIFORNIA, BERKELEY, BERKELEY, CA 94720-3840, USA
E-mail address: slaman@math.berkeley.edu

Contemporary Mathematics
Volume **257**, 2000

Extensions, Automorphisms, and Definability

Robert I. Soare

ABSTRACT. This paper contains some results and open questions for automorphisms and definable properties of computably enumerable (c.e.) sets. It has long been apparent in automorphisms of c.e. sets, and is now becoming apparent in applications to topology and differential geometry, that it is important to know the *dynamical* properties of a c.e. set W_e, not merely *whether* an element x is enumerated in W_e but *when*, relative to its appearance in other c.e. sets.

We present here, and prove in another paper, a New Extension Theorem (N.E.T.) appropriate for Δ_3^0-automorphisms of c.e. sets with simple hypotheses stated in terms of a dynamical concept called "templates" for the first time. The N.E.T. is needed implicitly or explicitly in almost every current automorphism proof for c.e. sets. We then present sketches of some known automorphism results using the N.E.T., and show how some definability results also derive their power from their ability to enforce a dynamic flow of elements into or away from a certain c.e. set.

To reveal the history and motivation for these results and questions, we give a detailed historical description in §1 and §2 which should be understandable to anyone who has read the first three or four chapters of Soare [1987]. The rest of the paper contains sketches of proofs which should be understandable to most computability theorists who have never worked in automorphisms. The open questions are chosen to illustrate a method or frontier we need to cross to advance our knowledge. Another paper on c.e. sets in this volume is Cholak [ta].

1. Introduction

1.1. The Early History of C.E. Sets. One of the great achievements of logic of the twentieth century was the development of the notion of a computable function. Foreshadowed by Gödel's use of the primitive recursive functions in his famous Incompleteness Theorem [1931], there soon emerged Gödel's definition [1934] of the general recursive functions. Shortly thereafter followed Turing's definition [1936] of functions computable by a Turing machine, and his argument that these constituted the intuitively computable functions. Gandy [1988, p. 82] observed,

1991 *Mathematics Subject Classification.* primary 03D25.

Key words and phrases. computability theory, computably enumerable sets, automorphisms, definable properties.

The author was supported by National Science Foundation Grant DMS 98-9802619. He had helpful conversations with Leo Harrington and Peter Cholak about topics presented in this paper.

"Turing's analysis does much more than provide an argument for" Turing's Thesis, "*it proves a theorem.*" This definition and demonstration only then convinced Gödel of the Church-Turing Thesis that these comprised the class of all functions calculable by a human being with unlimited time and space. We accept this thesis and identify the terms "computable" and "Turing computable."

Almost simultaneously, Church and Kleene proposed a formal definition of an "effectively enumerable" set (the informal term for this concept), one which could be listed by an effective algorithm. This concept is probably second only to that of computable function for its importance in computability theory. These sets appear in many branches of mathematics (number theory, algebra, topology), and computer science. Actual digital computers produce such sets during their computations. Turing reducibility [1939] of one set to another can be defined by an effectively enumerable set of axioms.

Church [1936] and Kleene [1936] formally defined this notion as follows. A set $A \subseteq \omega$ is *computably enumerable (c.e.)* (previously called *recursively enumerable*) if A is the range of a (Turing) computable function. (The empty set was added later as a c.e. set.) This is a *static* mathematical definition, like the definition of a function being continuous if the inverse image of every open set is open. Let $\{W_e\}_{e \in \omega}$ be the standard listing of the c.e. sets, as in Soare [1987].

Very little was heard about c.e. sets for half a decade. Then Post [1941] and [1943] introduced a *different* and unrelated formalism called a *(normal) production* system. Post's normal canonical system is a *generational* system, rather than a *computational* system as in general recursive functions or Turing computable functions. It led Post to concentrate on *effectively enumerable sets* rather than computable functions. He showed that every recursively enumerable set is a normal set (one derived in his normal canonical system) and therefore normal sets are formally equivalent to recursively enumerable sets.

Post used the terms "effectively enumerable set" and "generated set" almost interchangeably, for "the corresponding intuitive concept." Post [1944, p. 285] (like Church [1936]) formally defined a set of positive integers to be *computably enumerable* (formerly "recursively enumerable") if it is the range of a computable function. Post [1944], p. 286] explained his informal concept of a "generated set" ("effectively enumerable set") of positive integers this way,

> "Suffice it to say that each element of the set is at some time written down, and earmarked as belonging to the set, as a result of predetermined effective processes. It is understood that once an element is placed in the set, it stays there."

Post then [p. 286] restated his thesis from [1943] that *"every generated set of positive integers is recursively enumerable,"* [the italics are Post's] and he remarked that "this may be resolved into the two statements: every generated set is effectively enumerable, every effectively enumerable set of positive integers is recursively enumerable." Post continued, "their converses are immediately seen to be true." This is sometimes referred to as *"Post's Thesis."*

1.2. Post Initiates the Study of C.E. Sets in 1944. Explaining c.e. sets in terms of the quote above, and using his normal canonical system when necessary for mathematical precision, Post [1944] introduced several fundamental themes which have deeply influenced the subject ever since.

THEME 1.1. *Clear, informal style, using the Church-Turing Thesis.*

Up to that time, the papers by Kleene and others had been written in terms of the Gödel equational calculus or the Kleene schemata I–VI. (See Soare [1987, Ch. I].) Post explained everything in a very clear, informal style, using informal descriptions as we do today, and relying on Post's Thesis above or the Church-Turing Thesis to convert these, if necessary, to formal definitions of c.e. sets. This approach greatly stimulated interest in the topic of c.e. sets which had previously been largely ignored.

THEME 1.2. *(Dynamic) stage by stage computable construction of c.e. sets.*

Post moved significantly beyond the static conception of c.e. sets in Church and Kleene [1936] to a dynamic, stage by stage construction of c.e. sets, much like the process by which a modern digital computer performs its task. This is second nature to us now with our desktop computers, but it is not the way most mathematics was done in Post's time. It is far easier to follow than a formal mathematical definition in terms of Kleene's six schemata. From now on we use the term "dynamic" to refer to theorems which use a stage by stage construction in their proofs, but are not necessarily stated in dynamic terms as we discuss below, and we use the term "very dynamic" for those such as the New Extension Theorem 5.1 which use dynamic properties even in their statements.

THEME 1.3. *The study of reducibilities and relative computability.*

Post brought to a broad audience the definition of Turing reducibility from Turing [1939], where it had been obscurely placed in the middle of a long paper on another topic. Post augmented it by introducing several stronger reducibilities such as 1-reducible, m-reducible, tt-reducible, wtt-reducible, in addition to Turing's T-reducible. These were easier to understand than T-reducible, and Post related them to classes of c.e. sets he defined, such as creative, simple, h-simple, and hh-simple. His aim was to better understand how the information in a set A can be coded into and recovered from another set B.

THEME 1.4. *Post's Problem.*

Post posed his famous *Post's Problem,* of whether there exists a c.e. set A, noncomputable but such that A is strictly T-reducible to $K = \{e : e \in W_e\}$. In a subsequent paper, Post [1948], Post introduced the notion of two sets having the same *degree of unsolvability* if each is Turing reducible to the other. Thus, his problem can be stated as whether there is any degree of a c.e. set other than $\mathbf{0}$, the degree of the computable sets, and $\mathbf{0}'$, the degree of K. All this work on reducibilities, degrees, and coding of information had great impact.

THEME 1.5. *Relationship between set theoretic properties of c.e. sets and their degrees.*

Post approached his problem by trying to find a relationship between the (Turing) *degree* of c.e. set A and its structure as a *set*, measured either with respect to set inclusion, or equivalently with respect to the epsilon relation for "being an element of." For example, Post defined a c.e. set A to be *simple* if $\overline{A}$ is infinite, but contains no infinite c.e. set. Post gave a dynamic construction of a simple set and showed that such a set was incomplete with respect to a certain stronger reducibility, known as m-reducibility. We refer to *Post's Program* as the program to find connections between a c.e. set and its degree. This program had great influence, for example in the maximal set results by Yates, Sacks, and Martin in the 1960's and many later results.

It was only later that Myhill [1956] noticed that the collection of c.e. sets forms a lattice $\mathcal{E}$ under inclusion, but interestingly enough, almost all of the c.e. sets introduced by Post were later shown to be definable in the lattice $\mathcal{E}$. First, it is easy to see that the notion of being a finite set is definable (Soare, [1987, Ch X]), so it does not matter whether we consider $\mathcal{E}$ or $\mathcal{E}^*$, the lattice modulo finite sets. For $A \in \mathcal{E}$ let A^* denote its equivalence class in $\mathcal{E}^*$.

Post's simple set is clearly $\mathcal{E}^*$-definable. Of Post's other sets, hh-simple and creative are $\mathcal{E}^*$-definable (as later proved by Lachlan and Harrington, respectively, see Soare [1987]), but only h-simple is not. *Post's Program* launched in his 1944 paper was to find connections between the set theoretic structure (usually the $\mathcal{E}$-structure) of a c.e. set A and its (Turing) degree. This has had great influence on research ever since.

Convention. From now on all sets and degrees will c.e. unless otherwise stated.

2. Maximal Sets and Automorphisms

In 1956 Myhill called attention to the structure of the c.e. sets under inclusion, $\mathcal{E} = (\{W_e\}_{e\in\omega}, \subset)$, and noted that indeed it forms a lattice, although the lattice relations $\cup$, and $\cap$, and greatest element ω, and $\emptyset$, are all definable over $\mathcal{E}$ from $\subset$, so to build an automophism, it suffices to deal only with inclusion. Myhill also asked whether there is a maximal set, namely a set maximal in the inclusion ordering modulo finite sets, $(\mathcal{E}^*, \subset^*)$.

2.1. Maximal Sets and the Structure of $\mathcal{E}$.

DEFINITION 2.1. A coinfinite c.e. set A is *maximal* if there is no c.e. set W such that $W \cap \overline{A}$ and $\overline{W} \cap \overline{A}$ are both infinite. Equivalently, A is maximal iff A^* is a coatom (maximal element) in $\mathcal{E}^*$.

At the large and memorable Cornell logic meeting in 1957, Friedberg presented his construction of a maximal set [1958], done shortly after his solution of Post's problem [1957]. Maximal sets then became the object of intense study for several reasons: they were coatoms of $\mathcal{E}^*$ and hence the building blocks for more complicated lattices of supersets, they were the ultimate realization of Post's search for sets with thin complements; their degrees were very interesting, as we now see.

Tennenbaum conjectured that every maximal set is complete, but Sacks [1964] refuted this by building an incomplete maximal set. Yates [1965] built a complete maximal set. Martin completed this thread with a beautiful characterization of the Turing degrees (information content) of maximal sets.

DEFINITION 2.2. A c.e. set A is *high* (*low*) if its Turing jump $A' \equiv_{\mathrm{T}} \emptyset''$ ($A' \equiv_{\mathrm{T}} \emptyset'$), and the degree of a set A, written $\deg(A)$, is high or low according as A is. (See Soare [1987, Ch. III].)

THEOREM 2.3 (Martin, 1968). *The degrees of maximal sets are exactly the high degrees.*

Extending from maximal sets to those with the next thinnest complements, the hh-simple sets, Lachlan then proved a corresponding theorem.

DEFINITION 2.4. (i) Let $\mathcal{L}(A) = \{W : A \subseteq W\}$ the lattice of supersets of A.

(ii) Let $\mathcal{E}(S) = \{W \cap S : W \text{c.e.}\}$ the lattice of c.e. sets restricted to S, where S is not necessarily c.e.

Notice that for A c.e. there is a natural isomorphism between $\mathcal{L}(A)$ and $\mathcal{E}(\overline{A})$.

THEOREM 2.5 (Lachlan, 1968). *(i) A c.e. set A is hh-simple iff $\mathcal{L}(A)$ is a Boolean algebra.*

(ii) The degrees of hh-simple sets are exactly the high degrees.

For (ii) it had been known that hh-simple sets are high. Lachlan used Martin's method to prove that for every hh-simple set B and every high degree $\mathbf{d}$, there is an hh-simple $A \in \mathbf{d}$ such that $\mathcal{L}(A) \cong \mathcal{L}(B)$, which is stronger than (ii).

2.2. Felix Klein's Erlanger Programm. Another mathematical predecessor of the present paper is the emphasis on automorphism of a structure introduced by Felix Klein. When he was appointed to his professorship at the University of Erlangen his inaugural speech was devoted to a new approach to geometry. He argued that one should characterize a structure by those properties which remain invariant under all transformations which preserve the structure. This had a profound effect on the development of geometry.

In computability theory, by the later 1960's the structure of $\mathcal{E}$ was better understood, and people began asking about the automorphisms of $\mathcal{E}$. Several of these questions were stated in Rogers book [1967] such as: "Is every automorphism of $\mathcal{E}^*$ induced by one of $\mathcal{E}$?" "Is creativity invariant under $Aut(\mathcal{E})$?" Martin and Lachlan asked whether any two maximal sets are automorphic, namely are they in the same orbit?

2.3. Static Automorphisms in Mathematics. Automorphisms of a given structure now play a very important role in many branches of mathematics, for example, group theory, field theory and many others. When an algebraist builds an automorphism Φ of a countable object such as a group $\mathcal{G} = \{g_0, g_1, g_3, \dots\}$, he usually starts with an element g_0, picks out a suitable image $g_{h(0)}$ and continues. Even if he does not use a step by step approach, he still views the elements of $\mathcal{G}$ as indivisible whole objects. He usually does not collect some information on "part of" g_n during the process as we do here.

Since $\mathcal{E} = (\{W_n\}_{n \in \omega}, \subset)$ is so complicated, here we have no idea how to pick out for an arbitrary W_n an image $W_{h(n)}$ of the same elementary 1-type, much less how to continue the process to make Φ total, onto, and an automorphism. Hence, we are forced to take the partial information offered about $W_{n,s}$ during the construction and use it to build our approximation $\widehat{W}_{n,s}$ at stage s to the image $\widehat{W}_n = \Phi(W_n)$.

2.4. Dynamic Automorphisms for C.E. Sets. Every computable permutation of ω produces a trivial automorphism of $\mathcal{E}$, but the first nontrivial automorphism was constructed by Martin (unpublished, see Soare [1987, p. 345]) who showed that hypersimplicity is noninvariant, the only one of Post's properties to be noninvariant. A much more complicated method was necessary for more general automorphisms and was given by Soare who answered the preceding Martin-Lachlan question with the next result.

THEOREM 2.6 (Soare, 1974). *If A and B are any two maximal sets then there is an automorphism Φ of $\mathcal{E}$ mapping A to B.*

We defer a detailed discussion of the proof until §6, but we give here a little intuition. To approach this problem we cannot use a static automorphism, like the algebraist, but rather we need a *dynamic* approach to building the automorphism Φ. Given a set U we must build a set $\widehat{U}$ for $\Phi(U)$, enumerating elements in $\widehat{U}$ as more and more are enumerated in A, B, and U.

For each automorphism theorem we fix some simultaneous computable enumeration of all the sets under consideration, and let X_s denote the finite set of elements enumerated in X by the end of stage s. With respect to this simultaneous enumeration, define

$$X \setminus Y = \{x : (\exists s)[x \in X_s - Y_s]\}, \tag{1}$$

$$X \searrow Y = (X \setminus Y) \cap Y. \tag{2}$$

If elements x enter $U \setminus A$, threatening to make $|U - A| = \infty$, then we must enumerate some elements y into $\widehat{U} \setminus B$ toward making $|\widehat{U} \setminus B| = \infty$ also. Now the opponent can move such y into B ensuring $|\widehat{U} \searrow B| = \infty$. But we must match each such y with an $x \in U \cap A$ (and the opponent may arrange that $|A \searrow U| = \emptyset$.) Hence, in general we will need to guarantee that we can satisfy a *covering property* such as:

$$|\widehat{U} \searrow B| = \infty \implies |U \searrow A| = \infty. \tag{3}$$

To see the necessity of a condition like (3), suppose that B is simple and A is nonsimple with an infinite set $U \subset \overline{A}$. Now $U \setminus A$ infinite forces $\widehat{U} \setminus B$ infinite, but the opponent enumerates B to force $\widehat{U} \searrow B$ infinite, while $U \cap A = \emptyset$.

This is only a very weak version of the covering that is needed, and then after that we need an extension theorem like Theorem 5.1 to guarantee that the covering is sufficient to produce an automorphism, as we develop in §5.

The main thing to take away from this section is that dynamically building an automorphism from say, $U_1, \dots U_n$, to $\widehat{U}_1, \dots, \widehat{U}_n$, involves moving elements around on a giant Venn diagram with 2^n pieces, and trying to guarantee that one piece (state) is *well-resided* (occupied by infinitely many permanent residents) iff the corresponding piece on the other side is well-resided. The conditions like (3), (1), and (2), often involve the dynamic *flow* of elements from one well-visited state to another in the Venn diagram, where a piece (state) is *well-visited* if infinitely many elements rest there at least temporarily.

Definition 2.7. We say that a construction like Post's is *dynamic* if it involves a stage by stage construction, like most in computability theory, and *very dynamic* if it requires notions like (3), (1), and (2) which involve the flow of elements.

We will examine this in more detail later. However, with this brief glimpse at very dynamic properties, let us look at the surprising role they play in definability results.

3. Very Dynamic Aspects of Definable Properties.

For properties like Post's simple or hh-simple set, the definition and consequences are entirely static even though the construction may be dynamic. In the last two decades there arose a new class of proofs which relate a property $P(A)$, which is $\mathcal{E}$-definable (and therefore entirely static), to other static aspects such as $\deg(A)$, but the connection is entirely via certain *very dynamic* properties regarding the flow in a Venn diagram as above.

During the 1970's and early 1980's researchers had been trying to use the automorphism method to show that creative sets were not invariant, by taking a creative set A to a noncreative set B by an automorphism. Harrington analyzed the failure of these attempts and started an entirely new line of $\mathcal{E}$-definable properties, whose static definitions force some very dynamic behavior.

3.1. The Defining Property for Creative Sets. Post [1944] defined a c.e. set C to be *creative* if there is a partial computable function ψ such that

$$(\forall e)[W_e \subset \overline{C} \implies \psi(e)\downarrow \in \overline{C} - W_e]$$

It follows by Myhill's Theorem (see Soare [1987, p. 43]) that C is creative iff C is m-complete, *i.e.*, $W_e \leq_m C$ for every e, or equivalently $K \leq_m C$ for the Gödel complete set K. Although these properties of C at first appear to be very far from being $\mathcal{E}$-definable, Harrington (see Soare [1987, p. 339]) exhibited the following $\mathcal{E}$-definable property CRE(A) which defines C being creative.

THEOREM 3.1 (Harrington). *A c.e. set A is creative iff*

(4) CRE(A): $(\exists C \supset A)(\forall B \subseteq C)(\exists R)[R$ *is computable*

(5) & $R \cap C$ *is noncomputable* & $R \cap A = R \cap B]$,

where all variables range over $\mathcal{E}$.

We may represent the property CRE(A) as a two person game in the sense of Lachlan [1970] between the $\exists$-player (called RED, the definability player) who plays the c.e. sets C, R (the red sets) and the $\forall$-player (called BLUE, the automorphism player) who plays the c.e. set B (the blue set).

THEOREM 3.2 (Blue). *If* CRE(A) *then $K \leq_m A$ so A is creative.*

PROOF. (Sketch). Suppose CRE(A). We may visualize R as dividing the universe ω into two halves and on the R half we visualize in the Venn diagram the following states (corresponding roughly to e-states) $\nu_1 = R \cap \overline{C}$, $\nu_2 = R \cap (C - B)$, $\nu_3 = R \cap (B - A)$, $\nu_4 = R \cap A$. The static condition CRE(A) forces certain dynamic properties of the sets as follows. The condition that $R \cap C$ is noncomputable means that $R - C$ is not c.e. so there must be an infinite c.e. set of elements, say $\{x_n\}_{n \in \omega}$, which move from state ν_1 to state ν_2. Define $\psi(n) = x_n$. If n enters K, then enumerate $\psi(n)$ in B, from which the second conjunct of (5) eventually forces that $\psi(n) \in A$, so x_n passes from ν_1 to ν_2 to ν_3 to ν_4 in that order. If $n \in K$ then x_n remains in ν_2 forever. Hence, $K \leq_m A$ via ψ. For more details see Harrington-Soare [1998]. □

3.2. The Property $Q(A)$ Guaranteeing Incompleteness. In 1991 Harrington and Soare gave an $\mathcal{E}$-definable solution to Post's problem by producing an $\mathcal{E}$-definable property $Q(A)$ which guarantees that A is noncomputable and incomplete. It produces a much slower dynamic flow into A which prevents A from being prompt and hence complete.

DEFINITION 3.3. (i) A coinfinite c.e. set A is *promptly simple* if there is a computable function p and a computable enumeration $\{ A_s \}_{s\in\omega}$ of A such that for every e,

$$W_e \text{ infinite} \Longrightarrow (\exists s)\ (\exists x)\ [x \in W_{e,\text{ at } s} \cap A_{p(s)}]. \tag{6}$$

(ii) An c.e. set A is *prompt* if A has promptly simple degree namely, $A \equiv_T B$ for some promptly simple set B, and an c.e. degree is *prompt* if it contains a prompt set.

(iii) An c.e. set or degree which is not prompt is *tardy.*

By the Promptly Simple Degree Theorem, see Theorem XIII.1.7(iii) of Soare [1987], a set A being prompt is equivalent to the following property. Let $\{A_s\}_{s\in\omega}$ be any computable enumeration of A. Then there is a computable function p such that for all s, $p(s) \geq s$, and for all e,

$$W_e \text{ infinite} \Longrightarrow (\exists^\infty x)\ (\exists s)\ [x \in W_{e,\text{ at } s}\ \ \&\ \ A_s{\upharpoonright}x \neq A_{p(s)}{\upharpoonright}x], \tag{7}$$

namely infinitely often A "promptly permits" on some element $x \in W_e$.

DEFINITION 3.4. (i) A subset $A \subset C$ is a *major subset* of C(written $A \subset_{\text{m}} C$) if $C - A$ is infinite and for all e,

$$\overline{C} \subseteq W_e \Longrightarrow \overline{A} \subseteq^* W_e.$$

(Note that if $A \subset_{\text{m}} C$ then both A and C are noncomputable.)

(ii) $A \sqsubset B$ if there exists C such that $A \sqcup C = B$ (i.e. $A \cup C = B$ and $A \cap C = \emptyset$).

THEOREM 3.5 (Harrington-Soare, 1991). *There is a property $Q(A)$ which guarantees that A is tardy and hence, $A <_{\text{T}} K$, and which holds of some noncomputable set.*

Define the property: $Q(A): \quad (\exists C)_{A\subset_{\text{m}}C}\ (\forall B \subseteq C)(\exists D \subseteq C)(\forall S)_{S\sqsubset C}$ [

$$[B \cap (S - A) = D \cap (S - A)] \tag{8}$$

$$\Longrightarrow (\exists T)[\overline{C} \subset T\ \&\ A \cap (S \cap T) = B \cap (S \cap T)]]. \tag{9}$$

THEOREM 3.6. *If $Q(A)$ then A is incomplete (*i.e., *$A \not\leq_{\text{T}} K$).*

PROOF. (Sketch only, see Harrington-Soare [1991] for details.) We may visualize the property $Q(A)$ as a two person game in the sense of Lachlan [1970] between the $\exists$-player (RED) who plays the c.e. sets A, C, D and T and the $\forall$-player (BLUE) who plays the c.e. sets B and S. For simplicity ignore all the sets but C, D, B, and A, since the others are necessary only to give us a suitable domain on which to play the following strategy. Visualize $C \supseteq D \supseteq B \supseteq A$, and let $\nu_1, \nu_2, \dots \nu_5$ denote the differences of c.e. sets (called *d.c.e. sets*): $\omega - C$, $C - D$, $D - B$, $B - A$, A respectively, but viewed dynamically like e-states, so an element can pass from ν_i to ν_j, $i < j$. The oversimplified $Q(A)$ property now asserts that if BLUE plays: $(8)'$ $D = B$ on $\overline{A}$, then RED will play: $(9)'$ $B = A$. In particular, if both players are following their best strategies, then for an element x to enter A, it must pass through the ν-states in the order $\nu_1, \nu_2, \dots \nu_5$ as proved in Harrington-Soare [1991]. However, the set B acts like a wall of restraint, like the minimal pair restraint of Lachlan and Yates in Soare [1987, p. 153]. When presented with an $x \in D - B$, BLUE may hold x as long as he likes, but must eventually put x into B at which point RED is free to put x into A but not before. This implies that A is tardy (*i.e.*, not of promptly simple degree, see Ambos-Spies, Jockusch, Shore, and Soare in Soare [1987, p. 284] or [1987, Chap III], so A is incomplete. $\square$

Furthermore, Harrington-Soare [1996b] have discovered that $Q(A)$ imposes a much stronger tardiness property on A (called *2-tardy*) which helps us classify those sets which can be coded into any nontrivial orbit.

There are other $\mathcal{E}$-definable properties such as $T(A)$ in Harrington-Soare [1998] which guarantee that A is complete and which can hold of promptly simple sets. This works by ensuring the existence of a state ν_1 (of elements) in $\overline{A}$ which is well resided, but from which it is legal to move any element to some ν_s inside of A for coding $K \leq_{\mathrm{T}} A$. There is also a property $NL(A)$ which ensures that A is not low even though it can be low$_2$ and promptly simple. It achieves this by forcing an infinite stream of elements to move through a sequence of states to overcome any function which is a candidate to prove lowness. (See Harrington-Soare [1998a] on $NL(A)$ for details.)

The conclusion to draw is that not only do definable properties act in opposition to automorphisms, each limiting the power of the other, but now the battlefield on which they compete is that of very dynamic properties, an arena not obvious from the static definitions of $\mathcal{E}$-definable properties.

4. Building an Automorphism of $\mathcal{E}$

We fix two copies of the integers, ω and $\widehat{\omega}$ and a standard listing $\{U_n\}$ of the c.e. sets on the ω-side and $\{V_n\}$ on the $\widehat{\omega}$-side, which we view as being played by the opponent, called RED . During our dynamic construction we, Player BLUE, must construct sets $\{\widehat{U_n}\}$ for the $\{\widehat{\omega}\}$-side, and $\{\widehat{V_n}\}$ for the ω-side which meet the following condition, (10) and usually the stronger condition (11).

4.1. Definitions of e-states.

Definition 4.1. (i) Given two sequences of c.e. sets $\{X_n\}_{n\in\omega}$ and $\{Y_n\}_{n\in\omega}$, define $\nu(e,x)$, the *full e-state* of x with respect to (w.r.t.) $\{X_n\}_{n\in\omega}$ and $\{Y_n\}_{n\in\omega}$ to be the triple $\langle e, \sigma(e,x), \tau(e,x)\rangle$, where

$$\sigma(e,x) = \{i : i \leq e \ \&\ x \in X_i\}, \quad \tau(e,x) = \{i : i \leq e \ \&\ x \in Y_i\}.$$

(ii) If $x \in \omega$ we measure $\nu(e,x)$ with respect to $\{U_n\}_{n\in\omega}$ and $\{\widehat{V}_n\}_{n\in\omega}$, and if $\hat{x} \in \widehat{\omega}$ we measure $\widehat{\nu}(e,\hat{x})$ with respect to $\{\widehat{U}_n\}_{n\in\omega}$ and $\{V_n\}_{n\in\omega}$. If $\nu(e,x) = \nu$ we say x is *in* e-state ν and likewise for $\hat{x}$ and $\widehat{\nu}(e,\hat{x})$.

(iii) If $U_{n,s}$, $\widehat{V}_{n,s}$, $n, s \in \omega$ is a computable approximation to U_n, $\widehat{V}_n$, then we define $\nu(e,x,s)$ as above, but with U_n, $\widehat{V}_n$ replaced by $U_{n,s}$, $\widehat{V}_{n,s}$ and likewise for $\widehat{\nu}(e,\hat{x},s)$ with respect to $\widehat{U}_{n,s}$ and $V_{n,s}$.

Definition 4.2. (i) The *well-resided* e-states on the ω-side and $\widehat{\omega}$-side respectively are

$$\mathcal{K}_e = \{\nu : (\exists^\infty x)[\nu(e,x) = \nu]\}, \quad \widehat{\mathcal{K}}_e = \{\nu : (\exists^\infty \hat{x})[\widehat{\nu}(e,\hat{x}) = \nu]\}.$$

(ii) The *well-visited* states are

$$\mathcal{M}_e = \{\nu : (\exists^\infty x)[\nu(e,x,s) = \nu]\}, \quad \widehat{\mathcal{M}_e} = \{\nu : (\exists^\infty \hat{x})[\nu(e,\hat{x},s) = \nu]\}.$$

(iii) For the ω-side the *well-resided* states are $\mathcal{K} = \bigcup_{e\in\omega} \mathcal{K}_e$, the *well-visited states* are $\mathcal{M} = \bigcup_{e\in\omega} \mathcal{M}_e$, and similarly define $\widehat{\mathcal{K}}$ and $\widehat{\mathcal{M}}$ for the $\widehat{\omega}$ side.

The picture is now very simple. The e-states ν measure boxes in the Venn diagram of ω partitioned by U_n, $\widehat{V}_n$, $n \leq 3\ e$, giving 2^{2e+2} partitions or full e-states, and likewise for the $\widehat{\omega}$-side. To ensure that Φ is an automorphism it is necessary and sufficient to ensure that every full e-state ν is well-resided on the ω-side iff the corresponding full e-state $\widehat{\nu}$ is well-resided on the $\widehat{\omega}$-side, namely iff

$$\mathcal{K} = \widehat{\mathcal{K}}. \tag{10}$$

In practice, we achieve (10) by achieving the somewhat stronger condition,

$$\mathcal{M} = \widehat{\mathcal{M}} \quad \text{and} \quad \mathcal{N} = \widehat{\mathcal{N}}, \tag{11}$$

where the set $\mathcal{N}$ ($\widehat{\mathcal{N}}$) consists of those $\nu \in \mathcal{M}(\widehat{\mathcal{M}})$ which are "emptied out" during the construction in the sense that almost all elements in ν are moved, either by RED or BLUE, to another state. Clearly, the well-resided states $\mathcal{K}$ are exactly those states ν which are well-visited ($\nu \in \mathcal{M}$) but not emptied out ($\nu \notin \mathcal{N}$). Hence,

$$\mathcal{K} = \mathcal{M} - \mathcal{N} \ \ \& \ \ \widehat{\mathcal{K}} = \widehat{\mathcal{M}} - \widehat{\mathcal{N}}. \tag{12}$$

Thus, the primary strategy for the automorphism player, BLUE, is to *copy* the moves of RED, namely first to copy the well visited states to make $\mathcal{M} = \widehat{\mathcal{M}}$ and then to empty out a state if RED empties out the corresponding state.

4.2. Building an Automorphism on a Tree. If we proceed as in §4.1 as for Soare's original proof of Theorem 6.5 we produce an effective automorphism in the sense that the automorphism is presented by a computable map on indices of c.e. sets. It is more powerful to combine the automorphism method with Lachlan's method of trees to produce, not effective automorphisms, but Δ^0_3-automorphisms as introduced by Harrington-Soare [1996c] and Cholak [1995]. These are much more versatile, as we shall see.

The priority tree T consists of nodes defined roughly as follows. Suppose node $\beta \in T$. We put nodes $\alpha = \beta^\frown\langle k_\alpha, \mathcal{M}_\alpha, \mathcal{N}_\alpha\rangle$ in T where sets $\mathcal{M}_\alpha$ and $\mathcal{N}_\alpha$ have approximately the meanings above, and k_α is the least n beyond which all elements move only among well-visited states. We let f denote the true path of T and we use a computable approximation f_s such that $f = \liminf_s f_s$.

Each element x of the ω-side is placed at the end of stage s on a node $\alpha \in T$ denoted by $\alpha(x, s)$. Define the c.e. set

$$Y_{\alpha,s} = \{x : (\exists t \leq s)[\alpha(x,t) \supseteq \alpha]\}, \tag{13}$$

where $\gamma \supseteq \beta$ denotes that γ is an extension of β on T. We omit most of the details of T which can be found in Harrington-Soare [1996c, §2].

4.3. α-states. For conceptual simplicity we do as little action as possible at each node $\alpha \in T$. If $|\alpha| \equiv 1 \bmod 5$ ($|\alpha| \equiv 2 \bmod 5$), we consider one new U set (V set). If $|\alpha| \equiv 3 \bmod 5$ ($|\alpha| \equiv 4 \bmod 5$), we consider new α-states ν ($\hat{\nu}$) which may be non well-resided on Y_α ($\widehat{Y}_\alpha$). If $\alpha \equiv 0 \bmod 5$ we make no new commitments for the automorphism machinery but we may perform action for some additional requirement (such as coding information into B). We shall arrange that for all $n \in \omega$ that for $\alpha \subset f$,

$$|\alpha| = 5n + 1 \implies U_\alpha =^* U_n, \text{ and} \tag{14}$$

$$|\alpha| = 5n + 2 \implies V_\alpha =^* V_n. \tag{15}$$

We let U_α and $\widehat{U}_\alpha$ (V_α and $\widehat{V}_\alpha$) be undefined if $|\alpha| \not\equiv 1 \bmod 5$ ($|\alpha| \not\equiv 2 \bmod 5$). We let $e_\alpha(\hat{e}_\alpha)$ correspond to n in (14) (respectively (15)). Namely, define $e_\lambda = \hat{e}_\lambda = -1$ and if $|\alpha| \equiv 1 \bmod 5$ then let $e_\alpha = e_{\alpha^-} + 1$, and otherwise let $e_\alpha = e_{\alpha^-}$. Define $\hat{e}_\alpha$ similarly with $|\alpha| \equiv 2 \bmod 5$ in place of $|\alpha| \equiv 1 \bmod 5$. Hence, $e_\alpha > e_{\alpha^-}$ ($\hat{e}_\alpha > \hat{e}_{\alpha^-}$) iff $|\alpha| \equiv 1 \bmod 5$ ($|\alpha| \equiv 2 \bmod 5$).

DEFINITION 4.3. An *α-state* is a triple $\langle \alpha, \sigma, \tau \rangle$ where $\sigma \subseteq \{0, \ldots, e_\alpha\}$ and $\tau \subseteq \{0, \ldots, \hat{e}_\alpha\}$. The only λ-state is $\nu_{-1} =< \lambda, \emptyset, \emptyset\rangle$.

The construction will produce a simultaneous computable enumeration $U_{\alpha,s}$, $V_{\alpha,s}$, $\widehat{U}_{\alpha,s}$, $\widehat{V}_{\alpha,s}$, for $\alpha \in T$ and $s \in \omega$, of these r.e. sets which we use in the following definition.

DEFINITION 4.4. (i) The *α-state of x at stage s*, $\nu(\alpha, x, s)$, is the triple $\langle \alpha, \sigma(\alpha, x, s), \tau(\alpha, x, s) \rangle$ where

$$\sigma(\alpha, x, s) = \{e_\beta : \beta \subseteq \alpha \ \& \ e_\beta > e_{\beta^-} \ \& \ x \in U_{\beta,s}\},$$

$$\tau(\alpha, x, s) = \{\hat{e}_\beta : \beta \subseteq \alpha \ \& \ \hat{e}_\beta > \hat{e}_{\beta^-} \ \& \ x \in \widehat{V}_{\beta,s}\}.$$

(ii) The *final α-state of x* is $\nu(\alpha, x) = \langle \alpha, \sigma(\alpha, x), \tau(\alpha, x) \rangle$ where $\sigma(\alpha, x) = \lim_s \sigma(\alpha, x, s)$ and $\tau(\alpha, x) = \lim_s \tau(\alpha, x, s)$.

4.4. Templates as a Guide. For $\alpha \in T$ we refer to the various objects, associated with α, $\mathcal{M}_\alpha$, $\mathcal{N}_\alpha$, and k_α as *templates*[1] because from this information alone α can enumerate the appropriate sets (during those stages when α is accessible, namely $\alpha \subset f_s$). If α is on the true path then these sets will be correct.

The crucial point of the template-tree method is that each node $\alpha \in T$ at a given level works completely independently, acting only when it is accessible, and using its templates as a guide to its action.

To *relativize* the templates to A ($\overline{A}$) means to restrict to only those states in A ($\overline{A}$) respectively, and similarly for B and $\overline{B}$. For example, $\mathcal{M}_\alpha^{\overline{A}}$ and $\mathcal{N}_\alpha^{\overline{A}}$ refer to the well-visited and nonwell-resided α-states of $\overline{A}$, respectively.

4.5. Skeletons and Enumerations.

DEFINITION 4.5. An array of c.e. sets $\{U_n\}_{n\in\omega}$ is a *skeleton* (*basis*) (for $\mathcal{E}$) if

$$(\forall e)(\exists n)[U_n =^* W_e].$$

DEFINITION 4.6. (i) Given tree T with true path $f \in [T]$, an *enumeration* $\mathbb{E}$ for T is a simultaneous computable enumeration of c.e. sets U_α, V_α, and $\widehat{U}_\alpha$, $\widehat{V}_\alpha$, for $\alpha \in T$, such that $\{U_\alpha\}_{\alpha\subset f}$ and $\{V_\alpha\}_{\alpha\subset f}$ are both skeletons.

(ii) Let $\rho = f(0)$ and let $A = U_\rho$ and $B = \widehat{U}_\rho$ (the first sets to be matched as in the Automorphism Theorem, Theorem 6.5).

(iii)

$${}^{\mathbb{E}}\mathcal{M}_\alpha = \{\nu : (\exists^\infty x)(\exists s)[\nu(\alpha, x, s) = \nu \ \& \ x \in Y_{\alpha,s}]\},$$

and ${}^{\mathbb{E}}\widehat{\mathcal{M}}_\alpha$ is defined similarly on $\widehat{\omega}$-side.

---

[1]The Merriam-Webster Collegiate Dictionary defines "template" as a "gauge, pattern, or mold used as a guide to the form of a piece being made; a molecule (as of DNA) that serves as a pattern for the generation of another macromolecule (as messenger RNA)." We have carefully chosen the this word to explain our concept after exploring many others with similar meaning, such as "blueprint." This conveys an important intuition into the entire automorphism machinery.

(iv) An enumeration $\mathbb{E}'$ is an *extension* of an enumeration $\mathbb{E}$ if $S \subseteq S'$ for each set $S = U_\alpha, V_\alpha, \widehat{U}_\alpha, \widehat{V}_\alpha$ of $\mathbb{E}$ and the corresponding set $S' = U'_\alpha, V'_\alpha, \widehat{U}'_\alpha, \widehat{V}'_\alpha$ of $\mathbb{E}'$.

There is a deliberate ambiguity in the templates. For example, the template $\mathcal{M}_\alpha$ is used to represent an abstract blueprint which α uses to build the flow of elements into Y_α and it is also used to denote the actual flow,

$$^{\mathbb{E}}\mathcal{M}_\alpha = \{\nu : (\exists^\infty x)(\exists s)[\nu(\alpha, x, s) = \nu \ \& \ x \in Y_{\alpha,s}]\},$$

For $\alpha \subset f$ these will be the same. Strictly speaking, the templates have the former meaning and we must attach the left hand superscript $\mathbb{E}$ to denote the latter, but we will omit this superscript when there is no danger of confusion and write merely $\mathcal{M}_\alpha$ to denote the actual flow when the intended $\mathbb{E}$ is clear.

DEFINITION 4.7. Given an enumeration $\mathbb{E}$ with true path f define for each set of templates, $\mathcal{M}_\alpha, \mathcal{K}_\alpha, \mathcal{N}_\alpha$, $\alpha \in T$, the "join" along the true path f,

$$\mathcal{S}_f = \bigcup \{\mathcal{S}_\alpha : \alpha \subset f\} \quad \text{and } \widehat{\mathcal{S}}_f = \bigcup \{\widehat{\mathcal{S}}_\alpha : \alpha \subset f\}. \tag{16}$$

where $\mathcal{S}_\alpha$ denotes $\mathcal{M}_\alpha, \mathcal{K}_\alpha, \mathcal{N}_\alpha$, or other α-templates.

CONVENTION 4.8. For fixed $\mathbb{E}$ and f we often drop the subscript f from $\mathcal{S}_f$ in Definition 4.7 and simply write $\mathcal{S}$.

Note that $\mathcal{S}$ is not strictly a template but rather an infinite set of templates $\mathcal{S}_\alpha$, but it enables us to simplify the statements of conditions because

$$\mathcal{S} = \widehat{\mathcal{S}} \iff (\forall \alpha \subset f)[\mathcal{S}_\alpha = \widehat{\mathcal{S}}_\alpha].^2$$

DEFINITION 4.9. For a fixed enumeration $\mathbb{E}$ define two other sets,

$$\mathcal{G}^A_\alpha = \{\nu : (\exists^\infty x)(\exists s)[x \in A_{s+1} - A_s \ \& \ \nu = \nu(\alpha, x, s)]\},$$

$$\widehat{\mathcal{G}}^B_\alpha = \{\nu : (\exists^\infty \hat{x})(\exists s)[x \in B_{s+1} - B_s \ \& \ \nu = \nu(\alpha, \hat{x}, s)]\}.$$

Think of $\mathcal{G}^A$ as a "gatekeeper set" consisting of nodes $\nu \in \mathcal{M}^{\overline{A}}$ from which infinitely many elements x leave to enter A.

5. The New Extension Theorem

The main tool for almost our results on automorphisms is the following.

THEOREM 5.1 (New Extension Theorem (N.E.T.)). *If an enumeration $\mathbb{E}$ satisfies both:*

(T1) $\mathcal{K}^{\overline{A}} = \widehat{\mathcal{K}}^{\overline{B}}$, *[Static equality of states of $\overline{A}/\overline{B}$], and*

(T2) $\mathcal{G}^A = \widehat{\mathcal{G}}^B$, *[Covering],*

then $A \simeq B$ by an enumeration $\mathbb{F}$ which extends $\mathbb{E}$.

[2] The direction from right to left holds because for every $\beta <_L f$ the set U_β finite, and for every $\beta >_L f$ we discard each β-set infinitely often. Therefore, each β-set S_β, $\beta \not\subset f$, may be taken to be finite, where S_β is a variable ranging over $U_\beta, V_\beta, \widehat{U}_\beta, \widehat{V}_\beta$. Hence $\mathcal{M}_\beta$ and all its subset templates are empty for $\beta \not\subset f$. Therefore, the equation (16) could have been defined with $(\forall \alpha \in T)$ in place of $(\forall \alpha \subset f)$.

REMARK 5.2. (i) (T1) implies that $\mathcal{L}(A) \cong \mathcal{L}(B)$, and hence ensures the equality between the static (well-resided) states of $\overline{A}$ and $\overline{B}$.

(ii) (T2) asserts intuitively that A (exactly) *covers* B in the sense that if infinitely many elements enter B from a state $\widehat{\nu}$ of $\overline{B}$ then infinitely many must enter A from the same state ν and conversely.[3]

(iii) Of the two conditions, (T2), is much more recognizable, and of course, both are necessary, but satisfying (T1) is in general much more delicate and requires the more ingenious construction.

(iv) We normally satisfy (T1) above by satisfying the stronger *dynamic* condition,

$$\text{(T1)}' \quad (\mathcal{M}^{\overline{A}}, \mathcal{N}^{\overline{A}}) = (\widehat{\mathcal{M}^{\overline{B}}}, \widehat{\mathcal{N}^{\overline{B}}}). \qquad \text{[Dynamic matching of } \overline{A} \text{ and } \overline{B}\text{]}$$

(v) In many cases, for example if A and B are promptly simple or maximal, we satisfy (T2) by satisfying both (T1)$'$ and

$$\text{(T2)}' \quad \mathcal{M}^{\overline{A}} = \mathcal{G}^{A} \quad \& \quad \widehat{\mathcal{M}^{\overline{B}}} = \widehat{\mathcal{G}}^{B}. \qquad \text{[Autocover]}$$

(vi) For simplicity of exposition, we have deliberately not stated the N.E.T. in its strongest form. See the General Extension Theorem, Theorem 6.9, (G.E.T.) which has the same proof.

Note that Condition (T1)$'$ implies (T1) because $\mathcal{K}^{\overline{A}} = \mathcal{M}^{\overline{A}} - \mathcal{N}^{\overline{A}}$ as in (12). However, (T1)$'$ is much stronger than (T1) because it asserts that the correspondence is not merely a static one at the end of the construction, but a *dynamic* one describing corresponding states *during* the construction.

Since we always have $\mathcal{G}^{A} \subset \mathcal{M}^{\overline{A}}$ and likewise for B, condition (T2)$'$ asserts that $\mathcal{G}^{A}$ is as large as possible, which we call the *Autocover* case because A (or more precisely $\mathcal{G}^{A}$) is dynamically covering its own complement $\overline{A}$ (or more precisely $\mathcal{M}^{\overline{A}}$). In the presence of (T1)$'$, note that (T2)$'$ easily implies (T2) by transitivity of equality. Hence,

$$(17) \qquad [(T1)' \ \& \ (T2)'] \implies [(T1) \ \& \ (T2)] \implies A \simeq B.$$

In fact, most automorphisms, both effective and Δ^0_3, have been built to satisfy (T1)$'$ and (T2)$'$ rather than merely (T1) and (T2). One of the most tempting pitfalls in the subject, leading to several false conjectures, has been the following fact.

$$(18) \qquad (T1) \ \& \ (T2)' \not\Rightarrow \quad A \simeq B.$$

It is very tempting, for example, to assert that if two promptly simple sets, A and B, satisfy $\mathcal{L}(A) \cong \mathcal{L}(B)$ then $A \simeq B$, but that is false. For a counterexample see Harrington and Soare [1998, p. 123] where B is low, A is low$_2$ with $\mathcal{L}(A) \cong \mathcal{L}(B) \cong \mathcal{E}$, and A and B are promptly simple, but $A \not\simeq B$.

---

[3]In the earlier effective Extension Theorem machinery this was stated in a form that for such ν in $\overline{B}$ there must be $\nu' \geq \nu$ meaning that $\sigma' \supseteq \sigma$ and $\tau' \subseteq \tau$, but in the Harrington-Soare Δ^0_3-machinery we have equality of well visited states on $\overline{A}$ and $\overline{B}$ which makes the covering hypothesis (T2) much simpler than the old set of hypotheses in Soare [1987, 352–353].

5.1. Duality. One of the beautiful things about studying automorphisms of $\mathcal{E}$, especially with the Δ^0_3-method, is the duality between the ω and $\widehat{\omega}$ sides. The hypotheses (T1) and (T2) ensure that $\mathcal{K}^{\overline{A}} = \widehat{\mathcal{K}}^{\overline{B}}$ and $\mathcal{G}^A = \widehat{\mathcal{G}}^B$. These properties ensure that for any template $\mathcal{S}$ defined from them we have,

$$(19) \qquad \mathcal{S} = \widehat{\mathcal{S}} \qquad \text{[Duality]}$$

where the "dual" $\widehat{\mathcal{S}}$ is obtained from $\mathcal{S}$ by replacing ω, U_α, and $\widehat{V}_\alpha$ by $\widehat{\omega}$, $\widehat{U}_\alpha$, and V_α, respectively.

In the following we will define various templates, such as $\mathcal{W}_\alpha$, $\mathcal{C}_\alpha$, and $\mathcal{D}_\alpha$. We will show that various properties on A, such as noncomputability, promptly simplicity, promptness, simplicity, imply various corresponding properties for these templates, and will immediately conclude that the dual template has the same properties. This is exactly what is needed for the coding in Theorems 7.1, 7.2, 7.5, and others.

5.2. Advantages of the New Extension Theorem. There are a number of advantages of the New Extension Theorem over the old Extension Theorem in Soare [1987, p. 352].

1. Compared to the old Extension Theorem, this New Extension Theorem is more powerful and versatile, producing Δ^0_3-automorphisms. These apply, for example, to Theorem 7.1 that any noncomputable c.e. set is automorphic to a high one, which effective automorphisms do not. The fact that the New Extension Theorem can be done on a tree while the old one cannot makes it much more applicable.
2. The N.E.T. can be applied wherever the old Extension Theorem applies, except that now we produce a Δ^0_3-automorphism instead of an effective one. However, this slightly weaker conclusion is usually of little importance compared to extra versatility.
3. The two new hypotheses, while similar to the old ones, are much more compact and simple to state. This comes partly from the tree notation, and partly from the fact that the Harrington-Soare method ensures equality of states in $\mathcal{M}$ so we do not have to use inequalities to state the "cover" and "co-cover" properties as in Soare [1987, p. 352].

The New Extension Theorem has the following advantages over the plain Harrington-Soare Δ^0_3-automorphism method [1996c] on which it is based.

1. The Harrington-Soare Δ^0_3-method was designed for the applications in that paper and a few others, so the most general statement is the Automorphism Theorem [1996c, p. 633], which refers to the details of the construction there and says roughly that any variation of the construction which satisfies four conditions, stated in terms of that construction, produces an automorphism of $\mathcal{E}$. This makes it less portable than N.E.T. to another user or even to the authors themselves who need N.E.T. in their proof (Harrington-Soare [ta]) that $\overline{\mathbf{L}_1}$ is not invariant. To apply it one must get into the details of the construction.
2. Other users and the authors themselves who wished to apply a modified or extended version of the Harrington-Soare Δ^0_3-method had to write things like, "Go to page 631 of Harrington-Soare, change this to that, and then modify the lemmas as follows." Now the reference can just be the statement

of the N.E.T., a statement *external* to the the steps of the construction, and the reader need not go into the construction at all.

Of course, the N.E.T. is proved with the Harrington-Soare Δ^0_3-method, so the proof does not require a completely new method, but now we do not have to reapply the latter to build the automorphism taking A to B. However, we often still use the Δ^0_3-method to satisfy (T1), namely guaranteeing the isomorphism from $\overline{A}$ to $\overline{B}$.

6. Type 1 Automorphisms: Where We Are Given Both A And B

We now begin a number of applications of the New Extension Theorem to give simple proofs of theorems in the literature. These are only proof sketches, but are nearly full proofs. In these proofs the ideas of the proof may be the same as in the original author's papers, but the presentations are recast in the more general N.E.T. framework for greater perspicuity, and from the N.E.T. we can give shorter, cleaner proofs than from the original Soare Extension Theorem which all the authors used.

6.1. Promptly Simple Low Sets. Let's begin to translate normal properties of computability theory into the abstract template properties used in the N.E.T.

LEMMA 6.1 (Maass, 1981). *If we are given an enumeration $\mathbb{E}$ and a promptly simple set A then we can find an new enumeration $\mathbb{F}$ which speeds up the enumeration of A, leaves the enumeration of all other sets alone, and satisfies (T2)$'$: $\mathcal{G}^A = \mathcal{M}^{\overline{A}}$.*

PROOF. Let $\{\widetilde{A}_s : s \in \omega\}$ be the enumeration of A in $\mathbb{E}$. For each $\alpha \in \mathcal{M}^{\overline{A}}$ and each α-state ν we will define a c.e. set Z_ν which has Kleene index $W_{h(\nu)}$ for a computable function h. Whenever a new element x appears in state ν at stage s of enumeration $\mathbb{E}$, we put x into Z_ν at stage s, wait for the first t such that $x \in W_{h(\nu),t}$ and compute $\widetilde{A}_{p(t)}$. Let u be the maximum of these $p(t)$ over all such ν at s, and define $A_s = \widetilde{A}_{p(u)}$. □

THEOREM 6.2 (Soare, 1982). *If A is low (or even if $\overline{A}$ is merely semilow), then $\mathcal{L}^*(A) \cong \mathcal{E}^*$ by an isomorphism effective on indices.*

This theorem and other applications of lowness use the Lowness Lemma 6.3 and (T3) below whose explanation requires a bit more notation. In the full Harrington-Soare Δ^0_3 automorphism method [1996c, §2], the set $\mathcal{N}_\alpha$ of states being emptied out was split into the disjoint union $\mathcal{N}_\alpha = \mathcal{R}_\alpha \sqcup \mathcal{B}_\alpha$, where the nodes in $\mathcal{R}_\alpha$ are being emptied by RED and those in $\mathcal{B}_\alpha$ by BLUE . We need $\mathcal{R}_\alpha = \widehat{\mathcal{B}_\alpha}$ so that if RED empties state $\nu \in \mathcal{R}_\alpha$ then BLUE empties the same state in $\widehat{\mathcal{B}}_\alpha$ on the $\widehat{\omega}$-side. The heart of the whole method is to prove $\mathcal{R}$-consistency, namely

$$(20) \qquad (\forall \nu_0 \in \mathcal{R}_\alpha)(\exists \nu_1)[\nu_0 <_R \nu_1 \ \ \& \ \ \nu_1 \in \mathcal{M}_\alpha],$$

If this holds, then by (19) (duality),

$$(21) \qquad (\forall \hat{\nu}_0 \in \widehat{\mathcal{B}}_\alpha)(\exists \hat{\nu}_1)[\hat{\nu}_0 <_B \hat{\nu}_1 \ \ \& \ \ \hat{\nu}_1 \in \widehat{\mathcal{M}}_\alpha].$$

Hence, if BLUE must empty out a state $\nu_0 \in \widehat{\mathcal{M}}_\alpha$ there is a state $\nu_1 \in \widehat{\mathcal{M}}_\alpha$ into which he can move the elements.

To prove Theorem 6.2 we must essentially match $\overline{A}$ with $\overline{B} = \widehat{\omega}$ so *no* elements of $\overline{B}_s$ ever leave $\overline{B}$ even if elements of $\overline{A}_s$ leave to enter A. Relativizing all the previous templates to $\overline{A}$, lowness allows us to prove the next lemma.

LEMMA 6.3 (Lowness Lemma). *If A is low then the enumeration $\mathbb{E}$ can be arranged so that we have the following.*

$$(T3) \qquad (\forall \nu_0 \in \mathcal{R}_\alpha^{\overline{A}})(\exists \nu_1)[\nu_0 <_R \nu_1 \ \& \ \nu_1 \in \mathcal{M}_\alpha^{\overline{A}}].$$

This means that if a state ν_0 of $\overline{A}$ is emptied by RED, he cannot empty *all* its elements into A, but must empty infinitely many into some other state ν_1 of $\overline{A}$. Hence, on the $\widehat{\omega}$-side BLUE can empty the state ν_0 of $\overline{B}$ into state ν_1 still in $\overline{B}$.

PROOF. (Sketch) For each state $\nu \in \mathcal{M}^{\overline{A}}$ choose a marker Γ. Using the Robinson lowness trick in Soare [1987, pp. 224-228] ask whether $Q : Z \cap \overline{A} \neq \emptyset$ where Z is the set of future positions of Γ. Move Γ to some x in ν only when one exists and the $0'$-oracle says yes to Q. Hold x against any BLUE changes so if ν_0 is emptied by RED , he must move some such x into a state ν_1 of $\overline{A}$. Repeat with infinitely many markers to get infinitely many such x. □

Since this is a result on $\mathcal{L}^*(A)$ rather than on automorphisms of A, Theorem 6.2 is not affected by the N.E.T., although the Δ_3^0-method gives a more perspicuous proof for the noneffective case.

THEOREM 6.4 (Maass, 1981). *If A and B are low and promptly simple then $A \simeq B$.*

PROOF. By Lemma 6.1 we have (T2)$'$. The proof of Theorem 6.2 guarantees (T1)$'$. Hence, by N.E.T. we have $A \simeq B$. □

6.2. Automorphisms of Maximal Sets.

THEOREM 6.5 (Soare, 1974). *If A and B are maximal sets then $A \simeq B$.*

PROOF. We will modify the fixed enumeration $\{U_n\}$ to get a skeleton $\{U_n'\}$, and then measure states ν with respect to the latter. Since A is maximal we have for each W either $W \subset^* A$ or $W \supset^* \overline{A}$, namely that $\overline{A}$ obeys a *zero-one law*.

First, we can arrange that the enumeration of U_n' is *order-preserving* in the sense that if an element, while in $\overline{A}$, appears in both U_0' and U_1' then it appears in U_0' *first*, namely that $(U_1' \searrow U_0') \setminus A = \emptyset$.

Second, if an element $x \in \overline{A}_s$ in state ν_1 enters U_n, and if state ν_2 (ν_2') corresponds to ν_1 together with U_n (U_n'), then we withhold x from U_n' either forever, or until some new element y in state ν_2 enters A and put y into U_n' just before allowing y to enter A. This guarantees (T2)$'$ in a very strong way because the element x is not even allowed to enter state ν_2' while in $\overline{A}$ until a new witness in state ν_2 enters A. This construction guarantees (T1)$'$ and (T2)$'$, and hence the automorphism by applying the N.E.T. □

REMARK 6.6. Although maximal sets may be either prompt or tardy, the previous proof shows that *maximal sets are always promptly simple on a skeleton*, and hence satisfy (T2)$'$, the crucial property for automorphisms.

QUESTION 1. All the known examples automorphisms of type 1 (where A and B are given and satisfy some property) require the Autocover (a kind of prompt simplicity) condition (T2)$'$ either directly or indirectly to guarantee (T2). Can one find examples of type 1 automorphisms which do not satisfy (T2)$'$?

6.3. Automorphisms of Hyperhypersimple Sets. Lachlan (Theorem 2.5) proved that a set A is hh-simple iff $\mathcal{L}(A)$ is a Boolean algebra. Maass realized that the technique used for the maximal sets can be applied to hh-simple sets.

THEOREM 6.7 (Maass, 1984). *If A and B are both hyperhypersimple, and if $\mathcal{L}(A) \cong_{\Delta_3^0} \mathcal{L}(B)$, then $A \simeq B$.*

PROOF. There are Δ_3^0 sequences X_n and Y_n such that the correspondence $X_n \mapsto Y_n$ induces the isomorphism $\mathcal{L}(A) \cong_{\Delta_3^0} \mathcal{L}(B)$, and the X_n are disjoint on $\overline{A}$. The key point is that for each U and X, U satisfies the 0-1 law on $X \cap \overline{A}$. Hence, the same proof as in the maximal set case guarantees (T2)$'$. Namely, when an x in state ν_1 in $\overline{A}_s$ enters U, withhold it from state ν_2' until an element y in state ν_2 enters A. If $U \cap X \cap \overline{A} =^* \emptyset$ then the finite restraint will not matter. If $U \supseteq (X \cap \overline{A})$ then every such element will eventually become unrestrained. □

While the preceding theorem shows that for hh-simple sets the hypothesis $\mathcal{L}(A) \cong_{\Delta_3^0} \mathcal{L}(B)$ is sufficient for the automorphism, surprisingly it is also *necessary.*

THEOREM 6.8 (Cholak-Harrington, ta). *If A and B are hh-simple and $A \simeq B$ then $\mathcal{L}(A) \cong_{\Delta_3^0} \mathcal{L}(B)$*

This reduces the problem of classifying the automorphism types of hh-simple sets to that of classifying their $\mathcal{L}(A)$ which Lachlan [1968] has done. (See Soare [1987, p. 203].)

6.4. Hemi-Maximal Sets. Downey and Stob [1992, p. 237] defined a set H to be *hemi-maximal* if there are a maximal set M and disjoint noncomputable c.e. sets A_0 and A_1 such that $H = A_0$, and $M = A_0 \sqcup A_1$ is a Friedberg splitting. Let $\sqcup$ denote disjoint union. A disjoint splitting $A = A_0 \sqcup A_1$ is a *Friedberg splitting* if $W - A$ not c.e. implies $W - A_i$ not c.e., $i = 0, 1$. It is easy to see (Downey and Stob [1992, p. 239]) that any nontrivial splitting of a maximal set is a Friedberg splitting.

Downey and Stob showed that the hemi-maximal sets form an orbit, namely any two are automorphic. To deduce their theorem in the present context consider the following theorem whose proof is nearly identical to that of the N.E.T.

THEOREM 6.9 (General Extension Theorem (G.E.T.)). *Suppose that $A = \sqcup_{1 \leq i \leq n} A_i$ and $B = \sqcup_{1 \leq i \leq n} B_i$. If an enumeration $\mathbb{E}$ satisfies both:*

(T1) $\mathcal{K}^{\overline{A}} = \widehat{\mathcal{K}}^{\overline{B}}$, *[Static equality of states of $\overline{A}/\overline{B}$], and*

(T2)$_i$ $\mathcal{G}^{A_i} = \widehat{\mathcal{G}}^{B_i}$, *for all $i \leq n$ [Covering],*

then there is an enumeration $\mathbb{F}$ which extends $\mathbb{E}$ which witnesses $A_i \simeq B_i$, for all $i \leq n$.

The proof is exactly the same as for the N.E.T. except that elements entering A immediately enter a unique set A_i and play against those elements entering the corresponding set B_i. The N.E.T. Theorem 5.1 and its proof bear the same relation to G.E.T. as the original Friedberg-Muchnik Theorem bears to the trivial extension that there are infinitely many c.e. sets of incomparable Turing degree (see Soare [1987, p. 120]).

THEOREM 6.10 (Downey Stob, 1992). *The hemi-maximal sets form an orbit.*

PROOF. Let A and B be maximal and $A = A_0 \sqcup A_1$ and $B = B_0 \sqcup B_1$. Now by the proof of the Maximal Set Theorem 6.5, there is an enumeration $\mathbb{E}$ satisfying (T1)$'$ and (T2)$'$. By exactly the same method as in the Maximal Set Theorem 6.5, we can guarantee $(\mathrm{T2})'_i : \mathcal{G}^{A_i} = \widehat{\mathcal{G}}^{B_i}$, for all $i \leq n$, from which the theorem immediately follows by the General Extension Theorem 6.9.

Namely, as in the Maximal Set Theorem 6.5 if an element $x \in \overline{A}_s$ in state ν_1 enters U_n, and if state ν_2 (ν_2') corresponds to ν_1 together with U_n (U_n'), then we withhold x from U_n' either forever, or until some new element y in state ν_2 enters A_0 *and* a new element z in state ν_2 enters A_1. If $U \supseteq \overline{A}$, infinitely many such must enter A, and by the Friedberg splitting of A into A_0 and A_1, infinitely many such x must enter *each* of A_0 and A_1. □

Exactly the same method may be used to show $A_0 \simeq B_0$ for any two sets A and B which are automorphic using the 0-1 property as if the maximal set case and and which are then Friedberg into $A_0 \sqcup A_1$ and $B_0 \sqcup B_1$. This includes hemi-maximal sets, Friedberg splittings of hh-simple sets with $\mathcal{L}^*(A) \cong_{\Delta_3} \mathcal{L}^*)B)$, Herrmann sets, and others.

QUESTION 2. In Theorem 6.10 we have seen one case where Friedberg splittings of automorphic sets are also automorphic. Which orbits have the property that for any two sets A and B in the orbit and any Friedberg splittings $A = A_0 \sqcup A_1$ and $B = B_0 \sqcup B_1$ we have $A_0 \simeq B_0$?

QUESTION 3. Under what conditions on A *alone* is it true that any Friedberg splitting $A = A_0 \sqcup A_1$ satisfies $A_0 \simeq A_1$?

For example, if A is promptly simple and $A = A_0 \sqcup A_1$ is a Friedberg splitting, then is $A_0 \simeq A_1$? At first one might conjecture yes because: (1) it works for maximal sets, and both maximal and promptly simple sets satisfy (T2)$'$ and hence (T2) for N.E.T.; (2) if the splitting is a prompt splitting in the sense of Downey-Stob [1993a, p. 181] (as Friedberg's original splitting theorem was prompt) then the conjecture is clearly true. However, the Friedberg splitting condition only guarantees a flow from $\overline{A}$ to A_i and not a prompt one, so A_0 may never get the elements from $\overline{A}$ *while* they are in a desirable state ν, and by the time they arrive later, they may have changed state. Some of these questions and others on splittings were raised in Downey-Stob [1992], [1993], [1993b], who showed that all Friedberg splittings of creative sets are automorphic.

Russell Miller turned these doubts into a theorem as follows. First he proved that there is a promptly simple set A and a Friedberg splitting $A = A_0 \sqcup A_1$ such that A_0 is tardy and A_1 is prompt. Now by Theorem 7.5, A_1 is automorphic to a complete set, but we need something a little stronger to conclude that A_0 is not. Miller invented a new $\mathcal{E}$-definable property for splittings like the $Q(A)$ property of Theorem 3.5 which guarantees that A_0 is $\mathcal{E}$*-definably* tardy, and hence that every set in its orbit is incomplete.

THEOREM 6.11 (Russell Miller, 2000). *(i) There is an $\mathcal{E}$-definable property $R(A_0, A_1)$ which implies that $A_0 \sqcup A_1$ is a Friedberg splitting of $A = A_0 \cup A_1$ and implies that A_0 is tardy.*

(ii) There exists a c.e. set A with a Friedberg splitting $A = A_0 \sqcup A_1$ such that all of the following hold: A is promptly simple of high degree; A_1 has prompt degree; and $R(A_0, A_1)$ holds.

It follows that A_0 and A_1 cannot be automorphic.

QUESTION 4. If we cannot answer the preceding questions positively with automorphisms, what kind of $\mathcal{E}$-definable properties can we exhibit like Russell Miller's to refute their existence? What other new definable properties or variations on existing ones can we find to refute the existence of automorphisms?

6.5. Low Simple Sets. The most general question remaining for type 1 automorphisms is this.

QUESTION 5. Classify properties $P(X)$ (like those above) such that $P(A)$ and $P(B)$ guarantee that A and B are automorphic.

Since this is an immense task, we restrict it in this section just to simple sets and even more to low ones.

QUESTION 6. If A and B are low simple sets, classify conditions P which guarantee that they are automorphic.

The reason for starting with low sets A and B is that by Theorem 6.2 we have $\mathcal{L}^*(A) \cong_{eff} \mathcal{L}^*(B)$, and by (T3) we have great control over which states of $\overline{A}$ will empty out. Now if A and B are promptly simple then by Theorem 6.4 they are automorphic, so we are now looking at the nonautocover case. We still need to satisfy (T2). Hence, the key is to find a new way of approaching the following question.

QUESTION 7. If A and B are low, simple, and nonpromptly simple, how do we control the enumeration to achieve (T2) in the absence of (T2)$'$?

Namely, we begin by allowing some element y of $\overline{B}$ to enter a state ν, but since we are in a type 1 automorphism construction where the opponent controls both A and B, we must expect that he will put y into B immediately if it suits him. Hence, we must have anticipated this move by previously forcing some x of $\overline{A}$ in state ν into A. How do we select which states, and what property of A (analogous to maximal or promptly simple) can be used to force such a move by x? (Can d-simple or non-d-simple be used?) This is the next major front in the type 1 games, because all the known cases use (T2)$'$.

6.6. Atomless r-Maximal Sets.

DEFINITION 6.12. (i) A coinfinite set A is *r-maximal* if there is no computable set R such that both $R \cap \overline{A}$ and $\overline{R} \cap \overline{A}$ are infinite.

(ii) An r-maximal set A is *atomic* if A has a maximal superset, and *atomless* otherwise.

If A and B are atomic r-maximal then it follows that $\mathcal{L}(A) \cong \mathcal{L}(B)$ because they are major subsets in their respective maximal sets. (See Maass-Stob [1983]).

QUESTION 8. If A and B are atomic r-maximal sets, under what conditions is $A \simeq B$? The conclusion cannot always hold because the major subset may or may not be small. (See Soare [1987, p. 194].)

QUESTION 9. If A and B are atomless r-maximal sets, under what conditions is $A \simeq B$? Assume $\mathcal{L}(A) \cong \mathcal{L}(B)$ or even $\mathcal{L}(A) \cong_{\Delta^0_3} \mathcal{L}(B)$. What if A and B are also promptly simple?

Cholak-Nies [ta] have proved that there are infinitely many r-maximal sets $\{A_i\}$ such that $\mathcal{L}(A_i) \not\cong \mathcal{L}(A_j)$ for $i \neq j$. Hence, let us assume that $\mathcal{L}(A) \cong_{\Delta_3^0} \mathcal{L}(B)$. In the case of maximal or hh-simple sets we could achieve (T2)′ because we had promptness on a skeleton. Can we achieve something like that here, and if not, can we discard (T2)′ and achieve (T2) by a more delicate balancing of the flows into A and B? The following may be an easy question, not yet examined.

QUESTION 10. If A and B are promptly simple atomless r-maximal sets and $\mathcal{L}(A) \cong_{\Delta_3^0} \mathcal{L}(B)$, then under what conditions is $A \simeq B$?

Lempp, Nies, and Solomon [ta] have proved that there is an atomless r-maximal set A such that the set $\{e : W_e \cup A =^* \omega\}$ is Σ_3-complete. This implies that the set A has no uniformly c.e. (u.c.e) weak tower in the sense of Soare [1987, p. 196].

6.7. Pseudo-Creative Sets. A set A is *pseudo-creative* if it is not creative and for every $W \subset \overline{A}$ there is a infinite set V disjoint from $A \cup W$. These sets A stand in contrast to the simple sets where most of our automorphism results lie. It is difficult to see how to get started. Classifying the automorphism type of a pseudo-creative set requires a different approach from that for simple sets because many c.e. sets do not intersect A. Also E. Herrmann has isolated a very interesting class of pseudo-creative sets which form an orbit, surprisingly for the same reason as the maximal sets. Herrmann considers the class of sets which are pseudo-creative, r-separable, and $\mathcal{D}$-maximal (now called *Herrmann sets*), and proves they form an orbit. See Cholak-Downey [ta] for definitions and a proof in dynamic form, whereas the Herrmann proof was originally in static form but used the Maximal Set Theorem 6.5 for its dynamic component.

In the maximal set case we had a 0-1 law in which every new set W satisfied either $W \subseteq^* A$ or $W \supseteq \overline{A}$ (not merely $W \supseteq^* \overline{A}$), and so we got prompt simplicity on a skeleton and (T2)′. The remarkable thing about the Cholak-Downey-Herrmann proof is that, modulo computable sets, one gets essentially the same thing for the Herrmann sets. Hence, the ideas of the maximal set theorem and Extension Theorem can be applied.

QUESTION 11. The Cholak-Downey-Herrmann theorem gives just *one* orbit of pseudo-creative sets. How can we classify others? Herrmann looked at sets modulo computable sets. Can we mod out by other classes to study the pseudo-creative sets? Can we begin a direct approach to achieving (T1) and (T2)? For example, if A and B are pseudo-creative and low, then we have $\mathcal{L}(A) \cong \mathcal{L}(B)$ so we need merely achieve (T2). Can we achieve promptness at least on a skeleton?

If A is simple then $A \times \omega$ is pseudo-creative and r-separable, so we can form many pseudo-creative sets.

7. Type 2 Automorphisms: Given Only A

This section resembles the preceding section, except that we are given only A and are required to construct an automorphic copy B with certain properties. The theorems in this section are not simple applications of the New Extension Theorem because the procedures to satisfy (T1), or more usually (T1)′, are quite complicated.

Nevertheless, the principles of the New Extension Theorem still apply, and we still need to meet (T2) somehow. Here it is usually easier to satisfy (T2), and we rarely require that the hypotheses guarantee (T2)$'$. We control the flow of elements into B, so it is safer to allow an element y of $\overline{B}$ to enter a state ν of $\overline{B}$ because we control whether y later enters B and hence whether (T2) is satisfied. Suppose we wish to put more information into B than A contains, for example to prove either of the following theorems.

7.1. Mapping A to Some B Which Codes Information.

THEOREM 7.1 (Harrington-Soare, 1996c, and Cholak, 1995). *For every noncomputable c.e. set A there is a high c.e. set B such that $A \simeq B$.*

THEOREM 7.2 (Harrington, see Harrington-Soare, Theorem 9.1). *For all c.e. sets A and C such that $\emptyset <_T A$ and $C <_T K$ there is a c.e. set $B \simeq_{\Delta^0_3} A$ such that $B \not\leq_T C$.*

We use the noncomputability of A to show that there are certain "coding nodes" of $\overline{A}$ which will carry over to $\overline{B}$ by (19) (duality). Define

$$\mathcal{W}_\alpha = \{\nu : \nu \in \mathcal{K}^{\overline{A}}_\alpha \ \& \ \text{RED has a winning strategy to move any } x \text{ in state } \nu \text{ into } A\}. \tag{22}$$

Picture all the α states as a giant chess board containing the finitely many α-states arranged so RED can move from ν_1 to ν_2 by enumerating x in a red set U_β for some $\beta \subseteq \alpha$ and BLUE can move from ν_1 by enumerating x in a blue set $\widehat{V}_\beta$.

Now $\nu \in \mathcal{W}_\alpha$ means that RED can keep x in $\overline{A}$ forever if he wishes (because $\nu \in \mathcal{K}_\alpha$), but by a sequence of voluntary moves by himself and forced moves by BLUE, RED can slowly move x into A. A typical forced move for BLUE is this. Suppose we know that $U_\beta \subset \widehat{V}_\gamma$. Then RED moves x into U_β by a voluntary red move, and waits for BLUE to move x into $\widehat{V}_\gamma$, which is now forced. An easy precise mathematical definition of $\mathcal{W}_\alpha$ from $\mathcal{M}_\alpha$ and the other parameters is given in Harrington-Soare [1996c, Definition 6.1].

Define $\nu <_R \nu'$ to hold if $\nu = \langle\alpha, \sigma, \tau\rangle$ and $\nu' = \langle\alpha, \sigma', \tau\rangle$ where $\sigma \subset \sigma'$. Define $\nu <_B \nu'$ to hold if $\nu = \langle\alpha, \sigma, \tau\rangle$ and $\nu' = \langle\alpha, \sigma, \tau'\rangle$ where $\tau \subset \tau'$. The idea is that on the ω-side, any x in state ν is a RED (BLUE) move away from entering state ν', because RED (BLUE) can enumerate x in any set U_β (V_β) to raise the σ component of the state of x from σ to σ' (τ to τ'). (On the $\widehat{\omega}$-side it is the reverse.)

Define

$$\mathcal{P}_\alpha = \{\nu : \nu \in \mathcal{M}_\alpha \ \& \ \neg(\exists \nu' \in \mathcal{M}_\alpha)[\nu <_B \nu']\}. \tag{23}$$

These nodes are *opponent maximal* in the sense that the opponent, BLUE, cannot move any x on the ω-side from ν to another state. The following *coding nodes* $\mathcal{C}_\alpha$ play a key role and are defined as follows in Harrington-Soare [1996c, Definition 6.2]. For $\alpha \in T$ define,

$$\mathcal{C}_\alpha =_{\text{dfn}} \mathcal{W}_\alpha \cap \mathcal{P}_\alpha. \qquad \text{[Coding Nodes]} \tag{24}$$

This means that, in addition to $\mathcal{W}_\alpha$, we add the property $\mathcal{P}_\alpha$ that *BLUE* cannot move x out of a state $\nu \in \mathcal{C}_\alpha$. Hence, RED can keep x in ν and in $\overline{A}$ forever, because $\nu \in \mathcal{K}^{\overline{A}}_\alpha$, or at a later stage can begin to gradually send x on a series of forced moves ending up in A, hence the word "coding nodes" as we will see. The first key fact about $\mathcal{C}_\alpha$ is that it is nonempty if A is noncomputable.

THEOREM 7.3 (Harrington-Soare, 1996c, Lemma 6.4). *If A is noncomputable and $\alpha \subset f$, the true path of T, then $\mathcal{C}_\alpha \neq \emptyset$.*

PROOF. (Sketch) Fix $\alpha \subset f$ and suppose $\mathcal{C}_\alpha = \emptyset$. Then almost every $x \in \overline{A}$ must eventually enter some α-state $\nu \in \mathcal{K}_\alpha^{\overline{A}}$. Now RED has no winning strategy to force x into A from ν, because $\nu \notin \mathcal{C}_\alpha$. It is easy to convert this (see Harrington-Soare [1996]) to the assertion that $\nu \notin \mathcal{W}_\alpha$. But the game is obviously determined. Hence, BLUE must have a winning strategy to keep x *out* of A. Now since this applies to all $\nu \in \mathcal{K}_\alpha^{\overline{A}}$ BLUE can computably enumerate $\overline{A}$. Hence, A is computable, contrary to hypothesis. □

Hence, A noncomputable implies $\mathcal{C}_\alpha \neq \emptyset$ which, by (19) (duality), implies $\widehat{\mathcal{C}}_\alpha \neq \emptyset$. These nodes can be used for elements y of $\overline{B}$ to rest in $\overline{B}$ forever, or to be forced by BLUE into B at will. This is what is needed in proving Theorem 7.1 or 7.2.

7.2. Avoiding an Upper Cone. Theorem 7.2 uses the Sacks strategy of avoiding the downward cone as explained in Soare [1987, §4]. It is usually much easier to avoid an *upper* cone using the Sacks preservation strategy as in Soare [1987, p. 122], but here the following question is still open.

QUESTION 12. **[Avoiding an Upper Cone]**

$$(\forall A <_{\mathrm{T}} \emptyset')(\forall C >_{\mathrm{T}} \emptyset)(\exists B \not\geq_{\mathrm{T}} C)[A \simeq B]?$$

The point is that to avoid the upper cone we will have to put some restraint on the enumeration of B as Sacks did. Doing so may be impossible if A is complete as the properties CRE(A) and T(A) of §3 show, so we must require that A be incomplete. The noncomputability of A ensured that $\mathcal{C}_\alpha \neq \emptyset$ above. The key question here is to find the corresponding property for A incomplete.

QUESTION 13. Use the hypothesis $A <_{\mathrm{T}} \emptyset'$ to get a $\mathcal{C}_\alpha$-style property P on A which will translate on the B-side to a property which will allow elements to be restrained from B for the Sacks negative preservation method.

Namely, what does the incompleteness of A say about the enumeration of A which prevents elements from appearing too quickly? The only progress on these questions so far is by Russell Miller [2000], after considerable effort by senior people,

THEOREM 7.4 (R. Miller, 2000). *If A is low and $C >_{\mathrm{T}} \emptyset$, then there is a $B \not\geq_{\mathrm{T}} C$ such that $A \simeq B$.*

The idea is that the Lowness Lemma 6.3 and (T3) above prevent states ν on $\overline{A}$ from being emptied out unexpectedly. This, in turn, would force states in $\overline{B}$ to be emptied too fast and would interfere with the Sacks' restraint to avoid the cone above C.

7.3. Prompt Sets and Completeness. If we want to map a set A to a complete set, we need more than just the hypothesis that $\mathcal{C}_\alpha \neq \emptyset$, because by Theorem 3.5 there are noncomputable sets A which are not automorphic to any complete set. The first major progress in the completeness direction was the result by Cholak, Downey and Stob [1992] that if a set A is promptly simple then A is automorphic to a complete set. Harrington-Soare [1996c] then improved this by proving the following theorem.

THEOREM 7.5 (Harrington-Soare, 1996c, §10). *If A is prompt (or even almost prompt [1996c, §11]) then there is a complete set B such that $A \simeq B$.*

The key point in the proof is to define a set of prompt coding nodes. For $\alpha \in T$ define,

$$\mathcal{D}_\alpha =_{\text{dfn}} \mathcal{K}_\alpha^{\overline{A}} \cap \mathcal{P}_\alpha \cap \mathcal{G}_\alpha^A. \tag{25}$$

The intuition is that if $x \in \nu \in \mathcal{D}_\alpha$ then RED can hold x in $\overline{A}$ forever as before by the first two clauses, but now, whenever he likes, RED can enumerate x *immediately* into A and not through a series of moves as when we only have $x \in \nu \in \mathcal{W}_\alpha$.

Translated onto the $\widehat{\omega}$-side this means that we have a state ν of $\overline{B}$ with infinitely many permanent residents in $\overline{B}$ but such that we can move any temporary resident *immediately* into B for coding.

THEOREM 7.6. *(i) If A is prompt then $(\forall \alpha \subset f)[\mathcal{D}_\alpha \neq \emptyset]$.*
(ii)If the enumeration of A can be arranged so that $(\forall \alpha \subset f)[\mathcal{D}_\alpha \neq \emptyset]$, then A is automorphic to a complete set.

Why are the prompt coding nodes necessary? If an element x in state $\nu \in \widehat{\mathcal{C}}_\alpha$ enters B *eventually* why is that not good enough to show that every noncomputable set is automorphic to a complete one?

The reason is that Theorem 7.6 only holds for $\alpha \subset f$ and we cannot tell during the construction whether $\alpha \subset f$. If we place a coding marker on an element x in state ν' for $\nu' \notin \mathcal{D}_\alpha$ where $\alpha \subset f$, then when x begins its journey toward B it may never receive the next move by the opponent which it expects, and may be stuck forever outside of B. For a coding marker this is a fatal error. In the case of $x \in \nu \in \mathcal{D}_\alpha$ for $\alpha \subset f$, however, there are no opponent moves to wait for, since BLUE can immediately enumerate x into B whenever he pleases. This bears on the next key question.

QUESTION 14. Find a set of necessary and sufficient conditions for a set A to be automorphic to a complete set.

Previously, one looked at this problem from the point of view of two sufficient conditions: maximal (and its derivatives like hh-simple, and hemi-maximal) and promptness or almost promptness. One of the contributions of this paper is to point out that maximal sets are promptly simple on a skeleton. This suggests the following.

QUESTION 15. Can one prove that a necessary and sufficient condition for a set A to be automorphic to a complete set is that it be almost prompt on a skeleton?

This question is a kind of dual to Question 13. The evidence suggests at the moment that Question 15 is true for Δ_3^0 automorphisms, but it is harder to get a handle on all automorphisms.

7.4. Hitting a Cone. Another theme in moving elements around is to try to hit a particular cone, say a downward cone. This is a result in that direction.

THEOREM 7.7 (Wald, 1999). *If A is low (or even if $\overline{A}$ is semilow) and C is promptly simple, then there is a $B \leq_{\mathrm{T}} C$ such that $A \simeq B$.*

PROOF. (Sketch) By the Lowness Lemma 6.3 states ν of $\overline{A}$ do not empty into A so B need not seek C-permission to empty corresponding states into B. Rather B need only get C-permission move elements rapidly from $\overline{B}$ to B to ensure $\widehat{\mathcal{G}}^B \supseteq \mathcal{G}^A$, something C can easily permit by promptness. □

THEOREM 7.8 (Wald, 1999). *If A is low (or even if $\overline{A}$ is semilow) and promptly simple, and C is promptly simple then there is a $B \equiv_{\mathrm{T}} C$ such that $A \simeq B$.*

PROOF. (Sketch) In addition to the former part, we now need to put elements y_x into B to code when some element x enters C. To do this we need prompt coding nodes like $\nu \in \mathcal{D}_\alpha$ as in Theorem 7.5. □

QUESTION 16. Does Theorem 7.8 hold if A is merely prompt in place of being promptly simple?

At first it would seem so, but there are some delicate timing questions where the C-permitting does not get along with the approximation f_s to the true path. Any attempt to move elements around is limited by the following.

THEOREM 7.9 (Downey-Harrington, ta). *There is a prompt low degree* **a** *and a tardy (nonprompt) degree* **b** *such that*

$$(\forall A \in \mathbf{a})(\forall B \leq_T \mathbf{b})[A \not\simeq B].$$

A corollary of Theorem 7.7 of Wald's thesis [Wald, 1999] is that we cannot improve Downey-Harrington Theorem 7.9 to "$(\forall A \leq_{\mathrm{T}} \mathbf{a})$." This also also follows by the existence of hemimaximals of arbitary low degree. The no fat orbit theorem in its full generality shows that no member of the $\text{low}_n/\text{high}_n$ hierarchy is defined by a single orbit, except for the high_1 degrees. The next question was raised by Downey.

QUESTION 17. Which classes of degrees are definable by single orbits? We know the high degrees and the complete degree are such. Is there a nice characterization of any other?

Harrington and Cholak have shown that all double jump classes are definable.

7.5. Prompt high orbits. One of the most interesting threads in the subject has been that initiated by Martin's beautiful Theorem 2.3 that the degrees of maximal sets are the high degrees. This led to the conjecture that for every noncomputable A and every high c.e. degree **d** there is a $B \in \mathbf{d}$ such that $A \simeq B$.

Maass, Shore, and Stob [1981] refuted this by producing an $\mathcal{E}$-definable property, $SP\overline{H}$ (possessing a certain splitting property and not being hh-simple), which distinguished between some prompt and nonprompt sets. Cholak [1995] proved a version of the conjecture by showing that the weaker conclusion "$\mathcal{L}(A) \cong \mathcal{L}(B)$" is true. It would be very interesting to know whether the original conjecture for automorphisms holds in the form where everything is promptly simple or prompt so the above barrier is removed. Note that we also need to add the hypothesis that A is incomplete to avoid the properties $CRE(A)$ and $T(A)$ of §3.

QUESTION 18 (Prompt high orbit question). Given a promptly simple set $A <_{\mathrm{T}} \emptyset'$ and a high promptly simple set D, does there exist $B \equiv_T D$ such that $A \simeq B$?

The fact that A and D are both promptly simple allows us to achieve (T2)$'$, but not necessarily (T2) since we do not have (T1)$'$. The highness of D allows us to empty states of $\overline{B}$ into B to achieve (T1): $\mathcal{L}(A) \cong \mathcal{L}(B)$, as in Cholak, but is this enough? If we could achieve the dynamic property (T1)$'$ we would succeed, but this is not clear. This resembles the Harrington-Soare refutation of a similar conjecture described in the paragraph directly following (18) and immediately preceding §5.1 There we also had (T2)$'$ and (T1) only, but not (T1)$'$ and we failed to produce the automophism. This seems to be a subtle but crucial obstacle to building automorphisms and deserves a lot more attention.

7.6. Orbit Complete Classes.

DEFINITION 7.10. For classes $\mathcal{S}$, $\mathcal{T} \subseteq \mathcal{E}$ we say that $\mathcal{S}$ is *orbit complete* in $\mathcal{T}$ if

(26) $$(\forall X \in \mathcal{S})(\exists Y \in \mathcal{T})[X \simeq Y].$$

Cholak [1998] answered a question of Herrmann by proving that the simple sets are orbit complete in the hypersimple sets. The key step here is to prove that

(27) $$A \text{ simple} \implies \mathcal{W}_\alpha = \mathcal{M}^{\overline{A}} \text{ for } \alpha \subset f.$$

The proof is as in Theorem 7.3 for $\mathcal{C}_\alpha \neq \emptyset$ except that for *every* $\nu \in \mathcal{M}^{\overline{A}}$ RED must have a winning strategy to force any element $x \in \nu$ into A, for if not, then B has a winning strategy to keep such x in $\overline{A}$ and hence enumerate an infinite subset of $\overline{A}$. Wald [1999] answered a question of Jockusch by proving that the promptly simple sets are orbit complete in the effectively simple sets.

QUESTION 19. For which other pairs of classes $\mathcal{S}$, $\mathcal{T} \subseteq \mathcal{E}$ is $\mathcal{S}$ *orbit complete* in $\mathcal{T}$?

8. Summary of Template Properties

The Δ^0_3 paper by Harrington-Soare and the present paper highlight the desirability of translating standard computability-theoretic properties such as noncomputability, simplicity, promptness, and maximality into abstract properties about the templates. We now summarize some of these.

$(T1)$ $\mathcal{K}^{\overline{A}} = \widehat{\mathcal{K}}^{\overline{A}}$.

$(T2)$ $\mathcal{G}^A = \widehat{\mathcal{G}}^B$.

$(T1)'$ $(\mathcal{M}^{\overline{A}}, \mathcal{N}^{\overline{A}}) = (\widehat{\mathcal{M}}^{\overline{B}}, \widehat{\mathcal{N}}^{\overline{B}})$.

$(T2)'$ $\mathcal{M}^{\overline{A}} = \mathcal{G}^A \quad \& \quad \widehat{\mathcal{M}}^{\overline{B}} = \widehat{\mathcal{G}}^B$.

(T3) $(\forall \nu_0 \in \mathcal{R}^{\overline{A}}_\alpha)(\exists \nu_1)[\nu_0 <_R \nu_1 \;\&\; \nu_1 \in \mathcal{M}^{\overline{A}}_\alpha]$ [For A low].

- **N.E.T. Theorem** $[(T1)' + (T2)'] \implies [(T1) + (T2)] \implies A \simeq B$.
- A, B maximal (or hhs with $\mathcal{L}^*(A) \cong_{\Delta_3} \mathcal{L}^*(B)$) $\implies$ (T1)$'$ and (T2)$'$.
- A, B Herrmann $\implies$ (T1)$'$ and (T2)$'$.
- A_0, B_0 hemimaximal $\implies$ (T1)$'$ and (T2)$'$.
- A, B low $\implies$ (T1)$'$ and (T3).
- A, B promptly simple $\implies$ (T2)$'$.

- A simple $\implies \mathcal{W}_\alpha = \mathcal{M}_\alpha^{\overline{A}}$ (for (T2)′), where
  $\mathcal{W}_\alpha = \{\nu :$ RED has a winning strategy to move $x \in \nu$ into A $\}$.
- A noncomputable $\implies \mathcal{C}_\alpha \neq \emptyset$, $\alpha \subset f$, (toward (T2)′), where
  $\mathcal{C}_\alpha = \mathcal{K}_\alpha^{\overline{A}} \cap \mathcal{W}_\alpha \cap \mathcal{P}_\alpha$, and [Coding nodes]
  $\mathcal{P}_\alpha = \{\nu : \nu \in \mathcal{M}_\alpha \ \& \ \neg(\exists \nu' \in \mathcal{M}_\alpha)[\nu <_B \nu']\}$. [Blue maximal nodes]
- A prompt $\implies \mathcal{D}_\alpha \neq \emptyset$, $\alpha \subset f$, toward (T2)′, where
  $\mathcal{D}_\alpha = \mathcal{K}_\alpha^{\overline{A}} \cap \mathcal{G}_\alpha^{A} \cap \mathcal{P}_\alpha$. [Fast coding nodes]
- $A \mapsto B \leq_{\mathrm{T}} C$ p.s. $\implies \widehat{\mathcal{G}}^B \supseteq \mathcal{G}^A$, toward (T2)′, where $A \mapsto B$ denotes that there is an automorphism mapping A to B.

9. The Next Frontier

The New Extension Theorem 5.1 provides conditions which are sufficient to imply that $A \simeq_{\Delta_3^0} B$ and are virtually necessary, but we do not attempt to formally claim or prove it here. The following remarkable result suggests that something along these lines may be necessary and sufficient for *all* automorphisms of $\mathcal{E}$.

Define

$$\mathcal{S}(A) = \{W : (\exists V)[W \sqcup V = A\},$$
$$\mathcal{R}(A) = \{R : R \subseteq A \ \& \ R \text{ computable } \},$$

and $\mathcal{S}_{\mathcal{R}}(A)$ be the quotient structure of $\mathcal{S}(A)$ modulo $\mathcal{R}(A)$.

THEOREM 9.1 (Cholak-Harrington, ta). *If A and B are automorphic then*

$$\mathcal{S}_{\mathcal{R}}(A) \cong_{\Delta_3^0} \mathcal{S}_R(B).$$

The connection between Theorem 9.1 and the New Extension Theorem 5.1 and condition (T2) is this. For a fixed ν let X_ν (Y_ν) be the set of elements which enter A in state ν (not in state ν). Now X_ν and Y_ν give a splitting of A. Theorem 9.1 says that even for the most general automorphism of $\mathcal{E}$ mapping A to B these sets must be in a Δ_3^0-correspondence, just as the New Extension Theorem 5.1 says that a corresponding Δ_3^0-condition is sufficient. These connections are far from being worked out, but they point the way toward the ultimate automorphism theorem.

10. Reflections on Computably Enumerable Sets

For sixty-five years we have intensively studied the Church-Turing Thesis and its characterization of computability. However, during the same period we have not studied a c.e. set foundationally as separate object, but only as the ranges of computable functions on ω. The results for: (1) definable properties of $\mathcal{E}$; (2) automorphisms of $\mathcal{E}$: (3) c.e. sets applied to differential geometry, by Nabutovsky [1996a] and [1996b] and Nabutovsky and Weinberger [ta1] and [ta2]; suggest that a key factor in understanding c.e. sets is a dynamic approach which studies them from the point of view of properties like prompt simplicity or the $A >> B$ domination relation of Soare [Ta2] examining how fast elements enter one set W_e in relation to elements entering other sets.

Virtually all our practical computing processes as well as many theoretical ones (like the search for local minima on manifolds) work as c.e. processes looking for an output, which cannot always be guaranteed ahead of time. It seems that more direct conceptual effort should be devoted to c.e. sets and particularly to their dynamic properties.

Older references not listed below can be found in Soare [1987].

References

[1] [Cholak, 1995] P. A. Cholak, Automorphisms of the Lattice of Recursively Enumerable Sets, *Memoirs of the Amer. Math. Soc.* **113**, 1995.

[2] [Cholak, 1998] P. A. Cholak, The dense simple sets are orbit complete, Proceedings of the Oberwolfach conference on computability theory in 1996, *Annals of Pure and Applied Logic*, **94** (1998), 37–44.

[3] [Cholak, ta] P.A. Cholak, The global structure of computably enumerable sets, in: P. Cholak, S. Lempp, M. Lerman, and R. Shore, (eds.) Computability Theory and its Applications: Current Trends and Open Problems, American Mathematical Society.

[4] [Cholak and Downey, ta] P. Cholak and R. Downey, Some orbits for $\mathcal{E}^*$, *Annals of Pure and Applied Logic*, to appear.

[5] [Cholak-Downey-Stob, 1992] P. Cholak, R. Downey, M. Stob, Automorphisms of the lattice of recursively enumerable sets: Promptly simple sets, *Trans. Amer. Math. Soc.* **332** (1992), 555–570.

[6] [Cholak-Harrington, ta] L. Harrington and P. Cholak, On the definability of the double jump in the computably enumerable degrees, to appear.

[7] [Cholak-Nies, ta] P. Cholak and A. Nies, Atomless r-maximal sets, *Israel J. Math.*, to appear.

[8] [Church, 1936] A. Church, An unsolvable problem of elementary number theory, *American J. of Math.*, **58** (1936), 345-363.

[9] [Dekker, 1954] J. C. E. Dekker, A theorem on hypersimple sets, *Proc. Amer. Math. Soc.* **5** (1954), 791—796.

[10] [Downey-Harrington, ta] R. Downey and L. Harrington, There is no fat orbit, *Annals of Pure and Applied Logic*, **80** (1996), 227-289.

[11] [Downey-Stob, 1992] R. Downey and M. Stob, Automorphisms of the lattice of recursively enumerable sets: Orbits, *Advances in Math.*, **92** (1992), 237-265.

[12] [Downey-Stob, 1993a] R. Downey and M. Stob, Friedberg Splittings of recursively enumerable sets, *Annals of Pure and Applied Logic* **59** (1993), 175–199.

[13] [Downey-Stob, 1993b] R. Downey and M. Stob Splitting theorems in recursion theory, *Annals of Pure and Applied Logic*, **65** (1993), 1–106.

[14] [Harrington-Soare, 1991] L. Harrington and R. I. Soare, Post's Program and incomplete recursively enumerable sets, *Proc. Natl. Acad. of Sci. USA*, **88** (1991), 10242–10246.

[15] [Friedberg, 1957] R. M. Friedberg, Two recursively enumerable sets of incomparable degrees of unsolvability, *Proc. Natl. Acad. Sci. USA* **43** (1957), 236–238.

[16] [Friedberg, 1958] R. M. Friedberg, Three theorems on recursive enumeration: I. Decomposition, II. Maximal Set, III. Enumeration without duplication, *J. Symbolic Logic* **23** (1958), 309—316.

[17] [Gandy, 1988] R. Gandy, The confluence of ideas in 1936, in: *The Universal Turing Machine: A Half-Century Survey*, R. Herken (ed.), Oxford Univ. Press, Oxford, 1988, 55–111.

[18] [Gödel, *1931*] K. Gödel, Über formal unentscheidbare sätze der Principia Mathematica und verwandter systeme. I, *Monatsch. Math. Phys.* **38** (1931) 173-178. (English trans. in Davis *1965*, 4–38, and in van Heijenoort, 1967, 592–616.

[19] [Gödel, *1934*] K. Gödel, On undecidable propositions of formal mathematical systems, Notes by S. C. Kleene and J. B. Rosser on lectures at the Institute for Advanced Study, Princeton, New Jersey, 1934, 30 pp. (Reprinted in Davis *1965* [**?**, 39–74]).

[20] [Harrington-Soare, 1991] L. Harrington and R. I. Soare, Post's Program and incomplete recursively enumerable sets, *Proc. Natl. Acad. of Sci. USA*, **88** (1991), 10242–10246.

[21] [Harrington-Soare, 1996a] L. Harrington and R. I. Soare, Dynamic properties of computably enumerable sets, in: 'Computability, Enumerability, Unsolvability: Directions in Recursion Theory,' eds. S. B. Cooper, T. A. Slaman, S. S. Wainer, Proceedings of the Recursion Theory Conference, University of Leeds, July, 1994, London Math. Soc. Lecture Notes Series, Cambridge University Press, January 1996.

[22] [Harrington-Soare, 1996b] L. Harrington and R. I. Soare, Definability, automorphisms, and dynamic properties of computably enumerable sets, *Bulletin of Symbolic Logic*, **2** (1996), 199–213.

[23] [Harrington-Soare, 1996c] L. Harrington and R. I. Soare, The Δ^0_3-automorphism method and noninvariant classes of degrees, *Jour. Amer. Math. Soc.*, **9** (1996), 617–666.

[24] [Harrington-Soare, 1998a] L. Harrington and R. I. Soare, Definable properties of the computably enumerable sets, Proceedings of the Oberwolfach Conference on Computability Theory, 1996, *Annals of Pure and Applied Logic*, **94** (1998), 97–125.

[25] [Harrington-Soare, 1998b] L. Harrington and R. I. Soare, Codable sets and orbits of computably enumerable sets, *J. Symbolic Logic*, **63** (1998), 1–28.

[26] L. Harrington and R. I. Soare, [ta] Martin's Invariance Conjecture and Low Sets, in preparation.

[27] [Kleene, *1936*] S. C. Kleene, General recursive functions of natural numbers, *Math. Ann.* **112** (1936), 727–742.

[28] [Lachlan, 1968] A. H. Lachlan, Degrees of recursively enumerable sets which have no maximal superset, *J. Symbolic Logic* **33** (1968), 431–443.

[29] [Lachlan, 1970] A. H. Lachlan, On some games which are relevant to the theory of recursively enumerable sets, *Ann. of Math. (2)* **91** (1970), 291–310.

[30] [Lempp, Nies, Solomon] On the filter of computably enumerable supersets of an r-maximal set [ta].

[31] [Maass, 1982] W. Maass, Recursively enumerable generic sets, *J. Symbolic Logic* **47** (1982), 809–823.

[32] [Maass, 1983] W. Maass, Characterization of recursively enumerable sets with supersets effectively isomorphic to all recursively enumerable sets, *Trans. Amer. Math. Soc.* **279** (1983), 311–336.

[33] [Maass, 1984] W. Maass, On the orbits of hyperhypersimple sets, *J. Symbolic Logic* **49** (1984), 51–62.

[34] [Maass, 1985] W. Maass, Variations on promptly simple sets, *J. Symbolic Logic* **50** (1985), 138–148.

[35] [Maass-Shore-Stob, 1981] W. Maass, R. A. Shore and M. Stob, Splitting properties and jump classes, *Israel J. Math.* **39** (1981), 210–224.

[36] [Maass-Stob, 1983] The intervals of the lattice of recursively enumerable sets determined by major subsets, *Ann. Pure Appl. Logic* **24**, (1983), 189-212.

[37] [Martin, 1966] D. A. Martin, Classes of recursively enumerable sets and degrees of unsolvability, *Z. Math. Logik Grundlag. Math.* **12** (1966), 295–310.

[38] [Miller, 2000] R. Miller, Computability theory, Ph.D. Thesis, University of Chicago, 2000.

[39] R. Miller Definable speed of enumeration and Friedberg Splittings, [ta]

[40] [Muchnik, 1956] A. A. Muchnik, On the unsolvability of the problem of reducibility in the theory of algorithms, *Doklady Akademii Nauk SSR* **108** (1956), 194–197, (Russian).

[41] J. Myhill The lattice of recursively enumerable sets, *J. Symbolic Logic* **21** (1956), 220 (abstract).

[42] [Nabutovsky, 1996a] A. Nabutovsky, Disconnectedness of sublevel sets of some Riemannian functionals, Geometric and Functional Analysis, **6** (1996), 703-725.

[43] [Nabutovsky, 1996b] A. Nabutovsky, Geometry of the space of triangulations of a compact manifold, Communications in Mathematical Physics, **181** (1996), 303-330.

[44] A. Nabutovsky and S. Weinberger [ta1] Variational problems for Riemannian functionals and arithmetic groups, to appear.

[45] Nabutovsky and Weinberger [ta2] The Fractal Nature of Riem/Diff, to appear.

[46] [Post, *1941*] E. L. Post, Absolutely unsolvable problems and relatively undecidable propositions: Account of an anticipation. (Submitted for publication in 1941.) Printed in Davis *1965*, 340–433.

[47] [Post, *1943*] E. L. Post, Formal reductions of the general combinatorial decision problem, *Amer. J. Math.* **65** (1943), 197–215.

[48] [Post, 1944] E. L. Post, Recursively enumerable sets of positive integers and their decision problems, *Bull. Amer. Math. Soc.* **50** (1944), 284–316.

[49] [Post, *1948*] E. L. Post, Degrees of recursive unsolvability: preliminary report (abstract), *Bull. Amer. Math. Soc.* **54** (1948), 641–642.

[50] [Shoenfield, 1976] J. R. Shoenfield, Degrees of classes of r.e. sets, *J. Symbolic Logic* **41** (1976), 695–696.

[51] [Soare, 1974] R. I. Soare, Automorphisms of the recursively enumerable sets, Part I: Maximal sets, *Ann. of Math. (2)* **100** (1974), 80–120.

[52] [Soare, 1982] R. I. Soare, Automorphisms of the lattice of recursively enumerable sets, Part II: Low sets, *Ann. Math. Logic* **22** (1982), 69–107.

[53] [Soare, 1987] R. I. Soare, *Recursively Enumerable Sets and Degrees: A Study of Computable Functions and Computably Generated Sets,* Springer-Verlag, Heidelberg, 1987.

[54] [Soare, ta] R. I. Soare, An overview of the computably enumerable sets, in: *Handbook of Computability Theory*, ed. E. Griffor, North-Holland, Amsterdam, to appear.

[55] [Soare, ta1] R. I. Soare, Templates, Extensions, and Automorphisms, to appear.

[56] [Soare, ta2] R. I. Soare, Differential Geometry and Computably Enumerable Sets, to appear.

[57] [Turing, *1936*] A. M. Turing, On computable numbers, with an application to the Entscheidungsproblem, *Proc. London Math. Soc.* ser. 2 **42** (Parts 3 and 4) (1936) 230–265;

[58] [Turing, *1939*] A. M. Turing, Systems of logic based on ordinals, *Proc. London Math. Soc.* **45** Part 3 (1939), 161–228; reprinted in Davis [1965], 154–222.

[59] [Wald, 1999] K. Wald, Automorphisms and Invariance of the Computably Enumerable Sets, Ph.D. Dissertation, The University of Chicago, Chicago, 1999.

Department of Mathematics, University of Chicago, Chicago, Illinois 60637-1546, USA

E-mail address: soare@math.uchicago.edu

Contemporary Mathematics
Volume **257**, 2000

Open Problems in the Enumeration Degrees

Andrea Sorbi

ABSTRACT. We survey some open problems in the enumeration degrees. The problems fall into the following three categories: Global properties of the enumeration degrees and their local structure, the interplay between structural properties of sets and degree-theoretic properties, and the context of the enumeration degrees within the Medvedev lattice.

1. Introduction

Enumeration reducibility is a formalization of the notion of relative enumerability of sets of numbers; a set A is enumeration reducible to a set B if there is an effective procedure for transforming any enumeration of B into some enumeration of A.

We will refer to the formalization given in [**FR59**] (see also [**Rog67b**]), which makes use of the notion of enumeration operator. Every c.e. set W defines an *enumeration operator*, i.e. a mapping Φ, from sets of numbers to sets of numbers, taking a set A to the set Φ^A, where

$$\Phi^A = \{x : (\exists u)[\langle x, u\rangle \in W \,\&\, D_u \subseteq A]\}$$

(D_u is the finite set with canonical index u). A is *enumeration reducible to* B, denoted by $A \leq_e B$, if $A = \Phi^B$, for some enumeration operator Φ. We denote by $\equiv_e$ the equivalence relation on sets of numbers generated by the preordering relation $\leq_e$ (i.e. $A \equiv_e B$ if $A \leq_e B$ and $B \leq_e A$); the equivalence class of a set A, denoted by $\deg_e(A)$, is called the *enumeration degree* of A. The corresponding degree structure $\langle \mathfrak{D}_e, \leq\rangle$ is an uppersemilattice with least element; the least element $\mathbf{0}_e$ is the enumeration degree of any c.e. set.

Pioneering work on the enumeration degrees dates back to Case [**Cas71**], and Medvedev [**Med55**]. In particular, Case showed that $\mathfrak{D}_e$ is not a lattice, as a consequence of the exact pair theorem, and Medvedev proved the existence of quasi-minimal enumeration degrees. The *local structure* of the enumeration degrees, i.e. the degrees below the first jump $\mathbf{0}'_e$, or equivalently the Σ^0_2 enumeration degrees, will be denoted by the symbol $\mathfrak{G}$.

A different approach consists of thinking of enumeration reducibility as a notion of relative computability of partial functions. Given partial functions φ, ψ, let $\varphi \leq_e$

1991 *Mathematics Subject Classification.* 03D30.
Key words and phrases. Enumeration reducibility; enumeration degree.

ψ if the graph of φ is enumeration reducible to the graph of ψ. This reducibility coincides with the reducibility on partial functions defined in [**Myh61**]. It also turns out (see e.g. [**Coo90**]) that $\varphi \leq_e \psi$ if and only if φ can be computed by a nondeterministic oracle machine having as oracle the partial function ψ. Of course, when queries of the form "is $\psi(m) = n$?" are asked of the oracle, and m is not in the domain of ψ, then the computation is undefined. The degree structure induced by this reducibility on partial functions is called the *partial degrees*, which are clearly isomorphic to the enumeration degrees. It is sometimes convenient to work with partial degrees instead of enumeration degrees. In particular, the Turing degrees ([**Med55**], [**Myh61**]) are isomorphic, as an uppersemilattice with 0, to the partial degrees of total functions.

Our notation and terminology are standard and can be found in [**Rog67b**] and [**Soa87**]. When talking about first order aspects of degree structures (the first order theory of a degree structure, definability in the structure, etc.), we will always refer to the first order language with signature $\langle \leq \rangle$. The structure of the Turing degrees will be denoted by $\langle \mathfrak{D}_T, \leq_T \rangle$; the Turing degree of a set A will be denoted by $\deg_T(A)$. The Turing degrees are an uppersemilattice with least element.

1.1. Open problems. In his book, [**Rog67b**, p. 282], Rogers notes that the partial degrees have not been extensively studied, and raises some open questions. Although these questions have been settled by now, there still are a large number of interesting open problems.

Most of the problems we propose are well known in the literature. In particular, we acknowledge [**Coo90**] and the list of problems in [**AL99**, p. 230]. We believe that solutions to these problems would provide a deeper understanding of the general properties of the enumeration degrees and their local structure.

2. Towards a global structure theory for the enumeration degrees

The fact that the Turing degrees can be embedded into the enumeration degrees relies on the following result:

LEMMA 2.1. *For every pair of total functions f, g,*

$$f \leq_T g \Leftrightarrow f \leq_e g.$$

PROOF. See, for instance, [**Rog67b**, Corollary 9.XXIV]. □

As an immediate consequence, we have, for every pair A, B of sets of numbers,

$$A \leq_T B \Leftrightarrow c_A \leq_e c_B$$

(functions and partial functions will be often identified with their graphs). Thus the mapping $\iota(\deg_T(A)) = \deg_e(c_A)$ is well defined, and

THEOREM 2.2 ([**Med55**], [**Myh61**]). *The mapping ι embeds $\mathfrak{D}_T$ into $\mathfrak{D}_e$ as an uppersemilattice with least element.*

PROOF. See e.g. again [**Rog67b**]. □

Cooper and McEvoy ([**MC85**]) define a jump operation J_e on the enumeration degrees as follows: For every set A, first define $J_e(A) = K_A \oplus \overline{K_A}$, where $K_A = \{e : e \in \Phi_e^A\}$, and then let $J_e(\deg_e(A)) = \deg_e(J_e(A))$. (A warning on notation: To be consistent with the most commonly used notation (see e.g. [**Coo90**]), we will write $\mathbf{0}_e^{(k)}$ instead of $J_e^{(k)}(\mathbf{0}_e)$.) It follows that the embedding ι defined above preserves the jump operation as well, i.e. $\iota(\mathbf{a}') = J_e(\iota(\mathbf{a}))$, for every Turing degree $\mathbf{a}$. Moreover, one can show that $\text{range}(J_e) = \iota(\text{range}('))$; the inclusion $\subseteq$ follows from Soskov's jump inversion theorem in [**Sosta**], an immediate consequence of which is that for every enumeration degrees $\mathbf{a}$ there exists a total $\mathbf{c} \geq \mathbf{a}$ such that $J_e(\mathbf{a}) = J_e(\mathbf{c})$.

The Turing degrees can therefore be viewed as a substructure of the enumeration degrees. We define the *total degrees* to be the (images of the) Turing degrees in $\mathfrak{D}_e$, and let TOT be the collection of all total degrees. It is easy to see that $\mathbf{a} \in \text{TOT}$ if and only if $\mathbf{a} = \deg_e(f)$, for some total function f. Medvedev ([**Med55**]) defined an enumeration degree $\mathbf{a}$ to be *quasi-minimal* if $\mathbf{a} \neq \mathbf{0}_e$ and there is no total degree $\mathbf{b}$ such that $\mathbf{0}_e < \mathbf{b} \leq \mathbf{a}$, and proved that quasi-minimal enumeration degrees exist, thus establishing that $\text{TOT} \neq \mathfrak{D}_e$.

Recently, an important characterization of the total degrees was given by Jockusch, who shows that the total degrees are exactly the enumeration degrees of introreducible sets. This solves one of the problems in the Kazan list; see [**AL99**, p. 231] for the details of Jockusch's solution.

Surprisingly, the total degrees generate the enumeration degrees, and thus are an automorphism base of $\mathfrak{D}_e$ as shown by the following theorem. We recall that given a structure $\mathfrak{A}$, a subset $A \subseteq \mathfrak{A}$ is called an *automorphism base* for $\mathfrak{A}$ if

$$(\forall \alpha \in \text{Aut}(\mathfrak{A}))[\alpha \upharpoonright A = \text{id}_A \Rightarrow \alpha = \text{id}_{\mathfrak{A}}].$$

For more information about automorphism bases and sets of generators of the enumeration degrees see [**Sor98**].

THEOREM 2.3. *For every enumeration degree* $\mathbf{a}$ *there exist total degrees* $\mathbf{b}$ *and* $\mathbf{c}$ *such that* $\mathbf{a} = \mathbf{b} \wedge \mathbf{c}$.

PROOF. This result is due to Rozinas, [**Roz78**]. For a proof, see e.g. [**Sor98**]. □

In recent years, one of the most important lines of research in computability theory has been the study of the so called global properties of degree structures. In this regard, the Turing degrees have been extensively studied, whereas little is known about the enumeration degrees. (There are a few notable exceptions such as Theorem 4.1 below, and Theorem 3.1 and 3.2 which show that homogeneity in $\mathfrak{D}_e$ fails.)

One would expect that Theorem 2.2 and Theorem 2.3 would enable us to extend some of the well known results about general properties of the Turing degrees to the enumeration degrees. This would undoubtedly be the case if the total degrees were invariant within the enumeration degrees (i.e., every automorphism of the enumeration degrees carries total degrees to total degrees, a question which was asked by Rogers in [**Rog67b**, p. 282]). In particular, as pointed out in [**Sor98**], we would know in this case that $\text{Aut}(\mathfrak{D}_e)$ is a subgroup of $\text{Aut}(\mathfrak{D}_T)$, and every automorphism of $\mathfrak{D}_e$ is the identity on the upper cone generated by $\mathbf{0}_e''$; we would also know that the jump operation would be invariant in $\mathfrak{D}_e$ (even definable if the total degrees were definable in $\mathfrak{D}_e$).

Cooper, [**Coo99**], has recently announced a solution to the problem raised by Rogers on the invariance of the total degrees: there exists an automorphisms of $\mathfrak{D}_e$ which carries some total degree to a quasi–minimal enumeration degree. Thus the total degrees are not definable, nor are they invariant in $\mathfrak{D}_e$. In view of Cooper's claim, the following questions remain open.

QUESTION 1. *What is the cardinality of Aut($\mathfrak{D}_e$)?*

Slaman and Woodin have shown that there is a finite automorphism base for the enumeration degrees ([**Sla91**]); thus the cardinality of Aut($\mathfrak{D}_e$) is at most that of the continuum. Can one show that there are at most countably many automorphisms as in the case of the Turing degrees (see again [**Sla91**])?

QUESTION 2. *Is there* $\mathbf{a} \in \mathfrak{D}_e$ *such that every automorphism of* $\mathfrak{D}_e$ *is the identity on* $\mathfrak{D}_e(\geq \mathbf{a})$*?*

Of course, as in the case of the Turing degrees, this question might be related to the invariance of the jump operation:

QUESTION 3 (Cooper). *Is the jump operation definable in* $\mathfrak{D}_e$*? Is it invariant with respect to the automorphisms of* $\mathfrak{D}_e$*?*

A very natural question arising from Theorem 2.3 is the following (this question will be reconsidered in Section 6):

QUESTION 4. *Can every automorphism of* $\mathfrak{D}_T$ *be extended to an automorphism of* $\mathfrak{D}_e$*?*

What about local automorphism bases? For instance, it is known ([**Sla91**]) that the computably enumerable Turing degrees are an automorphism base for $\mathfrak{D}_T$: does this carry over to $\mathfrak{D}_e$ via the embedding? We recall that the images of the computably enumerable Turing degrees are the Π^0_1 enumeration degrees.

QUESTION 5. *Are the* Π^0_1 *enumeration degrees an automorphism base, or at least is* $\mathfrak{G}$ *an automorphism base for* $\mathfrak{D}$*?*

Although Theorem 2.2 and Theorem 2.3 do not seem to yield direct information about these questions, it can be expected, however, that answers to some of the questions about the global structure of the enumeration degrees can be somehow reduced to the context of the total degrees within $\mathfrak{D}_e$. More generally, the investigation of this context appears to be a very promising line of research, not only for studying the enumeration degrees, but also for obtaining a better understanding of the Turing degrees. A recent example in this direction is provided by Soskov's proof, using enumeration degrees, of the theorem by Coles, Downey and Slaman about the existence of a least element in $\{(\deg_T(X))' : A \in \Sigma^X_1\}$, for every set A: see [**Dowta**] for a full discussion of this result, its relationship with the enumeration degrees, and its relevance to algebra.

On the other hand, no nontrivial automorphism base for the local structure $\mathfrak{G}$ is known (apart from the trivial fact that, by the branching theorem [**NSta1**], every dense class of elements of $\mathfrak{G}$ generates $\mathfrak{G}$, and thus is an automorphism base for $\mathfrak{G}$).

In particular:

QUESTION 6. *Is* $TOT \cap \mathfrak{G}$ *an automorphism base for* $\mathfrak{G}$*?*

Of course it can not be hoped that TOT $\cap\, \mathfrak{G}$ generates $\mathfrak{G}$ under infima in the same way as TOT generates $\mathfrak{D}_e$, since, by [**CC88**], it is known that there exists an enumeration degree $\mathbf{a}$ such that no $\mathbf{b}$ with $\mathbf{a} \leq \mathbf{b} < \mathbf{0}'_e$ is Δ^0_2; in fact Cooper and Copestake show that there exists an enumeration degree $\mathbf{a} \in \mathfrak{G}$ which is incomparable with all intermediate Δ^0_2 enumeration degrees. Since every total degree in $\mathfrak{G}$ is Δ^0_2, it follows that not every enumeration degree in $\mathfrak{G}$ is the infimum of total degrees in $\mathfrak{G}$. Every attempt to show that $\mathrm{TOT} \cap \mathfrak{G}$ generates $\mathfrak{G}$ must also take into account the fact that there exist nonzero join-irreducible elements in $\mathfrak{G}$ (see [**Ahm89**], see also [**ALta**]); it is, however, conceivable that more complicated generating schemes are possible.

3. Intervals of enumeration degrees

As is known, there is a complete characterization of the countable ideals of the Turing degrees; up to isomorphism, they are exactly the countable uppersemilattices with least element (see [**LL76**]). Hence the countable ideals of $\mathfrak{D}_T$ realize all possibilities, i.e. all situations compatible with $\mathfrak{D}_T$ being a locally countable uppersemilattice with least element. In particular, the finite initial segments of $\mathfrak{D}_T$ with greatest element realize all finite possibilities, i.e. all finite lattices. This characterization has played an important role in the understanding of the structure of $\mathfrak{D}_T$ (see [**Ler83**] for a full discussion of these topics).

The situation for the enumeration degrees is radically different. In particular, a consequence of Gutteridge's proof of the nonexistence of minimal enumeration degrees is that there is no finite ideal.

THEOREM 3.1. *There is no minimal enumeration degree.*

PROOF. See [**Gut71**]; see also [**Coo82**]. Gutteridge's result is the consequence of the following two facts:

- No Δ^0_2 enumeration degree is minimal.
- If $A <_e B$ and $\deg_e(B)$ is a minimal cover of $\deg_e(A)$ then $B \leq_T A'$.

□

Gutteridge's proof, however, left open for many years the question of whether or not minimal covers do exist, in other words whether the enumeration degrees are dense. It did, however, give some evidence that they are dense since the second bullet of the above proof shows that there can be only very few (at most countably many) minimal covers of a given enumeration degree.

The density question was answered by Cooper (see [**Coo90**]); Cooper's result was subsequently improved by Calhoun and Slaman (see [**CS96**]):

THEOREM 3.2. *$\mathfrak{D}_e$ is not dense.*

PROOF. See [**Coo87**], [**CS96**]. □

Every attempt to carry out a systematic study of the countable initial segments of $\mathfrak{D}_e$ should of course take into account Theorem 3.1 and Theorem 3.2 above. In particular, by Theorem 3.1 we know that every nontrivial ideal of $\mathfrak{D}_e$ is downwards dense, so there is no nontrivial finite ideal; however, not every ideal is dense.

QUESTION 7. *Characterize the isomorphism types of the countable ideals of the enumeration degrees.*

On the other hand, the nondensity result of Theorem 3.2 shows that finite intervals of enumeration degrees do exist, thus raising the following natural question:

QUESTION 8 (Cooper). *What are the isomorphism types of the finite intervals of enumeration degrees? Can every finite lattice be embedded as an interval of $\mathfrak{D}_e$?*

Calhoun and Slaman's proof of the nondensity of $\mathfrak{D}_e$ shows in fact:

THEOREM 3.3. *There are Π^0_2 enumeration degrees* $\mathbf{a}$ *and* $\mathbf{b}$ *such that* $\mathbf{a} < \mathbf{b}$ *and the degree interval* $(\mathbf{a}, \mathbf{b})$ *is empty.*

PROOF. See [**CS96**]. □

We conjecture that the methods in [**CS96**] can be used to positively answer:

QUESTION 9. *Can every finite lattice be embedded as an interval of the Π^0_2 enumeration degrees?*

4. The local structure

There has been a considerable amount of work on the local structure $\mathfrak{S}$ in recent years. Recent achievements include:

- The solution of the embedding problem ([**LSta**]): Every finite lattice is embeddable in $\mathfrak{S}$ preserving 0 and 1.
- The work ([**LSSta**]) on the problem of the extensions of embeddings in $\mathfrak{S}$, towards obtaining an effective characterization of the pairs of finite posets $\mathfrak{P} \subset \mathfrak{Q}$ for which any embedding of $\mathfrak{P}$ into $\mathfrak{S}$ can be extended to an embedding of $\mathfrak{Q}$ into $\mathfrak{S}$.

A convenient way to show that a degree structure is "complicated" is to show that its first order theory is as complicated as possible. We know that the first order theory of $\mathfrak{D}_e$ is as complicated as possible:

THEOREM 4.1. *Th($\mathfrak{D}_e$) is recursively isomorphic to true second order arithmetic.*

PROOF. See [**SW97**]. □

Surprisingly, the corresponding problem for $\mathfrak{S}$ is still open, although Slaman and Woodin, [**SW97**], are able prove that Th($\mathfrak{S}$) is undecidable. The key degree theoretic fact they use is that there exist uniformly Σ^0_2 sequences of independent enumeration degrees (see [**SW97**]), which are definable in $\mathfrak{S}$ from parameters.

QUESTION 10. *Is Th($\mathfrak{S}$) recursively isomorphic to true first order arithmetic?*

The Π_1-theory of the poset $\mathfrak{S}$ is decidable, of course, since every countable poset (even every countable uppersemilattice) can be embedded in $\mathfrak{S}$; similarly, the Π_1-theory of $\mathfrak{S}$ in the language of uppersemilattices is decidable. The work on extensions of embeddings by Lempp, Slaman and Sorbi shows that some nontrivial fragments of the Π^0_2 theory of $\mathfrak{S}$ are decidable. It can be expected that the methods of Lempp, Nies and Slaman (see [**LNS98**]) will give the undecidability of the Π_3-theory of $\mathfrak{S}$. However, this is, at the moment, still open:

QUESTION 11. *What is the least n such that the Π_n-theory of the poset $\mathfrak{S}$ is undecidable?*

The density of $\mathfrak{S}$ seems to suggest that the situation regarding the ideals of $\mathfrak{S}$ should be more regular than that of the full structure. However, as a consequence of the work in [**LSSta**], not every isomorphism type compatible with density can be realized. For instance, the existence of a linearly ordered ideal can be ruled out.

QUESTION 12. *Characterize the isomorphism types of the initial segments of $\mathfrak{S}$.*

Perhaps, the main difficulty in answering this question consists in the fact that permitting is not available in the context of the enumeration degrees. Building enumeration degrees below other enumeration degrees is often a nontrivial matter (undoubtedly, one of the reasons why the problem of the existence of minimal elements in $\mathfrak{D}_e$ remained open for several years). Working with the enumeration degrees below $\mathbf{0}_e'$, the situation is even more complicated because of the Σ_2^0 nature of the sets with which one deals in the constructions. Gutteridge's proof of the nonexistence of minimal elements is of little help: the proof does yield the existence of enumeration degrees below any given nonzero Σ_2^0 enumeration degree, but in a nonuniform way. Of course, there are different proofs. Paradigmatic are, in this sense, the proofs of the density of $\mathfrak{S}$, see [**Coo84**] and [**LS92**], i.e. infinite injury priority arguments which (due to a delicate balance between the timing of the construction and the true stages of the Σ_2^0 enumerations of the sets involved in the construction itself) can not be easily combined with strategies for different requirements. For instance, it can not be combined with any splitting strategy, since it is known that there exist nontrivial join-irreducible Σ_2^0 enumeration degrees!

It is to be hoped that the techniques introduced in [**LSSta**] will mark considerable progress in this sense. It is also conceivable that a positive solution to Question 10 would have to introduce useful techniques for the investigation of the algebraic structure of $\mathfrak{S}$.

5. Enumeration degrees and structural properties of sets

There are several results relating properties of sets to properties of the corresponding enumeration degrees. The most striking ones regard characterizations of the total degrees: Case [**Cas71**] showed that an enumeration degree **a** is total if and only if **a** contains a retraceable set, if and only if **a** contains a regressive set. As already remarked, Jockusch has recently shown that **a** is total if and only if **a** contains an introreducible set.

Solon asks if there is a similar characterization of the nontotal enumeration degrees:

QUESTION 13 (Solon). *Is there a property, P, of sets such that an enumeration degree* **a** *is nontotal if and only if* **a** *contains some set satisfying P?*

In this regard, Solon (see e.g. [**Sol92**]) has proposed interesting properties (e.g. generalizations/modifications of the notion of hyperimmunity) which turn out to be sufficient for guaranteeing nontotality. See [**Sol99**], for a lengthy discussion of these problems.

The first result along these lines (i.e. relating properties of sets to properties of degrees) concerning the local structure $\mathfrak{S}$ is the sufficient condition for noncupping found by Nies and Sorbi, [**NSta2**]. We say that a Σ_2^0 set A is *$\mathbf{0}'$-hypersimple*, if $\overline{A}$ is infinite and no total $f \leq_T \emptyset'$ majorizes $p_{\overline{A}}$.

THEOREM 5.1. *If* **a** *is $\mathbf{0}'$-hypersimple then* **a** *is noncupping.*

PROOF. See [**NSta2**]. □

This result shows interesting similarities between enumeration reducibility and other reducibilities, such as *wtt*-reducibility. It is hoped (Nies, [**AL99**, p. 230]) that relativizations of other well known properties of c.e. sets (e.g. maximality) will lead to interesting degree-theoretic properties in the context of the Σ_2^0 enumeration degrees.

6. The Medvedev lattice

This section is dedicated to problems that relate the enumeration degrees to the Medvedev lattice. Interest in the Medvedev lattice stems mainly from the fact that it provides a very natural wider context for both the Turing degrees and the enumeration degrees. There are still many open questions about it, dating back to [**Rog67a**] and [**Rog67b**]. Most of the questions raised by Rogers, [**Rog67b**, p.288], have been answered (see [**Sor96**] for a recent survey on the Medvedev lattice. Still open are the questions of whether the Medvedev lattice is rigid and Question 14 below).

We begin with a review of the basic notions about the Medvedev lattice. One defines a reducibility relation on subsets of ω^ω, which are called *mass problems*: given mass problems $\mathcal{A}, \mathcal{B}$, define $\mathcal{A} \leq \mathcal{B}$ if there exists a recursive operator Ψ such that $\Psi(\mathcal{B}) \subseteq \mathcal{A}$ (notice that this entails that $\Psi(f)$ is a total function, for every $f \in \mathcal{B}$).

Define $\mathcal{A} \equiv \mathcal{B}$ if $\mathcal{A} \leq \mathcal{B}$ and $\mathcal{B} \leq \mathcal{A}$. As $\leq$ is easily seen to be a preordering relation on the collection of mass problems, one gets that $\equiv$ is an equivalence relation; the equivalence class of a mass problem $\mathcal{A}$, denoted by $\deg(\mathcal{A})$, is called the *degree of difficulty* of $\mathcal{A}$. The set $\mathfrak{M}$ of degrees of difficulty is therefore a partially ordered set: if $\mathbf{A}, \mathbf{B} \in \mathfrak{M}$, we have $\mathbf{A} \leq \mathbf{B}$, if $\mathcal{A} \leq \mathcal{B}$, for $\mathcal{A} \in \mathbf{A}$ and $\mathcal{B} \in \mathbf{B}$. Moreover, $\langle \mathfrak{M} \leq \rangle$ turns out to be a bounded distributive lattice, in fact a *Brouwer algebra*, i.e., dual to a Heyting algebra.

DEFINITION 6.1 ([**Med55**]). Given a set A, define $\mathbf{E}_A = \deg(\{f : \text{range}(f) = A\})$. The degrees of the form $\mathbf{E}_A$, for some A, are called *degrees of enumerability.*

Then:

THEOREM 6.2 ([**Med55**]). *Define $\iota_e : \mathfrak{D}_e \longrightarrow \mathfrak{M}$ by $\iota_e(\deg_e(A)) = \mathbf{E}_A$. Then ι_e is a well defined order-theoretic embedding preserving least element and join.*

PROOF. See e.g. [**Rog67b**]. □

We observe, by the previous theorem, that

$$A \leq_e B \Leftrightarrow \{f : \text{range}(f) = A\} \leq \{f : \text{range}(f) = B\},$$

in accordance with our intuition (see the Introduction) about enumeration reducibility, i.e. a set A is enumeration reducible to a set B, if there exists an effective procedure (i.e. a recursive operator Ψ) which transforms any enumeration of B (i.e. any function f such that $\text{range}(f) = B$) into some enumeration of A (i.e. a function g such that $\text{range}(g) = A$).

It is immediate that the degrees of enumerability corresponding to total degrees are of the form $\mathbf{A} = \deg(\{f\})$, for some function f; they are called *degrees of solvability.* Rogers's question in [**Rog67b**] about the invariance of the property of being a degree of solvability was answered by Dyment, [**Dym76**], who showed that the degrees of solvability are in fact definable in $\mathfrak{M}$ by the formula

$$\varphi(x) =_{def} (\exists y)[x < y \,\&\, (\forall z)[x < z \rightarrow y \leq z]].$$

The corresponding question about the enumeration degrees, or rather, about their copies within $\mathfrak{M}$, is, however, still open:

QUESTION 14 (Rogers). *Are the degrees of enumerability definable (or at least invariant) in $\mathfrak{M}$?*

6.1. The lattice of weak degrees of difficulty. In this section, we present a different approach to Question 14.

Given mass problems $\mathcal{A}$ and $\mathcal{B}$, define $\mathcal{A} \leq_w \mathcal{B}$ if

$$(\forall f \in \mathcal{B})(\exists g \in \mathcal{A})[g \leq_T f].$$

The relation $\leq_w$ is a preordering relation. Let $\equiv_w$ be the equivalence relation induced by $\leq_w$; the equivalence class of a mass problem $\mathcal{A}$, denoted by $\deg_w(\mathcal{A})$, is called the *weak degree of difficulty* of $\mathcal{A}$. The collection of weak degrees of difficulty $\mathfrak{M}_w$ is partially ordered by $\mathbf{A} \leq_w \mathbf{B}$ if $\mathcal{A} \leq_w \mathcal{B}$, for $\mathcal{A} \in \mathbf{A}$ and $\mathcal{B} \in \mathbf{B}$.

It is easy to see that $\mathfrak{M}_w$ is a complete distributive lattice, called the lattice of *weak degrees of difficulty*, or *Muchnik lattice*, see [**Muc63**]. The mapping $I_w : \mathfrak{M}_w \longrightarrow \mathfrak{M}$, with

$$I_w(\deg_w(\mathcal{A})) = \deg(\{f : (\exists g \in \mathcal{A})[g \leq_T f]\})$$

is an order-theoretic embedding preserving 0 and (all, including infinite) suprema. It is interesting to see that $F_w(\deg(\mathcal{A})) = \deg_w(\mathcal{A})$ defines an onto lattice-theoretic homomorphism $F_w : \mathfrak{M} \longrightarrow \mathfrak{M}_w$ satisfying

$$I_w(\mathbf{A}) \leq \mathbf{B} \Leftrightarrow \mathbf{A} \leq_w F_w(\mathbf{B})$$

and $F_w \circ I_w$ is the identity.

The degrees of difficulty in the range of I_w are still called *weak degrees of difficulty*. Furthermore, there is a copy of the enumeration degrees inside $\mathfrak{M}_w$, provided by the embedding

$$\iota_e^w(\deg_e(A)) = \deg_w(\{f : A \leq_e f\})$$

of $\mathfrak{D}_e$ into $\mathfrak{M}_w$. We use the terminology *weak degrees of enumerability* to denote the weak degrees of difficulty that are in the range of ι_e^w, whereas *weak degrees of solvability* denote the weak degrees of enumerability which correspond to the total enumeration degrees. It is not difficult to see that the weak degrees of solvability are definable in $\mathfrak{M}_w$ (by the same formula which defines the degrees of solvability in $\mathfrak{M}$), and that they generate $\mathfrak{M}_w$ (by infinite infima: each weak degree of difficulty is the infimum of the weak degrees of solvability above it). Thus,

LEMMA 6.3. *$Aut(\mathfrak{D}_T) \simeq Aut(\mathfrak{M}_w)$.*

PROOF. Immediate by the above remarks and the fact that $\mathfrak{M}_w$ is complete and completely distributive. □

Interestingly:

THEOREM 6.4. [**Dym76**] *The property of being a weak degree of difficulty is invariant in $\mathfrak{M}$.*

PROOF. Let $\mathbf{A} \in \mathfrak{M}$. $\mathbf{A}$ is a weak degree of difficulty if and only if $\mathbf{A}$ contains a mass problem $\mathcal{A}$ satisfying: $f \in \mathcal{A} \,\&\, f \leq_T g \Rightarrow g \in \mathcal{A}$. It is not difficult to see that this is equivalent to saying that there exists a set $\mathcal{S}$ of degrees of solvability satisfying

$$\mathbf{S} \in \mathcal{S} \,\&\, \mathbf{S} \leq \mathbf{S}' \Rightarrow \mathbf{S}' \in \mathcal{S},$$

and $\mathbf{A}$ is the least element satisfying:

$$(\forall \mathbf{S})[\mathbf{S} \text{ degree of solvability } \,\&\, \mathbf{A} \leq \mathbf{S} \Rightarrow (\exists \mathbf{S}' \in \mathcal{S})[\mathbf{S}' \leq \mathbf{S}]].$$

□

An interesting version of Question 14 would therefore be:

QUESTION 15. *Are the enumeration degrees (or rather, their images, the weak degrees of enumerability) invariant in* $\mathfrak{M}_w$*?*

A positive answer to Question 15 would imply a positive answer to Question 4. Moreover, it would imply a positive answer to Question 14, if one could prove that every automorphism of $\mathfrak{M}$ that carries weak degrees of enumerability to weak degrees of enumerability also carries degrees of enumerability to degrees of enumerability.

6.2. The Dyment lattice. On mass problems of partial functions, define $\mathcal{A} \leq_e \mathcal{B}$ if $\Phi(\mathcal{B}) \subseteq \mathcal{A}$ for some enumeration operator Φ. The degree structure induced by $\leq_e$ is once again a bounded distributive lattice, called the *Dyment lattice*, see [**Dym76**]. The elements of $\mathfrak{M}_e$ are called *partial degrees of difficulty.* The partial degree of difficulty of $\mathcal{A}$ is denoted by $\deg_e(\mathcal{A})$.

The Medvedev lattice can be seen as a substructure of $\mathfrak{M}_e$, in a way similar to that in which the Turing degrees can be viewed as a substructure of the enumeration degrees. In fact, things are much more natural here, as shown by the following theorem:

THEOREM 6.5. [**Dym76**] *The mapping* $I_e : \mathfrak{M} \longrightarrow \mathfrak{M}_e$, *defined by* $I_e(\deg(\mathcal{A})) = \deg_e(\mathcal{A})$ *(for every mass problem* $\mathcal{A}$ *of total functions) is a lattice-theoretic embedding. The mapping* $F_e : \mathfrak{M}_e \longrightarrow \mathfrak{M}$, *defined by* $F_e(\deg_e(\mathcal{A})) = \deg(\{f : (\exists \varphi \in \mathcal{A})[\text{range}(f) = \varphi]\})$ *(for every mass problem* $\mathcal{A}$ *of partial functions) is an onto lattice-theoretic homomorphism. Moreover,* $F_e \circ I_e$ *is the identity and*

$$\mathbf{A} \leq_e I_e(\mathbf{B}) \Leftrightarrow F_e(\mathbf{A}) \leq \mathbf{B}.$$

PROOF. See [**Dym76**]. □

Thus $\mathfrak{M}$ is a sublattice of $\mathfrak{M}_e$. The analogue of Rogers' question about the invariance of the total degrees is therefore:

QUESTION 16. *Is the Medvedev lattice invariant in the Dyment lattice?*

There are now two different embeddings of the enumeration degrees into $\mathfrak{M}_e$, namely the mapping $\iota^{e,p}(\deg_e(\varphi)) = \deg_e(\{\varphi\})$ is an embedding of the partial degrees into $\mathfrak{M}_e$ preserving 0 and suprema, and the mapping $\iota^{e,t}(\deg_e(\varphi)) = \deg_e(\{f : \text{range}(f) = \varphi\})$ (i.e. $\iota^{e,t} = I_e \circ \iota_e$) is another embedding preserving 0 and suprema. A proof similar to that for the degrees of solvability shows that the degrees in the range of $\iota^{e,p}$ are definable (see [**Dym76**]).

QUESTION 17. *Are the degrees in the range of* $\iota^{e,t}$ *invariant with respect to the automorphisms of* $\mathfrak{M}_e$*?*

In this regard, Dyment ([**Dym76**]) notes:

THEOREM 6.6. *The degrees in the range of* $\iota^{e,t}$ *are invariant under all automorphisms which carry* $\mathfrak{M}$ *onto* $\mathfrak{M}$.

PROOF. See [**Dym76**]. □

References

[Ahm89] S. Ahmad. Some results on the structure of the Σ_2 enumeration degrees. *Recursive Function Theory Newsletter*, 38, 1989.

[ALta] S. Ahmad and A. H Lachlan. Some special pairs of Σ_2 *e*-degrees. To appear.

[AL99] M. Arslanov and S. Lempp, editors. *Recursion Theory and Complexity*, de Gruyter Series in Logic and its Applications, Berlin, New York, 1999. W. De Gruyter.

[Cas71] J. Case. Enumeration reducibility and partial degrees. *Ann. Math. Logic*, 2:419–439, 1971.

[CC88] S. B. Cooper and C. S. Copestake. Properly Σ_2 enumeration degrees. *Z. Math. Logik Grundlag. Math.*, 34:491–522, 1988.

[Coo82] S. B. Cooper. Partial degrees and the density problem. *J. Symbolic Logic*, 47:854–859, 1982.

[Coo84] S. B. Cooper. Partial degrees and the density problem. Part 2: The enumeration degrees of the Σ_2 sets are dense. *J. Symbolic Logic*, 49:503–513, 1984.

[Coo87] S. B. Cooper. Enumeration reducibility using bounded information: counting minimal covers. *Z. Math. Logik Grundlag. Math.*, 33:537–560, 1987.

[Coo90] S. B. Cooper. Enumeration reducibility, nondeterministic computations and relative computability of partial functions. In K. Ambos-Spies, G. Müller, and G. E. Sacks, editors, *Recursion Theory Week, Oberwolfach 1989*, volume 1432 of *Lecture Notes in Mathematics*, pages 57–110, Heidelberg, 1990. Springer–Verlag.

[Coo99] B. Cooper. Hartley Rogers' 1965 agenda. In S. Buss and P. Pudlak, editors, *Logic Colloquium '98*. Springer-Verlag, 1999.

[CS96] W. C. Calhoun and T. A. Slaman. The Π^0_2 *e*-degrees are not dense. *J. Symbolic Logic*, 61:1364–1379, 1996.

[Dowta] R. Downey. Computability, definability and algebraic structures. In *Proceedings of the 7th Asian Logic Conference*. To appear.

[Dym76] E. Z. Dyment. Certain properties of the Medvedev lattice. *Mathematics of the USSR Sbornik*, 30:321–340, 1976. English Translation.

[FR59] R. M. Friedberg and H. Rogers, Jr. Reducibility and completeness for sets of integers. *Z. Math. Logik Grundlag. Math.*, 5:117–125, 1959.

[Gut71] L. Gutteridge. *Some Results on Enumeration Reducibility*. PhD thesis, Simon Fraser University, 1971.

[Ler83] M. Lerman. *Degrees of Unsolvability*. Perspectives in Mathematical Logic. Springer–Verlag, Heidelberg, 1983.

[LL76] A. H. Lachlan and R. Lebeuf. Countable initial segments of the degrees. *J. Symbolic Logic*, 41:289–300, 1976.

[LNS98] S. Lempp, A. Nies, and T.A. Slaman. The Π_3-theory of the computably enumerable Turing degrees is undecidable. *Trans. Amer. Math. Soc.*, 350(7):2719–2736, 1998.

[LSta] S. Lempp and A. Sorbi. Embedding finite lattices into the Σ^0_2 enumeration degrees. To appear.

[LS92] A. H. Lachlan and R. A. Shore. The *n*-rea enumeration degrees are dense. *Arch. Math. Logic*, 31:277–285, 1992.

[LSSta] S. Lempp, T. Slaman, and A. Sorbi. On extensions of embeddings into the enumeration degrees of the Σ^0_2 sets. To appear.

[MC85] K. McEvoy and S. B. Cooper. On minimal pairs of enumeration degrees. *J. Symbolic Logic*, 50:983–1001, 1985.

[Med55] Y. T. Medevdev. Degrees of difficulty of the mass problems. *Dokl. Nauk. SSSR*, 104:501–504, 1955.

[Muc63] A.A. Muchnik. On strong and weak reducibility of algorithmic problems. *Sibirskii Matematicheskii Zhurnal*, 4:1328–1341, 1963. Russian.

[Myh61] J. Myhill. A note on degrees of partial functions. *Proc. Amer. Math. Soc.*, 12:519–521, 1961.

[NSta1] A. Nies and A. Sorbi. Branching in the enumeration degrees of Σ^0_2 sets. *Israel J. Math.*. To appear.

[NSta2] A. Nies and A. Sorbi. Structural properties and Σ^0_2 enumeration degrees. *J. Symbolic Logic*. To appear.

[Rog67a] H. Rogers, Jr. Some problems of definability in recursive function theory. In J. N. Crossley, editor, *Sets Models and Recursion Theory, Proc. Summer School in Mathematical*

Logic and Tenth Logic Colloquium, Leicester, August-September 1965. North–Holland Publishing Co., 1967.

[Rog67b] H. Rogers, Jr. *Theory of Recursive Functions and Effective Computability.* McGraw-Hill, New York, 1967.

[Roz78] M. Rozinas. The semilattice of e-degrees. In *Recursive Functions*, pages 71–84. Ivanov. Gos. Univ., Ivanovo, 1978. (Russian) MR 82i:03057.

[Sla91] T. A. Slaman. Degree structures. In *Proceedings of the International Congress of Mathematicians, Kyoto, 1990*, volume I, pages 303–316, Heidelberg, 1991. Springer–Verlag.

[Soa87] R. I. Soare. *Recursively Enumerable Sets and Degrees.* Perspectives in Mathematical Logic, Omega Series. Springer–Verlag, Heidelberg, 1987.

[Sol92] B. Solon. E-hyperimmune sets. *Siberian Math. J.*, 33:211–214, 1992.

[Sol99] B. Solon. Enumeration reducibility and the problem of the nontotal property of e-degrees. In M. Arslanov and S. Lempp, editors, *Recursion Theory and Complexity*, de Gruyter Series in Logic and Its Applications, pages 173–191, Berlin, New York, 1999. W. De Gruyter.

[Sor96] A. Sorbi. The Medvedev lattice of degrees of difficulty. In S. B. Cooper, T. A. Slaman, and S. S. Wainer, editors, *Computability, Enumerability, Unsolvability - Directions in Recursion Theory*, London Mathematical Society Lecture Notes Series, pages 289–312. Cambridge University Press, New York, 1996.

[Sor98] A. Sorbi. Sets of generators and automorphism bases for the enumeration degrees. *Ann. Pure Appl. Logic*, 94, 3:263–272, 1998.

[Sosta] I.N. Soskov. A jump inversion theorem for the enumeration degrees. To appear.

[SW97] T. A. Slaman and W. H. Woodin. Definability in the enumeration degrees. *Arch. Math. Logic*, 36:225–267, 1997.

Dipartimento di Matematica, Univerità di Siena. 53100 Siena Italy
E-mail address: `sorbi@unisi.it`